THE ICE HUNTERS
A History of Newfoundland Sealing to 1914

Shannon Ryan

BREAKWATER

Breakwater
100 Water Street
P.O. Box 2188
St. John's, NF
A1C 6E6

The Publisher gratefully acknowledges the financial assistance of the Canada Council.

This book has been published with the help of a grant from the Social Science Federation of Canada, using funds provided by the Social Sciences and Humanities Research Council of Canada.

Canadian Cataloguing in Publication Data

Ryan, Shannon, 1941-

The ice hunters

(Newfoundland history series, ISSN 0831-117X ; 8)

Includes bibliographical references and index.
ISBN 1-55081-095-2 (pbk.).— ISBN 1-55081-097-9 (bound)

1. Sealing—Newfoundland—History. 2. Sealing industry—Newfoundland—History. I. Title. II. Series.

SH362.R92   1994   639.29'09718   C94-950264-2

Copyright © Shannon Ryan 1994

ALL RIGHTS RESERVED. No part of this work covered by the copyright hereon may by reproduced or used in any form or by any means—graphic, electronic or mechanical—without the prior written permission of the publisher. Any request for photocopying, recording, taping, or storing in an information retrieval system of any part of this book shall be directed in writing to the Canadian Reprography Collective, 379 Adelaide Street West, Suite M1, Toronto Ontario, M5V 1S5. This applies to classroom usage as well.

ice-hunter: (a) man who engages in the hunt or fishery for seals; (b) vessel so engaged.

[c1833] 1927 Doyle (ed) 15 "Come All Ye Jolly Ice-Hunters": Come all ye jolly ice-hunters and listen to my song; / I hope I won't offend you, I don't mean to keep you long: / 'Tis concerning an ice-hunter from Tilton Harbour sailed away, / On the fourteenth day of March, eighteen hundred and thirty-three.

[1837] 1906 Murphy 3 [They were] trying to reach the land in a boat after their vessel, an ice Hunter, was driven on the rocks.

1861 De Boilieu 198 One of the ship-wrecked men saved from the schooner had been an ice-hunter, and whiled away many an hour relating the mode of catching seals in the spring of the year on the coast of Newfoundland.

1866 Wilson 278 Few of the masters or skippers of ice-hunters knew anything of navigation.

1887 Howley MS *Reminiscences* 3 No one but an experienced ice hunter, used to treading his devious way through the ice floe and taking advantage of every small opening and lead of water would attempt it.

1928 FPU (Twillingate) *Minutes* 5 Oct. There was quite a talk over Mr. Ashbourne buying a new ice hunter and having a Bonavista Bay man for captain.

T80/1-64 Now grandfather, he always runned a ice-hunter, you know — his own schooner, see.

T183-64 The ice hunters used to moor up there winter time an'go to th'ice.

1972 Brown 119 Even in fine weather ice hunters never went anywhere alone.

ice-hunting: hunting seals amid the ice-floes.

1873 Carroll 20 The great object is when ice-hunting masters come up with old seals on loose ice to keep by them and remain quiet until the ice runs together.

1877 Tucker 229-30 In February the steamers were ready to start for the ice-fields in search of seals; the accustomed 'Ice-hunting Sermons' were preached.

[1900 Oliver & Burke] 12 It was his first spring out ice-hunting.

1905 Grenfell 120 Though we made very little by it, somehow we all looked forward to the 'ice-hunting,' as we called it.

T187/9-65 "This here ice hunting racket," he said, "is a hell of a hard racket."

T401/2-67 Fifteen year old an' when I went ice huntin', took a man's share.

Source: Story, Kirwin and Widdowson, *Dictionary of Newfoundland English* (Toronto,1982), pp.264-5.

For my mother, Lillian; and in memory of my father,
Bernard, and Dr. Cater W. Andrews.

CONTENTS

Key to Abbreviations ix
List of Illustrations x
List of Tables in Appendix xi
Acknowledgements xiii
Preface xvii
Introduction 25
    Fishing Ship Fishery 25
    Bye Boat Fishery 27
    Bank Fishery 27
    Early Settlement 29
    Post-1775 32
    War and the Migratory Fishery: 1793 35
    War and Newfoundland: 1793 37
    Napoleonic War to 1807 39
    American Embargo and the Peninsular War 41
    Anglo-American War 42
    Settlement to 1815 43
    Post-War Depression 46
    Seals 47
    Early Sealing 49
    Spring versus Winter Sealing 54
    Summary 59
    Notes 59

Chapter 1- Market Place 65
    Early Lighting 65
    Leather Currying 68
    Other Traditional Uses of Oils and Fats 69
    Seal Skins 70
    Industrial and Commercial Developments 70
    Oil Production Methods 71
    Demand for Oil Increases 72
        Advances in lighting 73
        Eighteenth-century British oil markets 76
        Growing use of seal oil 78
        Nineteenth-century British oil markets 79
    Decline in Demand for Traditional Oils 81
    Global Impact of Oil Trade 84
    Seal Skin Markets 85
    Summary 86
    Notes 86

Chapter 2 - Economy to 1914...**92**
  Economy on Land and Sea from 1814 to 1860................ 93
    Cod fishery..93
    Agriculture 94
    Seal fishery..98
    Early perceptions of the seal fishery
    in the economy...100
    Summary: 1814-60...104
  Economy on Land and Sea from 1860 to 1914.............. 104
    Seal fishery..105
    Cod fishery...106
    Other fisheries 108
    Agriculture 108
    Other strategies..109
    Later perceptions of the seal fishery: conservation
    issues and government responses...................... 111
    Summary: 1860-1914...117
  General Summary...117
  Notes..118

Chapter 3 - Vessels and Ports...**121**
  Age of Sail.. 121
    Conception Bay shows the way....................121
    St. John's expands...126
    Seal oil processing..129
    Conception Bay and the growth of
    Harbour Grace..130
    Changing role of St. John's..134
    St. John's versus Conception Bay......................135
    Ports to the northward..136
    Vessels 138
    Summary of the age of sail...143
  Age of the Wooden Walls 144
    Major companies...144
    Transitional 1860s 147
    Wooden walls..147
    Companies and wooden walls...............................151
    Mercantile changes and
    the decline of outport firms...155
    Decline of sail 158
    Larger firms and centralization.............................168
    Competition from outside Newfoundland...........174
    Problems and changes...178
    Expansion, rising prices
    and mercantile control..183
    Summary of the age of the wooden walls...........186

Age of the Iron Clads..187
    Iron clad S.S. Adventure 187
    Iron clads versus wooden walls...................190
    Summary of the age of the iron clads............200
General Summary 200
Notes 202

Chapter 4 - Fishermen - Ice hunters 213
Manpower..213
    Sail and manpower.......................................213
    Steam and manpower....................................218
    Steamer captains..219
    Changes in manpower...................................223
    Unusual ice hunters......................................226
    Stowaways 227
    Ice hunters and St. John's 229
Berths 230
Ice Hunters' Incomes..234
    Fishermen 234
    Captains 240
    Income and living standards 242
Living and Working Conditions....................................243
    General living conditions..............................243
    Work and diet on sailing vessels 247
    Work and diet on steamers...........................255
    Accommodations 261
    Discipline 262
Landsmen..264
Summary 271
Notes 271

Chapter 5 - Disasters...................................282
Early Disasters 283
Disasters during the Age of Expansion 284
Disasters during the Industry's Zenith.......................289
'Spring of the Wadhams' and the 1850s 295
'Green Bay Spring' and the Decline of Sail 297
Landsmen..301
Early Steamer Losses...304
Greenland Disaster..306
Other Steamer Losses 307
Erna Disaster..310
Southern Cross Disaster..310
Newfoundland Disaster...311
Support for Survivors and Dependents 317
Summary 318

Notes 319

Chapter 6 - A Sense of Identity 328
  Labour's Reaction.. 329
      Strikes in Harbour Grace and
      Carbonear in 1832 329
      Strikes in St. John's in the 1840s.......................330
      Strikes in Harbour Main and Brigus in 1845.......336
      Labour after mid-century......................................341
      St. John's strike of 1902 342
      Labour during post-1902 347
      Seal skinners 349
      Manuses 350
      Societies and unions 352
      Summary 252
  Community Reactions.. 354
      Government 354
      Evening Telegram in the 1880s and 1890s..........355
      Evening Telegram and the Greenland Disaster..359
      Governors take an interest..................................363
      Fishermen's Protective Union.............................365
      Summary 368
  Culture and Reputation.. 368
      Narratives and vignettes......................................369
      Honour the heroes 379
      Songs 385
      Other cultural aspects 386
      Flippers 387
      Cruelty to seals..388
      Arctic explorations..393
      Summary 395
  General Summary.. 395
  Notes.. 396

Conclusion 404
Appendix 417
Bibliography 509
Index 527

Key to Abbreviations

BT	Board of Trade
c.i.f.	cost, insurance and freight
CO	Colonial Office (Records), PRO
CNS	Centre for Newfoundland Studies
CUST	Records of the British Custom's Office, PRO
cwt	hundredweight of 112 lbs. avoirdupois [about 51 kg.]
DCB	*Dictionary of Canadian Biography*
DNE	*Dictionary of Newfoundland English*
DNLB	*Dictionary of Newfoundland and Labrador Biography*
ENL	*Encyclopedia of Newfoundland and Labrador*
f.o.b.	free on board
FPU	Fishermen's Protective Union
kg.	kilogram
km.	kilometre
JHA	*Journal of the House of Assembly of Newfoundland*
lb.	pound, avoirdupois
m.	metre
MHA	Maritime History Archives, MUN
MUN	Memorial University of Newfoundland
MUNFLA	Memorial University Folklore and Language Archive
PP	*British Parliamentary Papers*
PANL	Provincial Archives of Newfoundland and Labrador
PRO	Public Record Office, Kew, England
stg.	sterling
£ s d	pounds, shillings, pence: 12d = 1s; 20s = £1

List of Illustrations

In preparing the photographs and illustrations, I have made every effort to identify, credit correctly and obtain publication rights from copyright holders involved. In this respect, I would like to thank the collectors, repositories and photographers concerned, and the appropriate credit is given at the end of each illustration. (Notice of any errors and/or omissions in this regard will be gratefully received and necessary corrections made in any subsequent editions.)

Unfortunately, it is not always possible to identify the people who captured so much of Newfoundland's visual history on film. Therefore, to be certain that they receive recognition, at least of a general nature, I would like to introduce the principal photographers of the late nineteenth and early twentieth centuries. The first well-known local photographer whose work included photographs of aspects of the sealing industry was Simeon Henry Parsons (1844-1908), originally of Harbour Grace, who opened a studio in St. John's in 1875. Through exhibits and publication in newspapers, Parsons gained an international reputation as a photographer, and after his death, his studio continued to operate under the direction of his children. In 1874, Robert Edward Holloway (1850-1904) arrived from England to take up the position of headmaster of the Methodist Academy, where he also served as the chemistry teacher. He travelled throughout Newfoundland and took many photographs, from which he selected those to be published in his *Through Newfoundland with the Camera*. He died from tuberculosis shortly before it was published in 1905. His daughter, Elsie, expanded on her father's hobby, opened Holloway Studio in 1908 and carried on an extensive photographic business until her retirement in 1946. James Vey (about whom little is known) was also a well-known photographer c.1900, and some of his excellent sealing photographs survive. Finally, Reuben Parsons, the nephew of S. H. Parsons, was an amateur photographer in Harbour Grace, where he expertly recorded many historical events and scenes. See *DNLB*, *ENL* and Antonia McGrath, *Newfoundland Photography: 1849-1949* (St. John's, 1980).

Figure 1	Map of Canada	xxi
Figure 2	Map of the Province of Newfoundland	xxii
Figure 3	Map of the Island of Newfoundland	xxiii
Figure 4	Map of Conception Bay and St. John's	xxiv
Figure 5	Harp Seals on the Ice Floes	48
Figure 6	Pan Lamp and Cruise Lamp	66
Figure 7	Wall-Mounted Double Argand Lamp	73
Figure 8	Floor-Model Double Argand Lamps	75
Figure 9	St. John's Sailing Fleet c.1860	141
Figure 10	S.S. *Retriever*	148
Figure 11	S.S. *Terra Nova*	149
Figure 12	St. John's Wooden-Wall Fleet c.1890	151
Figure 13	S.S. *Adventure*	187
Figure 14	Governor Visits the Fleet	256
Figure 15	Men Ready to go Overboard	257
Figure 16	Men Sculping Seals	258

Figure 17   Men Hauling Pelts..260
Figure 18   The Dead from the *Newfoundland* Disaster..........313
Figure 19   A Survivor from the *Newfoundland* Disaster........314
Figure 20   The 1902 Strike..343
Figure 21   Seal Skinners at Work 348
Figure 22   Wooden Buckets 387

List of Tables in Appendix

Intro.1	Newfoundland: 1696-97	418
Intro.2	Newfoundland Saltfish: Quantity and Prices, 1793-1800	419
Intro.3	Prices in Newfoundland: 1802	420
Intro.4	Prices in Newfoundland: 1804	420
Intro.5	Wages in Newfoundland: 1804	421
Intro.6	Wages in Newfoundland: 1810	422
Intro.7	Prices in Newfoundland: 1810	422
Intro.8	Prices in Newfoundland: 1813	423
Intro.9	Value of Seal Oil produced by Inhabitants: 1723-1802	424
Intro.10	Newfoundland's Population by District: 1789	425
Intro.11	Newfoundland's Fishery and Trade by District: 1789	426
Intro.12	Seal Fishery: 1803	427
Intro.13	Spring Seal Fishery: 1804	428
1.1	Train Oil Imports into England and Wales: 1700-71	429
1.2	Train Oil Imports into England and Wales: 1772-1808	430
1.3	British Imports from Newfoundland: 1772-1808	431
1.4	British Imports of Seal Skins and Train Oil from Newfoundland: 1809-53	432
1.5	Comparative British Train Oil Imports: 1831-1914	433
1.6	Britain's Principal Oil Imports: 1854-1907	436
1.7	Newfoundland Train Oil Prices: 1854-1914	439
1.8	Oil Prices on the London Markets: 1832-80	440

2.1	Adult Population of Newfoundland and Occupations: 1911	441
2.2	Adult Population of Newfoundland and Occupations: 1857	442
2.3	Newfoundland's Principal Imports: 1856-60	442
2.4	Newfoundland Vessels: 1803-33	443
2.5	Newfoundland's Production of Seal Skins: 1803-60	445
2.6	Newfoundland Oil Exports: 1803-60	446
2.7	Newfoundland Seal Exports: 1861-1914	448
2.8	Prices of Seal Fishery Exports: 1861-1914	450
2.9	Percentage of Newfoundland's Exports Consisting of Seal Products: 1850-1914 (Value)	451
3.1	Newfoundland Seal Fishery: 1819-33	452
3.2	St. John's Sealing Fleet: 1819-62	453
3.3	Conception Bay Sealing Fleet: 1814-65	455
3.4	St. John's and Conception Bay Sealing Fleets: 1851-62	457
3.5	St. John's and Conception Bay Sealing Fleets: 1853	458
3.6	Harbour Grace Sealing Fleet: 1867-1900	466
3.7	Production Of Seal Pelts: 1863-95	467
3.8	Newfoundland Sealing Steamers: 1863-1914	468
4.1	St. John's Sealing Fleet: 1834	470
4.2	Conception Bay Sealing fleet: 1835	473
4.3	Carbonear and Harbour Grace Sealing Fleets: 1836	478
4.4	St. John's Sealing Fleet: 1838	481
4.5	Sealing Vessels Clearing from Brigus: 1838	484
4.6	Newfoundland Sealing Fleet: 1869	487
4.7	Newfoundland Sealing Steamer Captains: 1869/1909	495
4.8	Harbour Grace Sealing Steamer Captains	500
4.9	High Liner Captains and Cargoes	501
5.1	Newfoundland Sealing Steamers Lost: 1863-1914	504
6.1	Relief Subscriptions: 1830	506

Acknowledgements

My interest in the history of the Newfoundland seal fishery arose out of the research I carried out for my MA thesis under the supervision of the late Professor Keith Matthews, MUN, in 1969-71. Although my thesis examined the cod fishery, Matthews insisted that I study the seal fishery as well in order to place the cod fishery in its proper perspective. Since 1971, I have continued to devote some attention to my study of the seal fishery—although my main area of concentration has been the history of the Newfoundland cod fishery. Consequently, portions of this book have appeared as follows: "The Newfoundland Cod Fishery in the Nineteenth Century," MA thesis, MUN, pp. 14-26; "Introduction," in Shannon Ryan and Larry Small, eds., *Haulin' Rope and Gaff: Songs and Poetry in the History of the Newfoundland Seal Fishery* (St. John's, 1978); *The Seal and Labrador Cod Fisheries of Newfoundland*, XXVI (Canada's Visual History Series, Ottawa, 1978); "The History of the Seal Fishery," the *Evening Telegram* (St. John's), 26 March 1983; "Fishery to Colony: A Newfoundland Watershed, 1793-1815," *Acadiensis*, XII, no. 2 (Spring 1983), 34-62; "Seals Spelled Survival," *Horizon Canada* III, no. 27 (1985), 638-43; "A historical overview of Canadian/Newfoundland/world sealing and the part this industry played in the development of the Atlantic Canadian/Newfoundland economy," a technical report, *Seals and Sealing in Canada: Report of the Royal Commission*, (Ottawa, 1986); (assisted by Martha Drake), *Seals and Sealers: A Pictorial History of the Newfoundland Seal Fishery*, based on the Cater Andrews Collection, (St. John's, 1986); Ryan, ed., *Chafe's Sealing Book: A Statistical Record of the Newfoundland Steamer Seal Fishery, 1863-1941*, (St. John's, 1989); "Newfoundland: Fishery to Canadian Province," *Atlantic Canada: At the Dawn of a New Nation* (Burlington, Ontario, 1990), pp. 7-44; "The Industrial Revolution and the Newfoundland Seal Fishery," *International Journal of Maritime History* IV, no. 2 (December 1992), 1-43; "Newfoundland Spring Sealing Disasters to 1914," *The Northern Mariner/Le Marin du nord* III, no. 3 (July 1993), 15-48; and "Newfoundland Sealing Strikes: 1830-1914," *The Northern Mariner/Le Marin du nord* IV, no.3 (July 1994). Portions of this study have also been delivered as papers: the annual meeting of the American Folklore Society (Philadelphia, 1976); the Workshop of the St. Lawrence (McGill University, 1977); the University of London (1978 and 1987); the University of Bergen (1978); the annual meetings of the Canadian Historical Association (1982, 1984 and 1990); the First International Congress of Maritime History (Liverpool, 1992); and the New Dimensions in Maritime History conference (Fremantle, Australia, 1993). I would like to thank editors, referees, conference participants and all involved in these presentations and publications for the many valuable suggestions received.

This particular study has been in progress for a long time—actually since I began my MA degree on the Newfoundland cod fishery in 1969. Consequently, the number of archivists, researchers, collectors, academics, students, former students and friends (and many fall into two or more categories) to whom I owe thanks is inevitably substantial.

As indicated, the late Keith Matthews insisted that I examine the history of the seal fishery in the context of my MA thesis, and to him, I shall always be most grateful. Memorial University of Newfoundland (through the department heads, deans, vice-presidents and presidents) has been most cooperative in allowing me sabbaticals, rearranged schedules, funding and research assistance. This made it possible for me to continue my research and writing with the least interference. Similarly, the Social Sciences and Humanities Council of Canada has been generous with its funding, thus allowing me to spend my research terms of 1985 and 1989 in repositories in Great Britain. It was in the framework of this general assistance that my work was able to continue uninterrupted—except for my own obligations and commitments.

In the area of specific research assistance, I am especially grateful to Mrs. Cater W. Andrews and Mrs. Judith Mallam—the family of the late Professor Cater W. Andrews—for donating to Memorial University of Newfoundland the archival collection of Professor Andrews. This extensive collection of material on sealing has been most useful in my work, and *Seals and Sealers* and *Chafe's Sealing Book*, mentioned above, have been based on it. The collection will remain equally important to other scholars in their work on Newfoundland communities, local and regional sealing history, the Labrador cod fishery, Arctic and Antarctic exploration, family histories and photographic studies—to name just a few of its possible applications.

The Cater W. Andrews Collection was placed under the supervision of three literary executors—the late Professor Frederick Aldrich, President Emeritus (and then University President) Leslie Harris, and the late Henrietta Harvey Professor of English George M. Story—and Ms. J. M. Neeson (then attached to the History Department) catalogued it, which she did extremely well, using a system based on Arabic and Roman numerals. In 1982, the literary executors and Mr. Richard Ellis, University Librarian, arranged for the collection to be moved to the Centre for Newfoundland Studies and allowed me unrestricted access to it; for this I am most grateful. Also, on behalf of the late Professor Andrews I must thank the researchers he hired in 1968 and in 1969 (with assistance from the Canada Council): Mr. Wayne Andrews, Mr. Michael O'Connell and Mr. Jesse Fudge. The photocopies and research notes they accumulated from local newspapers make up a very important part of the Collection, and reference to these will be found in the Bibliography. Professor Andrews appealed to many sources for assistance in building his collection and, on his behalf I thank all who responded. (Relevant correspondence may be found in the Collection as 211 VI, 212 VI and 213 VI.) Dr. Andrews prepared outlines of his proposed work and drafted sections; these also are readily available in the Collection. I must reiterate my thanks to Mrs. Cater Andrews and her daughter, Mrs. Judy Mallam, for their enduring interest and assistance.

On behalf of both myself and Memorial University, I want to thank the late Mr. Cyril Cornick (former employee of Bowring Brothers) for his contribution of original documents to the Centre for Newfoundland Studies. The Cyril Cornick Collection, donated in 1983, was inventoried by the late Ms. Nancy Grenville in 1984 and contains Bowring Brothers material, notes, seal-spotting charts (drawn from the air), diaries and statistics.

Among the many scholars who were most generous with advice and assistance, I must thank the following in particular (members of history departments, except where otherwise noted): the late Professor George M. Story,

Henrietta Harvey Professor of English, MUN, who, in 1982, first suggested that I begin work on a history of the Newfoundland seal fishery to parallel my study of the saltfish trade, and who was most supportive right up until his death on 9 May of this year; Professor Glyndwr Williams, University of London and my former Ph.D. supervisor; the late Dr. Frederick A. Aldrich, Moses Harvey Professor of Marine Biology, MUN; Dr. Melvin Baker, historian and records manager, MUN; Professor David Buchan, Folklore Department, MUN; Dr. Valerie Burton, MUN; Dr. Sean Cadigan, MUN; Mr. Robin Craig, University of London; Professor Andy den Otter, MUN; Professor Lewis R. Fischer, MUN; Professor Kenneth S. Goldstein, Folklore Department, University of Pennsylvania; Professor Emeritus Herbert Halpert, Folklore Department, MUN; Professor Gordon Handcock, Geography Department, MUN; Dr. Freda Harcourt, University of London; Dr. Leslie Harris, historian and President Emeritus, MUN; Professor James K. Hiller, MUN; Dr. Robert Holland, University of London; Dr. Gordon Jackson, University of Strathclyde; Professor Gregory S. Kealey, MUN; Mr. Kenneth Kerr, retired historian, St. John's; Professor Emeritus William Kirwin, English Department, MUN; Professor Peter Marshall, University of London; Professor Emeritus W. P. McCann, MUN; Professor Peter Neary, University of Western Ontario; Professor Helge Nordvik, Bergen; Professor Patrick O'Flaherty, English Department, MUN; Mr. Gregory Palmer, writer and historian, Harwich Town, England; Dr. Sarah Palmer, University of London; Professor G. E. Panting, MUN; Professor Andrew Porter, University of London; Dr. John Reid, M.P. and historian, London; Dr. Judith Rowbotham, Nottingham Trent University; Professor Chesley W. Sanger, Geography Department, MUN; Dr. Larry Small, Folklore Department, MUN; Professor Odd Vollan, retired historian, Ålesund, Norway; Dr. Wilfred Wareham, folklorist, formerly of MUN; Professor William Whiteley, MUN; Professor John Widdowson, Centre for English Cultural Tradition and Language, University of Sheffield; and the readers appointed by the Aid to Scholarly Publications Programme administered by the Social Science Federation of Canada.

I also want to thank the personnel of the archives, libraries and repositories where my research was carried out: Mr. Richard Ellis, MUN University Librarian, and staff—especially the management and staff of the CNS—the late Dr. Agnes O'Dea (former Head), Ms. Anne Hart (Head), and the late Ms. Nancy Grenville, Ms. Joan Ritcey, Mr. Bert Riggs, Ms. Gail Weir and Ms. Linda White; Mr. Charles Cameron and staff, Newfoundland Provincial Reference and Resource Library; Ms. Heather Wareham, the late Mrs. Doris Pike, Mrs. Gertrude Crosbie, staff and volunteers, MHA, MUN; Mr. David Davis and archivists and staff, PANL; the late Dr. Bobbie Robertson, Mrs. Catherine Power, Mrs. Katherine Earle and Miss Mary O'Keefe of the Newfoundland Historical Society; Ms. Gail Hogan, Newfoundland Law Society Library; Miss Norma Jean Richards, Newfoundland Legislative Librarian; the late Mr. William Parsons, curator, Conception Bay Museum, Harbour Grace, Newfoundland; Mrs. Gerald S. Doyle, for permission to reprint selections from the Newfoundland song books published by the late Gerald S. Doyle; Mr. Michael J. Murphy for permission to reprint selections from the books of history, songs and stories published by his grandfather, the late James Murphy; Mr. Henry Collingwood, Baine Johnston and Company, for access to the company's records; Mrs. Jane Wilson, Mr. Albert

Brooks and Mr. Howard Cooper, Trinity House, London; Mr. Michael Harrington, local historian and journalist; the late Mr. Derek Suttling, rare book dealer, London, who helped me add to my personal library; the librarians, archivists, historians and staff of the Institute of Historical Research, University of London; the staffs of the Science Museum Library, Guildhall Library, British Library, British Library of Political and Economic Science, University of London Library, Victoria and Albert Museum, National Maritime Museum, National Register of Archives (London), the Public Record Office (Kew), the Board of Customs Library (Mark Lane, London), the Dorset Record Office, Liverpool Record Office, and Gwynedd Archives Service (Caernarfon, Wales). And finally, I must thank the Director, Wardens and staff of the London Goodenough Trust who made it possible for me to live and work in comfort and without distractions in Central London (and a special thanks to the former Warden of Willy G., Miss Jill Morrogh, and to the present staff of Fellowship House.)

I would like also to thank students, former students, informants, friends and family for their assistance: Ms. Denise French, history student, who, over a period of five summers on Challenge grants, researched the fleets of sailing ships in the Newfoundland nineteenth century seal fishery (and I hereby thank the Canadian Employment and Immigration Challenge Program for their financial assistance and the MUN administrators who managed the Program); students and former students—Ms. Sarah Barron, Mr. David Bradley, Mr. Geoff Budden, Mr. Edward Cole, Ms. Deborah Duke, Mr. Darryl Goodyear, Mr. Keith Hewitt, Ms. Catherine Horan, Ms. Trudy Johnson, Ms. Susan Oke-McCarthy, Mr. Andrew Parsons, Ms. Kathryn Pike, Mr. Leo Power, Ms. Janice Reid, and Mr. Frederick Winsor; others who assisted in various ways—Mr. Hal Andrews (formerly of "Land and Sea," CBC); Mr. Ronald Hynes, who identified Michael Hennessey as the St. Brendan's native who died in the *Greenland* disaster; Mr. Randy Earle, who provided information on his grandfather, Isaac R. Randell, captain of the S. S. *Bellaventure* in 1914; Mr. Robert Parker, whose assistance in the British Library was invaluable; and the following oral history informants—Mr. Clifford Andrews, the late Mr. Cyril S. Chafe, Mr. David Dawe, the late Captain William Gillett, Mr. Eric Gosse, the late Mr. Arthur Hurdle, Capt. Morrissey Johnson, the late Mr. Edward Russell, Mr. John M. Ryan, the late Mr. Joseph Ryan, the late Mr. Bertram Shears, Mrs. Andrew Short, and the late Mr. Andrew Short.

Finally, I would like to thank the people who helped in the physical preparation of this volume: Ms. Martha Drake, who assisted with the photographs; Ms. Willeen Keough, who proofread the manuscript and offered extensive editorial advice; Mr. Jack Martin and staff, Photographic Services, MUN; my sister, Brenda Parmenter, who assisted with the Index; my brother, William, who prepared the maps with advice from Dr. Gordon Handcock; Ms. Frances Warren, who prepared the Bibliography and provided editorial assistance; the staff of the History Department, MUN; the staff of Breakwater Books; and my wife, Margaret, for her constant support.

Preface

At the beginning of the nineteenth century, Newfoundland crossed the divide from fishing station to colony. Most historians, myself included, have tended to take a colonial-centred view of this transitional period, and we explain it by pointing out that the migratory cod fishermen discovered the seal herds and stayed to hunt them. This explanation is incomplete, and we must begin to appreciate the growing appetite of the urban and industrial world for oils: fish, sea mammal and vegetable. In my study of the Newfoundland saltfish trade, *Fish out of Water*, I examined the appropriate records from 1814 to 1914 and concluded that:

> Newfoundland's political independence was incompatible with the commercial reality of the international saltfish trade [because] its economy remained primarily dependent upon the resources, connections and support of Great Britain's international commercial arrangements and activities.[1]

Although this conclusion is valid while one remains focused on the important cod fishery, it invites further questions relating to the evolution of Newfoundland from a fishing station to a dominion and, finally, to its economic and political collapse in 1934. The principal question that remains is why Newfoundland sought increasing political independence if it was so dependent on Britain for its commercial and economic existence. Only by studying the history of the seal fishery can we fully understand what happened in Newfoundland during this period. In fact, the demand for oil, beginning in the late eighteenth century, had an impact on all underdeveloped parts of the world where this resource, in any form, could be found. For the next 100 years, societies, cultures and environments underwent transformations that are still apparent today. By the time their oil-producing resources had become depleted and, sometimes exhausted, petroleum and electricity had become available to urban and industrial customers, who were, for the most part, oblivious to the international changes they had wrought. In those few cases where resources remained viable, demand collapsed, with the same chaotic result on local economies. It is within this context that the present book examines the history of the Newfoundland seal fishery to 1914; exploring the extent to which Newfoundland's political and economic development in the nineteenth century (and the country's ultimate collapse in 1934) was directly linked to the expansion, and eventual decline, of the local sealing industry as it responded to the vicissitudes of the international oil trade.

In the past, the history of the Newfoundland seal fishery to 1914 has attracted the attention of several types of writers. There were the early contemporary observers, such as Rev. Lewis Amadeus Anspach, Sir Richard Bonnycastle, Joseph Beete Jukes, Rev. Charles Pedley and Rev. Philip Tocque,[2] whose observations are invaluable and whose views will be presented in the course of this study. In the late nineteenth and early twentieth centuries, a generation of popular writers adopted the seal fishery as a major topic: Levi George Chafe, Maurice A. Devine, Patrick Kevin Devine, Moses Harvey, Patrick Thomas McGrath, Harris Munden Mosdell, William Archibald Munn, James Murphy, and Henry Francis Shortis.[3] Several major historians, including Tocque, Pedley,

Harvey (mentioned above) and Daniel Woodley Prowse, analyzed the industry and its impact on the economy as objectively as the practice of history then allowed.[4] Finally, in the early twentieth century, there were the writings of observers and participants in the industry, who recorded their experiences and views. Included in this group were the famous captains Robert Bartlett and Abram Kean, who wrote informative autobiographies. Others in this group, while not so well-informed, did go to the ice and published their accounts. The most notable of these was the American writer, George Allan England, who went to the ice as a passenger on the S.S. *Terra Nova* in 1923; less well-known works were produced by Major W. H. Greene, John Harvey and Alexander A. Parsons. And finally, there were the memoirs of the Newfoundland planter, Nicholas Smith.[5] In recent years, the study of the Newfoundland seal fishery has attracted scholars and academics, and foremost among these are Briton Cooper Busch, James E. Candow, Michael Harrington, James K. Hiller, Chesley W. Sanger and Naboth Winsor.[6] While their studies are valuable, and will be referred to where appropriate, none is primarily concerned with a detailed examination of the questions proposed here.

Thematically, this study begins where *Fish out of Water* left off. It examines the extent to which a new industry—the seal fishery[7]—changed the Newfoundland cod fishing station, and explores the consequences of the growth, development, and decline of this industry. But it looks outward as well to a broader context: to the late eighteenth and nineteenth century industrial, commercial and urban requirements for fuel and lubrication, and to the effects those escalating demands had on resources and people.

The study of the Newfoundland seal fishery to 1914 is, to a great extent, the study of the colony itself in the nineteenth century. However, because the origins of this industry and the setting in which it first flourished are so important, the 'Introduction' to the present study is a substantial discussion of the early history of Newfoundland to c.1815. The central focus is an examination of the seal fishery from c.1815 to 1914, and this period is treated topically. In keeping with the fact that demand for seal oil was the critical factor in the growth of this industry, Chapter 1 examines the evolution of the international and British markets up to 1914. Chapter 2 discusses and analyzes the place of the seal fishery in the economy of Newfoundland—a complex issue because the seal fishery was not only an important component in the economy, but was sometimes the major player. The following chapter examines the ports involved in the seal fishery and the fleets that originated from these ports. Manpower and the conditions under which work was carried out are studied in Chapter 4. Chapter 5 examines the disasters in the seal fishery in order to better illustrate the conditions under which this industry was prosecuted. Because the living and working conditions discussed in Chapter 4 sometimes led to the disasters discussed in Chapter 5, readers will notice some repetition in the latter as points mentioned earlier are reiterated. However, given the number of lives lost and families so tragically affected, the disasters merited a separate chapter and some repetition was unavoidable. Chapter 6 takes up where Chapter 2, on the economy, leaves off and deals with the impact of the seal fishery on the major aspects of Newfoundland's nineteenth century development; it looks at the responses of the community, both sealing and non-sealing, to these developments—all in the context of the growth of a Newfoundland identity. The Conclusion attempts to clarify the extent to which the rise and growth of the colony of Newfoundland, and its eventual collapse as an independent dominion, resulted from the evolving industrialization of the western world.

NOTES

1. See Shannon Ryan, *Fish out of Water: The Newfoundland Saltfish Trade, 1814-1914* (St. John's, 1986), p. 257.
2. Rev. Lewis Amadeus Anspach, *A History of the Island of Newfoundland* (London, 1819); Sir Richard Bonnycastle (British army engineer), *Newfoundland in 1842* (London, 1842); Joseph Beete Jukes (geological surveyor), *Excursions in and about Newfoundland* (London, 1842); Rev. Charles Pedley, *The History of Newfoundland from Earliest Times to the Year 1860* (London, 1863) and Rev. Philip Tocque, *Wandering Thoughts* or *Solitary Hours* (London, 1846), and *Newfoundland: as it was and as it is in 1877* (London, 1878). Please note that full names are given when known using the *DNLB* and the *ENL* as the principal sources.
3. Levi George Chafe (1861-1942), *Report of the Newfoundland Seal-Fishery from 1863 ... to 1894* (St. John's, 1894), *Report of the Newfoundland Seal Fishery from 1863 ... to 1905* (St. John's, 1905); Chafe also provided the statistics for the Mosdell edition of *Chafe's Sealing Book* (see below); Maurice A. Devine (1857-1915), journalist, songwriter and local historian; Patrick Kevin Devine (1859-1950 and brother of Maurice), journalist and author of *Ye Olde St. John's* (St. John's, 1936), as well as other works; Moses Harvey (1820-1901 and in collaboration with Joseph Hatton), *Newfoundland, The Oldest British Colony* (London, 1883); Patrick Thomas McGrath (1868-1929), journalist and author of *Newfoundland in 1911* (London, 1911), but best known for researching the Newfoundland case in the Labrador boundary dispute; Harris Munden Mosdell (1883-1944), journalist and author of *When was That: 5000 facts about Newfoundland* (St. John's, 1923), and also editor of *Chafe's Sealing Book* (3rd. ed.; St. John's, 1923); William Archibald Munn (1864-1940), prominent historian and author of a number of works on Newfoundland and an early advocate of the theory that the island was the site of a viking settlement—wrote a lengthy introduction to *Chafe's Sealing Book* (St. John's: 1923), and a "History of Harbour Grace," which appeared as articles in the *Newfoundland Quarterly* (St. John's 1934-39); James Murphy (1868-1931), writer and collector of songs and ballads—many of which were published in *Songs and Ballads of Terra Nova* (St. John's, 1895); and Henry Francis Shortis (1855-1835), journalist, 'historiographer' of the Newfoundland Museum and author of numerous newspaper articles on Newfoundland history and folklore—many of which he transcribed in an unpublished volume he called his "Fugitive History." These writers romanticized the Newfoundland seal fishery, and some of their works are presented in Chapter 6; all except Harvey were Newfoundland-born. Harvey and Prowse (see following endnote) wrote for both the local and international readership. Chafe was largely concerned with statistics; (see also Shannon Ryan, ed., *Chafe's Sealing Book: A Statistical Record of the Newfoundland Steamer Seal Fishery, 1863-1941* (St. John's, 1989) for an edition of his complete statistical collection.) Full names and titles are used only in the first references. Please note that in many cases, only examples of each writer's work are mentioned; many of these and other contemporary writers contributed valuable articles to newspapers and periodicals.
4. Daniel Woodley Prowse, *A History of Newfoundland from the English, Colonial, and Foreign Records* (London, 1895). This is by far the most comprehensive and authoritative history of Newfoundland.

5   Captain Robert Bartlett, *The Log of Bob Bartlett* (New York, 1928); Captain Abram Kean, *Old and Young Ahead* (London, 1935); George Allan England, *Vikings of the Ice* (New York, 1924), and reprinted as *The Greatest Hunt in the World* (Montreal, 1969); Major William Howe Greene, *The Wooden Walls Among the Ice Floes* (London, 1933); Hon. John Harvey (a partner of Harvey and Co.), "With the Ice Hunters," *Newfoundland Quarterly* VI, no. 4, (March 1907), 3-8; Alexander A. Parsons, "Our Great Sealing Industry," *Newfoundland Quarterly* XIV, no. 4., (April 1915), 6-9 and editor of the *Evening Telegram* (1879-1904); and Nicholas Smith, *Fifty-two Years at the Labrador Fishery* (London, 1936).

6   Briton Cooper Busch, *The War against the Seals: A History of the North American Seal Fishery* (McGill-Queen's University Press, 1985); James E. Candow, *Of Men and Seals: A History of the Newfoundland Seal Hunt* (Canadian Parks Service, 1989); Michael Harrington, *Goin' To The Ice: Offbeat History of the Newfoundland Sealfishery* (St. John's, 1986); James K. Hiller, "The Newfoundland Seal Fishery: An Historical Introduction," *Bulletin of Canadian Studies* VII, no. 2, (Winter 1983-84), 49-72; Chesley W. Sanger, "Technological and Spatial Adaptation in the Newfoundland Seal Fishery in the Nineteenth Century," (MA thesis, MUN, 1973); Sanger, "The Evolution of Sealing and the Spread of Permanent Settlement in North-eastern Newfoundland," *The Peopling of Newfoundland: Essays in Historical Geography*, ed. John J. Mannion (St. John's, 1977); and Naboth Winsor, *Stalwart Men and Sturdy Ships: A History of the Prosecution of the Seal Fishery by the Sealers of Bonavista Bay North, Newfoundland* (Gander, Newfoundland, 1985).

7   The term 'fishery' was applied to the prosecution, in a general and all inclusive sense, of the sealing operation. At the same time, it was recognized that the men and vessels were 'ice hunters' while searching for the seals on and among the ice floes. For definition and exemplification of both terms see *DNE*.

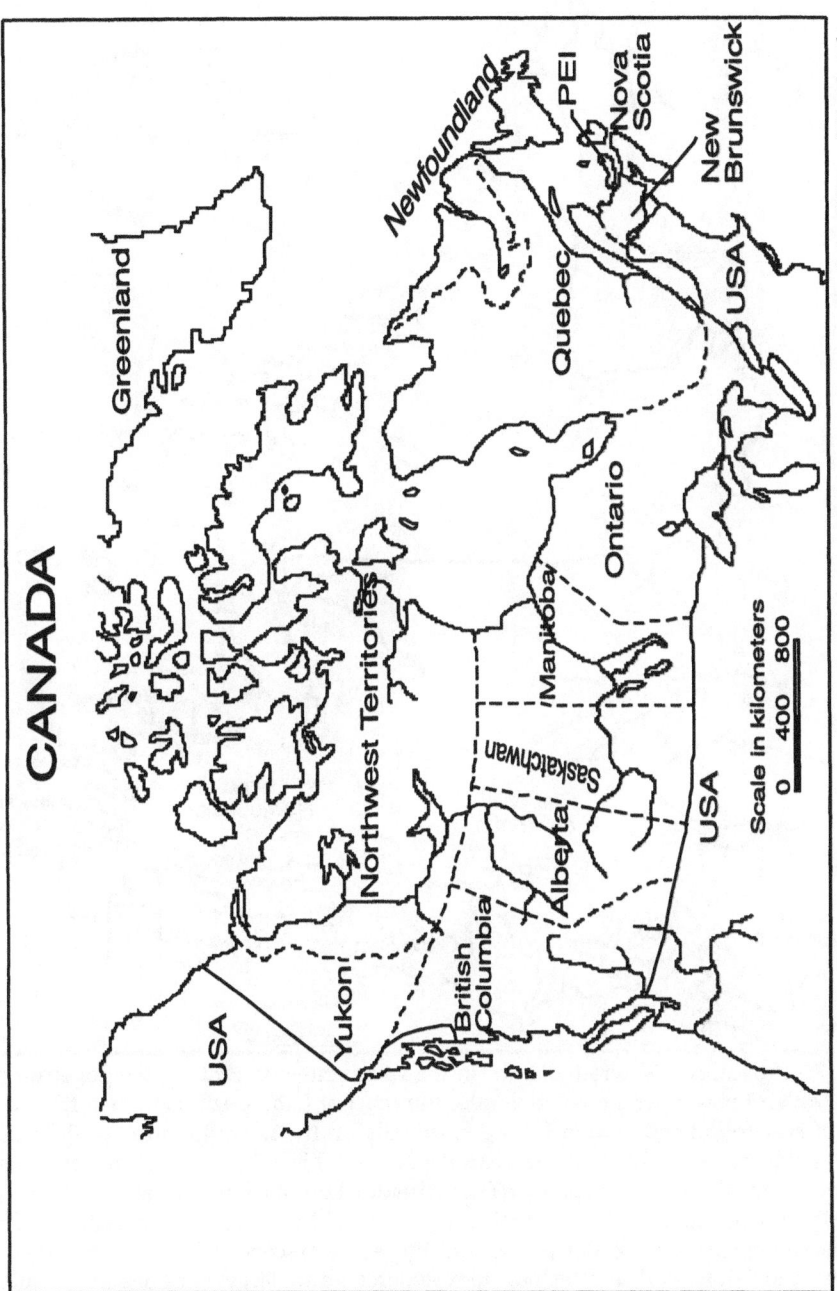

Figure 1 - Canada. England's exploration of North America began in the annual voyages of the West of England fishermen to the coast of the "New Found Land" (Terre Neuve) in the early 1500s - 100 years before the English

Figure 2 - Province of Newfoundland. In the late eighteenth century, the governor of Newfoundland was given jurisdiction over the coast of Labrador. Fishermen from the Island of Newfoundland started fishing seasonally on the Labrador coast in the early 1800s. Battle Harbour became the central point of this migratory fishery and also supported a local inhabitant fishery. The Labrador boundary was established by the British Privy Council in 1927 after a joint request by Canada and Newfoundland. The southern settlements are populated in general by people descended from Europeans who had stayed to fish for cod, and the northern settlements are largely populated by native people. However, many harbours, coves and islands were inhabited intermittently and do not have any permanent population; others that supported permanent population were abandoned in this century. (For example Battle Harbour, Ramah and Okak have been resettled and Davis Inlet is scheduled for resettlement.) Map by William Ryan.

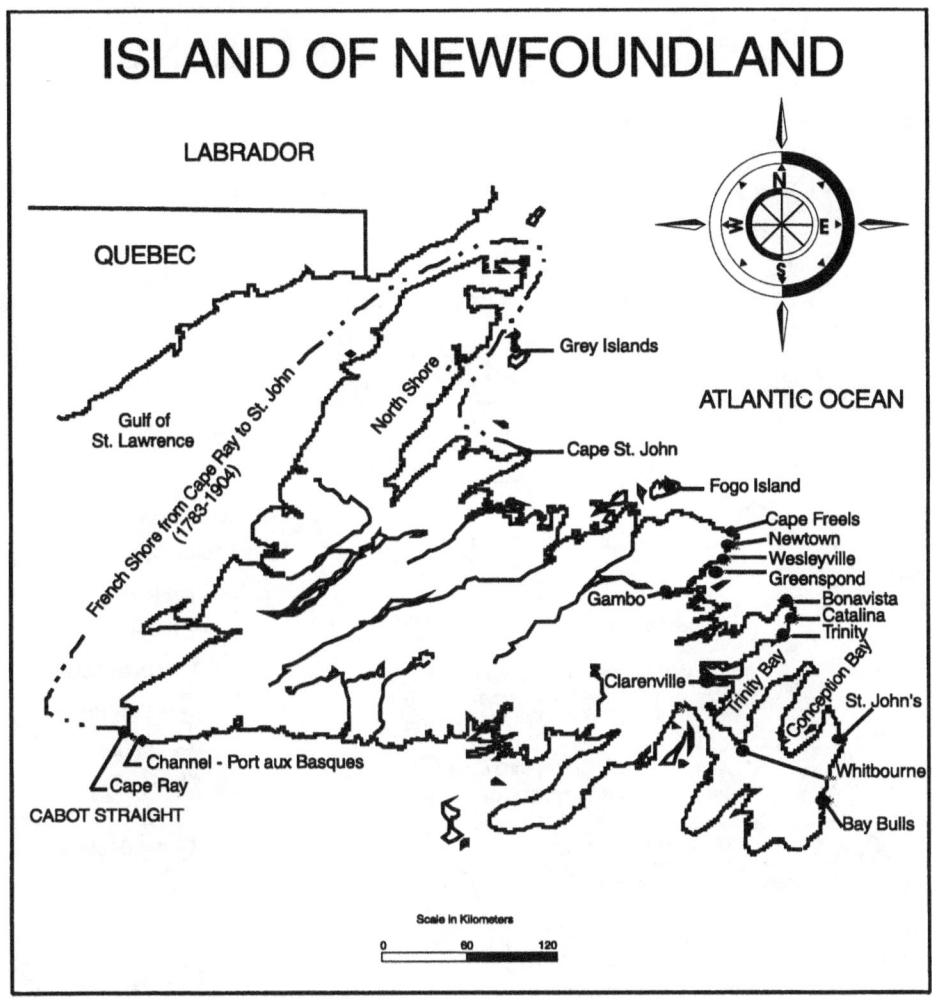

Figure 3 - Island of Newfoundland. Conception Bay fishermen led the way in establishing a migratory fishery on the 'North Shore' (i.e., The Great Northern Peninsula) during the absence of the French fishermen between c.1800 and 1815. When the French returned to the 'French Shore' after 1815, English fishermen from Conception Bay and St. John's began fishing on the Labrador coast further north. Most seals were found on the ice floes which accumulated in the area between the North Shore and Bonavista and extended many miles into the Atlantic Ocean. This area supported a 'landsman' seal fishery, in addition to the vessel fishery which was based in Conception Bay and St. John's. With the concentration of the sealing industry in St. John's in the late 1800s, many men travelled to the capital for employment in the seal fishery. After the building of the trans-island railway in the 1880s and 1890s, sealers walked to the main railway stations, particularly Whitbourne, Clarenville and Gambo, for transport by rail. As Conception Bay declined economically, sealers and their officers tended, increasingly, to come from the northerly ports - especially from the Cape Freels area. Map by William Ryan.

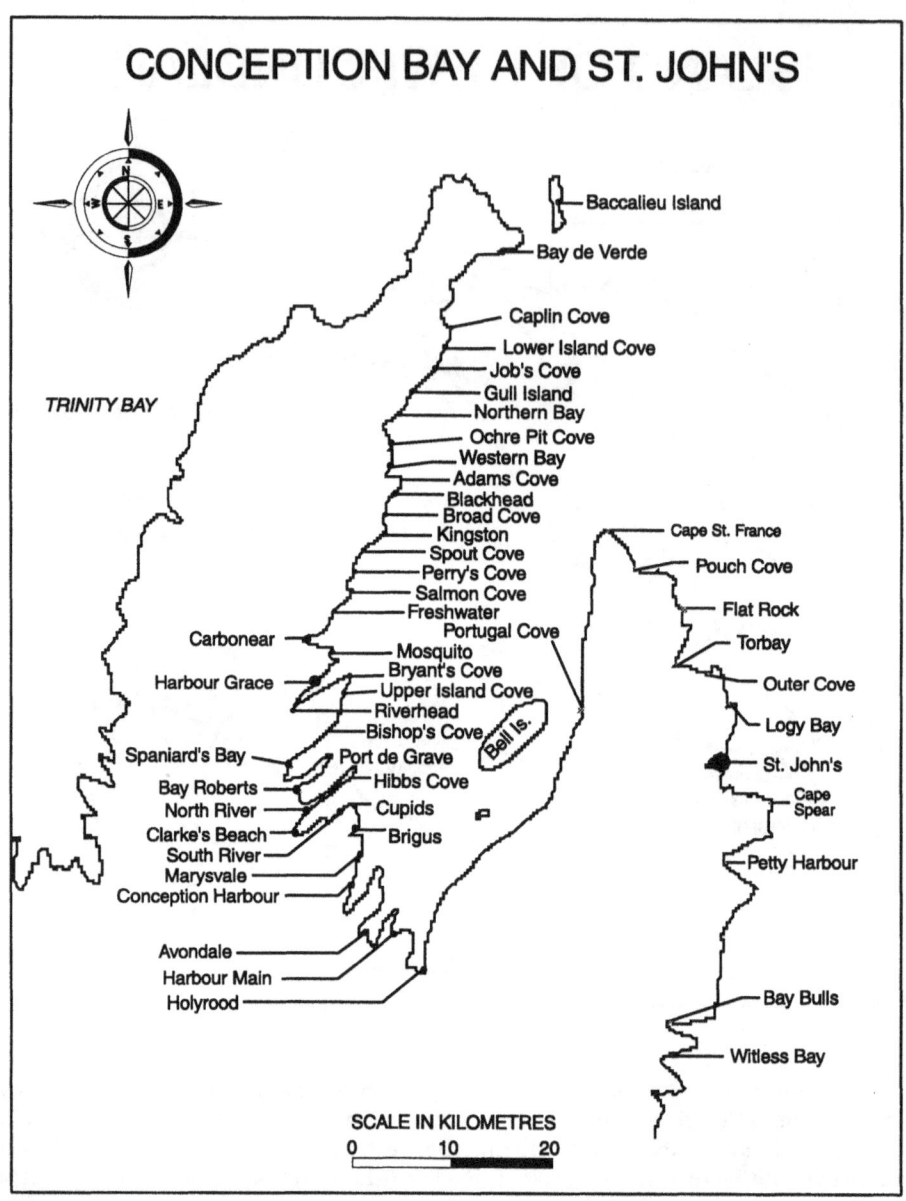

Figure 4 - Conception Bay and St. John's. The planters (or wealthier fishermen) from these two areas expanded their cod fishery operations during the Napoleonic Wars by sending small vessels to the North Shore for the summer fishing season. After 1815, these vessels sailed farther north to the Labrador coast; they were also used in the spring seal fishery. Harbour Grace, followed by Carbonear and Brigus, were the leading outports in Conception Bay, and most fishermen from the smaller coves and harbours journeyed to these ports for employment in the seal fishery and, often, in the Labrador fishery. The west side of Conception Bay enjoyed shelter from the prevailing west and northwest winds and was also protected, to some extent, from the offshore fog. With the decline of sailing ships in the seal fishery, the industry became dominated by St. John's. Map by William Ryan.

Introduction

Successful European settlement in North America was always heavily dependent on the resources that prospective settlers discovered. The waters around Newfoundland offered unlimited supplies of cod fish (but very little else) to European fishermen who visited annually during the summer months. However, while the migratory fishery encouraged and supported limited and scattered settlement in Newfoundland from the end of the sixteenth century, the short fishing season, combined with an inadequate agricultural base, prevented settlement from growing and acquiring an existence independent of the migratory fishery until the nineteenth century. However, the population of Newfoundland expanded during the early years of the nineteenth century because of the unusual circumstances of this period; and accompanying this expansion, there was the development of a new industry—the seal fishery. But before examining the history of the spring seal fishery, it is necessary to examine the internal developments in Newfoundland and the vicissitudes of its trade, fisheries and settlement in order to understand the local factors which allowed and even encouraged its appearance and growth.

Fishing Ship Fishery

In its infancy, the West of England-Newfoundland migratory fishery was easily defined.[1] From the beginning of the sixteenth century, fishing ships carrying fishing crews, boats, equipment and supplies came annually from the West of England ports to the harbours of Newfoundland's east coast to engage in the summer cod fishery. This summer fishery exploited the schools of cod that migrated from offshore waters to the coastal or inshore waters in June of each year in pursuit of the schools of caplin that arrived to spawn in the shallow waters near sandy beaches. The cod remained in these coastal waters in large numbers until about the middle of August. In certain areas, they lingered just offshore into the autumn, thus supporting a fall fishery. However, the sixteenth- and seventeenth-century fishermen tried to catch and dry as much fish as possible before September because of diminishing returns after that due to scarcer fish and poorer drying weather.

The ships were anchored in the harbours and the crews went ashore to build tilts, cookrooms, fishing stages and flakes. The fishing stage was a covered shed extending out over the water and served as a wharf for tying up the boats and as the place for splitting and salting the cod. The flake was a large platform (extending inland behind the stage) built of poles and covered with boughs on which the salted cod was dried. The men slept in the tilts and ate in the cookrooms. All structures were built from local timber and covered with tree rinds to keep out the rain. Such a fishing premises was usually referred to as a 'fishing room', and larger ones occupied by permanent (or semi-permanent) inhabitants or caretakers were labelled 'plantations'.

The fishermen fished from small boats holding three (but sometimes more) men, each using one or two baited hooks on long hand lines. Because three men in the boat and two men ashore curing the catch was the norm (once the fishery was in full operation), each ship usually carried five men for each boat. At the beginning of the season, the men used mussels for bait; later in the season, herring became available for this purpose; in late June, caplin were plentiful; and, finally, in mid summer, squid.[2] Once landed in the stage, the fish was headed, gutted and split and the backbone removed.[3] It was then laid flat in rows and covered with salt, and when a pile was about three feet high, another pile was started. After a week or two in salt (depending on the type of product desired), the fish was spread on the flakes to dry in fine weather, with great care taken to protect it from rain. It was piled in small mounds ('faggots') at night and, if necessary, covered with tree rinds. When completely cured, it was placed in large piles covered with rinds to await shipping at the end of the voyage. It was then known as 'saltfish'. Meanwhile, the livers were dropped into a vat, where they gradually rendered into 'cod oil' during the course of the warm summer.

This early English fishery was a small operation, but it increased in the later sixteenth century due to the problems that beset the larger French, Spanish and Portuguese fisheries, and by the beginning of the seventeenth century, the West of England migratory fleet numbered over 200 ships and supplied Spain and Portugal with most of their saltfish needs.

While the fishing ships initially transported their fish to market, it was impractical and expensive to take 50 or 60 men to Spain when a trading ship with 10 or 12 men could do this. Therefore, in the early seventeenth century, trading (or 'sack') ships came to Newfoundland late in the summer and brought the fish to market. This left the 'fishing' ships free to return directly to England, carrying the fishermen, the cod oil (usually referred to as 'train' oil), a little saltfish for local consumption and dried timber for firewood and other uses. This 'fishing ship' fishery reached its zenith during 1600-20, but in the latter part of the 1620s, it began to experience difficulties because of wars, depressions, market problems and a lack of naval protection from the Sallee Rovers.

In 1634, after much petitioning, the 'Western Adventurers', as the West Country merchants were called, managed to persuade the English government to grant them their own charter—the 'Western Charter'. The Charter contained clauses intended to allay the fears of the migratory fishermen that they would lose their rights in Newfoundland: provisions (including some from John Guy's 'Laws'—see below) attempting to put an end to abuses in the fishery; and clauses giving the first captain to enter a harbour in the spring the title of 'Admiral' and the responsibility to maintain law and order in the harbour during that particular summer. However, the migratory ship fishery continued to experience problems and the western adventurers began to blame the new bye boat keepers and the inhabitants for their difficulties. After abortive efforts to get rid of these two groups in the 1660s, 70s and 80s, the adventurers learned to adjust to the new situation, and soon, a different industry had evolved. By the end of Queen Anne's War in 1713, the old ship fishery had declined considerably, and some adventurers were supplying bye boat keepers and a small inhabitant fishery, while others were just beginning to prosecute the bank or deep sea fishery. Matthews concludes that there was no place for the old fishing ship fishery in Newfoundland after 1763.

Bye Boat Fishery

The next major change in the structure of the migratory fishery was the development of a 'bye boat' fishery. This branch of the industry began in the 1640s,[4] when the traditional ship fishery was experiencing problems. Briefly stated, a boat owner arranged transportation to Newfoundland for himself and his crew on board a fishing ship. Once he arrived in his chosen harbour, he and his men built a fishing room—albeit a smaller one than the usual fishing ship room—and then engaged in the fishery during the summer months. In the late summer or early autumn, the bye boat keeper sold his fish to a sack ship and returned to England with his crew, once again on board a fishing ship.

This industry had its limitations because the bye boat keeper was dependent on the fishing ships for transportation and dependent on the sack ships for the sale of his fish. However, these disadvantages became assets during periods of high costs and depression. The bye boat keeper did not have to invest in his own ship with 40 to 60 men and 8 to 12 boats, for example. When economic conditions dictated, he could take as few as four men and one boat, and increase his investment when times improved. In addition, he could stay out of the fishery for a year or two if necessary without any great loss—an option that the average ship owner did not usually enjoy. Furthermore, he could arrange for transportation on the earliest, fastest and safest fishing ships, and his men were not required to work as sailors on the voyage; neither were they required to go to southern Europe in the fall at short notice, which sometimes happened to the fishing ship crews. These characteristics made bye boat fishing attractive to fishing servants, and the bye boat keepers could hire from among the best. In addition, the bye boat keeper could arrange to arrive early in Newfoundland and leave later, thereby lengthening the fishing season; in fact, he could and did remain in Newfoundland over the winter when it was to his advantage. Matthews maintains that the bye boat keepers began the practice of paying their men wages instead of shares, which allowed them to attract the best fishing servants during the depressed 1640s and 1650s, when this branch of the fishery became established. Finally, the bye boat keeper enjoyed advantages over the few resident fishermen. He could buy his supplies in England, where prices were somewhat cheaper, and he could find employment in England during the winter months when it was necessary.

Given the wars and depressions of the seventeenth and eighteenth centuries, it is not difficult to see why this branch of the migratory fishery expanded, reaching its peak during the early 1770s, immediately prior to the American Revolutionary War, when the number of bye boat keepers and servants averaged 525 and 5,691 respectively per year. However, the bye boat keepers could easily become inhabitants or 'planters', and this is why this class of fishermen disappeared in the late eighteenth century.

Bank Fishery

Unlike the other major European fishing nations, England was slow to develop a bank or deep sea fishery. While the Spanish and Portuguese fleets ceased to be of significance by the end of the sixteenth century, the French continued to operate a large bank fishery during peace time and also operated an inshore fishery on the island's south coast, with headquarters at their colony, Placentia, after the 1660s. Early in the eighteenth century, England/Britain began to prosecute the bank fishery using the Newfoundland harbours as a base.[5]

British fishing ships began to frequent the offshore fishing banks after the Treaty of Utrecht in 1713—the peace treaty that saw the ending of France's colonization efforts in Newfoundland and the confining of that nation's fishing fleets to part of the island's north and east coasts on a migratory basis only.[6] Thus, beginning in 1713, the British migratory ship fishery began to expand slowly to include a bank fishery offshore. This development appears to have resulted from catch failures in the inshore fishery during the years immediately following the end of war. Many captains began to realize that they could engage in the bank fishery with smaller crews, and earn money by transporting bye boat keepers, their crews and supplies as well as by bringing supplies and crews out to the resident planters. After three or four weeks' fishing on the offshore banks, the ships returned to the island, where the fish was washed and dried by residents at an agreed price per cwt while the ship returned to the fishery. Therefore, the bank ships could concentrate their full attention on production, and the fishing period could be extended as well because the ships were not at the mercy of the migratory habits of the inshore cod stocks.

Statistically, it is impossible to distinguish bank ships from the traditional fishing ships for much of the eighteenth century because the annual reports which were collected on a more or less regular basis after 1698 classified all ships engaged in fishing in the same category until 1769.[7] However, it is worth noting that the total fishing fleet expanded during the peaceful interlude between King William's War and Queen Anne's War. Of the total number of ships visiting Newfoundland in 1699 (i.e., 236—figures for 1698 missing), 166 were classified as fishing ships and these employed 669 fishing boats. At an average of five men per boat, this would indicate that about 3,345 men were fishing. Similarly, in 1700, there were 171 fishing ships (out of 220) employed fishing with 800 boats and about 4,000 men. Thus, at the turn of the eighteenth century, the average fishing ship carried 20 men and four fishing boats. In 1720, out of a total of 151 ships, 73 were classified as fishing ships with an average burthen of 81 tons, and these carried, on average, 21 men and three boats. Because bank ships carried boats for obtaining bait, wood and water and not for the actual fishery, which was carried out from the side of the ship, they did not need as many boats. Therefore, the ratio of boats to men decreased as more of the 'fishing' ships were employed in the bank fishery.

Throughout the century, this became the trend. In 1769, when the types of ships were differentiated in the annual reports for the first time, there were 354 British fishing ships in Newfoundland waters, and of these, 222 were bank ships, leaving 132 employed in the regular inshore fishery. There is no information on the number of men or boats each of these two fleets carried, but a total of 430 boats were employed, and since few would have been carried by the bank ships, it must be concluded that nearly all of these were engaged in the inshore fishery—roughly three boats per fishing ship. From this and other information, Matthews concludes that even the old-style fishing fleet had changed and was also involved in the sack trade—fishing while awaiting cargoes to be cured by the bye boat keepers, residents and bank ships.

The traditional migratory fishing ship fishery had evolved to include various branches by 1763, and this trend continued in the succeeding years and decades. The bank fishery reached its peak in the years immediately following the American Revolutionary War. However, before continuing the discussion of the migratory fishery, it is necessary to examine other developments—particularly those concerning settlement and the establishment of a resident fishery using migrant labour.

Early Settlement

It is generally agreed that the first attempt by the English to live in Newfoundland all year round took place in 1610,[8] when John Guy was appointed by the London-Bristol Company to establish a colony in Newfoundland. Guy chose Cupids in Conception Bay as the site for this colony, probably because it was small and not too popular with the migratory fishermen.[9] Other similar attempts followed: the Bristol Company in Harbour Grace (Bristol's Hope); Vaughan's and Falkland's futile attempts in Renews; George Calvert's (Lord Baltimore) more ambitious attempt in Ferryland; and finally, the most important venture of all, the endeavour, also in Ferryland, carried out by David Kirke and his associates during 1637-52. Whether these colonies, or any one of them, should be classified as successes or failures has long been a matter of debate. Most professional historians agree that they failed as investments but that they succeeded as settlements because some fishermen continued to live there long after the organizers left. With Baltimore's original investment and Kirke's later administration, Ferryland stands out as the major site of these attempts, but more information is needed before one can successfully compare it with Harbour Grace (Bristol's Hope) and with unofficial and informal settlements, such as the one at Bonavista.

Meanwhile, permanent and semi-permanent settlement did not flow only from the colonization efforts of the first half of the seventeenth century. The migratory fishery also became involved in a big way. By 1600, the east coast of Newfoundland—probably much of the shoreline between Ferryland and Bonavista—was becoming an ecological mess. Hundreds of ships and tens of thousands of men visited the harbours annually during the latter decades of the sixteenth century. Many ships carried stones for ballast and these were usually thrown into the harbours thus making the water more shallow. Stages, flakes, cookrooms and tilts were needed every year and trees were cut for these purposes. Other trees were rinded to provide water-tight bark for roofing material for the buildings and to provide covering for the fish piles; this resulted in large stands of dead trees which fuelled disastrous forest fires. In Newfoundland, where forest regrowth was slow and generally did not occur at all because of soil erosion, the result of all this activity was extensive deforestation. Some idea of how rapidly and thoroughly this deforestation occurred in the latter sixteenth century can be ascertained from the differing observations of Stephen Parmenius and John Guy within a 30-year period. In 1583, Parmenius, a Hungarian poet accompanying Sir Humphrey Gilbert, wrote, in a letter to Richard Hakluyt in London, that they could not travel any distance from shore in St. John's because the trees came down to the water's edge.[10] In 1610, less than three decades later, John Guy was astounded at the destruction of the forests that had occurred and, in 1611, posted his 'Laws' prohibiting the traditional practice of destroying the fishing premises at the end of the season. This had always been the custom because the fishery operated on a 'first-come, first-served' basis with the first ships to arrive each spring being permitted to take the best fishing rooms—even if buildings had been left intact by others from the previous year. Thus, a fishing room that was built by one crew in any given year could be taken over by another crew the following year. Crews were known to burn their premises rather than leave them for others. In addition, the dried sticks from the fishing premises were worth carrying back to the West Country when space was available on the returning ships. However, the adventurers recognized that the practice of building new premises each spring was wasteful in terms of time and money—and increasingly so as the forest cover receded annually. Consequently, almost simultaneously with the publication of Guy's laws, it appears that the migratory captains began to leave caretakers behind to maintain their property during the winter and to have everything in readiness in the spring. These caretakers were often joined by their relatives along with some migratory fishermen, who decided to take up residence, sell their fish to the sack ships and buy their supplies and hire fishing servants from England. In this way, settlement became an integral part of the migratory fishery.

Therefore, in a parallel manner, harbours other than those settled by the colonizers were being gradually populated by a handful of resident fishermen during this period. By 1696-97, there were over 1,500 men (including planters and servants) living in over 38 separate harbours and coves along the English shore (see Table Intro.1). Because there were about 200 fishing establishments, or 'plantations' (including those north of Bonavista) one may assume that there were also about 200 men in charge of these plantations. Some were, no doubt, agents of migratory captains, while others, known as planters, were operating independent fishing businesses but were tied to the migratory ships financially and through family connections.

The fishery carried out by these resident fishermen was similar to that of the bye boat keepers and fishing ships; each resident hired a crew to fish and cure the catch. These crews were transported to Newfoundland in the spring and were returned to England in the fall by fishing captains, who charged them for their passage. In effect, all resident men (and the few independent women) were masters (and mistresses) of fishing operations that employed migrant fishing servants—many of whom spent one winter and two summers in Newfoundland as the conditions of their employment. A growing number of these men continued on to the English settlements on the mainland of North America at the conclusion of their contracts, where steady employment and the presence of women allowed them to establish homes and families. Thus, this resident fishery (or 'inhabitant' fishery, as the naval commanders began to call it) was essentially a migratory fishery dependent on an annual supply of migratory fishing servants.

All English fishing operations, except those in Bonavista, were destroyed by the French raid in 1696-97. People in the Harbour Grace/Carbonear area managed to escape the French by fleeing to Carbonear Island, but they lost all their property. An inhabitant population was re-established during the peace between 1697 and 1702, but during the following war, the French resumed their destructive winter raids from their headquarters in Placentia. However, in 1713, the French were forced to surrender their settlements in Newfoundland. Soon, English (and later, Irish) fishermen began to move back into the traditional harbours and coves of the old English shore, and also into St. Mary's Bay and Placentia Bay on the south coast, where the French had traditionally lived and fished.

Although the removal of the French from the south coast in 1713 brought an end to the threats from that quarter, the depressed state of the fishery after 1713 discouraged any growth in the migratory (including resident) fishery. In 1719, the first year for which the figures are complete (and seem reliable), the population of the island was as follows: 264 masters (of households or plantations); 1,346 men servants; 172 mistresses (of households or plantations); 81 women servants; and 466 children—a total of 2,329 people.[11] The resident population hardly grew at all during the 1720s and only very slowly during the 1730s. In 1738, the figures were as follows: 472 masters; 3,147 men servants; 354 mistresses; 144 women servants; and 861 children—a total of 4,978.

There were a couple of factors involved in the reasons for this slow population growth. In the first place, settlement was an integral part of the total migratory fishery and, under the circumstances, could only expand in tandem with the general growth of this industry. Related to this was the fact that it was easy to obtain passage to the mainland, where opportunities were so much greater.

However, the number of inhabitants increased gradually. Some moved farther north and became more involved in the salmon fishery, fur trapping, shipbuilding and the seal fishery; in other words the economy became slightly more diversified. In 1750, the population figures were as follows: 850 masters; 5,415 men servants; 609 mistresses; 310 women servants; and 1,043 children—a total of 8,225 people. Thus, approximately 850 fishing plantations located in hundreds of coves and harbours were dependent on 5,415 migrant fishing servants.

Two new developments brought further increases in the resident population. Firstly, in the 1750s, the potato was introduced into the island and became an important food staple; in addition, it proved to be a valuable source of vitamin C and, no doubt, led to the elimination of scurvy. Secondly, in 1763, New France was surrendered to the British, and Britain replaced France as that colony's trading partner. This resulted in more British shipping entering Newfoundland harbours, especially St. John's. These 'partial adventurers' (who were disliked by the West Country families established in the Newfoundland fishery and trade) would often visit Newfoundland on speculation—to fill some remaining cargo space, to sell supplies, to deliver barrel staves from Quebec, for example. In other words, while a trip across the Atlantic could not be justified for these purposes only, calling in at Newfoundland added very little expense, if any, to the cost of a voyage between Britain and Quebec. This created a more stable environment for inhabitant plantations, and by the late 1760s, the total population numbered 11,000 to 12,000 inhabitants—including 5,000 to 7,000 transient servants.

In the meantime, Newfoundland had become something of a problem for the British government. In a parliamentary act of 1699, "An Act to encourage the trade to Newfoundland," the government had granted inhabitants the right to live and fish and occupy fishing rooms on the island under certain conditions—the prime one being that they not interfere with the migratory ship fishery. The first captain to arrive in each harbour in the spring was to become the 'Fishing Admiral' for the season, as by previous custom, with the added stipulation that inhabitants could appeal to the naval convoy commander if they were dissatisfied with a decision of an admiral. In 1729, the convoy commander, whose primary responsibility was to patrol the waters along the coast, was appointed governor, with authority to appoint justices of the peace and constables. From this point on, the convoy commander was also the naval governor, with a small fleet at his disposal in Newfoundland waters each summer. He collected records, defended the fishery when necessary, protected Britain's sovereignty and helped maintain law and order. Only after the Napoleonic war were these governors required to serve in Newfoundland all year round. In all, these developments made life and property more secure and most certainly were factors contributing to the growth of population described above. But even with these developments, there were only about 1,300 households in Newfoundland by the outbreak of the American Revolutionary War.

Post-1775

The British government had been convinced since the latter seventeenth century that the Newfoundland cod fishery was valuable as a migratory ship fishery only. It employed British fishermen; it earned foreign exchange; imports from the fish markets paid tariffs to the treasury; the fishery consumed British goods; and it provided vast numbers of trained seamen who were available to the navy when needed. On the other hand, the migrant fishing servants hired by the inhabitants travelled to Newfoundland as passengers and did not learn the skills required in order to sail large ships. Also, the British government was aware that if Newfoundland developed into an independent colony—like its neighbours along the Atlantic seaboard—most of the benefits above would accrue to the colony rather than to Britain. Consequently, the modest growth in the inhabitant fishery alarmed some observers, and in 1775, a parliamentary act was passed strengthening the government's power to prevent the growth of a resident fishery. The preamble stated:

> Whereas the fisheries carried on by H.M. subjects of G. Britain and of the British dominions in Europe have been found to be the best nurseries for able and experienced seamen; always ready to man the Royal Navy when occasions require; and it is therefore of the highest national importance to give all due encouragement to the said fisheries, and to endeavour to secure the annual return of the fishermen, sailors, and others employed therein to the Ports of Great Britain etc. at the end of every fishing season.[12]

Thus, two and three-quarter centuries after the discovery of Newfoundland's fish stocks by Europeans, the British government could only conceive of Newfoundland as the 'Grand Cod Fishery of the Universe'. This picture was to change completely within 40 years, but in the meantime, Newfoundland was still viewed by the British government as a 'Fishing ship moored on the Grand Banks'.

This parliamentary act, known in Newfoundland as Palliser's Act, coincided with the beginning of trouble between Britain and its 13 principal colonies on the American mainland. The dispute came to a head in the spring of 1775, when the government closed the port of Boston.[13] In support of Boston, the other ports on the Atlantic coast stopped their trade with Newfoundland. The short-term effect was to create starvation conditions in Newfoundland, both among the inhabitants and the migratory fishermen, because the entire fishery had come to depend upon ships from the British colonies to supply flour and hard bread (both necessities); also, vegetables, fruits, livestock and small ships (needed but not indispensable); and rum, sugar and molasses from the West Indies. (In return, the colonial ships had bought large quantities of the poorest quality fish for sale to the West Indian plantations.) The outbreak of war also put an end to the regular migration of fishing servants and bankrupt planters to New England and the other colonies.

The number of 'Ships from America' that entered Newfoundland ports (especially St. John's) increased from 75 in 1750 to 89 in 1760, and to 138 by 1770; in 1774, the number peaked at 175. In 1775, the number declined to 66, as only those already en route managed to reach the island, and in 1776, the records show that only 3 ships made the trip. There were great shortages of bread and flour, in particular, and many migratory ships were forced to return to Britain in the summer of 1775 to obtain whatever bread was available. The following winter was extremely difficult for the inhabitants, but by the summer of 1776, the industry had learned to adjust.

Meanwhile, American privateers operated along the coast of Newfoundland, and the war soon changed its character from a local revolutionary war to an international one as France declared war on Britain in 1778; Spain did likewise in 1779, and Holland, in 1780. The exportation of saltfish from the Newfoundland fisheries as a whole declined from 600,220 cwts in 1775 to 495,350 in 1776, and plummeted to 132,780 cwts in 1779. And the price paid for saltfish fell to its lowest point in many years in 1779, when fish fetched only between 8s and 11s per cwt. Meanwhile, many inhabitants fled the near-starvation conditions on ships returning to Britain, and the resident population dropped from about 1,400 households in 1775 to about 1,000 in 1781.

With the end of hostilities in 1783, Newfoundland entered a new era in terms of both fishery and settlement. There was growth in local trade as some merchants made St. John's their headquarters in order to participate directly in the two-way trade between Newfoundland and the West Indies, from which the United States of America was now excluded. This led to a little diversification of the economy and society, and to the establishment of a unified court in the early 1790s. Increased employment opportunities—in shipbuilding, for example—and the reduction in the numbers of American ships visiting the island, which reduced the flow of fishermen to the mainland, resulted in an increase in the inhabitant fishery and in settlement in general. In addition, according to at least one writer, the increasing tolerance shown to Irish Roman Catholics encouraged more of them to remain on the island.[14] Population rose from about 1,000 households in 1784 to over 2,200 in 1789, but declined to less than 2,000 in 1792—the last year of peace. Part of the reason for this decline was the depression in the fishery beginning in 1789 (see below).

The migratory ship fishery also expanded after 1783, although the records are somewhat incomplete. In 1784, 346 ships entered Newfoundland harbours, including 236 fishing ships. The fishing ships had a total burthen of 22,535 tons (average—95 tons), and carried 2,603 men (average—11 men) and 572 boats (average—2.5 boats). In 1788, the migratory fishing fleet reached a peak of 389 ships—the largest number of ships in the history of the West of England-Newfoundland fishery. This fleet had a combined burthen of 38,846 tons and carried 4,306 men and 273 boats. Also, the fleet's efficiency improved a little because while the average crew size remained at 11 men, the average burthen increased to 100 tons. While the number of these fishing ships which fished on the banks is not recorded, most were bank ships, according to other indications: for example, in 1786, there was a total of 280 fishing ships, including 181 bankers, in Newfoundland; and in 1789, the fleet was made up of 182 bankers and 122 other fishing ships.

Also during this period, the production of saltfish reached its highest level in Newfoundland's history to that point. In 1784, the total catch amounted to 437,316 cwts, which was distributed as follows: British fishing ships—131,650 cwts; bye boat keepers—93,050; and inhabitants—212,616. In 1788, the total catch amounted to 948,970 cwts, allocated as follows: British fishing ships—412,580 cwts; bye boat keepers—79,285; and inhabitants—457,105. The markets collapsed under the glut, prices declined, and bankruptcies occurred, creating the legal problems which forced the British government to reform thoroughly the whole legal system with the establishment of a unified court.

A few changes were still taking place, but for the remainder of this period up to the outbreak of war in 1793, the migratory ship fishery operated at a fairly stable level, averaging 260 ships annually during the three-year period 1790-92. As already indicated, this was primarily a bank fishery, but also included ships that combined both shipping and fishing while waiting for the early cargoes of saltfish to be ready for market. For example, in 1792, there were 187 bank ships in Newfoundland, 89 other British fishing ships, 161 British sack ships and 57 ships from America. The fish produced was as follows: bankers—139,450 cwts; other British fishing ships—16,910; inhabitants (including bye boat keepers)—395,900. However, the bye boat fishery seems to have become largely irrelevant by the end of 1788, when the last statistics were recorded. As Matthews points out, in order to retain their fishing rooms, they were forced to reside on them all year round, and therefore they became inhabitants (or 'planters', which was the term commonly applied to them in the nineteenth century). On the eve of the French Revolutionary War, the inhabitants had managed to take over the larger portion of the Newfoundland migratory fishery.

It is important to remember that this inhabitant fishery was basically a migratory fishery centred around independent, property-owning inhabitants (or 'masters', as the statistics recorded them). Thus, in 1792, while it is obvious from the figures quoted above that the inhabitant fishery produced far more fish than the British banking and other fishing ships, these inhabitants were dependent upon a considerable number of migratory fishing servants. In that year, there were reported 1,996 masters, 1,602 mistresses (usually wives of masters), 5,306 children and at the same time, 6,726 male servants, 833 female servants and 697 dieters.[15] While, no doubt, a few servants married and established their own households,[16] the number to do so was limited by the availability of winter employment, which was scarce and consisted only of a few jobs provided by the mercantile establishments and wealthier planters. In fact, the planters themselves provided much of the labour needed in the winter. Thus, the single fishing servants in the inhabitant fishery were migratory workers in the same way as their relatives and neighbours who fished for the British fishing ships. This is a point that has been overlooked by historians, who have tended to equate the growing proportion of fish caught by the inhabitants as direct evidence of increasing settlement and a declining migratory fishery. More accurately, the increasing inhabitant share indicates that the migratory fishery was simply becoming more and more under the control of the inhabitants. In other words, if Newfoundland could be (and was) compared to a fishing ship anchored on the Grand Banks, then the inhabitant households were fishing stations in coves and harbours with fishing boats stored safe and sound and in readiness for their crews of migratory fishing servants every spring.

War and the Migratory Fishery: 1793

The beginning of the war in 1793 had the usual immediate effects on the European migratory fisheries in Newfoundland waters. The French fleet stayed home, putting an end to that nation's extensive fisheries on the French Shore, the banks and the fishing grounds around the islands of St. Pierre and Miquelon, which had become their possessions by the Treaty of Paris in 1763. France did not have the wherewithal to protect its fisheries, and it had always been its practice to mobilize fully its maritime assets in contests with Britain. In the case of the British fishery, the West Country fishing ships suffered from the impressment of large numbers of their fishermen into the Royal Navy and from the reluctance of fishermen to seek employment in the fisheries because of the fear of impressment.

Between 1792 and 1793, the British migratory fishing fleet declined from 276 ships to 148; within this general decline, the number of bankers dropped from 187 to 63. Some of the bankers may have decided not to return to Britain in the fall of 1792 because the following fall of 1793, 19 bankers were reported as belonging to the island. The end of the bye boat fishery and the business it had generated in the transportation of men and supplies, combined with the dangers of crossing the Atlantic, were the causes for this development. Like the bye boat keepers, these ships, and, in some cases, crews, simply chose not to return home. In 1794, there were 100 British bankers and 25 island-based bankers fishing on the banks, and in 1795, the numbers were 53 British and 62 island-based. In 1796, the war was extended to the banks by a French squadron of at least ten Men of War under Admiral Richery, and it seems that the bank fishery was cut short.[17] It recovered in 1797, and 34 British and 78 island bankers returned to the banks. In that latter year, 47 other 'British fishing' ships visited Newfoundland, making a total of 81 British migratory fishing ships in all.

The British and Newfoundland bank fisheries recovered following the signing of the Treaty of Amiens in March 1802, and in that year, 71 Newfoundland and 58 British bankers prosecuted the fishery. It declined again with the outbreak of war in 1803 and recovered somewhat after the Battle of Trafalgar in October 1805, but declined again during the Anglo-American war of 1812-14. For a variety of reasons which will become clearer below, the Newfoundland bank fishery did not become viable, and the last records show four ships operating in 1833. The British bank fleet remained stronger and numbered 28 in 1820, but the last records disclose that only nine were in Newfoundland waters in 1833. Of course, long before this, the fishery had been taken over by the residents, and such small numbers were no longer relevant to the total Newfoundland fishery.

Meanwhile, the fate of the other British fishing ship fishery (as distinct from the bankers) was much more swiftly sealed. It is impossible to differentiate bank ships from these others in terms of manpower and tonnage before 1793 because separate records were not kept. However, by examining the number of fishing boats carried on all fishing ships, certain conclusions can be drawn. During the 1784-89 period, an annual average of 318 British fishing ships, averaging 89 tons burthen and ten-man crews, sailed to Newfoundland. During the same period, an average of 134 British sack ships entered Newfoundland annually; the average burthen was 123 tons and the average crew contained ten sailors. Therefore, the fishing ships (including the bankers and those that fished inshore) carried only a few men beyond the number needed to sail the craft and between one and four boats each. The reason for the limited size of the bank ships seems to have been the desire to make a number of short trips to the fishing grounds in order to prevent the fish from remaining too long in salt. However, some of these 'fishing ships' continued to be somewhat of a mystery, and Matthews' explanation that they probably fished during the early part of the season while waiting for a cargo seems reasonable. Similarly, it is quite possible that these fishing ships engaged in the inshore fishery in conjunction with certain inhabitants whom they supplied with provisions, equipment and servants. However, by 1810, these fishing ships were no longer enumerated.

Finally, as mentioned above, the bye boat keepers were no longer considered a distinctive body of fishermen after 1788, but there are references to bye boats up to 1801 and to fish caught by bye boat men up to 1803. However, as already indicated, if these people wanted to participate in the cod fishery, they would have to live on their fishing rooms all year round. The decline of the bye boat fishery hurt the bank ships and, no doubt, other ships, financially, because they lost a source of income.

In 1793, the government appointed a committee to enquire into the problems of the Newfoundland fishery, and the major complaints received are contained in the evidence given by the Dartmouth merchants—the population of Newfoundland was growing, 'adventurers and hucksters' were moving into St. John's, and the fish trade was in danger. As already pointed out, after 1763, the West Country merchants were under a certain amount of pressure from the competition introduced into the fish trade by the partial adventurers from other ports of the British Isles because of the growth in the British-Quebec trade. It was only natural that some of these partial adventurers would become a permanent part of the fish trade. In his rebuttal to the Dartmouth merchants, John Reeves, the newly appointed chief justice of Newfoundland, made it clear that as saltfish was the only commodity produced in Newfoundland, the "Hucksters and Adventurers" and the growing population must be finding it profitable or they would not settle there. He went on to say:

> [Even if] ... the whole Town of Dartmouth sustains Losses that are not to be borne without Bankruptcy ... other Persons have been Bankrupts, and other Towns have been ruined, and the Trade has yet gone on. Biddeford and Barnstaple were once great Towns in this Trade, and have long ceased to employ any Ship at all. Perhaps Dartmouth rose upon the Fall of these Towns, and some others may rise upon the Fall of Dartmouth; and with all these changes, the Fishery, as a national Concern, may remain the same. We know that the Place of these decayed Towns has been supplied by Adventurers from other Ports of his Majesty's Dominions; Glasgow is one, Waterford is another; from both of these Towns there are very successful Trades carried on, sufficient to raise the envy of Dartmouth....[18]

Doubtless, Reeves did not appreciate the full scope of his prophecy. Drawing his conclusions from his observations of the current characteristics of the saltfish industry and trade, he did not realize that the Newfoundland fishing ports would be the successors to the West Country ports in the production of Newfoundland saltfish.

War and Newfoundland: 1793

The outbreak of war in 1793 was the beginning of a period of considerable disruption in Newfoundland. Total exports of saltfish from Newfoundland declined to about 450,000 cwts in 1793, but in the following year, there was a slight recovery. This was followed by another decline in 1795 which continued into 1796. Certainly, the steep decline in the size of the migratory fleet was probably the most important factor here, and the higher prices paid seems to point in this direction because less fish was entering the market.

The situation took a change for the worse in 1796. Admiral Richery's squadron destroyed fishing stations on the Labrador coast[19] and then carried out raids south of St. John's burning and sinking bankers and inshore shallops.[20] Governor Wallace remarked, "The fishery this year [1796] has been interrupted but the injury done to it is by no means equal to what might have been expected when we consider the Force of the Enemy that came to destroy it."[21] More seriously, in 1796, Spain ceased its neutrality and supported France in the conflict, thereby closing its markets to Newfoundland saltfish while at the same time cutting off the fishery from its major salt supplies. This resulted in the production of poor quality fish the following year. Governor Waldegrave wrote in August 1797: "Vast quantities of fish have been already brought into the different Ports of this Island, but the scarcity of salt and the unfavourable season for drying, has caused some part of it to be injured, and I fear some has entirely rotted."[22] The closure of the Spanish market, the scarcity of salt and the wet weather all resulted in making the 1797 fishery the most disastrous since the worst years of the American Revolutionary War period; only 374,940 cwts were produced and probably less than that exported (see Table Intro.2).

In 1798, the situation did not improve to any extent; the slight increase in production was offset by a decline in price. While it would appear that the closure of the Spanish market was the principal culprit, Governor Waldegrave believed that a glut created by the combined export of fish produced in 1797 with the 1798 catch was chiefly to blame.[23] The top price paid for fish was 11s per cwt, while the cost of provisions had risen considerably, according to Waldegrave, who pointed to the price of salt pork at nine to ten guineas per barrel in March of that year as an example.[24] By the fall of the year, the governor reported that there was a serious threat of starvation, and he requested and received £1,000 worth of copper coins from the British Treasury. These were put into circulation, and almost immediately, the price of bread declined from 6d (silver) per loaf down to one or two pence.[25] To illustrate the problem that the inhabitants were facing, Waldegrave computed the expenses of the average planter who employed four fishermen and two shoremen as follows:

four fishermen's wages	@ £21	= £ 84
provisions for above	@ 10	= 40
two shoremen's wages	@ 19	= 38
provisions for above	@ 10	= 20
bait for the boat		11
boat, lines, hooks and her outfit		20
Total		£ 213

He went on to point out that in 1798, the average catch for a crew this size was about 280 quintals, at an average price of 9s per quintal. This would mean that the voyage was worth only £126, leaving the planter with a loss of £87. Salt was not included in the cost of the outfit because, as Waldegrave explained, the train oil produced generally covered the price of the salt. However, he added, the price of salt was rising, while the price of oil remained the same. And to add to the distress, there was a partial potato crop failure that same year.[26]

Besides sending the copper coin (which became a common practice during the following years), the British government had earlier given Waldegrave permission to allow imports of bread, flour and other foods from the United States in order to keep prices down and prevent famine. (This permission was granted annually until the Americans imposed an embargo on trade in 1807, but by then, the European scene had changed.) In response to complaints from the merchants that crimes were being committed against their property during the winter by impoverished, unemployed fishing servants who had stayed on the island throughout the winter, in 1798, the British government ordered that the chief justice remain in Newfoundland all year round.[27] However, Waldegrave had very little sympathy for the merchants, and his reaction to their petitions for government assistance and to their complaints about the lack of judicial protection was expressed in his correspondence with the Colonial Office in February 1799, when he wrote:

> I must ... observe that in all the dealings I have had with the Merchants of St. John's on account of Government I have ever found them the most illiberal and rapacious body of Men I ever before met with, and for this I fear there is no remedy: a circumstance they but too well know themselves.[28]

Meanwhile, as already indicated, the nature of the fishery was changing fast, as one contemporary had already informed the Governor: " ... in a few years an extensive fishery may be carried on there [in Newfoundland] by the Inhabitants alone without ... Men from England and Ireland as here to fore...."[29]

The opening of the 1799 season was just as inauspicious as the beginning of the previous year. Governor Waldegrave forecast that the glut of top quality (merchantable) fish in Portugal would push the second quality (madeira) into the West Indies and leave the poor quality (West India) without any market.[30] Exports did exceed 450,000 cwts, but there was much poor quality fish, which was probably the result of the scarcity of salt and the low prices of merchantable fish the previous year. However, prices of merchantable improved, presumably because there were scant supplies of it. At the end of the season, the governor's concern over the plight of the fishermen had increased and he wrote:

> ... unless these poor wretches emigrate they must starve, for how can it be otherwise whilst the Merchant has the power of setting his own price on the supplies issued to the Fishermen, and on the Fish which these men catch for them. Thus we see a set of unfortunate beings working like Slaves, and even hazarding their lives, when at the expiration of their term (however successful their exertions) they find themselves not only without gain, but so deeply indebted, as to force them to emigrate, or drive them to despair.[31]

Waldegrave did not distinguish between planter fishermen and fishing servants in this report. However, the context in which he discusses merchants, fishermen and supplies indicates that it was the planter or inhabitant fisherman that he had in mind. He probably assumed that the fishing servants left the island at the end of the fishing season as a matter of course. Fishermen were not the only ones affected by the depression; merchants were as well because there were an unusual number of bankruptcies and forced property sales by well-established names in the fishery, including Harris and Roope, Thomas Stokes and William Henley.[32]

It seems that the first break in this depression took place in 1802, and although the quantity of fish exported was still less than 500,000 cwts, top prices reached a rather extraordinary 27s per quintal. This was, no doubt, due to the signing of the peace treaty in the spring of that year and the subsequent opening of the Spanish market. Governor Gambier, who had succeeded Waldegrave, was conscious of the Colonial Office's concern about the economics of the Newfoundland fishery and reported in detail on prices that year (see Table Intro.3). Also, according to Gambier, wages for seamen and fishing servants were within the £30 to £50 range per season.[33] The fishing operation and the people who depended upon it had gone through a severe depression and dislocation during this war, and by the time it ended, adjustments had been made and changes were apparent—the principal one being the fact that the fishery had become "... in a great degree sedentary [with] very few passengers [arriving] from His Majesty's European Dominions during the period...."[34]

Napoleonic War to 1807

When war was resumed in 1803, the Newfoundland fishery was better able to cope with the situation than it had been ten years earlier. Gibraltar began to import large quantities of fish, no doubt for re-export, taking, for example, 75,000 cwts in 1804 and 100,000 cwts in 1805.[35] In addition, American exporters began to purchase Newfoundland fish for sale in Spain; by 1805, the amount involved in this trade had risen to 100,000 cwts.[36] Also, in an attempt to depress food prices and wages, the British government, immediately on the resumption of hostilities, granted permission for Newfoundland to import provisions from the United States.[37] The demand for servants to work in the fishery increased: in 1787 and again in 1788, over 6,000 passengers had arrived in Newfoundland to work in the fishery; the numbers declined to 3,588 in 1800 and dropped further to 1,892 in 1802; but there was an increase to 2,732 in 1803. With the outbreak of the second phase of the war, the numbers fell again, to 646 in 1804, 755 in 1805 and 600 in 1806.[38] As a consequence, the governor reported in 1804 that the fishery was carried out almost exclusively by residents—both planters and servants.[39] In this same report, he pointed out that Newfoundland was continuing to lose servants to the United States because many fishermen used the fishery as a stepping stone to the continent. He feared that without a large annual influx of servants, wages would rise too high and the trade would suffer (see Tables Intro.4 and Intro.5 for prices and wages in 1804).[40]

By 1805, it appears that the centre of the Newfoundland fishery was shifting from the West Country to St. John's, and Gower described the situation as follows:

> This Harbour is no longer a mere fishing station, built round with temporary Flakes, Stages, and Huts of trifling value, but ... it is a port of extensive Commerce ... importing near two thirds of the supplies for the whole Island, and furnished with extensive Store-Houses and Wharfs for trade, containing a quantity of Provisions, Stores for the Fishery, British Manufacturers and West Indian Produce, as well as Fish and Oil ready for exportation, which together with the Buildings is computed to be worth more than half a million Sterling.[41]

For security reasons, the saltfish trade was beginning to use St. John's as the entrepôt for the whole British fishery. Gower also pointed out that food prices had risen and that bread cost 42s to 50s per cwt (up from a range of 18s to 28s in 1802) and flour was selling at £3 10s per barrel of 196 pounds (up from £2 to £3 in 1802). The fishery continued to be moderately productive and prices fell back from the highs of 1802. Gower suspected that the principal reason for the lower prices was the recent practice of price-fixing by the merchants. He wrote:

> It is remarkable that it [the price] varies greatly from year to year. Within these four years [including 1805] the price of the merchantable quality has fallen from twenty seven shillings to 14/6. The price is less affected by the demand for it abroad than by purchases in the Island. A custom prevails in St. John's for the merchants to meet together and settle the price of fish and oil, which is termed 'breaking the price'. This is done about the beginning of August, after having received advice from Europe and ascertained the state of the markets etc....[42]

This is the first indication that much of the decision-making, in terms of fish prices, had been moved from the West Country and London offices to St. John's and that the fishery was becoming a residential industry in terms of management as well as production.

There were no major changes during the early years of this second war. Spain imported Newfoundland fish via the Americans, and when the Portuguese market became glutted, Britain prohibited Barbados, St. Vincent and Grenada from purchasing fish from the Americans in favour of Newfoundland so that exports to the West Indies increased from an annual average of about 50,000 cwts prior to 1805 to 101,000 cwts in 1806.[43] Prices and wages remained fairly stable because of the continuing importation of provisions from the United States, and one of Gower's biggest concerns was the fact that the fishery was completely in the hands of the residents and they were supplying all available markets.[44] Exports of fish in 1806 totalled over 770,000 cwts, with merchantable fetching 14s per cwt. In 1807, maximum prices reached 13s 6d, and the quantity exported amounted to about 675,000 cwts. The period from 1803 to 1807 was one of moderation—in production, wages and prices—and the records reflect this. However, major and rapid changes were imminent.

American Embargo and the Peninsular War

The saltfish trade had finally been adjusted to compensate for the closure of the Spanish and Mediterranean ports to British shipping. Gibraltar's imports had reached 130,000 cwts in 1807, while the British West Indies bought over 100,000 cwts, and the Americans, over 155,000. However, the American Embargo Act of 22 December 1807 changed the situation. In 1808, the United States bought only one-third as much as it had purchased in 1807, and the price of provisions rose as both the British West Indies and Newfoundland were forced to turn to British North America for supplies. In addition, the French occupation of Lisbon closed off this major saltfish market for a while.

The following year, 1809, brought both good news and bad. The ports of Spain and Portugal were being reopened as a consequence of the arrival of a British army on the peninsula and the beginning of the Peninsular War, which was to see the British under Wellington assist the Spanish in driving the French out of Spain by 1814. At the same time, the huge American fishery on the Labrador coast ceased. This fishery had grown considerably in recent years. In the 1790s, it had employed 4,000 to 5,000 men and produced about 400,000 cwts for export annually.[45] A report in 1804 stated that about 1,360 American ships fished on the Labrador coast and in the Gulf of St. Lawrence and employed 10,600 men. This report went on to point out that the Americans sold their wet-salted cod to France and the American home market, their dry-salted product to the Mediterranean, and the worst of the latter to the British West Indies.[46] It was also reported, in that same year, that the Americans sent an average of 150,000 cwts annually to the British West Indies while Newfoundland sent only 50,000 cwts there.[47] One other report noted that in 1805, there were 1,500 ships involved in this fishery, and each one carried from 12 to 14 men.[48]

The ending of the French naval threat after the battle of Trafalgar (1805), the opening of the Iberian ports and the elimination of the American fishery gave Newfoundland a distinct advantage in the trade. However, the winter of 1808-09 was one of great hardship because of the lack of American supplies. Governor Holloway wrote:

> From the uncertain State of affairs with the United States of America and no supplies being received from thence this year, and Nova Scotia and Prince Edward Island not allowing any Stock to be exported, this Island is totally destitute of Beef for the Supply of His Majesty's Ships and the Inhabitants are deprived of it; I may therefore be under the necessity of allowing the Importation of Livestock, Grain and Fruit from the Azores or Western Islands, which I trust will meet your Lordship's Approbation.[49]

Despite the shortage of supplies, exports of saltfish exceeded 810,000 cwts, with almost 620,000 cwts going to Portugal and Gibraltar and top prices at a reasonable 13s 6d per cwt. The following year, exports reached 884,470 cwts and merchantable fish fetched 14s 6d per quintal. The exporters had, by this time, adjusted to the new trade pattern. They had increased their exports to the British West Indies, displaced the American market in southern Spain and were continuing to dominate the trade to Portugal. However, both wages and prices had risen somewhat since 1804 (see Tables Intro.6 and Intro.7).[50] In 1811, Spain and Portugal imported the enormous quantity of 611,960 cwts of Newfoundland fish, while 140,000 cwts went to Gibraltar and over 152,000 cwts to the British West Indies for a grand total of 923,540 cwts, with prices for merchantable fish fetching up to 22s per cwt.

Anglo-American War

In the meantime, another change in the international situation occurred in 1812, when Britain and the United States went to war, thereby forcing the Newfoundland governor to concentrate on the security of the island rather than on the trade. Total exports reached 711,256 cwts of saltfish, and the top price climbed to 22s 6d per cwt. However, there was a shortage of shipping, as many European ship owners were afraid to risk their shipping in the western Atlantic. At the end of the 1812 season, Governor Duckworth reported that, in his opinion, the resident population of Newfoundland was now so great and so completely in control of the fishery that the migratory fishery could never be re-instated. Furthermore, he recommended that most old ships' fishing rooms—"unoccupied places"—be given up to the inhabitants and that they also be granted land for purposes of cultivation.[51]

The fishery was unusually productive again in 1813, and 912,183 cwts were exported. However, the weather was unsuitable for drying and there was a lower proportion of merchantable fish, which forced the price of top quality to unprecedented levels: 32s per cwt in Newfoundland, according to one source; 42s per cwt in the market, reported another.[52] At the same time, there were inflationary pressures, as wages and prices were rising. Servants received an average of £70 per season in 1813,[53] and the price of bread increased from about 45s per cwt in 1810 to about 75s in 1813; flour, from about 65s per barrel to 120s; and pork, from about 140s per barrel to 195s (see Tables Intro.7 and Intro.8).[54] Before the year was out, Governor Keats, like his predecessors, commented on the changes taking place in Newfoundland:

> St. John's became the Emporium of the Island in consequence of this extended war, with a population of nearly 10,000 inhabitants, seems to have grown out of its character from a Fishery to a large Commercial Town, and

for a considerable time past has offered such advantages to the Farmer and Gardener as to surmount in a great degree all the restraints which Nature or the Policy of Government has laid on the Cultivation of a Soil, certainly less sterile than has generally been considered. But this character which it has latterly assumed, it is very doubtful it will be able to support on the return of Peace.[55]

Keats recognized that the war had turned St. John's into a commercial centre and that people were making a living from agriculture because food prices were high. However, he correctly realized that the cod fishery alone would not be able to sustain a centre of 10,000 people once peace was restored.

In 1814, peace was concluded in both Europe (April) and North America (December), and once again, the Newfoundland fishery flourished. Exports of saltfish rose to 947,811 cwts, of which 768,010 were sent to Spain, Portugal and the Mediterranean, and although prices dropped a little from the previous year, merchantable fish fetched 24s 6d per cwt. It was an unusually prosperous and secure year. There was full employment, high wages, a large catch and an almost insatiable market. Late that year, however, the first signs of trouble appeared as fish prices began to decline and information was received that the French fishery was being re-established at St. Pierre and Miquelon—though not as yet on the French Shore.[56] The year 1815 also experienced a good fishery and trade, uninterrupted by the short resumption of war, which ended at Waterloo in June with the final defeat of Napoleon. The largest quantity of fish ever produced in one year was exported—1,182,661 cwts—and fetched prices of up to 21s per cwt.

Settlement to 1815

Not only were the character of the fishery and the complexion of St. John's changed, but the composition of the resident population and the pattern of settlement were also affected by the war.[57]

Newfoundland's early inhabitant population originated in the West of England, particularly in the counties of Dorset and Devon. Although early seventeenth century colonization attempts involved participation by Bristol, London, Ireland and Wales, the non-West-Country settlers were soon assimilated by arrivals from the West Country. By the latter seventeenth century, the West Country fishing ships were calling at the port of Waterford to take on certain supplies which were cheaper and more readily available there than in England; these were 'wet' provisions, including barrels of salt pork and salt beef, butter, cheese and porter. It became common practice to hire young men to fill the crew's compliment and, no doubt, young women were brought out to Newfoundland to work for the more affluent inhabitants. (All the early accounts record small numbers of women servants.) Thus the Irish became acquainted with the Newfoundland cod fishery.

The second stage in Irish emigration to Newfoundland occurred after the Treaty of Utrecht in 1713. For a number of reasons, including the post-war depression and the presence of the British military, many West of England fishermen were reluctant to settle or even fish in Placentia Bay after the French vacated it. The Irish gravitated to this area, and their presence here and in St.

John's and Conception Bay was often noticed and remarked upon during this century. In 1754, the naval governors began to take an annual census of these Irish Roman Catholics (and this annual record was kept until 1830). Irish passengers to Newfoundland numbered between 3,000 and 4,000 annually during the late 1760s and throughout the 1770s. During the remainder of the century, the number declined; only about 2,000 came annually during the depressed early 1790s, and when the war began in 1793, very few ventured across the Atlantic—83 arriving in 1797. Like their English counterparts, the Irish either returned home each fall or continued on to the mainland, because in 1803, it was reported that there were only 11,021 Protestants and 8,008 Roman Catholics on the island, including planters and servants living in fewer than 3,300 houses.

When war resumed in 1803, the scene was set for an increase in Irish emigration to Newfoundland. English servants were more difficult to obtain because many were being pressed into the navy, and others would not take a chance by going anywhere near the West Country sea ports. Then, in 1803, the Passenger Act was passed by the British Government in order to protect the lives and health of British subjects emigrating to the United States (and possibly to slow emigration). The act stipulated that certain quantities of food and water had to be carried by all immigrant ships and certain sanitary and health conditions maintained—all of which increased the cost of transportation and discouraged many would-be emigrants. Naturally, the ships clearing British ports for the Newfoundland fishery were excluded from the provisions of the Passenger Act and, consequently, transportation costs to Newfoundland were much lower.

A situation developed which seemed to please all concerned. Ships' captains calling at Waterford could make extra money transporting Irish men and women to Newfoundland. (Gower reported in 1805 that the cost of a fisherman's passage from Ireland to Newfoundland was £6 to £7, while the return passage cost £5.)[58] The planters in Newfoundland were pleased because it kept down wage costs and allowed them to expand their operations. And the Irish were relieved to obtain a passage to Newfoundland, whence they could always continue on to the mainland if they so desired. It was reported that 83 Irish passengers arrived in 1797; 93 in 1801; and 1,642 in 1803. Numbers declined for several years, but in 1811, 1,368 Irish arrived; 1,943 in 1812; 1,842 in 1813; 2,254 in 1814; 5,838 in 1815; 2,636 in 1816; and 441 in 1817. As can be seen, the numbers arriving increased dramatically beginning in 1812 and were abnormally high in 1815. The difference between the period before the war and the early nineteenth century is the fact that so few Irish stayed in the eighteenth century while so many stayed during the later period. For example, during the 11-year period from 1765 to 1775, a total of almost 40,000 Irish passengers arrived in Newfoundland (out of a grand total of 73,000 passengers). Fewer, but significant numbers, arrived during the 1780s. And yet, as already pointed out, there were only 8,000 Irish men women and children on the island in 1803—and many of them were, no doubt, transient. This situation changed during the latter years of the Napoleonic War.

During these few very prosperous years, planters in larger centres, particularly St. John's, but also Harbour Grace, Carbonear, Brigus and other ports in

Conception Bay and Trinity Bay, expanded their operations. They could not do this very well in the harbours in which they lived because of the lack of space, but they could send fishing shallops and small schooners north to the French Shore, which was completely vacant during this period. Therefore, they built the necessary craft and began to participate in what became known in the records as the 'North Shore' fishery—the 'North Shore' being the English term for the eastern side of the Great Northern Peninsula. Governor Gower described the North Shore fishery as it existed in 1804 as follows:

> These [North Shore] vessels are, in general, the same that in the months of March and April are employed in the Seal fishery; after their return from which service they are again fitted out, and proceed to the harbours on that part of the north-east coast of the Island that lies to the northward of Cape St. John.... The chief part of this fishery is carried on from Conception Bay, where the Planters are more independent than in the other districts. From thence whole families remove in the spring of the year to the coast before mentioned, and carry on their fishery in the same manner as in their own Harbours, the men going out in the boats to catch the fish, while the women and children employ themselves on shore to split and cure it. The activity and enterprise of these industrious people is so great, that their women, even in advanced pregnancy, rather than stay at home, take midwives with them on this expedition.[59]

At the same time, the planters had discovered that seals could be slaughtered on the ice fields off the northeast coast of the island (and in the Gulf of St. Lawrence) and began to send their shallops and vessels in search of this source of valuable oil. The newly arrived Irishmen were not necessarily skilled fishermen and certainly not in any position to invest in their own fishing rooms, even if vacant ones were conveniently located. They built their dwellings away from the sea, sought employment from the established planters and were satisfied to spend months away from home at the ice fields and on the North Shore engaged in these two fisheries, welcoming the relatively high pay they received. Governor Gower explained the situation as well as any. He wrote in 1805:

> [There was a] further decline of the British Ship Fishery this year; nor is the number of Bankers fitted out in the Island so great as usual, while the number of Vessels employed on the North East shore is considerably increased. It is scarcely necessary to say that the latter mode of fishing is attended with less expense and risque, and generally better success than the former—It is also more convenient to the Planters as lying within the Compass of their limited means, and being connected with the Seal Fishery in the spring of the year, as both are carried on by the same Vessels, the former not commencing till after the latter is finished. But though the Cod Fishery on the above Coast begins late (usually not before the middle of June) it generally yields, during its continuance, the greatest abundance of fish that is found on any part of the Island.[60]

The fishing servants were now able to settle in Newfoundland because there was winter employment in the seal fishery and by 1815, it is most likely that many of the 10,000 men servants recorded in the reports were beginning to establish households and families. Thus, there developed around the better and wealthier ports, especially St. John's and the Conception Bay ports, a concentration of capital and labour which was not dependent on the inshore fishery but which could be applied to two more distant fisheries—the seal and North Shore cod. These concentrations of people were now engaged in farming and market gardening with the active encouragement of the governor; churches, including the Roman Catholic church, were being built and expanded, a newspaper had been established in St. John's, and this unofficial, quasi-colony had now become home to advocates of representative government.[61] Newfoundland had changed from a fishery based on an island to a string of coastal communities centred around two fisheries.

Post-War Depression

By the end of 1815, the boom had ended and a financial crash was imminent.[62] Fish prices plummeted in 1816. Prowse states that in the winter of 1815-16, St. John's and the outports were in a state of actual starvation, with numerous losses and insolvencies and inadequate provisions.[63] The winter of 1816-17 was worse, with a major fire occurring in St. John's in February to add to the distress. The following winter, 1817-18, which Prowse calls the "Hard Winter" or the "Winter of the Rals," was worse again. He writes:

> In the former season [1816-1817], starvation alone had to be contended with; now [1817-1818] famine, frost, and fire combined, like three avenging furies, to scourge the unfortunate Island. A frost that sealed up the whole coast commenced early in November, and continued almost without intermission through the entire season, and on the nights of the 7th and 21st of November, three hundred houses were burnt, rendering two thousand individuals, in depth of that cruel winter, homeless. Nothing can add to the simple pathos of the grand jury's presentment:
>
>> Calamities so extensive would have been in our most prosperous times productive of severe distress, but on retrospecting to our situation for the last three years, during which period we have alternately suffered by fire, by famine, by lawless outrage, and numerous mercantile failures, which have greatly injured the commercial reputation of the town, the recent conflagrations seemed only wanting to consummate our misfortunes. Several hundred men in the prime of life, without money, or the means of being employed, without adequate clothing or food, are at the hour of midnight wandering amidst the smoking ruins to seek warmth from the ashes, and food from the refuse of the half-consumed fish. In dwelling-houses the misery is little less. Many families, once in affluence, are now in absolute want. Within these two days, two men have been found perished of cold, and many hundreds must inevitably experience a similar fate if humanity does not promptly and effectually step forward to their relief.
>
> To add to this misery, gangs of half-famished, lawless men everywhere threatened the destruction of life and property; vigilance committees were formed in every settlement.... During these three unhappy years [the winters of 1815-16, 1816-17, 1817-18] everything was against us.... [In the meantime] the sufferings of the Newfoundland colonists called forth humane response from all quarters. The British Government sent £10,000, Halifax contributed liberally, but the most touching and generous gift was from the large-hearted people of Boston. A few years before they were fitting out privateers to destroy our fishing vessels, now they sent us a welcome ship-load of

provisions. In the most inclement month of the year, January, the Brig *Messenger* came into St. John'... [with] 174 barrels of flour; 125 barrels of meal; 11 tierces of rice; and 27 barrels and 963 bags of bread.

However, finally the distress let up and things began to improve. In Prowse's words—

> In the spring of 1818 this period of calamity came to an end. The seal fishery was unusually productive, in less than a fortnight scores of little vessels returned, loaded to the scuppers with fat; hope revived, the fisheries were good, the markets improved, and the poor old Colony again began to lift up her head.

The worst of the depression was over and the Newfoundland fishery was firmly in the hands of the residents. This resident population of 40,000 people, however, was not as homogeneous in 1815-18 as had been the population of 17,000 in 1792, nor was the economy as 'saltfish-centred' in the latter period as it had been earlier, and Prowse's acknowledgement of the importance of the seal fishery returns the discussion to the main subject.

It is obvious that the cod fishery alone was not able to support the planters and the fishing servants that had taken up permanent residence. If the cod fishery had remained the sole industry, the island's population of 40,000 (including 4,600 planters and 10,000 men servants) in 1815 would have declined at least to the levels of the early 1790s. The surplus population would have been forced to move to the mainland and planters would once again have depended on migratory servants. In fact, given the extent of the post-war depression, many of the old established resident fishing operations would have closed permanently. However, the fishery did not return to its pre-war state; it became entirely residential and, furthermore, population actually increased—all because of the discovery of the seal resources.

Seals

The north Atlantic harp seal came to the attention of European settlers on the northeast coast of the island of Newfoundland early in the eighteenth century. Up to then, this sea mammal had escaped the notice of the Europeans because the herds visited coastal Newfoundland only in the winter and spring months, while the summer fishermen spent these seasons in their home ports in Europe. The Beothuks on the island and the Innu and Inuit on the coast of Labrador had always utilized the flesh, skin and oil of the seal for food, clothing, housing, heat and light, but it was left to the English population that had begun to winter on Newfoundland's northeast coast to discover that the seal was commercially valuable as well.

Harp seals (phoca groenlandica)[64] congregate in large herds and spend their summers in the high Arctic. When the northern seas begin to freeze over in the autumn, the seal herds migrate south, where they feed for several months. A small proportion of the seals enter the Gulf of St. Lawrence but the majority aim for the waters off Newfoundland's east coast. Many seals swim near the coast of northern Newfoundland and eastern and southern Labrador, and historically have provided wintering fishermen with an opportunity to net them in narrow passages and off headlands. In February, the seals begin the return journey to the Arctic at the same time as the Arctic ice is drifting south. The herds find this mass of ice in late February and climb upon it, and the females give birth. For two or three weeks, the young are nursed while the adults live on these ice 'fields', entering the water only to feed. Meanwhile, the ice continues to drift, driven by wind and tide—sometimes far out to the southeast towards the Grand Banks, while at other times near and, occasionally, into the bays and harbours of Newfoundland's northeast coast. By late March, when the young are ready to travel, the herds resume their northward journey.

Figure 5 - Harp seals on the ice floes.

The harp seal is a hair seal measuring over five feet (1.6 m.) in length. Adult females weigh about 265 pounds (120 kg.), while the males are larger and weigh about 300 pounds (135 kg.). In Newfoundland terminology, the young pups are known as 'white coats', and they become 'raggedy jackets' as their coats change and they learn to swim. Young adolescents are called 'beaters', while those approaching breeding age are known as 'bedlamers'. Finally, the adults are referred to as 'dogs and 'bitches'. The skin with the fat attached is known as the 'pelt' or 'sculp' (and less frequently, 'scalp').[65] Whitecoat pelts weigh from 30 to 50 pounds (14 to 23 kg.), depending on when the young are killed, and occasionally weigh up to 60 pounds. Whitecoats yielding pelts of less than 30 pounds are referred to as 'cats'.

A second species of seal, the 'hood' or 'hooded' (cystophora cristata), follows a similar migration and feeding pattern and whelping practice, but travels in scattered family groups, which has always made them difficult to harvest in commercial numbers. Furthermore, they have always been fewer in number, and although some are always found in the vicinity of the harps, the two species do not mix. The hoods are much larger than the harps; 'bitches' weigh about 770 pounds (350 kg.) and measure over seven feet (2.2 m.), while 'dog' hoods weigh about 880 pounds (400 kg.) and average over eight feet (2.6

m.) in length. The young hoods are known as 'blue backs' because of their distinctive colour and are somewhat larger than whitecoats.

Historically, the skin of the adult harp was a useful source of leather; however, it was the two- to three-inch (5 - 8cm.) layer of fat attached to the skin that was by far the most commercially valuable. As in the case of the harps, the fat of the hoods was more valuable than the skin, although the latter produced a stronger and heavier leather.

Early Sealing

Leaving demand aside for the moment, the first prerequisite for a commercial seal fishery or hunt in Newfoundland was the establishment of a permanent population with commercial connections and the necessary technology. The second requirement was the establishment of fixed settlements in certain specific locations. However, there was a lapse of about 100 years between the establishment of a resident cod fishery in about 1600 and the beginnings of a local commercial seal fishery.

The first settlement of English fishermen and women—both official and unofficial—took place in Conception Bay, St. John's and in the harbours of the eastern Avalon peninsula (the 'Southern Shore'). This was the area where the earliest English cod fishery was prosecuted, and settlement followed from that. During the seventeenth century, settlement spread into Trinity Bay and north to Bonavista. For example, in 1696-97, it was reported that Bonavista was home to 300 men, although most were, no doubt, living there temporarily during the winter because of the threat from the war between England and France.[66] Any people wintering in this area would have seen many seals annually as these sea mammals were carried along on the ice floes. Furthermore, it would have become apparent by this time that by moving north, residents' opportunities to benefit from this fishery would be greatly enhanced. Indeed, the report mentioned above states that in 1696-97, there were an estimated 150 resident English fishermen scattered throughout the area of [present-day] Fogo and Twillingate and other northern points. It would have been a natural inclination for the wintering fishermen to go out among the loose ice pans in the spring and shoot seals and to go out on the solid ice fields, which occasionally appeared, in search of these sea mammals. The flesh would provide a welcome change from salt beef and pork, and the skins could be used for a variety of local needs. In addition, the use of cod and sea mammal oil for lighting was not unknown.

The first statistical account to include the amount of seal oil produced by Newfoundland inhabitants was submitted at the end of 1723 and records that the quantity thus produced was worth £6,025. Since train (or cod) oil was valued at between £10 10s and £14 per tun during that year, it is very likely that seal oil was worth at least £12 per tun, given the usual price pattern. This would mean that the seal fishery in 1723 yielded about 500 tuns of seal oil compared with 811 tuns of cod oil—a comparatively large proportion of Newfoundland's total oil production. However, unlike the cod oil industry, which remained relatively stable from year to year (except in war time) because it was totally dependent on the quantity of saltfish produced, the seal oil industry was most erratic because it was so dependent on the weather, the direction of the wind and the general ice conditions. Thus, the catch of seals often varied considerably from one year to the next.[67] This can be seen clearly in the early reports. For example, the value of seal oil produced in 1726 amounted to £6,305, whereas in 1734, only £1,310 worth was produced (see Table Intro.9).[68]

Another feature of the eighteenth century seal fishery was the fact that it was dominated by the thinly populated northern ports. In 1726, for example, Trinity Bay and Bonavista accounted for £3,500 worth of seal oil, while the more heavily populated ports of Conception Bay, St. John's and the Southern Shore accounted for the remainder—less than £3,000. In 1734, this trend was even more strikingly illustrated when it was reported that Trinity and Bonavista districts together produced seal oil to the value of £1,200—nearly all of Newfoundland's total production.[69] Similarly, in Governor Hugh Palliser's conscientious report for 1768, it was recorded that nearly the total production of seal oil, worth £12,664, originated from the northern ports, including the 'Coast of Labrador'.[70]

In fact, the seal fishery on the Labrador coast predated the 1763 treaty and continued to be an important component of the economy of that area until well into the twentieth century. However, it was rarely reported upon in the eighteenth century. One interesting contemporary report on this fishery was included in a petition from the Labrador coast to the 'King', complaining about the attempt, begun by Governor Palliser in the 1760s, to reserve the British fishery on the Labrador coast for the use of migratory British fishing ships only; for a while, Newfoundland governors tried to enforce this regulation. The petition explains in detail the nature of the winter seal fishery on the coast of southern Labrador in the vicinity of the present-day boundary between Quebec and Newfoundland:

> ... we beg leave humbly to represent to Your Majesty, that while Canada remained in the hands of the French, and the Coast of Labrador was considered a Dependency thereupon, a fishery for seals was, amongst other objects of national concern, attempted and brought to a degree of perfection, and the returns from this branch of the Exports of Canada amounted annually to about £10,000 sterling ... The [seal] fishery cannot be prosecuted in the open seas and made general like those for cod and whales, but is practised in a manner widely different from any other fishery in the gulph or river St. Lawrence and requires much judgement and circumspection; it is chiefly formed by the contiguity of small islands or rocks to the mainland, which occasion strong currents called Passes where only such fisheries can be exercised and to which the make and contexture of the netts [sic] must be particularly fitted; it is chiefly followed in the winter season only and the immediate operation of catching these animals commences in December and lasts only about fifteen days [as a proportion of the herd migrates into the Gulf], but the fishers employed in the business must be at their stations in the course of the month of September and cannot get away from it before the end of May [because ice prevents shipping].... The seal fishery being of necessity a sedentary fishery, requiring great expense, nets of a particular quality of construction immediately fitted to the Pass they belong to, employing materials of a bulky nature, and requiring houses for wintering of fishers, cannot be made open and general in the manner of the fisheries above mentioned [seal and whale], nor can the posts occupied by Adventurers in this business be quitted at pleasure or transferred to first comers as

is practised in the Newfoundland [cod] fishery.... [Therefore the rules and regulations of the Newfoundland Governor which may be applied to the encouragement of the British migratory cod and whale fisheries] have been found absolutely incompatible with the principles on which the seal fishery can alone be conducted ... [and they request that this part of the coast of Labrador be restored to the jurisdiction of the Governor of Quebec.][71]

As can be seen, not only does this report the manner in which the winter seal fishery was carried out; it also brings attention to the fact that it could be prosecuted only by fishermen who remained in Newfoundland and/or Labrador during the winter. While this description of netting seals applies to the Straits of Belle Isle, where seals passing into and out of the Gulf of St. Lawrence could be taken, similar practices were engaged in along various islands and points of land on the north and northeast coasts, the west coast and along the Atlantic Labrador coast.

The erratic nature of the seal fishery during the eighteenth century can be seen most clearly by a cursory examination of reports from these northern districts. For example, it was reported that Trinity Bay produced seal oil valued as follows: in 1725—£2,600; 1726—£2,000; 1727—£960; 1734—£600; 1735—£480; and 1768—£840. The reports for Bonavista during these same years were as follows: £1,036; £1,500; £200; £600; £2,334; and £1,620.[72] Nevertheless, with the ending of the American war of independence and the slow but steady growth in the resident fishery, the seal fishery in the north emerged as a constant component of the local economy (and encouraged the local shipbuilding industry). In 1786, Governor Elliot's report included the following comment:

> The cod fishery in White Bay [well beyond the northern range of English settlement] is very inconsiderable. Our Bye Boat keepers and Planters used to send their shallops without the Bay; presumably while the French were absent during the recent war] to the Northward and on the Coast of Labrador for almost all the Fish they took, but in the winter they killed a great number of Beasts yielding valuable Furs and saved a quantity of Seal oil, which with the advantage of the largest and finest Timber on all Newfoundland with which they built Vessels and a number of Shallops rendered it to those who frequented there a Valuable place.[73]

Even a cursory examination of the governors' reports during the remainder of the 1780s and 1790s will demonstrate the predominance of the northern ports in the seal fishery. Fishermen residing in these harbours and coves hunted seals among the ice floes with guns. Occasionally, the sheet ice[74] drifted in against the shore, carrying thousands of nursing whitecoats; then they could be slaughtered by the hundreds or even thousands. This was an unusual occurrence, but it happened occasionally, and when it did, it brought a bonanza to the residents involved. Men, women and children would rush out and do their share to kill and haul ashore as many as possible (see Chapter 4). The people who engaged in hunting seals on and among the ice floes near their home communities were referred to as 'landsmen' during the nineteenth century although in the beginning (i.e., the eighteeth century) they and the net seal fishermen were usually described by the governors as "engaged in the winter seal fishery." Furthermore, the principal and most dependable early method to capture seals in this northern region was by the use of nets, as described in the petition above and by John Bland (see below). (In both cases, this was in order to distinguish this fishery from the nineteenth century 'spring' seal fishery.) In any case, it is useful to look briefly at individual regional reports to identify the areas from which this fishery was launched. In 1787, the number of seal skins exported from Newfoundland was valued as follows:[75]

	£
St. John's	440
Harbour Grace and District	6,077
Trinity and District	32,425
Fogo and Twillingate	2,100
Total	£41,042

In 1793, the value of seal oil exported from the island was as follows:[76]

	£
St. John's, etc.	400
Old Perlican	200
Bay de Verde	100
Trinity and Bonaventure	800
Bonavista	3,500
Greenspond	2,000
Kings Cove	200
Keels	250
Fogo	500
Harbour Grace	60
Carbonear	400
Total	£8,410

As one can see, St. John's, Harbour Grace and Carbonear—the oldest, most populous and, indeed, most important English centres (see Tables Intro.10 and Intro.11)—had a very limited involvement in the seal fishery up to 1793, and the old cod fishing centres in the Ferryland area were not involved at all. On the other hand, the northern ports, including Trinity, Bonavista, Greenspond, Fogo and Twillingate, were employed in, comparatively speaking, a fairly substantial seal fishery. In essence, the peak of the 'winter' seal fishery had been reached.

This situation, however, was to change completely during the following years. A letter from Magistrate John Bland, Bonavista, to Governor Gambier, St. John's, in September 1802, describes in great detail the winter seal fishery and introduces the reader to the development of the new 'spring' seal fishery:

> Your Excellency has been pleased to request of me some information respecting our seal fishery, and as far as my own experience and general observation can lead, I shall endeavour to comply with that request.
> This adventurous and perilous pursuit is prosecuted in two different

ways—during the winter months by nets, and from March to June in ice-skiffs and decked boats, or schooners. The fishery by nets extends from Conception Bay to the Labrador, and in the northern posts there is most certainty of success. About fifty pounds weight of strong twine will be required to make a net, the half worn small hawsers, which the boats have used in the summer fishery, serve for foot ropes; new ratline is necessary for head ropes, and each net is required to be about forty fathoms [about 72 m.] in length, and nearly three [5.4 m.] in depth. I am thus minute [sic] that your Excellency may form some idea of the expense attending this adventure, as well as of the mode in which it is conducted.

Four or five men constitute a crew to attend about twenty nets, but in brisk sealing this number of nets will require a double crew in separate boats. The seals bolt into the nets while ranging at the bottom in quest of food, which makes it necessary to keep the nets to the ground, where they are made to stand on their legs, as the phrase is, by means of cork fastened at equal distances along the head ropes. the net is extended at the bottom by a mooring and killock fixed to each end, and it is frequently placed in forty fathoms water [sic], for we observe that the largest seals are caught in the deepest water. To each end of the head rope is fixed a line with a pole standing erect in the water to guide the sealers to the net.... On the Labrador coast the seal fishery begins in November and ends about Christmas, when the nets are taken up. With us it begins about Christmas and continues through the winter, the ice in this quarter being seldom stationary for any considerable length of time.

The seals upon this coast are of many species ... but ... our dependence rests wholly upon Harps and Bedlemers, which are driven by winds and ice from the north-east seas. The harp in its prime will yield from ten to sixteen gallons of oil, and the bedlamer, a seal of the same species, only younger, from three to seven ....

I will now, sir, proceed to bring into view the produce of this fishery in Bonavista Bay for some years back ... In the winter of 1791-2, a succession of hard gales from the north-east brought the seals in great numbers, before the middle of January, unaccompanied by any ice, a circumstance that rarely occurs. In Bonavista about two hundred men night have been employed in attending the nets, and the number of seals caught amounted to about seven thousand. The entire catch at Bonavista Bay may be taken at ten thousand, and two-thirds of the whole reckoned Harps. The Harps yielded thirteen shillings each, and the bedlemers seven shillings and sixpence, which at that time afforded the merchant a large profit in the English market.

The winter of 92-3 the nets were unproductive, but the ice-skiffs in the spring were generally successful, killing from five to six thousand seals, principally young, which yielded five shillings each.

For the winter and spring 93-4, we may reckon about five thousand....

In 95 the nets were successful in the prime part of the season, and the numbers taken about six thousand, two-thirds of which were harps.

The five succeeding years may be averaged at six thousand each, and three-fourths of these may be reckoned young seals. In the spring of 1801 we may count about twenty thousand, the greatest part of which were dragged on the ice by men, women, and children, with incredible labour. These seals were principally of the hooded kind, and about nine-tenths of them young. The old yielded about fourteen shillings and the young seven. The last season may have produce six thousand, yielding on an average about seven shillings each.... I have confined myself to the district in which

I live.... The ports on the coast of Labrador, on the northern shore, now possessed by the French, and southward to Fogo, have been out of all comparison more successful in this fishery by nets.

The sealing-adventure by large boats, which sail about the middle of March has not been general [sic] longer than nine years.... From two to three thousand men have been employed in this perilous adventure, and it may excite surprise that so few fatal accidents have happened. It is, however, in my opinion, for various reasons, upon the decline.... Out of the harbour of Saint John's this adventure has been followed with uncommon spirit, and your Excellency, it is likely, may obtain useful information on this subject from the merchants of that quarter.[77]

Bland's summary of the eighteenth century winter seal fishery is the best contemporary description of this activity on record. And his failure to foresee the future development of a new activity is an oversight that has not been uncommon among commentators on Newfoundland's economic future.

Exports of seal and cod oil averaged 2,823 tuns annually during 1787-92, inclusive, and 2,619 tuns during 1793-97, inclusive. Exports of seal skins (usually a more accurate indicator of the extent of the seal fishery) averaged 31,705 during the earlier period and 42,091 during the latter.[78] As one can see from the statistics on exports of skins, the seal fishery increased during this period. The decline in total oil exports was, no doubt, due to the decline in the production of cod oil which resulted from the decline in the cod fishery because of the war.[79]

Nevertheless, the early seal fishery contributed little to the total fishery economy: during the five-year period ending in 1727, the average annual value of seal oil exports was less than £5,000; and during the five-year period ending in 1792, it was still less than £8,000, as Table Intro.9 indicates. Meanwhile, the value of saltfish produced annually during the earlier period amounted to almost £100,000, and the value during the latter period averaged over £350,000 annually. Nevertheless, it had become obvious by the 1790s (see Chapter 1) that there was a reliable and growing market in Britain for fish oils, and the planters and merchants prosecuting the cod fishery in the major harbours on the northeast coast began to address the problem of how to harvest this available and under-utilized resource.

Spring versus Winter Sealing

It is generally accepted that the first attempt by planters on the east coast of Newfoundland to seek out the seal herds on the ice fields occurred in 1793. Bland's report written in 1802 stated that "sealing adventure by large boats, which sail about the middle of March has not been general [sic] longer than nine years." Governor Le Marchant's report written in 1848 explained that this industry was started in 1793 by a St. John's merchant who sent two small vessels of about 45 tons each to search for seals, and that both were successful in bringing in a total of about 1,600 of these mammals.[80] W.A. Munn wrote in 1923 of the origins of this industry as follows:

> The earliest records of our modern Seal fishery date from 1795 [sic], when two small fore-and-afters were fitted out from St. John's the first week in April. They were both successful, and one returned with 800 seals.

In 1796 four schooners went from St. John's and several from Conception Bay, some of which met with good success.[81]

Munn described how the industry began to grow as enterprising planters went farther and farther afield in search of the seals:

> At first these [ships] were shallops, decked in fore and aft with movable deck boards in the centre of the boat. The shelter cuddies at each end gave the crews some protection when they remained out over night. It is generally stated in all the Histories that these shallops never went past the headlands at the mouth of the Bays [i.e. Trinity and Conception Bays], but we have one authentic instance of Capt. William Bartlett of Brigus, an uncle of Capt. Abram Bartlett, the grandfather of our well known Arctic Navigator, Capt. Robert Bartlett. He was out seal hunting about the year 1800 in his open boat and not finding the ice at Baccalieu [at the mouth of Conception Bay], he decided to "Follow On", to Cape Bonavista, and then to the Funks, and still there was no ice and he "Followed On", till he sighted the Spotted Islands on the Labrador, and off Hiscock Island he met the ice and there were the seals. He loaded his shallop and returned safely to Brigus.[82]

Lewis Anspach, who lived in Harbour Grace from 1801 to 1813, gave the Conception Bay inhabitants most of the credit for the successful establishment of the spring seal fishery.[83]

This new spring seal fishery was finally recognized by the record keepers in 1803, during Governor Gambier's administration. In that year, he reported on the extent of both the winter and spring seal fisheries. His report states that during the previous winter, 493 men in Fogo, Bonavista and Trinity districts, using nets, took 4,461 seals valued at £2,230 and produced 58 tuns of seal oil. The same report describes the spring seal fishery of that year as follows: 77 ships and boats employing 998 men took 49,007 seals valued at £23,152 which yielded 616 tuns of oil (see Table Intro.12). Even allowing for the fact that this may have been an unusually unproductive year for the winter net fishery and also that Gambier, most likely, had not received complete reports from the north, nor reports from the coast of Labrador, it is apparent that the spring vessel fishery was firmly established by now. Moreover, given the fact that out of the 77 vessels ('ships and boats'—the latter were no doubt shallops) involved, 35 (45 per cent) came from Conception Bay and 19 (25 per cent) were out of St. John's. The remaining 23 vessels (30 per cent) of the fleet came from the Trinity and Bonavista districts. The report for the following year is more complete and also shows a considerable growth in the sealing fleets of St. John's to 35 schooners, while the fleet in Conception Bay increased to 68 shallops and schooners (see Table Intro.13). This spring seal fishery was to continue to expand, and the dominance of Conception Bay and St. John's was to increase as well.

The entry of Conception Bay and St. John's into the sealing industry was closely associated with the developments in the cod fishery during this period, as discussed earlier. A significant proportion—if not most—of the Irish immigrants who flocked to Newfoundland during the Napoleonic War settled in the largest, most prosperous harbours, partly because these were the centres which provided most employment and also partly because it was to these harbours that most of the ships on which the Irish travelled were bound.

Thus, the majority of the immigrants arriving during this period settled in St. John's, near the major harbours in Conception Bay—especially Harbour Grace, Carbonear, Brigus and Bay Roberts as well as others, and along the Southern Shore—where fishing rooms, which had been vacated by the British bank and bye boat fishermen because of the war, were now available. After 1808, as has already been discussed, there was secure employment in the major harbours, as merchants and planters expanded production to meet the growing demand for Newfoundland saltfish in all markets. Along the Southern Shore, this expansion was confined to the shore fishery because of the more thinly spread population, and the same situation applied to the northern harbours.

However, in St. John's and in the old, long-established Conception Bay ports like Harbour Grace, population expansion took a new direction in terms of Newfoundland history. Here, the merchants and planters built vessels and sent them to the 'North Shore' for the summer fishery. That coastline was part of the 'French Shore' and reserved, for the most part, for the use of the French migratory fishery. However, the French fishermen had not visited the area since 1792 because of the continuing wars, and therefore, there were hundreds of vacant fishing rooms that were now available for the use of the fishermen of the over-crowded harbours in Conception Bay and in the St. John's area. Almost simultaneously, the merchants and planters discovered that they could employ their vessels in the seal fishery in the spring and at the North Shore fishery during the summer months. Because neither industry was confined to the harbour where the merchants and fishermen resided, there were really no local physical limits on the extent to which expansion could take place, as had always been the case with the old shore fishery, where the inhabitants had lived along the shore and fished, literally, in front of their homes, processing and curing their product on the beach and strip of land in the immediate vicinity. Thus it was that fishing servants who had been seasonal migrant workers could now obtain winter employment, marry, raise families and remain in Newfoundland. They were the first 'employee' fishermen to do so. The Irish formed the majority of this body of fishing servants because they were the people available when the need arose for year-round workers.

Beginning in 1803, as described above, the governors paid more attention to the seal fishery as a separate industry and, consequently, one can follow the progress of this fishery from their reports. For example, the reports for 1803 show that 53,468 seals in all were taken by inhabitants, as discussed above. In 1804, this figure had risen to 106,739, and during the period from 1811 to 1816, inclusive, the average annual kill amounted to 128,618 seals.[84] The total kill declined in 1817 (to maybe as few as 50,000 seals) during the disastrous depression of that year,[85] but, as pointed out, it rose to 165,622 in the spring of 1818 to help save the local economy from annihilation. In that latter year, 1,809 tuns of seal oil valued at over £63,000 were exported.

The growth, during this period, of the spring seal fishery and the 'North Shore' fishery, as prosecuted by residents from St. John's and Conception Bay, is well documented.[86] In 1807, 47 vessels from St. John's, 78 from Conception Bay, 2 from Trinity Bay, 3 from Bonavista Bay and 1 from Twillingate fished on the North Shore. Therefore, out of a fleet of 131 ships, 125 came from St. John's and Conception Bay. That same year, fishermen from St. John's brought into port

33,950 seal skins, while fishermen from Conception Bay brought in 58,950. At the same time, Keels (in Bonavista Bay) produced 18,000 and Fogo and Twillingate together another 25,800. An additional 8,519 pelts were brought in by Trinity Bay fishermen and 5,000 by fishermen in Bonavista and King's Cove. St. John's and Conception Bay fishermen relied exclusively on the spring vessel fishery; Trinity Bay, Bonavista Bay, Fogo and points north relied on both the spring and winter fisheries, with the latter becoming more important in the more northerly ports. The evidence also suggests that the winter fishery was most vulnerable to the forces of nature. In 1809 (for which complete statistics are available) the overall number of vessels on the North Shore had declined to 90, composed of 30 from St. John's, 56 from Conception Bay, 3 from Trinity Bay and 1 from Fogo. Meanwhile, it was reported that the sealing fleet that year was made up as follows:

Harbour	Shipping			Seals	Tuns of Oil
	No.	Tons	Men		
St. John's	37	1,778	484	30,477	468
Conception Bay	66	3,156	916	41,433	625
Trinity Bay	6	332	79	3,776	25.5
Bonavista Bay and Greenspond	13	516	150	7,409	93
Fogo	8	375	93	1,330	13
Ferryland	5	270	73	2,191	40
Total	135	6,427	1,785	89,669	1,264.5

At the same time, the returns of the winter seal fishery of 1808-9 which are included in this report show that only 2,538 seals were taken and 43 tuns of oil produced by this branch of the industry. Thus, the vessel fishery was much more valuable and almost exclusively concentrated in the south. In 1811 (statistics for 1810 are incomplete), St. John's sent 48 ships to the seal fishery, while Conception Bay sent 81. (For the first time, it was reported that 13 vessels from St. John's and 11 from Conception Bay were lost in the ice, but all the crews were saved, with the exception of one individual from the latter.)[87]

The cod fishery on the North Shore continued throughout this period. St. John's sent 29 vessels there in 1812; 26 in 1813; 45 in 1814 (as demand for saltfish grew); 36 in 1815; 17 in 1816 (during the depression); 24 in 1817; and 34 in 1818. Conception Bay's numbers for this same period were: 77 in 1812; 83 in 1813; 81 in 1814; 70 in 1815; 86 in 1816; 85 in 1817; and 118 in 1818. Only these two places sent vessels to the North Shore in 1815, although for the next three years, one ship sailed annually from Trinity Bay. Thus a migratory cod fishery involving St. John's and Conception Bay fishermen and reminiscent of the old West of England fishery in Newfoundland had become established. Moreover, although governors continued to report 'Vessels on the North Shore', it must be remembered that the North Shore was part of the French Shore and was returned to the use of the French fishermen by the treaties of 1814 and 1815. Other sources indicate that the vessels which had been fishing on the North Shore began the practice of sailing on to the Labrador coast after 1815, although it is unclear just how rapidly the English North Shore fishery became a Labrador coast fishery. In 1826, the governor reported 239 vessels on the North Shore, while the last report of this nature, in 1827, reported 73 "British Vessels on the North Shore and Labrador."[88] Governor Hamilton reported in 1820 that "An unusual number of people ... resorted from this Island [to Labrador] when compared with former years."[89] And a final report in 1853 pointed out that the North Shore fishery concluded in 1821, when the fishermen "with few exceptions abandoned the fishery and betook themselves to the Labrador."[90]

The threat of the French returning to the French Shore and displacing the English-Newfoundland fishermen had been recognized as early as 1804 by Governor Gower, who reported that the fishermen of Conception Bay had established quite a significant cod fishery north of Cape St. John and expressed concern that they would have to make the longer voyage to the Labrador coast once a peace treaty returned the North Shore fishery to France.[91] Although the North Shore fishery quickly and smoothly became a Labrador fishery, nobody involved was pleased with the return of the French to the area, and one of the first petitions of the newly established representative government assembly in St. John's addressed this concern to the Colonial Office as follows in 1834: "The Labrador cod fishery was generally considered to be a precarious venture and most agreed that losing the right to fish at Petit Nord (North Shore) was very unfortunate to the Newfoundland fishermen."[92]

In looking at the establishment of a spring seal fishery during the Napoleonic War, one cannot overemphasize the interdependence of the spring seal fishery and the North Shore (later Labrador) cod fishery. The sealing ships were fitted out in February, they sailed in March and returned to port in April or early May, at which time the vessels were outfitted for the cod fishery. Governor Gower summarized the new industry in a report on the 1804 season:

> This fishery commences about the middle of March and continues till the early part of May. The Merchants or owners of the vessels, who are at the whole expense of the outfits, receive one half the proceeds of the voyage, and the other half is divided in equal proportions among the crew, who generally clear from £5 to £25 each man; and though the success of the voyage is precarious, yet the personal interest which the men feel in it, stimulates them to encounter the most inclement weather, and expose themselves to the most imminent dangers, and instances last spring occurred of Crews, who were taken off the wrecks of vessels that were crushed between the ice, and brought home, having procured other vessels and made a successful voyage. The owners' profits are generally sufficient to defray the expense of fitting out the same vessel in the Cod Fishery, which commences about the time of her return from sealing.... It appears that upwards of 1600 men were last spring employed in the Seal Fishery, and it is certain that there is no employment so well calculated to form hardy and intrepid seamen, and for this reason it appears very desirable that every encouragement should be afforded it [by the British government].[93]

Back in port, 'seal skinners' went to work on the pelts and skilfully removed the fat from the skin. The fat was placed in vats to render into oil and the skins were tanned for export. The ships were then refitted to sail to the North Shore or Labrador coast in June for the cod fishery, returning with their cargoes of saltfish in September or October. After the fish was culled according to the demands of the various markets, the larger vessels were dispatched to these markets with cargoes of saltfish, while the smaller vessels engaged in the coastal trade during the fall. Thus, the same ports, ships and men were engaged in a round of activities which occupied them for almost the entire year and allowed, indeed encouraged, expansion and growth in the ports concerned.

Summary

During the period from 1500 to 1800, Newfoundland was the site of a major English/British migratory fishery, valued for its contributions to the economy and to the navy. Its development evolved from a fishery prosecuted solely by migratory fishing ships coming annually to Newfoundland for the summer months to one whereby resident planters maintained substantial operations on a year-round basis, using migratory labour, while at the same time, supporting the migratory ships by securing and maintaining the latter's premises during the winter months. The migratory ships reciprocated by supplying these planters with the provisions, equipment and labour needed to carry on their fishery and bringing the final product to market. The work force on which the resident fishery depended could not earn enough in the short season to become full-time residents of Newfoundland. This all changed during the Napoleonic War period, when the price of saltfish rose dramatically and the Irish were encouraged to join the fishing ships and sack ships in the port of Waterford for employment in the Newfoundland cod fishery. They became the employees of the resident planters in the major harbours, who decided to expand their operations to include the cod fishery on the North Shore. The vessels used for these purposes were then available to hunt seals among the ice floes the following spring. Newfoundland had finally discovered what it needed for permanent and independent settlement—a second industry.

NOTES

1. For a thorough discussion of the early migratory fishery, see Keith Matthews, "History of the West of England-Newfoundland Fishery" (D. Phil. thesis, Oxford, 1968).
2. See C. Grant Head, *Eighteenth Century Newfoundland* (Toronto, 1976), pp. 2-6.
3. The term 'fish' is singular when used in this context.
4. See Matthews, "West of England," for further information.
5. See Matthews, "West of England," for the best discussion of the origins and growth of the bank fishery. See also Head, *Newfoundland*, for an excellent account of all aspects of Newfoundland history during this century. The term 'England' is generally used when referring to that country during the period up to the union of England and Scotland in 1707. After 1707 the term 'Great Britain' is most appropriate. While one must try to be consistent, the term 'England' is most fitting when discussing Newfoundland's development up to the last quarter of the eighteenth century.
6. The French were, at first, restricted to that part of the coast extending from Cape Bonavista on the northeast coast around the tip of the northern peninsula south to Point Riche, about one-third the distance to Cape Ray. With the conclusion of the

American Revolutionary War in 1783, the boundaries of the 'French Shore' were adjusted to remove the French fishermen from that part of the northeast coast between Cape Bonavista and Cape St. John and to allow them to fish on all the west coast of the island. However, the French were refused the right to build permanent structures on the coast and were required to vacate the coast at the end of each fishing season. Furthermore, the British never agreed to give the French the rights to an exclusive fishery on the French Shore, and consequently, there was often conflict between fishermen of both nations until the French finally surrendered all rights to the French Shore in 1904. The eastern side of the French Shore was usually referred to as the 'North Shore' by the British fishermen.

7   Shannon Ryan, comp., "Abstract of C.O.194 Statistics" (Unpublished manuscript, Memorial University of Newfoundland, 1969). See the CO 194 volumes for the original reports.

8   See Gillian Cell, *English Enterprise in Newfoundland: 1577-1660* (Toronto, 1969); Gillian Cell, ed., *Newfoundland Discovered: English Attempts at Colonization, 1610-1630* (London, 1982); W. Gordon Handcock, *Soe longe as there comes no women: Origins of English Settlement in Newfoundland* (St. John's, 1989); Head, *Newfoundland*; Matthews, "West of England"; Peter Pope, "The South Avalon Planters, 1630 to 1700: Residence, Labour, Demand and Exchange in Seventeenth-century Newfoundland" (Ph. D thesis, MUN, 1992; and Prowse, *Newfoundland*, for discussions of early settlement in Newfoundland.

9   Official colonization attempts in Newfoundland encountered two unique problems: when colonizers—from Guy in 1610 to Kirke in 1637—arrived in Newfoundland, they found thousands of English fishermen engaged in the cod fishery, like their fathers, grandfathers and often great grandfathers before them, and these fishermen claimed their traditional rights; and colonies in Newfoundland were never able to support themselves with food.

10 David B. Quinn and Neil M. Cheshire, eds., *The New Found Land of Stephen Parmenius* (Toronto, 1972), pp. 169-73.

11 Unless otherwise stated, the following statistics have been taken from Ryan, "Abstract."

12 Quoted in Prowse, *Newfoundland*, p. 344.

13 See Matthews, "West of England"; Keith Matthews, *Lectures on the History of Newfoundland: 1500-1830* (St. John's, 1988); and Prowse, *Newfoundland*, for complete discussions of the following summaries.

14 Prowse, *Newfoundland*, p. 365.

15 In 1789, the single men who remained in Newfoundland during the winter and worked for their room and board only were first described as 'dieters', referring, no doubt, to the fact that they received no other recompense. See *DNE*, p. 140.

16 It is very likely that there were huts in the vicinity of the 'households', inhabited by servants who managed to survive from one fishing season to the next.

17 CO 194/39, fols. 25-44. Governor Wallace to CO, 29 September 1796.

18 Sheila Lambert, ed., *House of Commons Sessional Papers of the Eighteenth Century*, (Wilmington, 1975), 90: 404. "Third Report from the Committee appointed to enquire into the State of the Trade to Newfoundland, 17 June 1793—Evidence of John Reeves, Chief Justice of the Island of Newfoundland...."

19 CO 194/40, fols. 17-34. Report from Admiral Crofton, Commander of the *HMS Pluto*, to Governor Waldegrave. Information was collected in the summer of 1797 and the report is dated 10 January 1798.

20 CO 194/39, fols. 27-45. Wallace to CO, 24 November 1796.

21 CO 194/39, fols. 27-45. Wallace to CO, 24 March 1796.
22 CO 194/39, fols. 86-8. Waldegrave to CO, 14 August 1797.
23 BT 6/92, fol. 164; and CO 194/40, fols, 135-7. Waldegrave to CO, 30 October 1798.
24 CO 194/40, fol. 65. Frederick Warren, R.N., to Waldegrave, 24 March 1798.
25 CO 194/40, fol. 105. Waldegrave to CO, 18 October 1798.
26 CO 194/40, fols. 135-7. Waldegrave to CO, 30 October 1798.
27 CO 194/40, fol. 83. Waldegrave to CO, 11 June 1798. It was not until 1817 that the governor was ordered to remain in Newfoundland throughout the winter. In the meantime, the governor's annual tour of duty in Newfoundland was confined to the fishing season, and sometimes only for part of that.
28 CO 194/42, fol. 37. Waldegrave to CO, 25 February 1799.
29 CO 194/40, fol. 31. Captain Crofton of the *HMS Pluto* to Waldegrave, 10 January 1798.
30 CO 194/42, fols. 59-60. Waldegrave to CO, 9 May 1799.
31 CO 194/42, fols. 112-3. Waldegrave to CO, 22 October 1799.
32 CO 194/42, fol. 148. Report of the St. John's Magistrates to Waldegrave, 24 October 1799.
33 CO 194/43, fol. 122. Gambier to CO, 22 January 1803.
34 CO 194/43, fols. 24-8. Lieutenant Governor Robert Barton to CO, 2 January 1802.
35 BT 6/92, fol. 164. "Computed by deducting 30,000 cwts for British and Irish consumption."
36 CO 194/45, fols. 155-9. Governor Gower to CO, 9 November 1806.
37 CO 194/43, fols. 284-9. Order in Council, 2 March 1803.
38 BT 6/92, fol. 164. These figures vary slightly from those reported in the CO 194 (see Ryan, "Abstract").
39 CO 194/44, fols. 40-3. Gower to CO, 28 March 1804.
40 CO 194/44, fols. 198-200. Gower to CO, 28 November 1805. Gower's reports are among the most comprehensive of all reports submitted by Newfoundland governors. Not only does he provide information on prices by item and region and wages by occupation and region, but he also includes many other facts and observations. For example, in the report above, he states that it cost a fisherman £6 to £7 to travel to Newfoundland from Ireland in 1804 and £5 to return home. Also, as will be seen later, he reports on the fledgling seal fishery.
41 CO 194/44, fols. 115-7. Gower to CO, 18 July 1805.
42 CO 194/45, fols. 17-47. Gower to CO, 18 March 1806. This is a very comprehensive report on Newfoundland. Gower explained that it was late because "... for want of some Accounts from the Out Harbours, [it] could not be completed sooner." This particular report is referred to below as "Gower's Report for 1804." Please note that in addition to their despatches, reports and replies to enquiries the governors were required to submit a consolidated report on the fisheries, trade and inhabitants at the end of each year.
43 CO 194/45, fols. 155-9. Gower to CO, 9 November 1806.
44 CO 194/45, fols. 61-9. Gower to CO, 29 April 1806.
45 BT 6/91, fol. 3. "Report laid before the House of Representatives by the Secretary of the Treasury respecting the fishermen of the United States on the 29th of January, 1802."
46 CO 194/44, fols. 24-5. Lieutenant Morrison R.N., Commander, HMS *Charlotte*, to Gower, September 1804.

47 CO 194/44, fols. 24-5. This was before the increase in exports from Newfoundland which began in 1805 (see above).
48 BT 6/91, fol. 116. Report from Captain James Murray, R.N., HM Sloop *Curlew*, on the Labrador coast, 23 July 1805.
49 CO 194/48, fol. 23. Holloway to CO, 19 July 1809.
50 CO 194/48, fols. 51-129. Governor Duckworth to CO, 25 November 1810.
51 CO 194/53, fols. 3-8. Duckworth to CO, 2 November 1812.
52 Ryan, "Abstract"; and Edward Chappell, *Voyage of the HMS Rosamund to Newfoundland and the Southern Coast of Labrador* (London, 1818), pp.245-7.
53 CO. 194/54, fols. 55-66. Society of Merchants, St. John's, to Governor Keats, 23 June 1813. According to Pedley, *Newfoundland*, p. 284, fish splitters received £90—£140 in 1814.
54 CO 194/54, fols. 55-66. Society of Merchants, St. John's, to Governor Keats, 23 June 1813.
55 CO 194/54, fols. 159-74. Keats to CO, 18 December 1813.
56 CO 194/55, fols. 95-104. Keats to CO, 29 December 1814.
57 See Prowse, *Newfoundland*; John Mannion, *Irish Settlements in Eastern Canada: A Study of Cultural Transfer and Adaptation* (Toronto, 1974); Mannion, ed., *Peopling of Newfoundland*; Handcock, *Origins of English Settlement*; Head, *Newfoundland*; and Matthews, "West of England," from which the following short section has been summarized.
58 CO 194/44, fols. 198-200. Gower to CO, 28 November 1805.
59 CO 194/45, fols. 17-47. "Gower's Report for 1804."
60 CO 194/45, fols. 199-205. Gower to CO, 23 December 1806. This is Gower's valuable report on all aspects of Newfoundland's trade and population for the year 1805.
61 See Prowse, *Newfoundland*, Chapter XIII.
62 For an examination of the fish trade in post-1814 Newfoundland, see Ryan, *Fish out of Water*. The two factors that affected exports almost immediately were the introduction of prohibitively high import tariffs on Newfoundland (British) saltfish imported into Spain and the entry into the marketplace of cheaper Norwegian klippfisk. Of course, the usual post-war depression compounded the problem.
63 The following short discussion of the distress in Newfoundland during this period is summarized from Prowse, *Newfoundland*, pp. 402-11.
64 Most of the information about the terminology used to describe the various stages of the development of adult seals has been taken from an interview on 17 June 1976 with Cyril S. Chafe, former manager of Bowring Brothers seal operations, St. John's (since deceased). Tape in author's possession. See also *Seals and Sealing in Canada: Report of the Royal Commission on Seals and the Sealing Industry in Canada* (Ottawa, 1986), chaired by Albert H. Malouf; and *DNE*.
65 The skin of the seal with the fat attached is nearly always called the 'pelt' in the written records; similarly, it was and is used as a verb. Sculp is a synonym for pelt but it was not used in the written records as much and rarely in government or business records. However, 'sculp', as both a noun and a verb, was widely used by the ice hunters. 'Scalp' was used very infrequently. *DNE*.
66 Prowse, *Newfoundland*, p. 698.
67 The term 'catch' was often used to describe the number of seals taken in a given year or on a given occasion. *DNE*, p.90.

The Ice Hunters

68 Unless otherwise stated, all statistics have been taken from Ryan, "Abstract," and in some cases, approximate calculations have been made, using prices and quantities from the recorded statistics.

69 See Commodore's "Answer to the Heads of Enquiry for the year 1726," CO 194/8, fol. 43; and Commodore's "Answer to the Heads of Enquiry for the Year 1734," CO 194/9, fol. 262. These reports have been chosen to illustrate the point being made and also, unlike some others, they appear to be reasonably complete.

70 CO 194/18, fol. 41. Although Labrador fishermen produced a considerable amount of seal oil in the year ending in October, 1768, it is unclear whether later governors always included oil from this area in their annual reports. However, for the present writer's purposes this was not an important point; the fact remains that the seal population was gradually becoming recognized as a valuable resource.

71 BT 6/90, fols. 50-6. "Representation to the King relative to the Seal Fisheries on the Coast of Labrador." Signed by John Roberts, Bamber Gascoyne and Robt Spencer, Hillsborough, 24 June 1772.

72 CO 194/8, fols. 15, 43 and 164. Commodore's "Answer to Heads of Enquiry ..." for the years ending in October 1725, 1726 and 1727; CO 194/9, fol. 262 and CO 194/10, fol. 16. *Ibid.*, for years ending in October 1734 and 1735; and CO 194/18, fol. 41. "A General Scheme of the Fishery and Inhabitants of Newfoundland for the year 1768."

73 CO 194/21, fol. 39. Elliot to CO, 14 November 1786.

74 Sheet ice: " a stretch of flat, thin ice ... frequented by seals." *DNE*, p. 469.

75 CO 194/21, fol. 45. "Account of Trade and Shipping for Year ending October 10, 1787."

76 CO 194/21, fol. 425. "Annual Report on the Fishery and Trade of Newfoundland for the year ending October 10, 1793." Except when quoting original sources, the *Gazetteer of Canada: Newfoundland and Labrador* (Ottawa, 1968), will be used as the guide to the correct spelling of the names of communities; please note, however, many inconsistencies in this source. Thus, readers will find King's Cove and Bryants Cove, Crockers Cove and Bishop's Cove as well as many other examples of the use and non-use of the apostrophe 's'.

77 Prowse, pp. 419-20. Quoted from CO 194/43.

78 CO 194/23, fol. 426. Tables of shipping and trade outwards ... from Newfoundland.

79 Ryan, "Abstract," pp. 66 and 93.

80 CO 194/129, fol. 147. Le Marchant to CO, 4 May 1848.

81 Mosdell, *Chafe*, p. 20.

82 Mosdell, *Chafe*, p. 20.

83 Anspach, *Newfoundland*, pp. 421-2.

84 'Kill' was often used as a verb in the sense of "to successfully hunt," and as a noun as in to go "over the side [of the ship] for another kill." *DNE*, (2nd ed.), p. 707. It often means to 'catch' fish and leading fishermen are known as 'fish killers'.

85 Prowse, *History*, pp, 404-5. Ryan, "Abstract," records a catch of 38,288 seals for that year without noting that there are no returns available for St. John's. However, given the fact that Conception Bay's seal fishery produced 103,358 pelts in 1816, 32,939 in 1817, and 118,228 in 1818, while St. John's produced 27,485 in 1816 and 24,735 in 1818, one can conclude that the 1817 voyage amounted to no more than 50,000 seals.

86 See the annual "Return[s] of the Fishery and Inhabitants of Newfoundland for the Year ..." for each year from 1807 to 1818 inclusive in CO 194/47, fol. 54; CO 194/48,

fol. 7; CO 194/49, fols. 14 and 50; CO 194/51, fol. 36; CO 194/54, fol. 39; CO 194/55, fol. 29; CO 194/55, fol. 107; CO 194/57, fol. 12; CO 194/59, fol. 27; CO 194/61, fols. 22 and 157-8. Of particular interest are the reports of "Vessels on the North Shore" from St. John's and Conception Bay and other ports of origin, information on the catch of seals, the production of seal oil (less regularly reported) and the number of ships that participated in the seal fishery.

87 Losses to lives and shipping is an important topic in itself and will be discussed in a later chapter.
88 Ryan, "Abstract."
89 CO 194/63, fol. 128. Hamilton to CO, 14 November 1820.
90 CO 194/139, fols. 327-8. Hamilton to CO, 28 September 1853.
91 CO 194/45, fols. 17-47. "Gower's Report for 1804."
92 CO 194/87, fol. 87. 1 May 1834. Annual "Return."
93 CO 194/45, fols. 17-47. "Gower's Report for 1804."

CHAPTER 1
Market Place

THE IMPETUS FOR the development and rapid growth of the West of England-Newfoundland cod fishery was provided by the ready market for saltfish (salted, dried codfish) in France in the 1560s, when that nation became involved in a civil war which put an end, temporarily, to its own fishery. Within a few years, the Spanish and Portuguese Newfoundland fisheries were permanently eliminated—with a few exceptions—and the export of English-Newfoundland saltfish to these countries commenced and eventually developed into a major trade. Thus, it was the demand in foreign markets for saltfish that stimulated the growth of the English cod fishery in Newfoundland waters.

At the beginning of the nineteenth century, another industry developed in Newfoundland—a seal fishery—and the market forces which stimulated the expansion of this industry emanated from England/Great Britain itself. In order to discuss the growth of the Newfoundland seal fishery, it is necessary to examine briefly the demand in England for the two products obtained from this sea mammal—oil and leather, the former originally of much more importance than the latter. This examination will proceed by first discussing the growing number of uses of oils (and fats) and the changes in supply and demand that affected Newfoundland's product. Skins (including seal skins) will be considered briefly within this context as well.

Early Lighting

With the centralization and strengthening of governmental powers in northern Europe at the end of the Middle Ages, curfews were no longer enforced and people began to participate in evening activities.[1] They began to hang lights outside their houses on festival days, and gradually, the town authorities took the initiative and commanded householders to do so on special occasions in order to make the streets safer for pedestrians—safer from thieves and safer for walking.[2] In the seventeenth century, as part of this new era, churches ceased to be the only major centres where people were permitted to mingle at night. Tea houses, coffee houses, eating houses and taverns became popular. In the eighteenth century, clubs provided further attractions.[3] Under these circumstances, it became increasingly necessary to provide regular street lighting.

In 1668, the inhabitants of London were ordered to hang candles outside their houses[4], a practice that the authorities in Paris had instituted two years earlier.[5] This requirement triggered a demand for improved street lighting, and in 1684, Edward Wyndus was granted a patent for a new street oil lamp with a convex lens to reflect the light. He also received the contract to provide street lighting in certain parts of central London. He fulfilled his contract with his new oil lamps burning rape seed (also called colza) oil, and, although Wyndus

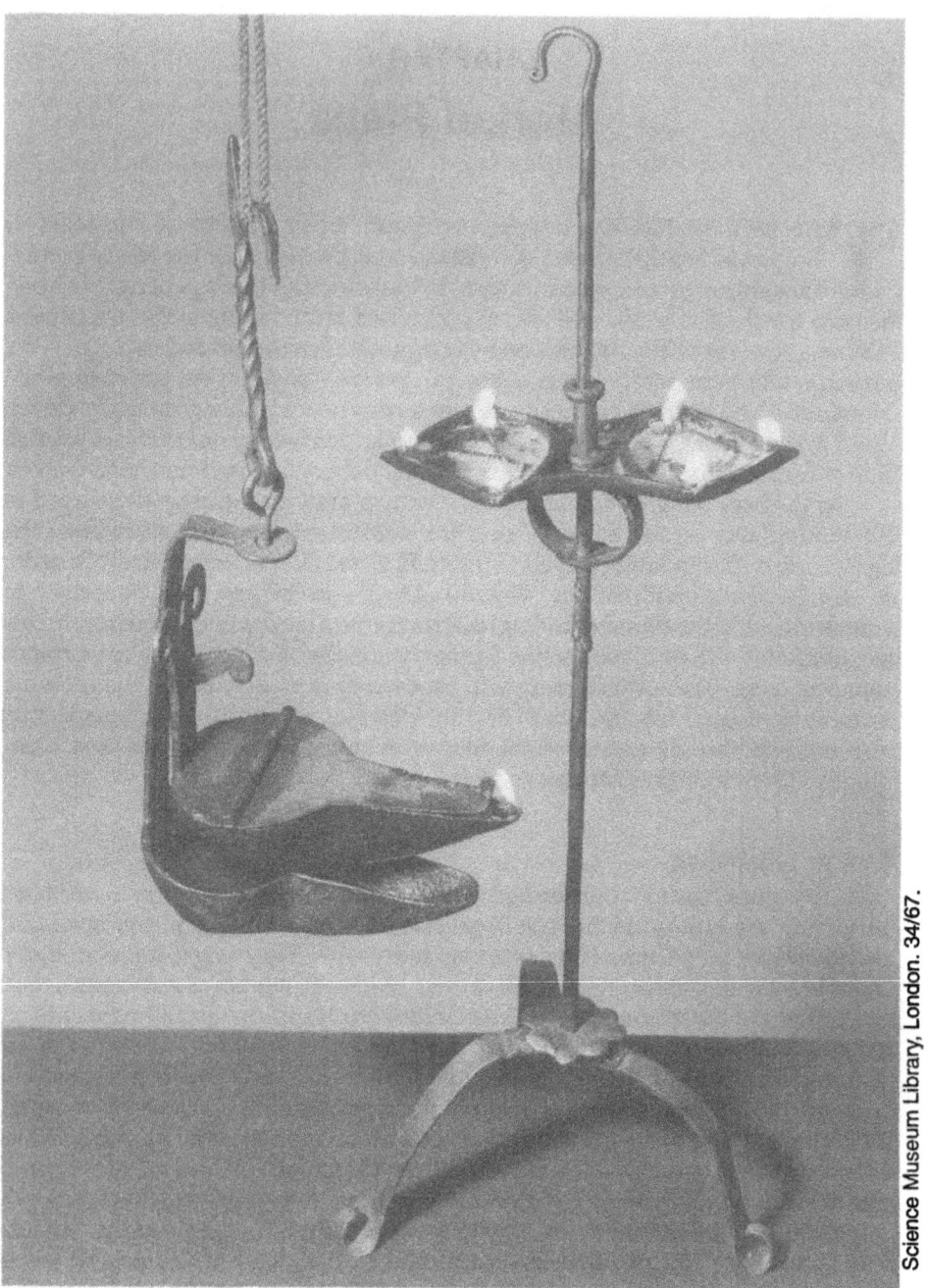

Figure 6 - On the right, a double-pan lamp with six wicks; pan lamps were the earliest oil lamps. the development of the cruise lamp (left) was a major improvement over the pan lamp. The lower vessel on the cruise lamp was designed to catch oil drips. The betty lamp (not shown), with an enclosed spout, was a further development.

dropped out of the picture, his successors built on this advancement. In 1691, a report stated that London had "three notable Conveniences ... not elsewhere to be found; viz. the New Lights, the Penny-Post, and the Insurance-Office for Houses in case of Fire."[6] Like many inventions, the new lamp had its critics, and in 1690, three city companies (or guilds), whose businesses were directly and adversely affected (i.e. the Tallow Chandlers, the Tin Plate Workers and the Horners—who supplied candles and lanterns), tried to prevent the new contractors from fulfilling their contracts, but to no avail.[7] The new lighting company then proceeded to gain a contract to provide street oil lamps over the whole City of London, collecting a sum of money from each householder. The effectiveness of these lights is supported by the evidence of a woman who witnessed a murder on the night of 6 January 1692. She testified:

> I well observed Mr. Harrison but do not know the other Man; there were two Lamps burning; one in Brownlowe Street, and the other in Holborn ... and they lighted quite through the Coach.[8]

By 1736, it was no longer practical for householders to pay contractors to provide street lighting in London, and in that year, an Act of Parliament transferred the responsibility for street lighting to the aldermen in each ward and set rates that householders were required to pay. Before 1736, there were about 1,000 lamps in the City and surrounding area, and these were lit for about 750 hours per year; by 1739, this number had increased to nearly 5,000 lamps maintained by 17 contractors. As de Beer writes: "In one stride the City had become the best lit of the great urban areas of Europe."[9]

A similar pattern of development occurred in Bristol. Inhabitants were at first allowed to walk outdoors at night after the ringing of curfew, provided they carried lights. Then, in the sixteenth century, the Bristol Corporation provided a few lanterns with candles to be hung out at night. In 1606, a man was hired, and paid half-a-crown per quarter (three months), to look after the street lanterns. By 1748, oil lamps were in use,[10] and in 1788, one report stated that Bristol was exceeded only by London in terms of the quality of its street lighting.[11] By the latter years of the eighteenth century, the town of Liverpool was also lighted with oil lamps, and we know for certain that at least some of these lamps burned Newfoundland cod oil.[12] The fact that smaller ports were also supplied with street lighting is illustrated by the report of an incident concerning Yarmouth during the war with Revolutionary France:

> ... fish-oil was available in almost unlimited quantities [for street lighting]. The Russians evacuated with the British troops after the Netherlands campaign of 1799 ... astonished the people of Yarmouth, where they landed, by drinking the oil out of the street lamps.[13]

Therefore, by the latter part of the eighteenth century, street lighting had become firmly established in the major commercial centres of Britain, and responsibility for maintaining this service was accepted by the local political authorities. One contemporary report from Bristol summarizes this development:

> London was first lighted with lamps in the reign of Charles 2nd [1660-1685]. The act for lighting Bristol was not procured till the latter part of the reign of King William 3d [1689-1702]. This act obliged Citizens to hang out their

lamps: subsequently public lamps were lighted for half the year only, but for many years past, have been kept burning every night during their proper hours. Many of the principal streets and places of Bristol are well lighted [in 1794]....[14]

These developments in street lighting created a growing demand for lighting oil.

Meanwhile, other developments in lighting were taking place. The English labourer had traditionally depended on the rush dipped in grease for the brief and feeble light that was sufficient to see his family through the short period of darkness during which they sat up after curfew. However, as evening activities became common, candles and lamps were required to provide longer and more dependable light. Candles were made from bees wax and from tallow. The former were moulded, expensive and much in demand by the Church and by the wealthier classes. The latter could be moulded but were usually made by dipping. The English made exceptionally good tallow candles, using equal parts of mutton fat and beef fat.[15] However, even the best tallow candles needed constant attention because the tallow and the wick generally burned at uneven speeds. While this made them useless for street lighting purposes, they were adequate for lighting the homes of all but the wealthiest classes. Foul-smelling oil lamps remained the preferred source of light for streets, docks, wharfs and all outside purposes.

The first oil lamps were low flat pan lamps with the wick laying on the oil. Later, lamps were developed with channels up the side to hold the wick. These were known as 'cruises', and they were usually provided with a second pan underneath to catch the surplus oil. In the next development, the channel was converted into a spout (like that of a tea kettle) through which the wick was threaded, and the final refinement was the addition of a cover which produced the 'Betty' lamp.[16] Thus, as one writer stated: "During this [eighteenth] century the oil lamp began to emerge from a long dormant period."[17] Nevertheless, despite these developments, and for a variety of reasons, there was "virtually no improvement in the techniques or efficiency of lighting, but only increasing refinement and embellishment of its vessels."[18]

Leather Currying

Before continuing with a chronological treatment of lighting, it is necessary to discuss the traditional importance of oil in the manufacture of leather. The origins of this industry, like the origins of lighting, are lost in the distant past. However, leather was used by all early inhabitants for clothing and tents and, later, to meet a wide variety of household and factory needs.

To handle the preparation, sale and utilization of leather, a number of guilds (or gilds) evolved in England after the Norman conquest; and the political and social prominence of certain guilds associated with the leather trades and crafts illustrate the importance of this product. For example, the saddlers' and skinners' guilds were among the most important in fourteenth century England and ranked as equal with the goldsmiths, merchant taylors and mercers (dealers in certain textiles). In 1351, the thirteen principal guilds (which were beginning to be referred to as misteries or companies) included three which were associated with leather— the skinners, saddlers and cordwainers. Also, the cordwainers, saddlers and girdlers were granted Royal Charters before the end of the fourteenth century; and the saddlers and cordwainers were among the thirteen guilds to send representatives to the Common (Guild) Council in 1351.[19] Furthermore, it should be noted that of the 111 groups of craftsmen recorded in London in 1422, fourteen were directly concerned with leather and/or leather products. These were as follows: skinners, saddlers, cordwainers, leathersellers, girdlers, loriners (or lorimers—makers of bridles, bits and spurs), tanners, curriers, pouchmakers, cofferers (chest makers who used leather coverings), whitetawyers (those who prepared white leather), leather dyers, glovers and malemakers (those who made leather trunks and packs).[20] Thus, one can appreciate the importance of leather in European life on the eve of the age of expansion.

In the preparation of leather hides and skins,[21] the raw product was 'tanned', resulting in a dried hide or skin which was usually stiff. This new product was then sent to the leather curriers, who worked oil or fat into it and produced 'leather'. This was an essential stage in the production of leather, and the curriers' guild became quite important and expanded with the increasing demand for leather. As the demand for leather grew, so too did the demand for oil and fat to be used in the currying process. The leather industry did not undergo any revolutionary changes until advances in chemistry in the late nineteenth century resulted in the development of new techniques. Then, the major change was the virtual disappearance of the currying process, which was absorbed into the tannage process, and 'oil tannage' became only one of several methods used. Nevertheless, as late as 1946, it was reported that "present-day oil-tanned leathers are produced by the aid of fish and marine animal oils such as cod-liver, seal or whale...."[22] Thus, the leather industry was to become a very important consumer of fish and seal oils.

Other Traditional Uses of Oils and Fats

In the meantime, oil was utilized in other ways as well. Soap making, using olive oil, was introduced into Marseilles in the thirteenth century and into England in the next century.[23] Basically, two soaps evolved: hard soap, which required superior tallow (ox, hog, sheep) and soda; and soft soap, which could be made from a range of fish oils and potash. The latter was cruder, contained impurities and was used primarily for washing certain fabrics in the textile industry.[24] Lubrication also required oil and fat, and both were used to keep metal parts operating smoothly. In addition, olive oil had long been used in the manufacture of textiles. One writer describes the trade in olive oil as follows:

> Olive oil, a vital raw material of the woollen industry, came in great tonnage from southern Spain, but the import began to be supplemented, before the middle of the seventeenth century, by an Italian supply.[25]

Another writer points out that it was necessary in the manufacture of linen.[26] Thus, the traditional uses of seal oil included lighting, leather currying, soap making, lubrication and textile manufacturing.

Seal Skins

As indicated at the beginning of this section, the seal fishery involved the production of two products—oil and skins. While the former was much more important, the latter certainly found a ready market in England and elsewhere. The skins from the older harp seals produced a light leather that was very 'close'—strong and tough with a bold grain.[27] They were used to make upholstery, hats, waistcoats, jackets, and boot and shoe uppers.[28] Another source indicates they were also used by 'trunk-makers' and for making breeches and saddle covers.[29] Thus, the Newfoundland seal fishery provided oil and leather: two products which could be absorbed immediately into England's traditional trades and crafts in the context of expanding industrialization and urbanization.

Industrial and Commercial Developments

By the middle of the eighteenth century, England (and Great Britain as a whole) was being affected by a confluence of industrial and commercial developments. The 'commercial revolution' that had begun with the expansion of the English cod fishery in Newfoundland waters, had led to major developments in international trade and colonial expansion. For example, in 1570, when the West of England-Newfoundland cod fishery had begun to expand, woollen cloth sales to the continent of Europe had made up four-fifths of the total value of English exports. By the middle of the eighteenth century, that pattern had changed, and Britain was engaged in an extensive trade with its American colonies—a trade that continued to grow after the end of the American Revolutionary War. Meanwhile, trade with Asia and Europe expanded and included the increased exportation of manufactured goods and the importation, and to some extent the re-exportation, of colonial and tropical and oriental products, including sugar, tea and coffee.[30] It is within the context of England's commercial and industrial growth that the demand for oil and leather must be studied.

As the demand for oil in eighteenth Britain increased, the nation actually faced an oil shortage at times.[31] Among the vegetable oils, there were drying oils, such as linseed, which were needed in the paints and varnishes in growing demand in the new urban centres;[32] semi-drying vegetable oils, such as rape, needed for lubrication and lighting; and the non-drying vegetable oils, particularly olive, used for lighting, in soap and perfume, as well as for food. In the animal oils, there were: fish oils pressed from salmon, herring and sardines, and used for currying leather; and cod oil (from the fish livers) also used for currying leather and for lighting. Seal oil was used both for lighting and currying leather, and whale oil was used for these purposes as well as for soap making and fibre dressing in the manufacture of certain textiles. In addition, animal tallow was used in the manufacture of candles and hard soap. However, there was considerable overlap involved; a shortage of whale oil, for example, would force consumers to substitute cod or seal oil. In 1839, it was reported as follows:

> The accounts received from the Davis Straits [whale] Fishery being of an unfavourable character, it being estimated that the produce will not exceed 1000 tuns, have occasioned holders to demand advanced rates for all descriptions of common fish oils.[33]

In addition, fish oils could be used interchangeably with rape, as a report in 1842 indicates:

> The uncertainty of a continental demand for Fish Oils this season, owing to the reported abundant crop there of Rape, induces the dealers here [Liverpool] to hold back from purchasing pale Seal [oil] for arrival.[34]

Similarly, fish oils and/or tallow could be used in the manufacture of soap;[35] in 1841, for example, there were complaints that cheap South American tallow was being used as a substitute for oil.[36]

In addition to whale oil, another, more valuable product was sperm oil, which was classified as an animal oil during the nineteenth century, but by the twentieth had been reclassified as a liquid wax. It was often referred to as spermaceti oil, but the spermaceti itself was the solid wax that separated and solidified after the carcass of the whale had cooled. This spermaceti (or 'headmatter', as it was sometimes called) was used along with beeswax in the manufacture of superior candles. Sperm oil was very popular as a fuel for lamps, especially in lighthouses, but it was much more expensive than the common fish oils, which limited its use.

There was no real limit to the degree to which common fish oils and other oils and tallow could be used as substitutes for each other, and, as one can see, all these products—from the solid waxes to the tallows, vegetable oils and animal oils, drying and non-drying—overlapped in their uses.

Newfoundland was a supplier of oil to England's markets from the beginning of the cod fishery in c.1500. The cod fishery not only produced saltfish for sale in Spain and Portugal, but also cod oil (train oil) for sale in England. A report in 1616 indicates that a 200 ton ship with 44 men could catch and cure about 200 tons of saltfish worth £2,100 sterling, 10,000 salted (but not dried) cod worth about £100, and 12 tuns (48 hogsheads) of cod oil worth £96.[37] It is apparent that a strong demand for cod oil existed in England, and one of the earliest records indicates that in 1701, migratory fishing ships brought home 1,268 hogsheads of this product, while resident fishermen sent to England 2,533 hogsheads, for a total of about 950 tuns.[38] Therefore, Newfoundland was recognized, from an early period, as a source of oil.

Oil Production Methods

The production methods of these various eighteenth century oils need some elaboration. Vegetable oils, such as olive, rape seed, and linseed, were produced in mills by pressing and crushing. Oils of different qualities were produced, and those from the first pressings were superior to oil from the later pressings. Thus, 'pale rape' was the choice rape seed oil and 'Gallipoli' was the best olive oil.[39] The common fish oils, which were cod and whale during most of the eighteenth century, were produced by rendering; cod livers were allowed to decay naturally in vats on fishing premises in Newfoundland; whale blubber was brought back to Britain in casks by the Arctic whaling fleet, and the process of natural decay was speeded up by heating the oil.

Again, when it came to sales, 'fine' whale oil was superior to 'good'. Cod oil seems to have been of a uniform quality because there were never any distinctions made with this product. The fact that the cod livers were allowed to render out slowly over the course of the summer is what probably produced a uniform product. (Also a small amount of blubber, which was the residue from the rendering of the livers, was usually exported to Britain.) Seal oil was always produced in various qualities. The seal blubber or fat was cut into small pieces and put in large vats, which were set over receptacles. The weight of the fat pressed out a certain amount of oil, which ran out through small holes in the vats. This 'cold drawn' or 'pale seal' oil was the most valuable. Continued decay produced 'straw' or 'yellow' and, later, 'brown'. Finally, the residue was then heated to make 'boiled' seal oil. One authority states that whale oil had an unpleasant odour and was, therefore, best suited for the lighting of streets, wharfs and docks.[40] This was possibly the case with seal oil as well, or at least with the yellow and brown qualities, if a mid-nineteenth report is accurate. The writer described the rendering process and then continued as follows:

> As pressure is applied and time elapses, decomposition takes place, and the [seal] oil becomes darker. The operation is exceedingly disagreeable from first to last, on account of the stench that accompanies it, and it makes St. John's during July, August, and September, a most undesirable residence.[41]

However, the pale seal oil, according to an observer in 1842, "burns equally bright ... with the best spermaceti, and is not half the expense ...[and] the odour ... is very faint."[42] Therefore, it is obvious that the different qualities of seal oil could be used for both indoor and outdoor lighting. Also, it is obvious that the production and supply of oil to the British consumers would increase as demand increased.

Demand for Oil Increases

Whether pressed or rendered, drying or non-drying, animal or vegetable, the oils were in considerable demand and competed with one another and with tallow and wax, to some extent; and in the climate of industrial growth and urbanization in eighteenth- century Britain, there was a strong and increasing demand for these products. Consequently, any new industry, such as sealing, which could guarantee new sources of oil, would enjoy considerable advantage in the market place.

By the beginning of the last quarter of the eighteenth century, several developments were in the offing—particularly in Britain. The demand for oil was increasing because of its many and growing urban and industrial uses; available sources, for Britain, were not unlimited. In addition, lighting technology took a giant step forward, resulting in even greater increases in demand for this purpose. Advances in lighting technology came none too soon; the lamps in use in the lighting of streets, wharfs, docks, large buildings and lighthouses were no longer adequate for the requirements of the period, as there had been little improvement in the technology up to the latter eighteenth century. One writer describes the breakthrough in lighting as follows:

> The industrial revolution might have foundered on the hazards and inadequacy of the lighting that attended its birth. Its success was in no small measure dependent on the researches that provided better illuminants and

brought to civilization relief from the darkness with which it had formerly been encumbered.[43]

Advances in lighting

The major advance in lighting arrived with Argand's invention of a new lamp between 1780 and 1784.[44] This invention solved several major problems that had plagued lamp lighting for centuries. All oil lamps, including the pan, the cruise and the betty, burned very smokily and were only practical for outdoor use (although in parts of northern Europe oil lamps were burned indoors). This was the case because the oils in general use were of such a viscosity that smooth capillary action through the wick was considerably impeded. Argand, a Swiss inventor living in France, solved this problem by designing a lamp with an oil reservoir located at a higher level than the burning wick. The oil flowed down through a tube to the wick-holder and rose up through the wick, pushed by the

Figure 7 - Double-burner Argand lamp on a wall bracket. The invention of the Argand lamp c. 1780 was a revolution in lighting technology and was adapted for homes, factories and shops, streets, wharves and docks, theatres and other public places as well as lighthouses. It was improved upon, but only slightly, and remained an important source of light until the refining of kerosene (or paraffin) from mineral oils and the invention of the electric light. The valve below the oil reservoir was needed in order to adjust the flow of oil.

weight of oil in the reservoir. In order to provide sufficient oxygen for a bright efficient light, Argand designed a wick-holder and wick in the shape of a tube or cylinder. Thus, the wick was enclosed in two metal tubes, one inside the other, so that the flame at the top was completely circular. The two tubes were sealed in such a way as to prevent any oil leakage, but the centre was open so that air could flow up through the wick-holder and provide oxygen to the inside of the burning wick. A regulating valve adjusted the flow of oil to the wick. Argand then designed a glass globe or shade to enclose the flame and protect it and also to facilitate the flow of air up both the inside and outside of the wick. This design reduced the smoke considerably because the oil burned brighter, hotter and more efficiently. One report states that the light produced by one lamp was as bright as twenty candles.[45] This lamp could use whale oil, rape seed oil and seal oil.

One drawback to the Argand lamp was the fact that the regulating valve constantly needed to be adjusted as the level and weight of oil in the reservoir dropped. This problem was solved by the inventor Carcel, who, in 1798-1800, designed a clockwork pump which forced oil into the wick-holder at a constant pressure.[46] In 1836, another inventor, Franchot, added a piston driven by a spiral spring, which forced the oil up through a narrow tube.[47] Nevertheless, the original Argand lamp remained a favourite. However, it and its successors were expensive; thus, their use was confined to the better-class homes, shops and public places, including the Drury Lane theatres and the theatres on the continent.[48] A chandelier with 40 Argand lamps was installed in the newly-built Comédie-Française in Paris by the end of the century, and in 1810, the Berliner Opernhaus was lighted with Argand lamps as follows: 72 footlights, 256 wing lamps and an audience chandelier of 32 lamps.[49] In addition, the Argand lamp was invented "just in time for the great increase in the demand for light to guide trans-ocean traffic as the Industrial Revolution got into its swing."[50]

It was in the area of lighthouse lighting that the Argand lamp and its successors created their own revolution. The first warning lights in northern Europe had been signal fires on the ground. Later, stone towers were erected, which made these wood and coal fires more visible. Later still, torches were used, and finally, candles and flat-wick oil lamps were adopted.[51] Britain took the lead, among the commercial powers, in the construction of new lighthouses, using the newly invented lamps—although it took some time for the older lights to switch to the new technology. For example, the famous Eddystone light used tallow candles until 1800, and the Isle of May light in the Firth of Forth depended on a coal fire until 1816. Some idea of what the new lighthouses looked like can be obtained from a description published in 1869:

> The Argand lamp used with a Fresnel lens of the first order, has four concentric wicks, the largest being three and one-half inches in diameter, and produces a flame six inches in height. Protection against the immense heat thus generated is afforded by the superabundant supply of oil [?], which is pumped up by clock-work on the plan of the Carcel lamp. The annual consumption of oil by a lamp of this description is about 800 gallons.[2]

Lighthouse illumination improved dramatically. Most of the larger lighthouses were about 200 feet (61 m.) high and could be seen for 15-20 miles (24-32 km.), while some were even higher and more visable; for example, the Lundy light-

Figure 8 - Two elegant floor-model Argand lamps.

house was 540 feet (165 m.) in height and could be observed for 30 miles (48km.).[553]

With trade increasing, the British moved quickly to rationalize and expand their lighthouse system. In the early nineteenth century, Trinity House Corporation was assigned responsibility for all the royal lighthouses and was given the right to purchase those in private hands. At the same time, construction continued unabated, and in 1868, England was reported as having a lighthouse for every 14 miles of coastline.[54]

That the new Argand lamp (and its successors) improved lighting and increased the level of artificial lighting in use is unquestioned. In fact, as H. Parrott Bacot writes in his excellent illustrated study of nineteenth century candle lighting, "... more inventions were developed and more improvements made in lighting during the period from 1783 [Argand] to 1879 [Edison] ... than during the entire previous history of man."[55] It is obvious that the advances in lighting technology resulted in an increased consumption of oil (see Tables 1.1 and 1.2). However, the kinds of oil in use during any particular time are not clearly differentiated, although the most common lighting oil in use in England in the eighteenth century was that produced by the whale fishery. The English had been whaling with the help of Basque whaler-men in the sixteenth century. By 1611, they were harvesting the favoured 'right' whale in Greenland waters,[56] so that when street lighting became common, whale oil was the fuel adopted.[57] While the Argand lamp could burn rape seed, whale, seal and cod oils, the more refined the oil, the whiter the light and the lesser the odour.[58]

Eighteenth-century British oil markets

In addition to the British whaling industry, by the mid-eighteenth century, the New England colonies were very heavily involved in supplying British markets with whale oil. As Gordon Jackson points out in his major study on British whaling, "imports of non-Greenland oil into England rose from 2,419 tuns per annum in 1725-29 to 3,067 tuns in 1745-49 and 6,494 tuns in 1765-69."[59] He also indicates that a portion of the non-Greenland train oil imported came from the Newfoundland cod fishery (no doubt, some of this was actually seal oil).

In 1725-29, train oil production in Newfoundland averaged 1,014 tuns; in 1745-49, 2,482 tuns; and in 1765-69, 2,641 tuns.[60] Thus, when one examines Jackson's figures above, one can determine that a fair proportion of Britain's imports of train oil actually came from the Newfoundland fishery. Even this, however, does not provide the complete picture, because the fledgling Newfoundland seal fishery was beginning to become important. While statistics are incomplete, and because the nature of the early industry created a situation where numbers and quantities fluctuated widely, it is evident, nevertheless, that the seal fishery must have been contributing to the growth of the train oil trade. As already discussed in the Introduction, in 1726, seal oil valued at £6,305 was produced in Newfoundland. Records available for four years of the 1745-49 period indicate that seal oil produced was worth, on average, £2,324 per annum. For the 1765-69 period, when records are complete, the value of annual production averaged £7,113[61]. As the average price for a tun of seal oil in that period was about £16, one can calculate that the quantity of seal oil produced during the latter 1760s averaged 445 tuns per annum—a not insignificant amount.[62] These figures indicate that Britain was importing an increasing quantity of fish oils during this period. At the same time, the British whale fishery was slowly expanding[63] (assisted by government bounties of 30s per ton—raised in 1750 to

40s), and in 1783 this fleet numbered 83 ships. However, colonial-produced oil continued to serve many of the nation's requirements.

The outbreak of the American War of Independence led to a crisis situation because of Britain's growing dependence on New England whale oil and sperm oil. Imports from New England declined from 4,093 tuns in 1775 down to 52 tuns in 1776, and then ceased completely. Following the peace treaty of 1783, imports of oil from the new United States faced heavy tariffs, and imports stopped entirely in 1793. As a consequence, the British whaling fleet expanded rapidly to 102 ships of 73,000 tons burthen carrying 10,000 men in 1784, and to 250 ships in each of 1787 and 1788. Oil produced by Britain's Greenland whale fishery averaged 1,791 tuns per annum in 1775-79, 3,148 tuns in 1780-84, and 7,732 tuns in 1785-89. Production increased to such an extent that prices declined from £28 per tun in 1781 to £22 in 1785, and down to £17 in 1788. This latter year proved to be a disastrous one for the fleet in Arctic waters, and losses totalled almost £200,000. The Government bounty of 40s per ton was lowered to 30s, then to 25s and, finally, to 20s in 1795. Meanwhile the number of ships in this fishery declined, dropping to an average of 68 annually in 1793-95 and 65 annually in 1796-1800. However, production of whale oil increased even while the number of ships was declining. As Jackson points out, between 1790 and 1794, an average of 97 ships brought back 3,309 tuns of whale oil from the Arctic, whereas during the second half of the decade, 61 ships brought back 4,872 tuns. "Moreover", he continues:

> because of expanding domestic consumption in industry and lighting, the increasing quantities of oil did not have an adverse effect upon its price, which continued to rise more or less steadily from [£17 in] 1788, to £22 in 1790, £31 in 1795, and £38 10s in 1800.[64]

There is no doubt about the extent of the use and public awareness of whale oil. One writer reported that:

> It is largely used in the lighting of streets of towns, and the interior of places of worship, houses, shops manufactories, etc.; it is extensively employed in the manufacture of soft soap, as well as in the preparing of leather and coarse woollen cloths; it is applicable in the manufacture of coarse varnishes and paints; in which, when duly prepared, it affords a strength of body more capable of resisting the weather, than paint mixed in the usual way with vegetable oil; it is also extensively used for reducing friction in various kinds of machinery; combined with tar, it is much employed in ship-work and in the manufacture of cordage; and either simple or in a state of combination, it is applied to many other useful purposes.[65]

The increasing demand for fish oil by the 1790s is obvious from the records, but the whale fishery continued to be an expensive and risky venture. Meanwhile, the production of cod oil was tied to the amount of cod caught and cured, and any decrease in the catch, for whatever reason, resulted in a reduction in the amount of cod oil produced. Consequently, there was room for a new product that was not tied to the whale fishery or to the production of codfish—and one that could be produced without a major investment.

Growing use of seal oil

By the late eighteenth century, seal oil was becoming recognized as a lighting fuel. For example, in 1780, the agent at the Portland lighthouse reported to the Trinity House Brethren that he could not buy seal oil at Poole under £34 per tun of 280 gallons.[66] He was ordered to buy a stock at that price "as soon as the Newfoundland ships arrived." In June of that same year, Pugh and Son, brokers, were told to send to the Portland light one tun of seal oil and 14 pounds of cotton. Similarly, a brief perusal of the Trinity House Corporation records indicates that 322 gallons of seal oil were burned in the Lowestoft light in 1778, and 557 gallons during part of 1778 and 1779; that 154 gallons were burned in the Caster light; and that 156 gallons were burned in the Well light during this same period. In 1780 and 1782, Lowestoft, Portland and Caster lights burned seal oil; however, by the late 1780s, whale oil was being burned almost exclusively in these particular lights. Thus it would seem that these lighthouses turned to Newfoundland seal oil when their supplies of American whale oil ceased during the American Revolutionary War.[67]

As the Argand lamp began to replace the older lamps, sperm oil became the preferred fuel (and this was later supplanted by rape seed oil). The outstanding British lighthouse engineer Robert Stevenson (1772-1850) reported as follows in 1801:

> In the present construction of Argand's burner only spermaceti oil can be used, not on account of its giving a better light but because it is much less liable to coagulation.[68]

He also reported that it cost half as much to buy oil for a common burner as it did for the Argand type, because sperm oil cost two-thirds more than "the very best common oil."[69] However, as indicated, the expensive sperm oil eventually gave way to rape seed oil, and in 1878, it was reported:

> The oil employed in the lighthouses of Great Britain, Ireland and France is the colza, which has of late years entirely superceded spermaceti oil, as producing an equal quantity of light at little more than half the expense.[70]

Nevertheless, the modifications to the Argand lamp and its successors were quite sufficient to allow them to burn seal oil, and, in fact, Newfoundland lighthouses burned seal oil until at least the late 1880s.[71] Furthermore, it is obvious that seal oil was viewed as a common lighting fuel in the 1820s and 1830s, and one that was competitive in price. A report dating from 1823 compares the various costs of lighting as follows:

Argand burner with gas:	3/4 penny per hour
Argand burner with sperm oil:	1 and 3/4 pence per hour
Argand burner with seal oil:	1 penny per hour
Tallow candle; 8 to the pound:	1 and 1/2 pence per hour
Wax candle; 8 to the pound:	6 pence per hour

Another report compares lighting costs in 1838:

Argand burner with gas:	2 and 3/4 pence
Argand burner with sperm oil:	6 pence
Argand burner with seal oil:	4 pence
Wax candle; 8 to the pound:	1 shilling[72]

As reported earlier, various other references state emphatically that seal oil was used in lighting[73] and for other purposes. One contemporary writer, in 1884, stated:

> The immense consumption of seal-oil in the United Kingdom is known to everybody. The increasing demand for it in the United States, where only the great cities are lighted with gas, may be supposed.[74]

Another writer, in 1861, reported that seal oil "is extensively used for machinery, both in Europe and in the United States...."[75] Similarly, A Newfoundland newspaper quoted an American newspaper, in 1871, as follows: "[Seal oil] is used largely in lighthouses, for machinery, and in the manufacture of the finer kinds of soap."[76] Another contemporary, who viewed the seal fishery as a cruel activity, stated that "as long as people burn seal-oil, and wear 'kid' boots made of seal skins, will the slaughter of the innocents continue."[77] This writer went on to point out that the inferior oil was used in the manufacture of soap, while the "finer qualities were used for mines, machinery, and lubricating purposes generally."[78] A local report in a Newfoundland newspaper, in 1886, begrudged the fact that seal oil had lost most of the lubrication market to petroleum and vegetable oils and was now in the process of losing the market formerly provided by the Scottish jute industry.[79] Market problems towards the end of the nineteenth century will be discussed later; the point here is that most evidence indicates that whale oil and seal oil were similar "in origin and properties" and were often used interchangeably.[80]

The demand for oils as lamp fuel, as a component in the manufacture of soap and textiles, and as a lubricant guaranteed the growth of the British and American whale fisheries and also the Newfoundland seal fishery. One writer summed up the developments of the latter eighteenth century as follows:

> Towards the end of the century, however, the rapid growth of population in south west Lancashire, and the factorisation [sic] of that area, created wider demand for the aquatic products of Newfoundland, i.e., codfish, cod-oil, and seal-oil, or "train oil" as it later became known.[81]

It was the Newfoundland seal fishery that produced the seal oil that is the subject of this study.

Nineteenth-century British oil markets

There was an extensive and growing demand for oil in Britain during the nineteenth century. The demands in that market place, and in other industrializing nations to a slightly lesser degree, were met by importing a wide range of oils as well as by the local production of rape seed. This demand, which was apparent by the beginning of the nineteenth century, stimulated the importation of Newfoundland seal oil.[82] Although it is impossible to separate seal oil from cod oil in the available British import records, the significant increase in the importation of Newfoundland seal skins beginning in the late 1790s is a reliable indicator that seal oil imports had also begun to increase. During 1791-95, imports of Newfoundland seal skins averaged 33,141 annually (see Table 1.3); during 1804-08 the average had increased dramatically to 128,550. Similarly, imports of train oil (i.e., seal and cod) from Newfoundland increased from an average of 2,400 tuns annually to an average of 5,667 during the same period.

Another source states that the amount of cod oil produced in Newfoundland during these same two periods amounted to 1,863 and 2,705 tuns respectively.[83] Meanwhile, the latter source shows that the annual value of seal oil produced in Newfoundland averaged about £7,987 annually in 1791-9 and about £38,036 in 1804-08. One can calculate from these figures that Newfoundland produced approximately 400 tuns of seal oil annually during 1791-95 and 1,900 tuns annually during 1804-08. The fact that 1,900 tuns made up only a small proportion of British imports, which consisted of over 22,000 tuns of train oil by 1808, does not detract from the point that the Newfoundland seal fishery was growing rapidly.

Meanwhile the consumption of all oils was on the increase in British markets. For example, in 1800, the value of British oil imports were as follows:[84]

Castor (from America)	£937 7s 9d
Olive (mostly from Italy)	£121,476 3s 10d
Palm (mostly from Africa)	£4,528 16s 2d
Train (from America)	£57,642 13s 9d
Train (Greenland whale)	£76,349 8s 9d
Train (Southern whale)	£77,605 9s 4d
Total Train	£211,597 11s 10d
Grand Total	£338,540 11s 5d

As is readily obvious, train oil made up 63 per cent of total imports by value, while most of the other imports consisted of olive oil. (Unfortunately, it is difficult to compare quantities of oils from the records because of the use of different weights and measures. Similarly, the figures for the values of the various imports are useful for comparison purposes only, and are of little use otherwise because these were simply 'official' prices which were set about a century earlier. This situation continued until 1854, when the practice of assigning 'real' values to imports commenced.) A later report, in 1839, illustrates how the oil trade had evolved by then. In addition to the considerable decline in the proportion of oil imports consisting of train oil, there was an enormous increase in the importation of palm oil and, in all, a greater diversity in the oil trade, as demonstrated by the following:[85]

Oils	1837(Tuns)	1838(Tuns)
Train	10,477	13,861
Sperm	5,420	6,366
Olive	4,970	7,455
Cocoa Nut	1,495	1,905
Palm	10,100	14,020
Total	32,432	43,607

Not only was the consumption of oil increasing in Britain during the nineteenth century, but the trade was continuing to become more diversified. In 1857, for example, Britain imported oil as follows:

Coconut	207,239	cwts	Rosin	1,058	cwts
Lard	£6,224		Other seed	4,381	tuns
Linseed	2,039	tuns	Train	15,765	tuns
Olive	18,862	tuns	Spermaceti	5,410	tuns
Palm	854,791	cwts	Turpentine	108,336	cwts
Rapeseed	6,689	tuns	Animal	3,603	cwts
Rock	56,544	cwts	Castor	40,621	cwts
			Other	£ 7,757	

As one can see, the 'lard' and 'other' have been assigned values in sterling, while the remaining types have been measured in cwts and tuns. However, even with this lack of consistency, it is possible to see that the industry has become more diversified and to appreciate the comparative importance of olive, rape seed, train and spermaceti (all in tuns), and coconut and palm (both in cwts). It must be noted, also, that of the total imports of train oil, 11,819 tuns, or 58 per cent, were supplied by Newfoundland.[86]

Meanwhile, British demand for Newfoundland seal and cod oil to meet much of its train oil requirements is fully documented in the custom records, as Table 1.4 illustrates. During the five-year period 1814-18, imports of train oil from Newfoundland averaged 5,447 tuns annually, and by the five-year period 1824-28, this figure had risen to 6,982 tuns annually. (About 1,200 tuns of this increase was made up of seal oil imports, and about 300 tuns consisted of cod oil.) By 1831-35, Britain's imports of Newfoundland train oil had increased to an annual average of 9,195 tuns, and to 9,944 tuns during 1841-45 (see Table 1.4). But cod oil production continued to stagnate, as the cod fishery itself had ceased to expand.[87] However, British imports of train oil from all sources continued to grow, as did the country's total imports of oil as well. As Table 1.5 illustrates, Newfoundland provided a sizeable proportion of this trade, especially if one sets apart figures on spermaceti imports. It is equally clear from Table 1.6 that after mid-century, train oil continued to decline within the parameters of the total oil trade, and Newfoundland's share itself declined steeply during the last half of the century (see Table 1.5). By 1914, Britain imported only 3,452 tuns of train oil from Newfoundland—a mere 6.57 per cent of its total importation of 52,537 tuns.

Decline in Demand for Traditional Oils

The traditional oil industry in general, and the train oil trade in particular, had undergone a metamorphosis. For example, during the last quarter of the eighteenth century, Liverpool had been lit with oil lamps.[88] In 1816, efforts were begun to light Liverpool's streets with gas, and that city's 'Oil Gas Company' began operations in 1823.[89] Newman and Company, which operated extensive fishing and trading operations in Newfoundland, reported in the early years of the nineteenth century as follows: "April 1816, the price of cod oil continues low at £28 per tun. The introduction of gas light for the streets and shops [in Liverpool and other cities] lessens the consumption very considerably and will be worse."[90] Thus, Liverpool had switched from "the old oil lamps which, few and far between, used to twinkle in the distance and just to make darkness visible," to gas for its street lighting.⁹¹ This impression of the oil lamps is in considerable contrast to the portrait of 1794, which pointed out that "Castle-street is a principal and most elegant [street in Liverpool] ... well lighted by a regular arrangement of lamps on each side."⁹² The oil lamps of which writers had been so proud in the latter eighteenth century were, by the early years of the nineteenth century, considered greatly inferior to gas, and between 1825 and 1833, other reports stated that "chop and eating houses [were lit] after the most approved London fashion with gas...."⁹³ Beginning with lighting of the Pall Mall in London in 1807, the use of gas for street lighting spread rapidly throughout Britain and across the Atlantic, with Baltimore being the first American city to use this new technology in 1816, and Montreal being the first city in Canada in 1840.⁹⁴ One writer summed it up as follows in 1864:

> Towards the end of the last century the only materials used for lighting were animal fats, such as tallow and fish oils, the former being used as candles, the latter burnt in lamps.... In towns the use of gas soon prevailed over the use of other lighting materials, but in the country, and especially in remote places, candles and lamps still continue to be largely used.⁹⁵

One obvious reason for the rapidly increasing preference for gas can be found in the comparative costs of lighting materials. As indicated earlier, gas was the cheapest lighting fuel in Britain, as reported in 1823 and in 1838 (in both cases seal oil was second). An American writer, in 1843, explained that lighting by tallow candle was 7.1 times as costly as lighting by gas; cocoa candle was 7.3 times as costly; palm candle—10.5 times; spermaceti candle - 16.2 times; wax candle - 14.4 times; sperm oil lamp - 8 times; and whale oil lamp - 5 times.⁹⁶ However, it took a considerable period before gas could be refined to such an extent that it was acceptable to householders. Furthermore, the sulphur produced odours and was suspected of ruining book bindings, drapes and furniture,⁹⁷ and consequently, the Argand lamp and tallow and other candles held out in homes until the introduction of the kerosene lamp, further improvements in gas and, finally, the electric light.

Similarly, lighthouses continued to use fish and vegetable oils, but in this case, the determining factor was that lighthouses were too remote from gas lines. As already pointed out, Newfoundland lighthouses used seal oil until late in the nineteenth century; another, more general report written in 1857 related how olive oil was formerly used, but no longer; that gas was not practical, nor electricity reliable; and it concluded as follows:

> Practically, the chief materials used for lighthouse illumination, are sperm oil and colza, or rape seed oil. The former is still used in this country [USA in 1856]; the latter, which is derived from a species of wild cabbage, is used entirely in France, chiefly in Great Britain, and is, indeed altogether the main reliance of the European lights.... Colza oil gives the intensest [sic] light; produces less charring of the wick; is less affected by cold, and is, in most places, very much cheaper than sperm oil.⁹⁸

However, as late as 1852, a report quoted from the *Woburn Journal* in the United States referred to the importation of Newfoundland seal oil in 1851 as follows: "The present season is the first introduction into the Market of the Seal Oil, which is said to be a first rate article, and fully equal, if not superior to the Whale

Oil...."[99] Nevertheless, at the beginning of the second half of the century, many experiments were begun to discover a substitute for fish and vegetable oils for use in lighthouses. Electricity was introduced in 1857, when a Professor Holmes received permission from Trinity House Corporation to establish an electric light with a steam engine generator at the South Foreland lighthouse; this meant that Britain became the first nation to adopt electricity for lighthouse illumination.[100] Later, oil gas was given the approval of the Trinity House Corporation, and the new fuel was adopted in the 1870s.[101]

In the early 1870s, traditional oils were under severe competition. Nevertheless, they still seemed to have a viable future. A report in 1872 made this point as follows:

> The oil trade is another large and important branch of industry. In London and its Suburbs alone, which has one-tenth of the entire population, there are about 270 oil merchants, 1,440 oil and colour men, 30 oil brokers, 50 oil-refiners, 80 melters and tallow chandlers, 65 wax chandlers, 10 wax-vesta makers, and 100 soap-makers.[102]

Another observer wrote two years later how petroleum was becoming the accepted lighting fuel, but did not foresee its replacing other oils as a lubricant. He wrote as follows:

> It is now some years since, when the discovery of new sources of petroleum added immensely to the supplies of oil for lighting purposes at a low price, that it is was generally thought that the cultivation of oleaginous seeds could not profitably be continued. On the contrary the cultivation of the colza for the oil has never been conducted on so large a scale as it has been during the last few years; and the quantity of colza oil now employed for the lubrication of machinery only, is greater than the total produce in earlier times.[103]

However, by the early 1880s, the revolutionary changes brought about to the oil trade by the increase in petroleum were almost complete, and lubrication had also been conquered by this new product. An observer summarized the situation as it existed in 1883 by showing what had happened to reduce the demand for the traditional oils. He pointed out that train, olive, rape seed, palm and coconut oils and tallow had all been used for lubrication (although palm and coconut were mostly used for soap making). He continued and explained how and why mineral oil had invaded this arena:

> Mineral oils do not decompose either at very low temperatures or at degrees of heat which far exceed those prevailing in the steam cylinders, etc., where they are employed. They do not undergo any change either on contact with the air or with water or with steam; they do not attack metals, even the most easily oxydisable such as potassium or sodium, and are as little changed or decomposed by the metals themselves.
>
> This chemical indifference is the principal advantage possessed by mineral oils over all fat oils, whether they are of vegetable or animal origin. All those fat oils decompose in time on exposure to the air, at high temperatures, on contact with metals or their oxides, and thus destroy, sometimes more quickly, sometimes more slowly, the parts of the machinery which they are intended to preserve.... Lubrication with mineral oils has, within the short space of the last five years, made such progress that it may justly be called, not only the lubricant of the future, but that of the present day.... The

> displacement of fat oils for lubricating machinery by mineral oils is a great technical progress. But the use of mineral oil is a great advantage also from the point of cheapness. The best mineral oils are now only half the price of fat oils.[104]

Therefore, by the mid 1880s, the demand for traditional oils had been seriously affected by the introduction of petroleum into the market for illumination and lubrication oils. In addition, gas lighting continued as a major competitor and electric lighting was beginning to be utilized.

Thus, it is obvious that the number of uses for Newfoundland train oil had declined. In the late eighteenth century, it was used for practically every purpose for which oil was needed. By the late nineteenth century, it was no longer used for lighting, nor lubrication nor many other industrial purposes. Nonetheless, it was still used for leather currying, in the making of soap and in the jute industry. Jackson points out that the Dundee sealing fleet (which will be discussed more fully later) survived in the latter nineteenth century because of the growth in Dundee of the jute industry, which included a process called 'batching', whereby the jute was soaked in oil to soften the fibres before spinning.[105] In the soap industry, all train oil was, no doubt, suffering severe competition from tallow imports. Tallow had found markets in both the candle-making industry and in soap making. With gas lighting, kerosene (paraffin in Britain) lamps, and electricity, tallow candles had become a rarity, and the tallow trade was forced to depend on the demand of the soap industry. Indeed, this well-established traditional British trade declined in value by over 50 per cent between 1860 and 1890.[106] In addition, the palm oil and coconut oil trades, creations of the nineteenth century, were also dependent upon the soap industry, particularly with the decline in the candle industry. Thus, Newfoundland train oil was subjected to considerable competition.

Consequently, there was a steady decline in the market price for train oil during the second half of the century. As Table 1.7 demonstrates, the average price of seal oil imports into Britain declined from about £41 per tun in the 1854-63 period down to £22 during the 1894-1914 period, with the steepest decline beginning in the mid-1880s (see also Table 1.8). However, beginning in 1900, seal oil increased in value because of the decline in the Norwegian cod fishery.[107] This decline resulted in a shortage of Norwegian cod oil and led to an increase in the demand for both cod and seal oil from Newfoundland. In fact, it was reported that Newfoundland seal oil increased in value in Britain from £18 to £28 per tun in 1903 and that it was being mixed with cod oil by the manufacturers of medicinal cod liver oil.[108]

Global Impact of Oil Trade

The oil marketplace was a capricious one, and it consumed oils at an enormous rate, oblivious of the dramatic changes that were inflicted on the world. Colonies from Newfoundland to Australia to the Falkland Islands were created and prospered because they were near major seal and whale populations and became centres for oil production. Established and sophisticated, but non-industrial societies—as in Hawaii, for example—were changed forever once American whalers began to use them as bases. African societies with relatively well balanced systems of agriculture were encouraged to produce palm oil and coconut oil for export. Dahomey, for example, under pressure by British business interests, became the centre of an expanded palm oil industry during the period leading up to the mid nineteenth-century; this economy declined in the latter half of the century with irreparable results.[109] Arctic native societies, living along the coasts of Alaska and Siberia, dependent at a subsistence level on their harvest of walruses, were reduced to starvation after the commercial hunters destroyed the walrus herds.[110] And huge areas of farmland devoted to rape seed were forced to change to other cash crops. Finally, species of sea mammals were slaughtered to the point where it was no longer economically viable to hunt them, and in some cases, to the extent that they are now facing extinction.[111]

Seal Skin Markets

In the meantime, Britain had been an importer of seal skins from the beginning of the seal fishery. Skins and hides (and furs, of course) were traditional imports, and the growth of this trade, in many ways, paralleled the growth of the oil trade. Both were traditional trades supplying the needs of pre-industrial England; and both were necessary for the growth of urbanization and industrialization. The many uses of skins and hides and the guilds and crafts that depended on them have already been discussed, and Newfoundland seal skins had their own niche within this market place. A contemporary writing from Harbour Grace in c.1810 reported on the uses of Newfoundland seal skins as follows:

> To the European, the skin of the seal has sometimes made muffs; it covers his trunks, and supplies him with shoes and boots. When it is well tanned, the grain is not unlike that of Morocco leather: it is not quite so fine, but it preserves its colour longer; even waistcoats in the Greenland style are not infrequently seen in the metropolis of Great Britain.[112]

With the developments of industrialization in the nineteenth century, the demand for these items increased dramatically. The expanding middle class required larger quantities of hats, boots, gloves, cases, trunks, harness, carriage upholstery and the many other items that were needed in their personal, business and leisure activities as well as for their local travels and foreign tours and voyages. The British military and naval needs continued to increase, as did the number and variety of accessories in the factories and mills. Also, as in the case of oil imports during the nineteenth century, the imports of skins and hides expanded and became more diversified. At the same time, Britain imported an increasing quantity and variety of hides, skins and leather as well as furs during this period; and unlike fish and vegetable oils, the hides and skins were not to be supplanted by more modern products until well into the twentieth century.[113] For this reason, no doubt, the value of skins was maintained during the last half of the century, while the value of oil declined. For example, the average price of seal skins imported into Britain from Newfoundland during 1854-58 amounted to 4s 5d each, while in 1908-12 Newfoundland seal skins fetched, on an average, 6s.[114] In other words, the skins were becoming more valuable to the Newfoundland economy as the value of the oil declined (although they would never compensate for the loss in production and value of the oil trade, as will be shown later).

Summary

In conclusion, the international trade in oils during the period from about 1780 to 1914 was very complex, rapidly expanding and very competitive. In the beginning, fish oils, together with vegetable oils, enjoyed a considerable boom. New lamps and their application to the lighting of homes, public places, wharfs, streets and lighthouses increased the demand for oils and had a direct impact on the growth of the Newfoundland seal fishery. The further demand for oil as a lubricant, in leather currying, in textile manufacturing and soap making was also the result of the increasing influence of the industrial revolution. Even the changes in technology during the nineteenth century that reduced the uses for seal oil did not immediately affect the demand for this product. As one market closed—as in the case of gas lighting—another expanded—as in lubrication, leather currying and textile manufacturing. Producers of such oils enjoyed an enormous advantage in the world of growing urbanization and industrialization. However, in the last one-third of the nineteenth century, circumstances changed. Petroleum became a major factor in the oil equation and was soon to wreck the markets for the traditional oils. Then, the general expansion in world trade and the growing ease and speed of transport resulting from the use of steam engines combined to depress the prices of many commodities, thereby bringing further pressure on producers. In addition, as Table 1.5 shows, Britain was importing fish and sea mammal oils from Norway, Denmark, Iceland, Portuguese possessions, Japan, the United States, Chile, Argentina, Natal and the Falkland Islands by the end of the century. Therefore, by the end of the nineteenth century, Newfoundland's export trade in train oil, especially seal oil, was under severe pressure from international competition from all oils, including petroleum.

NOTES

1. Alastair Laing, *Lighting: The Arts and Living* (London, 1982), p. 56.
2. Laing, *Lighting*, p. 56; and F.W. Robbins, *The Story of the Lamp and the Candle* (Oxford, 1939), p. 140.
3. Laing, *Lighting*, pp. 56-7.
4. Robins, *Lamp*, p. 140.
5. Laing, *Lighting*, p. 56.
6. E.S. de Beer, "Early History of London Street-Lighting," *History*, New Series, XXV (1941), 316-8.
7. de Beer, "Early History," 318.
8. Quoted in de Beer, "Early History," 320-1.
9. de Beer, "Early History," 323.
10. Elizabeth Ralph, *The Streets of Bristol* (Bristol Branch of the Historical Association, Pamphlet #49, 1981), p. 5. The above information on Bristol is taken from this source.

11 Ralph, *Bristol*, p. 5. Quoted from Daniel Defoe, *A Tour through the Whole Island* (4 vols., London, 1788), II, 269-70.
12 Richard Brooke, *Liverpool as it was During the Last Quarter of the Eighteenth Century: 1775-1800* (Liverpool, 1853), p. 451; and J. Aspinall, *Liverpool: A Few Years Since* (Liverpool, 1885), pp. 88-9. See Prowse, *Newfoundland*, p. 403, for references to cod oil.
13 W. T. O'Dea, *The Social History of Lighting* (London, 1958), p. 97.
14 George Heath, *The New History, Survey and Description of the City and Suburbs of Bristol* (London, 1794), p. 50.
15 Laing, *Lighting*, pp. 13-36.
16 Leroy Thwing, *Flickering Flames: A History of Domestic Lighting through the Ages* (London, 1959), pp. 17 and 33-5. The above information on early lamps is taken from this source.
17 W. T. O'Dea, *Darkness into Light: An Account of the Past, Present and Future of Man-Made Illumination* (London, 1948), p. 6.
18 Laing, *Lighting*, p. 9.
19 George Unwin, *The Gilds and Companies of London* (London, 1963), 4th edition, pp. 54-82. The Cordwainers were involved in the manufacture and sale of leather goods. Eventually, they lost important components of this industry to newer more specialized guilds and the Cordwainers became confined to the manufacture and sale of footwear.
20 John W. Waterer, *Leather in Life, Art and Industry* (London, 1946), p. 54. Other leather crafts evolved later and went out of existence or combined with other crafts in larger guilds. These include, for example the Harness Makers and Pursers. See also H. H. Ditchfield, *The City Companies of London and their Good Works* (London, 1904).
21 Hides referred to larger animals—cattle, horses and buffalo; skins referred to smaller animals—reptiles, birds and fish. See Waterer, *Leather*, p. 132.
22 Waterer, *Leather*, p. 145.
23 *Encyclopedia Britannica*, 11th ed., "Soap."
24 For further information, see Abraham Rees, A Selection from "The Cyclopedia; or Universal Dictionary of Arts, Science and Literature," *Rees's Manufacturing Industry (1819-20)*, ed. Neil Cossons, (London, 1819-20; reprint, 1972), 14-16; Professor [?] Church, "The Manufacture of Soap," *British Manufacturing Industries*, ed. G. Phillips Bevan, (13 vols., London, 1876),IV, 75-83; J. T. Jenkins, *A History of the Whale Fisheries* (London, 1921), p. 40; Gordon Jackson, *The British Whaling Trade* (London, 1978), p. 11; C.R. Alder, "The Manufacture of Toilet Soaps," *Journal of the Society of the Arts* (London), XXXIII (1884-85), 1073-88; and William Hawes, "On the Manufacture of Soap," *Ibid.*, IV (1856), 325-33.
25 Ralph Davis, *The Rise of the English Shipping Industry in the Seventeenth and Eighteenth Centuries* (London, 1962), p. 228.
26 Jackson, *Whaling*, p. 74.
27 *Encyclopedia Britannica*, 11th ed., "Leather."
28 Robert Hunt, *Ure's Dictionary of Arts, Manufacturers and Mines, containing a clear exposition of their principles and practices*, (6th ed., London, 1875), 84-5 and 757.
29 "Seal Hunting," *The Penny Magazine* (London), IV (1835), 100-03.
30 See Ralph Davis, *A Commercial Revolution: English Overseas Trade in the Seventeenth and Eighteenth Centuries* (London, 1967). For an overview of the growth of English (and British) trade during the eighteenth century see Elizabeth Boody

Schumpeter, *English Overseas Trade Statistics: 1697-1808* (Oxford, 1960), pp. 1-14. For a similar overview of the oil trade during this period see Tables 1.1 and 1.2.

31 Derek Hudson and Kenneth W. Luckhurst, *The Royal Society of Arts* (London, 1954), p. 160.

32 See Hunt, ed., *Ure's Dictionary of Arts* and the *Encyclopedia Britannica*, 11th ed., "Oils."

33 *Mark Lane Express* (London), 21 October 1839. The term "common fish oils" is used quite frequently in this publication. These include cod, seal and whale (but not sperm).

34 *Mark Lane Express*, 27 June 1842.

35 *Mark Lane Express*, 26 November 1838.

36 *Mark Lane Express*, 6 September 1841. Russia was the traditional supplier of superior tallow, which was called 'Petersburg Yellow Candle' or 'PYC' tallow.

37 Quoted in Cell, *English Enterprise*, p. 150. 'Trayne', 'traine' and 'train' oil are the terms used to describe cod oil and whale oil, and sometimes seal oil.

38 Ryan, "Abstract." Although cod oil was sometimes measured in hogsheads and barrels and other similar containers, the usual measurement was by the tun. The tun contained 252 wine gallons (the wine gallon was to become the standard United States gallon) or eight barrels (31.5 gallons each), six tierces (42 gallons each), four hogsheads (63 gallons each), three firkins or puncheons (84 gallons each), or two pipes or butts (126 gallons each). In 1824, the Imperial gallon became the legal measurement, and this was set at 120 per cent of the wine gallon. After 1824, the Newfoundland tun contained 252 imperial gallons. See also R.D. Connor, *The Weights and Measures of England* (London, 1987).

39 The oil market reports in the *Mark Lane Express* beginning in 1832 usually indicate the qualities of oil on the London market.

40 Jackson, *Whaling*, p. 34.

41 "Impressions From Seals," *Chambers's Journal* (Edinburgh), (4 February 1854), 75-7.

42 Bonnycastle, *Newfoundland in 1842*, vol. 2, p. 134.

43 O'Dea, *Darkness*, p. 1.

44 Laing, *Lighting*, pp. 60-1.

45 Laing, *Lighting*, pp. 60-1. Argand's contemporaries Quinquet and Lange were the first to perfect the globe with its narrow top to accelerate the flow of air. In fact, there was considerable controversy in France over who actually invented the lamp in question, and in that country, this lamp became known as the Quinquet lamp. See Gösta M. Bergman, *Lighting in the Theatre* (Stockholm, 1977), p. 199.

46 Laing, *Lighting*, p. 61.

47 Laing, *Lighting*, p. 61.

48 O'Dea, *Social History*, p. 108.

49 Bergman, *Lighting*, pp. 199-202.

50 O'Dea, *Social History*, p. 108.

51 See Edward Abbott, "Lighthouses," *Galaxy* (New York), VII (1989), 237-47; "Lighthouses," *Harper's New Monthly Magazine* New York), XXXVIII (1868-69), 405-14; and "Light-House Construction and Illumination," *Putnam's Monthly Magazine of American Literature, Science and Art* (New York), VIII (1857), 198-213.

52 Abbott, "Lighthouses," 240.

53 Abbott, "Lighthouses," 241.

54 "Lighthouses," *Harper's*, 409.
55 H. Parrott Bacot, *Nineteenth Century Lighting: Candle-powered Devices, 1783-1883* (West Chester, Pennsylvania, 1987), "Introduction."
56 Jackson, *Whaling*, p. 5.
57 Jackson, *Whaling*, p. 34.
58 Bergman, *Lighting*, p. 201.
59 Jackson, *Whaling*, p. 51.
60 Ryan, "Abstract." Quantities have been rounded off to the nearest whole number.
61 Ryan, "Abstract."
62 The £16 has been calculated as follows: Seal oil was usually slightly more valuable than cod oil, and the value of this oil hovered between £12 and £18 per tun during 1765-69. Also Jackson (*Whaling*, p. 268) puts the value of seal oil at £18 in 1770. Incidentally, all oil was valued at £18 per tun in 1719, and at £16 per tun in 1720. Ryan, "Abstract."
63 The following information has been taken from Jackson, *Whaling*, pp. 55, 67, 70-7 and 264.
64 Jackson, *Whaling*, p. 77.
65 William Scoresby Jun., *An Account of the Arctic Regions with a History and Description of the Northern Whale Fishery* (Edinburgh, 1820), pp. 420-1.
66 D. Alan Stevenson, *The World's Lighthouses Before 1820* (London, 1957), p. 291. This appears to have been an unusually large 'tun' measurement.
67 Trinity House Archives, London, "Trinity House Cash Book, 1778-1784," 10 October 1778; 22 May 1779; 26 February 1780; 23 December 1780; and 5 January 1782.
68 Robert Stevenson, *English Lighthouse Tours: 1801, 1813, 1818.* (London, 1946), p. 28.
69 Stevenson, *Tours*, p. 28.
70 W. H. Davenport, *Lighthouses and Lightships* (London, 1878), p. 81.
71 See *JHA*, Appendices. See also Malcolm Macleod, "Lighthouses," *ENL* (St. John's, 1991), III, 295-303.
72 Both reports are cited in O'Dea, *Social History*, pp. 53-4. Candles were classified by the number required to make one pound (avoirdupois). While the second report does not indicate whether lighting costs were 'per hour', it seems safe to assume they were. If this is correct, then costs rose considerably during this 15-year period.
73 *Encyclopedia Britannica*, 11th ed., "Oils."
74 "Impressions From Seals," *Chambers's*, 75-7.
75 "The Seal Fishery of Labrador," *Hunts' Merchants Magazine* (New York), XLV, no. 1 (1861), 463-4. The title is misleading because this short article describes the traditional Newfoundland spring seal fishery.
76 *Royal Gazette*, 18 April 1871. Quoted from the *Boston Traveller*.
77 "Newfoundland," *Blackwood's Magazine* (Edinburgh), CXIV (1873), 53-4.
78 "Newfoundland," *Blackwood's*, 53-4.
79 *Evening Mercury*, 11 March 1886.
80 Mattieu Williams, "Oils and Candles," *British Manufacturing Industries*, ed. Bevan, 118. Church, "The Manufacture of Soap," *Ibid.*, 75.
81 Arthur Wardle, "Liverpool and the Newfoundland Trade," *Liverpool Nautical Research Publications*, I (1933-44), 3.

82 See Table 1.3 for information on the importation of seal skins and train oil from Newfoundland during the period from 1772 to 1808. In 1808, the total value of imports from Newfoundland into Britain was greater than the combined value of all imports from New Brunswick, Prince Edward Island, Nova Scotia and Cape Breton. See CUST 17, Vols. 1-30.

83 See Ryan, "Abstract," for these statistics and those below.

84 CUST 5, Vol. 1B.

85 *Mark Lane Express*, 14 January 1839. This does not include rape seed and some lesser oils. There are some discrepancies between these figures and those recorded in *Trade of the United Kingdom* (London, 1837 and 1838).

86 CUST 5, Vol. 57. 'Rock' was an early name for petroleum. Note that the total from Newfoundland as stated in Table 2.6 amounts to 11,988 tuns and the total as found in Table 1.5 is given as 12,237 tuns. These are not significant differences given the fact that shipments could be made from parts of the Labrador coast—and indeed, from parts of the island—without being recorded.

87 See Ryan, *Fish out of Water*.

88 Brooke, *Liverpool*, p. 451.

89 Thomas Baines, *History of the Commerce and Town of Liverpool and of the Rise of Manufacturing Industry in the Adjoining Counties* (London, 1852), p. 569; and *Co-Partners' Magazine* (Liverpool), IX, no. 2 (April 1948), 26.

90 Prowse, *Newfoundland*, p. 403.

91 J. Aspinall, *Liverpool: A Few Years Since* (Liverpool, 1885), pp. 88-9.

92 James Wallace, *A General and Descriptive History of the Ancient and Present State of the Town of Liverpool* (Liverpool, 1794), p. 79.

93 *Liverpool Fifty Years Ago* (a series of articles appearing in the *Albion* between 1825 and 1833). Reprinted in the *Liverpool Daily (Evening) Albion*, November 1878-79.

94 M. Baker, J. Miller Pitt and R.Pitt, *The Illustrated History of Newfoundland Light and Power* (St. John's, 1990), pp. 4-5.

95 B. H. Paul, "Artificial Light and Lighting Materials," *Journal of the Franklin Institute* (Philadelphia), XLVIII (July-December 1864), 124-32.

96 Andrew Fyfe, "On the Comparative Expense of Light Derived from Different Sources...," *Journal of the Franklin Institute*, V (1843), 260-70.

97 J. Scoffern, "Artificial Illumination," *St. James Magazine* (London), XV (1866), 45.

98 "Light-House Construction and Illumination," *Putnam's*, 207-8.

99 *Patriot*, 13 September 1852.

100 "Lighthouse Illuminants," *Van Nostrand's Eclectic Engineering Magazine* (New York), XXXI (June-December 1884), 309-11.

101 John Tyndall, "A Story of our Lighthouses," *Nineteenth Century* (London), XXIV (1888), 79.

102 P.L. Simmonds, *Science and Commerce: Their Influence on our Manufacturers* (London, 1872), p. 516.

103 "Lubricants For Machinery," *The Practical Magazine* (London), IV, no. 19 (1874), 113.

104 Herr Lux, "Lubricants: Proceedings of the Society of German Engineers," *Van Nostrand's Eclectic Engineering Magazine*, XXIX (July-December 1883), 486-8. This article introduces another interesting point: the increasing conversion of good agricultural land to the production of seed oils—a development which troubled the author. He wrote: "...the ultimate introduction of lubrication with mineral oils is of importance also from an economical point of view. Our population is steadily

increasing and the difficulties of gaining a livelihood are growing with it. On the other hand, there is an abundance of suitable material for lubricating with mineral oils, for, without reckoning the almost inexhaustible stores in America, petroleum is now found in large quantities also in Russia, Galicia, and Germany. Our food supply would be greatly extended, either directly, by appropriating large quantities of fats and oils for the maintenance of the people, or indirectly, by restoring the areas now used for the cultivation of oil-producing seeds for raising cereals."

105 Jackson, *Whaling*, p. 150.
106 *Trade of the United Kingdom* (London, 1860-90).
107 Ryan, *Fish out of Water*, p. 88.
108 *Evening Herald*, 27 May, 18 and 22 June 1903.
109 See John Reid, "Warrior Aristocrats in Crisis: The Political Effects of the Transition from the Slave Trade to Palm Oil Commerce in the Nineteenth Century Kingdom of Dahomey" (Ph.D. thesis, University of Stirling, Scotland, 1986).
110 For example, see Peter Murray, *The Vagabond Fleet: A Chronicle of the North Pacific Schooner Trade* (Victoria, B.C., 1988).
111 See Busch, *Seals* and Jackson, *Whaling* for many examples of sea mammal populations that were decimated during the nineteenth century.
112 Anspach, *History*, p. 417.
113 See CUST 5, vols. 12, 57 and 161 for the record of imports of skins and hides into Britain during the years 1823, 1857 and 1899, for example.
114 *Trade of the United Kingdom* (London, 1854-1860).

CHAPTER 2
Economy to 1914

THE SALTFISH INDUSTRY, according to the report of the Select Committee appointed to enquire into the state of the Newfoundland trade in 1817, could not support the population of approximately 40,000 inhabitants without substantial help from the British Government.[1] This report was based on evidence of saltfish trade representatives from St. John's, Poole, London, Liverpool and Teighnmouth, and the consensus was that Great Britain should provide a bounty or subsidy to the fish exporters of not less than 2s per cwt, remove at least 5,000 people from the island, and obtain a reduction in the import tariffs in the various markets, especially Spain. The alternative, as these exporters envisioned it, was the withdrawal of all capital investment from the island, followed by the collapse of the trade and the starvation of the inhabitants. John Henry Attwood, representing the St. John's merchants before this committee, saw the need for other measures besides those emphasized above: "... we seriously recommend to your Lordship's consideration the propriety of encouraging and promoting the cultivation of the soil, a measure calculated to assist the labouring fishermen in the support of himself and his family."[2] As early as 1812, a writer had pointed out that "every inducement and encouragement should be held to stimulate greater exertion of more general attention to the growth of Potatoes."[3] In fact, during 1813-15, Governor Keats had encouraged agriculture in an effort to reduce the population's dependence on imports and to curb the rising price of food.[4] By 1817, the general feeling was that the quantity and value of saltfish produced in Newfoundland during the short fishing season could never—on its own—be expected to sustain such a large permanent population of 40,000 people. A writer to a St. John's newspaper in that year summarized the prevailing conviction among those involved in the cod fishery when he suggested that the only solution to Newfoundland's economic problems lay in the exploitation of new resources.[5]

Nevertheless, by 1911, the last year for which a census was taken during the period under study, the population had increased to 242,619. In order to support this six-fold increase in population, either the saltfish industry would have had to have grown correspondingly, or considerable and radical changes would have had to have occurred within the economic system. Saltfish production did increase, but not substantially; during the ten-year period beginning in 1814, about 930,000 cwts of saltfish were exported annually, while the annual average exports for the ten-year period ending in 1914 amounted to about 1,407,000 cwts.[6] However, while exports (by 40,568 residents and 14,716 summer fishermen) in 1815 reached 21.4 cwts of saltfish per capita, this figure had dropped considerably to 5.8 cwt per resident at the end of the period under study.[7]

There are several possible explanations for the fact that such a large population could exist when the saltfish industry had grown so little, and these include: the cessation of the migratory fishery; the inflow of capital from Newfoundlanders living and working in the United States of America; government and business borrowing; the development of import substitutes; an increase in the value of the fish produced; and the diversification of the local export economy. The first element was not a significant factor throughout the period—in fact, summer or migratory fishermen ceased coming to Newfoundland in the early 1820s. Remittances from Newfoundlanders in America made up a significant, but as yet unexplored, inflow of money to the colony; however, this did not really become an important factor until after the mid-1880s. The government went heavily into debt beginning in the 1880s, especially in order to build the trans-island railway, but this provided comparatively little employment, as Table 2.1 illustrates. Furthermore, there was only a little development in the field of import substitution prior to 1914 (this will be discussed later). And there was no increase in the price of saltfish.[8] There was, however, from the very beginning of this period, some economic diversification, as under utilized resources were exploited. While not ignoring the overall picture, this chapter intends to study the extent to which the economy became diversified during the century up to 1914 as well as the role of the seal fishery within the total context of this diversification and the overall economy.

The report of the parliamentary committee, examined above, did not result in any effective action by the British Government on Newfoundland's behalf; no bounties were granted to support the fishery (unlike France's fishery subsidization program), and no effort was made to reduce the population or impede emigration to the island. Only in the sphere of the market place did the British Government promote Newfoundland's saltfish trade through genuine and persistent efforts to acquire for the colony more favourable trading agreements with Spain, Portugal, Italy, Brazil and, later, Greece.[9] These efforts were, for the most part, futile; however, the suggestions concerning diversification did bear fruit.

A study of the Newfoundland economy from c.1814 to 1914 can be approached in a variety of ways. However, for the purposes of a general overview, this writer will examine this subject in two chronological units using 1860 as a turning point.

Economy on Land and Sea from 1814 to 1860

The economy during the period from 1814 to 1860 was dominated by three major components: the dominant saltfish industry; the agricultural sector; and the seal fishery.

Cod fishery

The Newfoundland saltfish trade remained the mainstay of the local economy during this period (as it did during the later period. Exports of this product were as follows (five-year averages):[10]

Year	cwts	Year	cwts
1811-15	935,450	1836-40	841,525
1816-20	883,387	1841-45	961,260
1821-25	920,607	1846-50	980,340
1826-30	927,993	1851-55	959,126
1831-35	737,805	1855-60	1,236,868

As is readily obvious, production declined after 1815 and recovered somewhat beginning in the 1820s; it declined substantially in the early 1830s, and it was not until the early 1840s that production exceeded the 1811-15 level. From 1841 to 1855, the export of saltfish was relatively stable, and there was considerable expansion in the trade in the late 1850s. Meanwhile, fish prices (f.o.b.) declined from 24s 6d per cwt for the superior quality product (merchantable) to 14s by 1820, 12s by 1830 and 10s by 1849. Prices recovered slowly during the early 1850s to 14s by 1857; and the late 1850s experienced a brisk increase in price to 19s in 1860.[11] However, it is obvious that the stagnation in the Newfoundland cod fishery—which was certainly the case to at least the mid-1850s—was not offset by any increase in the value of the product. In fact, the saltfish trade was unusually depressed throughout the 1820s and 1830s and only began to partially regain lost ground during the 1840s and 1850s. At the same time, the population of permanent residents rose from 40,568 in 1815 to 60,088 in 1830; 74,993 in 1836; 96,296 in 1845; and 124,228 in 1857.[12]

The indisputable conclusion is that Newfoundland's resident population could not possibly have been totally dependent on the saltfish trade unless there had been an appropriately large decrease in the cost of living, which was not the case. One must look, then, for another explanation: the exploitation of other resources—in this case, land resources: the soil and the forests.

Agriculture

Agriculture, on a small scale, was practised in Newfoundland from the beginning of the first settlement attempts. Richard Whitbourne reported in 1620 that "it is well knowne to me, and diverse that trade there yeerely, how that Cabbage, Carrets, Turneps, Lettice, Parsley, and such like, prove well there."[13] However, the history of early agriculture is not well-documented, and it was not until 1729 that the governors began reporting on the amount of 'improved' land, which was the description given to land that had been cleared and/or cultivated (the reports do not specify). Moreover, the reports do not indicate whether the land so described was the total amount improved or the amount improved since the previous report. For example, the first report to mention this topic—the report of 1729—states that two acres of land were improved which, given the length of permanent residency in the island, could not have been the total amount cultivated.[14] Also, while the report of 1731 states that 250.5 acres of land on the English shore were improved (a reasonable figure), the report of 1733 mentions only 50 acres. Nevertheless, the governors continued to report on this topic, and by the mid-1820s, the amount of improved land was reported to be over 10,000 acres, while the final report in 1833 states that there were 16,813 acres of improved land in the island (excluding, presumably, the French Shore).

While Newfoundland's northern and southern coasts contain little soil, the traditional English shore, which, after 1713, included Placentia Bay (formerly French), is comparatively favoured. Head says that the fishermen of Harbour Grace and Carbonear were engaged in "considerable agricultural production" in the late seventeenth century and goes on to state: "In fact, the whole group of Conception Bay communities showed a marked interest in agricultural matters."[15] He also points out that the first major expansion in local agriculture occurred during the American War of Independence, when Newfoundland fishermen—both migratory and resident—were cut off from trade with New England, whence much of their food supplies had come. He quotes Governor Edwards, who wrote, in 1779: "The Ground which is cultivated produces vast quantities of (and very fine) Potatoes and other Vegetables."[16] Head also describes the growth in the amount of land improved during this period and points out that the greatest increases took place in St. John's, Harbour Grace and Carbonear. Head attributes much of the agricultural development in St. John's to the presence of the military, because the officers and troops were engaged in a considerable amount of farming. But he says:

> On the other hand, the district including Harbour Grace and Carbonear, in which the activity was 'civilian' boasted of more than the usual amount of agricultural production even in the seventeenth century, and its soils and climate were probably superior to those of most other coastal areas. During the eighteenth century its population grew at a very high rate, and it was essentially a resident one. These factors considered, it was logical that agriculture would be emphasized....[17]

By the end of the eighteenth century, it is obvious that the residents of St. John's and Conception Bay were engaged in farming—certainly, in the production of potatoes.

During the early years of the nineteenth century, this activity increased for a number of reasons. Once again, war disrupted trade, and the governors began to grant land to applicants for the sole purpose of farming, which was a departure from the traditional practice. Governor Keats published the following notice on 13 June 1813:

> All residents and industrious inhabitants desirous of obtaining small grants of land for the purposes of cultivation in the neighbourhood of St. John's, subject to very moderate quit-rents, are desired to give in their applications to the office of the Secretary to the Governor, before the last day of July.[18]

This was a new development; vacant land had always been reserved by law for the use of migratory fishing ships, and residents had generally cultivated land they had simply enclosed.

It was during the early years of the nineteenth century that others began to advocate the development of agriculture. Dr. William Carson, whom Prowse calls "the real founder of agriculture and constitutional government in the Colony,"[19] not only advocated agriculture, but engaged in extensive farming himself. Another prominent reformer, Patrick Morris—businessman and politician—encouraged the expansion of agriculture during the 1820s and 1830s, and in 1837 wrote:

> ...agriculture in this Island...has made rapid progress within the last ten or fifteen years. It is now a source of employment and subsistence to a considerable portion of the inhabitants in almost every part of the island [excluding the French Shore]. It has been found a great auxiliary to the fishery and has been the principal means with a large portion of the people of enabling them to subsist and remain in the country; and it is questionable whether the British fisheries could at all be maintained against foreign competition were it not for the additional resource afforded to the people by the cultivation of the soil.[20]

In addition to the generally widespread subsistence farming described by Morris, a recent study by Robert MacKinnon has shown that in the St. John's area (and no doubt in the major Conception Bay centres as well), there was a considerable number of commercial farms.[21] By 1840, there were approximately 400 farms in the St. John's area, and, although land-use data is available on only 135 of these, the average farm consisted of 40 acres of land, of which 15 were cultivated.[22] However, 4 of these farms had between 51 and 100 acres under cultivation, while another 16 had 21 to 50 acres under cultivation; at the same time, 32 cultivated only 1 to 5 acres and 35 from 6 to 10. MacKinnon's study summarizes this commercial agriculture as follows:

> Irrespective of farm size or acreage under cultivation, the basic mix or balance of goods produced and the methods of production changed little. Hay, potatoes, turnips and cabbage were the main crops and a variety of other vegetables, including carrots, beets, parsnips and savory were produced in a "kitchen garden." Fruits, such as apples, raspberries, gooseberries and rhubarb, as well as a variety of flowers were grown in a "front garden." Usually, about three quarters of improved acreage on all farms was under hay and the remainder under arable crops. Because milk was the commercial staple, the main objective of the crop rotation scheme was to maximize hay production. Potatoes were followed by oats and hayseed (timothy and clover) and then hay was cultivated, perhaps for four years, before the land was ploughed again and planted with turnips and cabbage (Mannion, 1974:63). During the summer, livestock grazed in the woods and when field crops were harvested, cattle were allowed in the meadows. Female calves were normally retained for milking and breeding and males sold for veal.[23]

Certainly by 1860, agriculture in Newfoundland, especially on the Avalon peninsula, had become well established, with commercial farming meeting some of the needs of those without access to land—which would include most clergymen, doctors, lawyers, mechanics, merchants and traders, and the workers on the docks, wharves and shore premises of the fish merchants of the bigger commercial centres, such as Harbour Grace, Carbonear and, especially, St. John's (see Table 2.2 for occupations in 1857). Meanwhile, most fishermen would have been engaged in subsistence farming.

Farming, however, was only one use to which local land was put. The forests provided timber for building purposes—fishing boats and vessels, and houses, sheds, barns, stages and fish flakes as well as all kinds of commercial buildings, from the humble blacksmith shop to the major mercantile enterprises of the fishing and sealing merchants of the 'Water Streets' of St. John's, Harbour Grace and Carbonear.[24] With settlements along the coast and a large unoccupied interior, the Newfoundland environment provided every community with a plentiful supply of wood—for building, as described above, but also for heat. Only when one recognizes the fact that the vast majority of immigrants to Newfoundland had no access to free wood in their English and Irish communities of origin can one understand the impact that the sight of unlimited forest resources must have made on those who came here to fish.

Therefore, it is difficult to envision how Newfoundland would have developed, if at all, as a local community, without the resources of soil and forest. As Table 2.3 illustrates, imports of potatoes and vegetables as well as imports of board and plank seem small in comparison to other items brought from outside. Nevertheless, an enormous quantity of agricultural products were imported. In 1826, the colony imported bread, flour, salted pork and beef, butter and cheese to the value of £243,391, or 28 per cent of its total imports of £862,453; and it exported saltfish valued at £481,970.[25] In 1838, the colony imported bread, butter, flour and salted pork to the value of £226,258, or 35 per cent of its total imports of £639,268; and it exported £484,649 worth of saltfish. In 1839, imports of these same four items totalled £284,885 in value and made up 40 per cent of the total import trade of £710,557; saltfish exports were worth £508,157. In 1840, the products in question were worth £307,534, or 39 per cent of the total import of £784,045; and saltfish exports were worth £576,245. Finally, as Table 2.3 indicates, Newfoundland imported bread, flour, butter and salted pork worth about £528,000, or about 42 per cent of the total value of its imports of about £1,272,000; meanwhile, exports of saltfish amounted to £789,124. If one adds the smaller imports of bacon, ham, various vegetables and grains and farm animals, one can conclude that in 1856, at least 50 per cent of Newfoundland's imports— in terms of value—were made up of agricultural products that were being produced in the other northern mainland colonies. Newfoundland's inability to supply itself with bread, flour, salted pork and butter—to name just the most significant agricultural products imported—reveals all too strikingly the severe limitations of the agricultural component in the local economy.[26]

Meanwhile, the population expanded from about 55,000 in 1826 to over 124,000 in 1857. It is obvious that both the population and the economy had moved away from what had been a total dependence on the saltfish trade. It is also evident that agriculture had not been able to make a significant reduction in Newfoundland's dependence on imports of basic foods, although both statistical and literary evidence indicate that there was a strong reliance on local potatoes. However, only in the smaller outports could people grow their own potatoes; those living and working on the 'Water Streets' of St. John's, Harbour Grace and Carbonear, to name the largest centres, and particularly those in the capital, were forced to purchase potatoes, either from local producers or from importers. To these people, the increase in local agricultural production made very little immediate difference to their lives. Therefore, as the import figures show, the expansion of local agriculture was not the decisive economic achievement of the 1814-60 period. In order to explain the development of the Newfoundland economy during this period and to account for the growth of population in the midst of a stagnating and problem-ridden cod fishery, one has to study the impact of diversification in the fisheries and examine the industry which was developed by residents during the Revolutionary and Napoleonic Wars—the spring seal fishery.

Seal fishery

Once the seal fishery had become established and the sealers discovered that the female seals bore their young around the end of February, that young seals several weeks old provided a source of excellent oil and, furthermore, could be found in large herds or 'patches' on the ice floes, the industry rapidly increased. From what is already known about the saltfish trade, it is fairly certain that neither the North Shore nor Labrador fisheries could have survived the post-1815 depression without the additional seal resources. Therefore, while the number of ships engaged in the seal fishery during the earliest period is unknown, the rapid expansion in the North Shore fishery beginning in 1818 (see Table 2.4) is a good indication that the seal fishery was beginning to expand as well. As the table shows, 263 vessels were engaged in the North Shore fishery in 1820, and 300 ships were engaged in the Labrador fishery in 1828. However, starting in 1827, reports show that nearly 300 ships were engaged in sealing that year, and this number rose to over 400 in 1832. At the same time, the original reports state that the sealing fleet of 1827, which was made up of 290 vessels, had a combined burthen of 17,445 tons and carried a total of 5,418 men. In 1832, there were 407 vessels employed and these had a total burthen of 27,241 tons and carried 8,649 men in all. When one realizes that the total population of adult men on the whole island was reported as being just over 16,000 in 1830 (the latest statistics available for this period), it is obvious that at least one half of them were engaged in the spring seal fishery in 1832.

The rapid expansion of this fishery is also evident from the reports of the number of seal skins exported. As Table 2.5 illustrates, exports of seal skins rose from under 170,000 prior to 1819 to almost 280,000 in that year. The number climbed to over 368,000 in 1822, to over 536,000 in 1830, and over 601,000 in 1831. However, production was very erratic, and, as can be seen in the above table, the results of this fishery were rather disappointing in 1824, 1828 and 1829. Nevertheless, despite the unpredictable nature of this industry, merchants and fishermen were only too willing to participate because the financial returns were considerable. For example, in 1831 and 1832, just over 700,000 cwts of saltfish were exported, which meant that, given the price of 8s to 12s per cwt, this industry earned only about £350,000 in each of these years—and by then the population was over 60,000 people. In these same two years, returns from the seal fishery averaged well in excess of £200,000. When one allows for the minor exports of pickled herring and salmon and a few other items, it is obvious that the seal fishery accounted for at least one-third of Newfoundland's exports during these successful early years. Furthermore, one must remember that the seal fishery helped to support the Labrador cod fishery, and that without the latter, production of saltfish would undoubtedly have been lower.

There are considerable difficulties in trying to quantify the value of the seal fishery in a consistent manner during the period before the newly-installed local assembly began to publish annual reports in the *Newfoundland Blue Books* and, later, in the *Journals of the House of Assembly*. However, the growth of the sealing fleet and the increase in the employment of sealers combined with the considerable expansion in the production of seal oil are all the evidence one needs to appreciate the growing importance of this industry to the well-being of the local economy. By using the figures that are available, one can better appreciate the developing situation.

Certainly, as Table 2.6 indicates, exports of seal oil surged forward in 1819, the same year that there was a considerable increase in the size of the North Shore fleet and an increase in the number of seal skins exported. As one can see, seal oil production rose from less than 1,300 tuns, worth about £36,000, in 1814 to 4,253 tuns, worth over £95,000, in 1819; there was a major increase in this industry in 1830, when 7,110 tuns were produced, worth about £160,000 (taking £22 10s as an average price), and a further expansion the following year, when at least 601,742 seal pelts were brought in and 8,761 tuns of oil produced, worth about £205,000 (taking £23 10s as the average price). As Table 2.6 shows, it is impossible to distinguish seal oil from cod oil during 1834-44, inclusive, but the large numbers of seals killed in the early 1840s coincide with large quantities of oil being produced, especially in 1840, 1841, 1843 and 1844. There was a decline in 1846, but exports rose again in 1848 to 6,508 tuns, worth just over £160,000, and averaged over £230,000 annually during 1851-58, inclusive—although the figures are missing for 1852 and 1855. This average would be somewhat less if the missing information was available because, as Table 2.5 illustrates, production was lower in 1855 than in any of the other years. On the other hand, although the 1852 seal fishery was racked with losses to shipping, literary evidence indicates that despite "such destruction among the vessels ... we understand the average catch at this time is equal to that of last year."[27] At the same time, it must be pointed out that there was a slight dip in the seal fishery in 1858 and 1859, as shown in Tables 2.5 and 2.6, and indeed, decline in price is also evident in Table 2.6, as seal oil dropped from a peak of £43 per tun (as valued in Newfoundland) in 1856 to £30 in 1859 and 1860. One must remember, also, that while the fat was the most valuable part of the seal 'pelt', the skin was also in demand in the British market place.

To place the seal fishery in the context of the overall economy one must return to the discussion of imports during 1856-60; from Table 2.3, it can be seen that the value of total imports for each of the four years in question were: £1,271,604; £1,413,432; £1,323,288; and £1,254,128. To pay for these imports, Newfoundland was primarily dependent on exports of saltfish, cod oil, seal oil and seal skins, valued below in pounds sterling:

Exports	1856	1857	1859	1860
Saltfish	£789,124	1,006,129	894,966	846,338
Cod Oil	162,313	159,130	111,839	118,949
Seal Oil	216,006	265,131	166,970	145,959
Seal Skins	71,386	99,217	57,607	51,631
Total	£1,238,829	£1,529,607	£1,231,382	£1,162,877
All Exports	£1,338,797	£1,651,171	£1,357,113	£1,271,712

The above demonstrates the extent to which the Newfoundland trade was dependent on both the cod and seal fisheries up to the end of this period. Most of the remaining exports were comprised of varying amounts of pickled cod, herring, salmon and trout.

Early perceptions of the seal fishery in the economy

Contemporaries were not unaware of the importance of the seal fishery to the newly-evolving colony. One contributor to a St. John's newspaper in 1829 was of the opinion that this industry was perhaps even more important than the cod fishery. He wrote in March:

> For several days past the greatest bustle and activity has prevailed in our streets, and on the merchants' wharves, among the sealers, in preparing to start upon their hazardous but important expedition. In consequence of the favourable state of the weather during this month, the vessels have been got in readiness for sea at a much earlier period than we ever before recollect; and it is gratifying to us to remark that, in the opinion of persons best acquainted with this main branch of our trade, the prospects for the present sealing voyage are unusually flattering.[28]

Similarly, in July 1830, the same newspaper reported on a recent meeting and dinner of the Commercial Society (in St. John's), which was presided over by the president, Thomas H. Brooking—a prominent merchant. The reporter wrote as follows:

> He [Mr. Brooking] took occasion to compare the result of the seal fishery of 1829 with that of the present year, and congratulated the Society upon the unprecedented success which had attended the exertions of our hardy and enterprising fishermen this season. In 1829, about 300 sail of vessels, manned with 5,000 fishermen, were employed in the Seal-Fishery, and they produced what was considered a tolerable result. The number of seals taken by them amounted to about 280,000, which was estimated to be worth something more than £100,000. This year about the same number of vessels, but with some little increase in tonnage, and carrying 400 to 500 men more, have been occupied in this fishery; they have taken about 554,000 seals, which may be fairly valued at a sum exceeding £210,000 and when it is considered this has been drawn from the ocean at a season when no other profitable employment could be found for our fishermen, this branch of our fisheries is of vast importance, and is entitled to all the encouragement and support which can be given to it. Besides advantages which are immediately derivable by those persons actually employed in catching the seals, this branch of the fishery gives to thousands of artisans and labourers employment in the course of every year; and, could we trace it through all its ramifications, the benefits are almost incalculable.[29]

The following year, in February, another St. John's newspaper pondered the coming voyage and the results of "the prosecution of so highly important a branch of the Newfoundland trade."[30] In March of that year, a local correspondent pointed out that it was the seal fishery which had saved the colony from disaster following the failure of the cod fishery in 1830:

> The unprecedented failure in the Cod fishery of last year, led to forebodings of a winter of more than usual distress; happily, however, it has not been exemplified in this vicinity so as to meet the public eye, in the degree anticipated. It is, doubtless, the extraordinary outfit which has all the while been in operation for the Seal Fishery that has warded off the evil so much

apprehended, and by affording employment in various ways, diffused as a means of relief among hundreds of those who would, otherwise, immediately suffer by the afflicting deficiency in the staple article of support.... We turn, however, from a scene of much relative inactivity and stagnation, to the contrast now presented on the wharves and business part of this and the neighbouring harbours, so cheering in every point of view; for the period has arrived when thousands of the population are again called into active employment; our harbours were never more beautifully studded with vessels, the streets are filled with crowds of men who are preparing to rally from our shores in quest of a means of subsistence annually afforded them by the bounty of Providence,—the contents of shops and stores are in pretty general request, and amidst the general bustle former losses are forgotten in the hope of future success....[31]

And another account published in Britain and quoted in a local newspaper demonstrates that this new industry was becoming widely recognized and appreciated:

There is another department of the colonial fishery, which has originated within no distant period, and is now becoming of great extent and importance. The large fields of ice, which in the months of March and April drift southward from the Polar seas, are accompanied by many herds of seals: these are found sleeping in what are called *seal meadows* of the ice, and are there attacked with fire-arms or bludgeons, and slaughtered in great numbers. For this purpose the fishermen of Newfoundland, from which island these voyages are principally made, without waiting till the return of spring shall have opened their harbours, saw channels through the ice for their vessels, and set sail in quest of those drifting fields, through the openings of which they work a passage, attended with great difficulties and dangers, till they encounter their prey on the seal meadows. This bold and hazardous enterprise seems well compensated by its success. The number of seals thus taken is almost incredible, and is greatly on the increase.[32]

In his book published in 1846, sometime after he observed and experienced the seal fishery, Philip Tocque wrote:

The seal fishery of Newfoundland has assumed a degree of importance far surpassing the most sanguine expectations of those who first embarked in the enterprise, and is now become one of the greatest sources of wealth to the country. The interest of every individual, from the richest to the poorest, is interwoven with it, and the prosecution of the voyage causes more anxiety, excitement, and solicitude, than any other business in Newfoundland, or probably in the world.[33]

A later correspondent wrote in praise of the seal fishery and with great optimism on 5 March 1846—innocent of the difficulties that lay ahead, especially for St. John's:

The Sealers have made the last week one of much excitement and interest. The preparations for the voyage are now completed, and our fleet only await a favourable breeze, to waft them to the scene of their springs operations, There is something, we think, in the fine and imposing appearance of our harbour at the present moment, in which the people of Newfoundland, both the Trade and the working population, might be pardoned if they indulged feelings of no ordinary pride. If we were asked for a striking instance of the importance of this country, or sought to enforce a conviction of her superiority as a place of trade, and of the value of her resources, we should at once point for our illustration to the port of St. John's with its hundred and twenty five or thirty sail in which are employed nearly four thousand men, now fully equipped for sea. The considerations which induce so extensive an outlay in this fishery are based upon the experience of its value, and though 'tis too undeniable that many have sunk large amounts of capital in such enterprises, 'tis equally demonstrable we imagine, that to this source we owe a considerable proportion of the wealth of our Merchants, as well as of the many comforts and comparative independence of the industrial classes.[34]

However, within two months, it was obvious that the 1846 season was not going to be a successful one, and one writer once again drew attention to the importance of this industry when he wrote on 11 May:

A failure of the seal fishery is undoubtedly a heavy blow levelled at the means of the people's subsistence, and at the vitality of the trade. It is an evil of superior magnitude in itself, and in its immediate and inevitable effects. But it produces besides, a very sensible depression in the arrangements for the summer's operations. Men who have been severe sufferers in one adventure, are naturally enough ill-disposed to hazard still more in what may prove an equally unproductive or perhaps ruinous speculation, and our resources are so limited, that inactivity, which is for a time partial or confined to a particular branch of trade, soon extends itself, and becomes augmented into a too general torpor.[35]

Unfortunately, 1846 experienced not just a mediocre seal fishery but saw much of St. John's burned to the ground, including the sealing premises, with their vats of oil and oil-soaked wharves and wooden structures. This was followed by a blight which destroyed much of the potato crop in the neighbourhood and in the outports.

Some idea of the value placed on the seal fishery in Newfoundland at this time can be obtained from a newspaper commentary on 4 March 1847, which stated:

The prospects so far are favourable, if indications may be at all relied on. The severe weather which has so often been the principal cause of the failure of this enterprise, it may be reasonably hoped, is in great part past, and the wind considerably cleared the ice off the coast. With all our heart we hope these appearances may not prove to have been delusive, for never we believe in the history of the country was a successful seal fishery a matter of greater moment than now.[36]

A more forceful commentator described the situation that spring as being on the verge of a calamity similar in severity, if not is scope, to the Irish famine. He wrote:

The accounts we continue to receive from various parts of the Bay are truly distressing. There may, perhaps, be a small stock of potatoes in the settlements north of Brigus, but with this exception the destitution complete [sic]—is awful. A gentleman from the North Shore assures us they will really [sic] have to bury the people there unless some relief is immediately extended. Of course, we have no inducement to exaggerate. We reiterate—the prospects are truly alarming. The sealing craft, we admit, will relieve the district of some five or six thousand mouths, but unfortunately these are they upon whose daily exertions the bulk of the population are dependent for support. Who, for the most part will be left behind? Women, the aged, the decrepit, and the helpless! How are those to be fed? How are 15 or 20,000 people to be fed during the months of March and April? ... Immediately on the sailing of the ice-hunters, supplies must be had from St. John's by some means or other.[37]

In 1848, a reporter in St. John's described the results of the seal fishery in terms that once again confirmed the local conviction that a healthy seal fishery was a vital element in the economy. He described and commented on the voyage of 1848:

Most of the Sealers [sealing ships] from this port have already returned, and the issue of the spring's fishery is decided at an earlier period than usual. It has, indeed, been an excellent fishery—the best, we believe, since the spring of 1831....

Never, in our time, never, perhaps, in the history of the country, was a good fishery of such vital importance, or hoped for with such feverish longing expectation [sic]. All our resources strained to their utmost endurance, short fisheries, short crops, a treasury drained almost to exhaustion, and unprecedented commercial embarrassments, formed a combination of agencies, which, if left to their own working, uninterrupted by some special counteraction, might have defied the most energetic struggle of which the country was capable; and we looked forward to the result of this adventure as to something which should either give assurance of stability, or warn us of inevitable ruin. We have great cause of thankfulness that the wish, and not the fear, has been thus realized.[38]

This attitude, that the seal fishery was so essential to the Newfoundland economy, was widely held in the colony up until the 1850s. Thus, one writer, when reporting on the people of Brigus in 1849, stated: "The people are badly off in this neighbourhood, and if we don't get a good seal fishery I don't know what they will do."[39] Another pointed out in 1852 that the seal fishery was the "mainstay of Newfoundland."[40] And finally, it must be remembered that despite its significance to the local economy, the seal fishery was carried out only during a brief period in the spring—a point that was summarized by one New York writer in 1853:

The catch of seals, probably, reaches half a million which at a fair valuation, may be set down as worth one million of dollars [about £200,000 stg.]; and considering the length of the voyage—a month, rarely extending to two months—the result will appear the more astonishing. Indeed we know of no venture—our own whale fishery not excepted—which yields so profitable a return in proportion to the capital invested, and the time occupied in the undertaking.[41]

Although, strictly speaking, his was not a contemporary voice because he was writing in the 1890s, Judge Prowse, Newfoundland's most famous historian, was a firm believer in the utilization of oral history and had talked with observers from the 1840s; he should have the last word on this subject. Writing of that period, he said:

All the country was prosperous—the seal fishery had immensely increased; exports of all kinds were very large; in 1840 over nine hundred thousand quintals of fish and over six hundred thousand seals rewarded the labours of our hardy fishermen; taxation was very light, the revenue for 1840 being only £43,863, and the whole civil expenditure £39,347—dear, delightful days of Arcadian simplicity! when we had no debt, and port wine was a shilling a bottle!

The Governments of Captain Prescott and Sir John Harvey [1834-46; Prowse uses political periodization points] marked the parting of the ways, the interregnum between the last days of the cookroom and the [migratory] fishing ships and the new departure.[42]

The above writings demonstrate that the importance of the seal fishery was recognized by observers within the colony and beyond.

Summary: 1814-60

It is obvious that the cod fishery alone could not have supported a comparatively large and increasing population after 1815; in fact, the state of the markets was such that even the traditional migratory cod fishery would have had difficulty surviving. Also, despite efforts to cultivate the soil, Newfoundland was never able to feed its population and remained a food importer. The spring seal fishery was the essential ingredient needed to create an economy that could support a colony. As one can see from Table 2.5, the seal fishery was a growing enterprise up to about 1845. The annual average production (and/or exportation—there is some confusion here) of seal skins rose from about 196,640 in 1816-20, to 260,750 in 1821-25, to 321,468 in 1826-30, to 498,633 in 1831-35, to 457,142 in 1836-40, and to 526,804 in 1841-45. Not all years were recorded, so it is impossible to say how accurate these figures are. However, it is obvious from those that are available and from the contemporary evidence already quoted that this was a growing and very important industry up to c.1845. During the following 15 years, production remained considerable, but growth had ceased. Annual exports (and by the 1830s the figures refer to exports) amounted to an average of 394,101 during 1846-50, to 442,150 during 1851-55, and to 407,688 during 1856-60. Newfoundland's economy survived this stagnation without great difficulty because the price of oil increased significantly, particularly during the early 1850s, as Tables 1.8 and 2.6 indicate. The seal fishery provided the necessary winter/spring employment that was necessary for large numbers of employee cod fishermen to remain in Newfoundland all year round. Also, as discussed more fully in the following chapter, the seal fishery forced the West Country firms to transfer operations to Newfoundland and, more importantly, encouraged the growth of local firms. The seal fishery, during 1814-60, was crucial to the development and survival of the Newfoundland colony.

Economy on Land and Sea from 1860 to 1914

By 1860, Newfoundland had grown considerably since its days as a fishing station at the beginning of the century, but it was dangerously dependent on only two industries: the production of saltfish and seal oil; and the production of two other products for export directly dependent on these industries—cod oil and seal skins. With the political advances that had been made and with the development of an infrastructure to meet the needs of a population of about 125,000 people, the future of Newfoundland in 1860 was much more secure that it had been fifty years earlier. The pressures to expand the fisheries and diversify the economy would intensify in the second half of this period.

Seal fishery

The second half of the period under discussion got off to a disastrous beginning, with the decline in the seal fishery in the first half of the 1860s, as Table 2.7 clearly shows. During the five-year period ending in 1865, annual average exports of seal skins amounted to 259,896—lower even than the early 1820s. During 1866-70, average exports recovered to 320,815 annually; they rose further to 386,028 during 1871-75 and remained firm at 382,250 during 1876-80. The 1880s experienced catches almost as low as those in the early 1860s, with exports averaging 282,956 seal skins annually during 1881-85, and 269,193 during 1886-90. There was a small recovery in the early 1890s, but the latter half of this decade saw a drop in production to an average annual export of only 232,738 skins during 1896-1900. There was a modest increase in the average to 351,206 during 1901-05, but a further decline followed, so that by the last years of the period leading up to 1914, exports of skins averaged less than 250,000 per year. With the exception of the 1870s, and exceptional catches on occasion, the last half of the nineteenth century experienced an overall decline in the seal fishery, which was especially obvious in the early 1880s.

This general trend was exacerbated by the overall decrease in the market price of seal oil during the second half of the century, as demonstrated by the figures in Table 2.8. The price rose in the early 1860s to a high of £47 10s per tun (f.o.b.) in 1864, and this, no doubt, helped to compensate for the low catches. However, after 1865, there was a slow and steady decline in price for reasons already discussed in Chapter 1. From $230 (about £47 stg.) per tun in 1865, the price dropped to $160-$170 in the latter 1860s, to an average of $144 per tun during 1870-74 and to about $122 during 1875-79. It rose a little to an average of about $128 during 1880-84—possibly in response to the decline in the catch during these years. However, beginning in the mid 1880s, there was a rapid and critical decline in price as the uses for fish oils became more restricted. During the five-year period ending in 1889, the average price declined to about $81 per tun; it dropped further to $79 during 1890-94 and fell to about $72 during 1895-1900. Prices rose to about $100 during 1900-04 (in response to the reduction in the Norwegian cod fishery); they dropped to $87 in 1905-09 and, finally, rose a little to $92 during 1910-14. These prices were determined by the declining demand for oil in the market place, as already discussed, as well as by general economic developments, which included a depression in the latter 1880s and throughout the 1890s, and a recovery during the early years of this century. However, on the whole, the market situation continued to deteriorate, as already discussed in Chapter 1. One commentator wrote in a local paper as follows in 1897:

> It is but a poor outlook for the coming seal fishery that the price of both oil and skins is so low as to leave little or no profit. Our oil has been for years fighting for its very existence against the inferior compounds which are

being brought into use, and now a substitute has been found for the fur seal of the Pacific and the south Sea, so that the price of these skins, the most valuable of their kind has dropped to $6., which figure renders our skins unsaleable for their fur, and they can only find a market to be manufactured as leather. The skin substituted for the costly skin garments affected by wealthy ladies is a monkey skin got in west Africa, and seal sacques and caps are now so cheap, in consequence, that they have ceased to be a badge of fashion and the wealthy will no longer wear them.[43]

In that particular year, the price for skins plummeted to $0.60, and it reached a record low of $0.49 two years later, in 1899 (see Table 2.8). While the price for seal skins recovered in the remainder of the period under study, the price for oil never regained its former high. It is obvious that the decline in the value of this product could only make the situation worse when combined with the stagnation and subsequent decline in production.

The inevitable result of this combination is presented in Table 2.9, where it can be seen that the contribution of the seal fishery to Newfoundland's export economy declined. This industry provided at least one-third of the colony's exports at times during the first half of the century—probably more during peak years. In 1850, 1851 and 1853, it provided 27 per cent, 30 per cent and 31 per cent, respectively, as the table indicates. By 1859 and 1860, these percentages had declined to 17 per cent and 16 per cent, respectively, probably because of the drop in price during these years. There was a fluctuation in the place of the seal fishery in the overall economy during the following years—with 1864 being an exceptionally poor year due to the low catch. After 1871, the percentage dropped below 20 per cent and stayed below that level. During the five-year period 1871-75, the annual average percentage of Newfoundland exports made up of seal products amounted to only about 17 per cent; by 1881-85, 12 per cent. It declined further to 7 per cent by 1896-1900, and during the last four years of the period under study, 1911-14, the proportion sank to 5 per cent. Certainly, stagnating and decreasing production, combined with a declining price, especially beginning in the mid 1880s, resulted in the extraordinary decrease in the importance of the seal industry to the Newfoundland economy by the end of this period.

Cod fishery

Meanwhile, the production of saltfish remained Newfoundland's principal industry, and saltfish its chief export. The industry recovered firmly between the early 1830s, when less than 750,000 cwts were exported on average annually, to over 1,200,000 cwts annually in the late 1850s. The 1860s was a decade of depression, principally because of market problems,[44] and average exports declined to about 1,100,000 annually. There was an increase in the early 1870s to over 1,300,000, followed by a decline in the latter part of that decade to an average of less than 1,200,000 cwts annually. A major increase in the first half of the 1880s brought the average of the colony's annual exports to almost 1,450,000 cwts, but a rapid decline reduced the trade to less than 1,150,000 during the last half of this decade, and there was only a slight recovery to just over 1,200,000 cwts annually during the 1890s. After 1900, there was a modest recovery to an annual export of about 1,300,000 cwts during 1901-05, a stronger increase to an annual average of over 1,500,000 cwts during 1906-10, and a drop to about 1,300,000 on average during the last four years of this period—ending in 1914.

As already pointed out, the actual economic situation was rather worse than these statistics suggest. The population of the colony continued to grow—to 146,536 in 1869; 161,374 in 1874; 197,335 in 1884; 202,040 in 1891; 220,249 in 1901; and 242,619 in 1911—with the obvious result that by 1914, per capita exports of saltfish had declined to about 6 cwts.

Furthermore, as in the case of the seal fishery, the situation was aggravated by a decline in the price of fish. At the beginning of the 1860s, there was a severe drop as the price declined from 19s per cwt in 1860 to 13s per cwt in 1861.[45] There was brief recovery to $5.09 per cwt in 1866, but this was followed by another fall in 1867 and 1868, when average prices declined from to $3.62 and $3.45, respectively.[46] (In fact, it is widely acknowledged that this decline encouraged people to look to confederation with Canada as a solution to the economic problems, and the price recovery in 1869 to $4 was a major factor in the defeat of the confederation party in the polls that autumn.) During the period from 1871 to 1875, the average price of saltfish exported was $4.18 per cwt, including a rise to $4.54 in 1875 and a further rise to $5.32 in 1876. Not until 1876 did published records begin to report the exported value of Labrador-cured and shore-cured fish separately. Prices prior to that year were calculated by including both qualities of fish, and since the former always brought a lower price than the latter, the recorded prices were averages and actually reflected a price lower than that obtained for shore-cured. However, the point to remember is that prices were lower than average during the 1860s and were to fall again later in the century. For example, as indicated, the value of shore fish in 1876 was set at $5.32, while the value of Labrador was set at $4.16. By 1880, Labrador fish had declined to $2.17 per cwt and shore fish to $3.33. Prices recovered to over $3 and $4, respectively, in the early 1880s, but in 1885, they declined again to $2.05 and $3.20, and in 1886, down to $2 and $3.15. There was a partial recovery in the late 1880s and early 1890s, but this affected shore fish more than Labrador fish—the latter coming under severe competition from French-caught fish in the Spanish and Italian markets. By the latter 1890s, the saltfish trade had nearly collapsed, as prices decreased to the all-time low of $2 and $2.56, respectively, in 1897.

Complicating the Labrador fishery was the fact that these particular fishermen had always depended on the seal fishery to provide a major part of their annual earnings. Without this supplementary industry, the Labrador fishermen were forced to find other work, which usually meant travelling to New England, especially Boston, and working there for long periods at a time. Many maintained their homes and families in Newfoundland in this manner, but many others remained and eventually became American citizens—particularly the young, unmarried workers who simply did not return. However, the money remitted to Newfoundland during this period was probably quite considerable, and references to lost letters containing money become common in the late nineteenth century, despite the usual practice of sending money orders. Nonetheless, many of these migrant workers were stubbornly committed to the Labrador fishery, and remained commuters to Boston all their lives.

Economic recovery began in the early years of the twentieth century, primarily because of problems in the fisheries of Newfoundland's competitors, especially Norway and France; by the final four years of this period, 1911-1914, the export value of Labrador fish averaged $4.09 per cwt and that of shore fish averaged $5.52. Nevertheless, it is obvious that, even with this recovery in value, the increasing population was outstripping the total returns from the cod fishery. And the major crisis occurred in 1894, when the two local banks 'crashed', several major fish businesses went bankrupt and others were forced to reduce their operations.

Other fisheries

Newfoundland had always exported small quantities of other fish products, with pickled salmon being the most important traditionally. However, rarely in the nineteenth century did the value of exports of this item amount to $100,000 annually, and in the last five years of this period, considerably less—about $78,000. The exportation of pickled herring was a more important activity, and exports rose from about $150,000 worth in 1860 to a very brief peak of over $400,000 in 1883; but were worth only between $200,000 and $300,000 annually by the end of the period. The production of canned lobsters began in the 1870s, and this industry was worth about $500,000 per year in exports by the end of the century but had declined to just over $400,000, on average, by 1910-14. One fishery—whaling—did not begin until the end of the century, and at first, it looked as if it would become a valuable, long-term industry. In fact, exports of whale products grew quickly and were worth $535,101 by 1905 (more than seal oil, which was worth $374,974 that same year). However, this fishery declined rapidly as well, and by 1914, annual exports were not worth above $100,000. In the meantime, the government had encouraged the development of a bank (or deep sea) fishery, and assisted by government subsidies, this industry had grown during the 1870s and 1880s. However, cuts in government funding and the continuing market problems caused a decline in this industry—especially on the northeast coast, although a smaller bank fishery became established on the south coast and continued into the twentieth century. The saltfish produced by this branch of the fishery was cured and exported as either shore fish or Labrador fish. In all, however, the diversification into other fisheries did not produce any really successful ventures of the sort that could compensate for the decline in the seal fishery.

Agriculture

From a slow informal beginning in the first half of the century, agriculture attracted people's attention in the early 1860s, especially after the disastrous seal fishery of 1862. A Select Committee of the House of Assembly on agriculture was struck, and meetings were held throughout March 1863.[47] Thirteen witnesses were examined, among them: W. F. Rennie, Secretary of the Agricultural Society (founded in St. John's in 1842); Justice Bryan Robinson, a member of the Agricultural Society since 1842 and, at times, its president; Justice Philip Little, the colony's first prime minister (1855-58); and Lawrence O'Brien, president of the Legislative Council and one of the most important sealing ship owners and suppliers of the 1850s. Rennie advocated the establishment of government-financed Agricultural Society branches throughout the colony. Robinson promoted agricultural fairs and markets, free grants of land to people who promised to cultivate it and the raising of sheep. Little supported the establishment of Agricultural Society branches, the distribution of improved seeds and breeding stock and went on to add:

> The population of Iceland—a bleak and sterile country, not to be compared to Newfoundland in point of fishery or agricultural resources—is about from 65,000 to 70,000 [people], and there are in that Island more than 50,000 cattle, and more than half a million sheep. Besides exporting fish and oil, the inhabitants send to markets abroad over a million pairs of stockings and mittens, frocks and jackets, manufactured with the wool of the sheep by the families at their homes; while our population import all such articles for their use, and only knit a few mittens and stockings with imported yarn, or pulled wool. Our people are comparatively idle for several winter months, for want of employment, while the Icelanders are engaged in looking after their cattle and sheep, or in spinning, weaving, knitting, preparing hides for shoes, shoemaking, or some mechanical employment.[48]

As a result of these hearings, breeding stock and seeds were distributed, agricultural societies formed, financial subsidies granted, and more land cleared, but progress was slow. In 1912, for example, the Agricultural Board reported that there were 77 agricultural societies operating in Newfoundland, and these had received from the government as follows: 56 bulls, 500 sheep, 284 pigs, 1 stallion, 1,233.5 barrels of potatoes, 776 boxes of potatoes and $7,600 in cash grants.

Nevertheless, by the end of the period, Newfoundland was importing as follows: flour—393,688 barrels worth $1,989,866 in 1913, and 395,729 barrels worth $1,823,551 in 1914; tens of thousands of barrels of salted beef, pork, pigs' hocks, pigs' jowls and pigs' heads worth hundreds of thousands of dollars; a wide variety of grains, farm animals, sizeable quantities of leatherware and forest products; and even potatoes—127,922 bushels worth $52,446 in 1913 and 146,874 bushels worth $75,177 in 1914; as well as other vegetables. However, imports of hard bread (ship's biscuit) declined towards the latter part of the century as local bakeries began to produce their own from imported flour; similarly, other investors began to manufacture butter and cordage from imported raw materials.[49] However, the importation of new 'necessities'—India rubber ware and kerosene oil—increased throughout the period, and added to the problems in Newfoundland's balance of trade on current account. Indeed, by the end of the period, the trade imbalance on current account was becoming a problem. For example, in 1913, total imports and exports were $16,012,365 and $14,672,889, respectively, although there was a slight balance in favour of exports in 1914. By this time, however, the public debt exceeded thirty million dollars.

Other Strategies

The Newfoundland government discovered, in the latter 1860s, that it could not depend on confederation with Canada to save it in the event of a major economic and trade crisis—the electorate had demonstrated its displeasure with a political party that took this approach. Therefore, the governments were forced to develop new strategies, and these became centred around the building of a railroad, industrialization, acquisition of the French Shore and reciprocity with the United States of America.

The collapsing fisheries (both seal and cod) forced the governments to borrow and make arrangements to have a railroad built (at first to Hall's Bay but eventually to Port aux Basques, with a ferry terminal by 1898). William Whiteway, the prime minister for much of this period, believed that the railroad would open up vast agricultural, mineral and forest resources in the interior of the island. While construction provided much-needed employment, the government was forced to increase its national debt from about $1.5 million in 1883 to $30 million by 1914, primarily to finance its railway-building scheme.[50]

Turning to the problem presented by the presence of the French on the French Shore, the local government's constant agitation finally prompted Britain to negotiate an agreement with France to end the freedoms and privileges the French fishermen enjoyed. Thus, beginning in 1904, Newfoundlanders were able to take advantage of this sparsely settled territory. However, the immediate economic returns were not readily obvious.[51]

Reciprocity with the United States had its immediate origins in the free trade agreement negotiated between British North America and the United States in the 1850s. The idea was kept alive during the period of the Washington Treaty during 1873-85, but in 1890-91, a separate deal between Newfoundland and the United States failed to be concluded at the last moment. Various subsequent governments saw reciprocity as a solution to Newfoundland's economic woes—especially those under Prime Minister Bond in 1900-08—however, efforts in this direction were unsuccessful. Many observers have since noted that Newfoundland's biggest problem was finding large, reliable markets for its saltfish, and since the United States was not a saltfish importer, the concept that free trade with that country would help to ease Newfoundland's economic situation was largely illusory.[52]

In the area of industrialization, Newfoundland's earliest experiences were all confined to the latter part of the nineteenth century, and not all strategies failed. Copper mining was begun and exports of copper became part of the colony's trade in the 1860s, briefly peaking at a value of over $1,264,000 in 1877. An erratic decline followed, with exports worth only $99,000 in 1884, up to $816,000 in 1888, but down to $227,000 in 1890. During the final years of the period, after a similar series of high and low export points, copper exports were worth only about $200,000 annually.

In 1895, an iron ore mine was opened on Bell Island, in Conception Bay, and this provided an important export product for the country and wage labour for the many unemployed sealers and Labrador fishermen. In 1905, a pulp and paper mill was begun in central Newfoundland—largely because a railroad existed. This mill produced both pulp and paper for export and provided employment for other fishermen. In 1913 and 1914, respectively, exports of iron ore were valued at $1,367,520 and $1,370,375; pulp exports were valued at $436,352 and $372,676; and paper exports, at $1,990,229 and $1,795,488. Meanwhile, during these two years, exports of all fishery products amounted to $10,242,556 and $10,907,677, respectively; and total exports were $14,672,889 and $15,134,543. By the end of the period, the pulp and paper industries and the iron ore mine were, to some extent, playing the same role, in terms of the export trade, as that played by the seal fishery during the first half of the century. However, this was not completely true, because the latter industry was an integral part of the overall fishery and highly labour intensive; the newer industries, on the other hand, were not labour intensive. For example, the 1911 census reports that 42,846 males were engaged in catching and curing fish and 22,472 females were engaged in curing fish. At the same time, there were only 2,260 males engaged in mining and 2,821 engaged in lumbering. While the number of mill workers was not specifically recorded, if mining is any guide, it is obvious that the pulp and paper industries actually employed as lumbermen and mill workers a comparatively small number of men when compared with the number employed in the fisheries.

Later perceptions of the seal fishery: conservation issues and government reactions

By mid century, observers realized that the continual annual increase in the seal catch was ending because the resource was being over exploited. In April 1853, the House of Assembly went into committee on a bill for the protection of the seal fishery, but this did not result in any legislative action.[53] A St. John's newspaper reported in 1855 on the diminishing seal resource:

> We learn with much satisfaction that arrangements are in progress for sending a vessel from this port to the northern seal fishery in the spring. It is the general opinion of the most experienced that the seal fishery on the Newfoundland coast is over-done, the number of old seals taken diminishing annually and the total catch not keeping pace with the increased number, larger tonnage and more numerous crews of the sealing vessels. Consequently the business is becoming less remunerative. It is therefore of the greatest importance to make an effort for opening up the northern fishery for the larger class of vessels—a fishery from which vessels from Hull and other British ports ... have carried out large trips.[54]

Although this experiment was not successful—in fact, Scottish ships began coming to Newfoundland to prosecute the Newfoundland seal fishery later in the century—it does show that contemporaries were conscious that the 1850s had become a turning point in the history of the productivity of the spring seal fishery. The Legislative Council made the first effort to pass legislation protecting the seal fishery (and other fisheries) in 1858.[55] Another attempt was made in 1861 to prohibit the wasteful practice of panning seal pelts, which often resulted in losses overnight through changes in the wind and weather,[56] but again, nothing concrete resulted. In its annual report of 1865, the St. John's Chamber of Commerce declared: "The result of the seal fishery of this past spring tends to confirm the opinion that resource of the colony has materially declined."[57] The following year, after a better catch, the same body expressed its opinion: "The more general success of the seal fishery this year is calculated to revive hope that the resource is still of considerable worth."[58]

Even in participation in international exhibitions, Newfoundlanders' own perception of the seal fishery had changed as well. In 1853, for the exhibition in New York, Newfoundland had sent a display which included "amongst the articles to be sent" an exhibit described as follows:

> ... a representation of a vessel at the Seal Fishery, executed by Mr. William Knight, of this town. We believe that nothing of the sort has ever yet been attempted here; and we would direct attention to this specimen of handiwork, not merely as an elaborate and skilful production, but as giving so complete an ideal of seal catching as to be true almost to life. Moored to the ice, in the center of a surface seven feet square, is a brig rigged craft, (named the *Governor Hamilton*) built of mahogany, chocked and sheathed; she is three feet long on deck, eight inches beam, built on a scale of 1 inch to 3 feet, giving a brig 108 feet long, 24 feet beam and about 160 tons. She is fitted out with the usual accessories of a sealer, having ten punts, with oars, batts and gaffs, iron cambouse [sic], water casks in the hold, pound boards etc. She is an excellent model and the exact counterpart of a superior class of sealing vessels. Seals (unmistakeable) are scattered in all directions and are of every variety and the "hands" are busily employed in killing, sculping, hauling and hoisting on board the brig, whilst a few unfortunate fellows are being treated to a cold bath between the ice, their ship mates helping them out. The ice is represented by plaster of Paris, upon blocks of wood and the whole is fastened upon a wooden bed divided into three parts ...[59]

In 1867, on the other hand, the articles sent to the exhibition in Paris were much more modest and consisted of two much smaller ship models, some stuffed seals, seal skin clothing and boots, and sealing gear.[60]

Because of the low yields on several occasions during the 1860s, people were becoming increasingly concerned about the seal stocks. This concern was reinforced by the increase in the spread of global information. For example, in 1871, a local newspaper published an excerpt from the *Montreal Daily News* which pointed out that in Newfoundland, "Great fears are entertained that the employment of so many steamers will over work the [seal] fishery and in the end exterminate the seals as has happened in the South Sea Seal Fishery."[61] While, the original writer did not believe such a fate would befall the Newfoundland seal fishery immediately and that "the present generation will catch all they please," he was not so sure about the future and concluded that, in any event, "... posterity will be left to take care of themselves."

Again, the Newfoundland seal fishery attracted international attention when, in 1873, the *Times* (London) published a letter which was reprinted in a local paper. This letter referred to the "yearly diminution of seals ... observable off the coast of Newfoundland."[62] Moreover, the writer blamed this diminution on specific practices. One of these major causes, he wrote:

> ... is the habit sealers have of panning—that is killing and leaving on floes of ice, numbers of seals which they are unable afterwards through their vessel drifting away, to get on board, the result being, boundless waste; ... [and another cause is] the custom many steamers have lately adopted of making the second trip in the same spring to the ice in which usually only old seals are met with, so that both young and old seals are destroyed together.

The writer suggested that "if legislation could be applied to these two causes there is no doubt the supply of seals would continue almost illimitable."

The panning of seal pelts was viewed as a major reason for the decline in the stocks around this time. A writer in the *Newfoundlander* described all the ways in which panning was wasteful:

> No greater injury can possibly be done to the Seal Fishery than that of bulking seals on pans of ice by the crews of Ice hunters. Thousands of seals are killed and bulked and never seen afterwards ... and very often the men pile from five hundred to two thousand in each bulk, which are from one to two miles apart. Care is taken that flags are stuck up as a guide to direct the men where to find such bulked seals. So uncertain is the weather, the shifting about of the ice, as well as heavy falls of snow and drift, that such bulked seals are never seen again by the men that killed and bulked them, as the vessels and steamers are often whirled or driven by gales of wind out of them and wheel'd or driven into another spot where the men again commence killing and bulking as before. In many instances it has happened that the crews of vessels have killed and bulked *twice* their load....[63]

The writer continued in the same vein to detail: how vessels were blown off course, away from their pans of seal pelts; how pelts left out on pans often became frost burnt or sun burnt and lost value; how the ice often raftered and turned over, sending all the pelts to the bottom. The author concluded that "ten thousand pounds cy. [local currency; about £8,000 stg.] worth of seals are lost to the country each sealing voyage by the present system carried on by sealing masters and crews."

Although legislation to resolve some of the problems observed in the seal fishery had been initiated as early as 1866,[64] it was not until 5 May 1873 that the Newfoundland Government finally responded to the problems of the seal fishery by passing "An Act to Regulate the Prosecution of the Seal Fishery."[65] This act was primarily concerned with allowing the white coats to grow as large as possible (but still remain helpless on the ice) before slaughtering them. Therefore, sealing steamers were prohibited from leaving port for the ice fields before 10 March, and sailing vessels were prohibited from leaving before 5 March. The original bill had included provisions aimed at preventing the panning of pelts and the harvesting of small white coats (or 'cats'), but these clauses did not survive the legislative process. In fact, the question of 'property rights' to seal pelts left on the ice became a very contentious issue, with strong views being expressed on both sides.[66] However, although the bill had strong support and passed, there were opponents to this modest measure as well. A local newspaper, the *Times*, argued, beginning in 1867, when the issue was first raised in a serious way, that prohibiting vessels from leaving port until a certain date could result in their being barred in the harbours by ice.[67] In other words, vessels had to be free to leave port at the first available opportunity. There was a certain amount of logic to this argument, but the advantages to be gained by the legislation far outweighed the disadvantages.

In fact, the inadequacy of this legislation and the growing fears of the Chamber of Commerce were both readily apparent in the Chamber's annual report in 1874. Apparently, the seals had given birth later than usual, and in spite of the fact that ships were prevented from leaving port until the dates established by legislation, the white coats were found in a very immature state, and cargoes of under-sized seals were harvested. The Chamber concluded that other measures would be needed to prevent this from happening again and that "further legal protection must be afforded to prevent this valuable fishery from becoming a thing of the past."[68] The Chamber was also quite concerned about the slaughter of the older, breeding seals:

> The destruction of old seals, particularly by steamers on second trips, is considered by many to be more injurious to the future of the fishery than killing the young in the first of the season; and the startling and exceptional features of this spring, already referred to, are by some attributed to the great number of old seals destroyed in the two immediately preceding seasons.[69]

This refrain was heard on a regular basis, with the Chamber emphasizing it again and again, pointing out that long-term losses arising from killing old seals far outweighed the short-term gains.[70] Then, in 1878, there was a re-occurrence of the 1873 season, when the white coats were found in an immature state late in the season and many were slaughtered, resulting in a considerable potential loss to the industry.[71] Finally, the government moved, and in 1879, an act was passed which included the clauses of the previous act but which specifically banned the killing of cats. Any seal pelt weighing less than 28 pounds (avoirdupois—about 12.7 kg.) would be considered as that of an immature or cat seal. No more than 5 per cent of the total pelts in any ship's cargo was to consist of cats, and vessels were to be fined $4 per pelt for every cat pelt over the 5 per cent limit. It was hoped that this would solve the problem, because just a few days difference in the age of the white coats could make a considerable difference to the weight of the final pelt.

Although the catch was above average in 1878, a general air of pessimism was beginning to permeate the industry. In May of that year, the editor of the *Newfoundlander* felt compelled to respond to an anonymous letter writer who had advocated terminating the industry altogether. The editor wrote a serious reply:

> Observer's [the letter writer's] doubts of the value of the Seal fishery to the country are, we believe, shared by not a few; but the experiment of abandoning it from a conviction of its worthlessness is probably remote. A lottery it may be and is; but whilever it tempts with large prizes so quickly secured, when secured at all, it will command capital and labour, whatever the result of any other pursuit.[72]

However, after the low catches of 1880, 1882, and 1884 and in response to the weakening price, the steamer owners attempted in 1885 to come to an agreement to terminate the seal fishery on 30 April. Besides protecting the old seals, it was felt that this would allow the fishermen more time to prepare for the summer cod fishery; delays in returning to port at a reasonable date caused regular complaints among the sealers/fishermen. Some owners also decided that if the agreement was not followed by all owners, they would approach the government with a request for the necessary legislation.[73]

As the economics of the sealing industry continued to worsen, the government moved to prohibit the wholesale, indiscriminate slaughter by passing legislation in 1887 dealing with the question of property rights to seal pelts. Basically it stipulated that seal pelts which were not under the actual charge of a particular individual or individuals were to be considered abandoned. (However, this proved to be a very contentious provision and it was repealed in 1889.) The 1887 legislation also re-affirmed the ban on killing seals before 12 March and added a ban on killing after 20 April. In addition, in response to the prevailing mood of the majority of participants, this legislation prohibited steamers from going on a second trip to the seal fishery, with the exception of cases where steamers were forced to return to port prematurely.

This was an important point, because although all vessels could stay out until late April, the pelts would begin to render out about then. The weight of the pelts, combined with the temperature, would hasten or delay this process, and crews often iced the pelts well to delay it; but it would begin to happen sooner or later, and once it started, much of the cargo could be lost. The oil would mix with the bilge water and be pumped out. Therefore, no vessel that had been successful in acquiring even a small load of white coats could afford to stay out long in search of the old seals because "when young fat commences to run it goes like butter before the sun."[74]

In the broader context, the Newfoundland government was trying to find new markets for its seal and cod products and had been negotiating with the United States of America over a Reciprocity Treaty which would lead to freer trade between the two. However, opposition from Canada, which convinced Great Britain that such an agreement by Newfoundland would undermine the Canadian cause, resulted in Britain's applying its veto. The *Evening Telegram*, supporting the government of Robert Bond, brought the issue to the attention of the sealing interests:

> The sources of prosperity for a country are a kind Providence, an industrious people, and a wise Government. This year a kind Providence has blessed us with abundance of seals. An industrious people has brought in big loads. A wise Government has succeeded in opening up a splendid market for seal oil and seal skins.
>
> But Canadian meddling and Imperial tyranny, aided by the treachery of a "kept" agent of Canada, have largely neutralized the blessings of a bountiful Providence and the labors of an industrious people. Men of Bonavista, one of your representatives, the gallant Capt. Blandford, has been on the ice-floes leading you to the big loads of fat. Another of your representatives, the unscrupulous Morine, has spent the spring, not in trying to find good markets for you, but in depriving you of markets. He has done his best to keep you from doubling your bill. If Bond's Reciprocity Convention had gone through, your bill would be twice as big as it is. Will you put up with the Canadian spy and informer forever? If you do, don't you think you deserve the treatment you get from him?[75]

Although the newspaper's partisanship is blatant, there was considerable disappointment, both within Bond's Government and among the public, over the failure of the reciprocity negotiations. As Chapter 1 points out, the decreasing demand for fish oils in the latter part of the nineteenth century led to ever lowering prices. Bond was anxious to come to an arrangement that would allow Newfoundland's products to enter the United States on more favourable terms and stimulate the colony's export trade. He would waste considerable time and effort, throughout his whole public life, in a futile effort to see this objective through to a successful conclusion.

In the meantime, panning continued to focus people's attention. This was probably because there was nothing quite as obvious and concrete as a pan of pelts going to waste. Eye witness reports were difficult to overlook (especially by the media) and seemed to make a greater impression on the general public than the more abstract explanations given for the decline of the seal population. For example, Captain Thomas Dicks, "one of the principal schooner-holders out of Channel" who engaged in sealing in the Gulf, wrote:

> I left here on Wednesday, the 20th inst. [March], for the ice, and shortly afterward spied a large bulk of seals off Duck Island, panned, and I also saw the smoke of the steamer *Panther* off Cape Ray. After I passed them, a couple of men from Mouse Island, went and took a tow from the bulk which they said, contained upwards of 600 seals. A few more bulks were seen by the people here but the ice rafted over them. If the law permits the steamers to destroy the seals in that manner, our Seal Fishery in the Gulf will soon be a thing of the past.[76]

Nothing was done about this problem, although further legislation was passed in 1892 to delay the departure of steamers until 6:00 a.m. on 12 March and to establish a legal time frame for the hunt—14 March to 20 April. Once again, the prohibition on second trips was re-confirmed. In the following year, it became illegal to kill seals on Sunday. (While conservation was a factor in the passage of this legislation, the maintenance of cordial working relationships was just as important—see Chapter 4). During that same year, 1893, the time and date of departure of the sealing steamers was changed to 2:00 p.m. on 10 March, or 9 March if 10 March should be Sunday. Then, in 1895, the legislature repealed the prohibition on a second trip (but no man could be forced to return to the ice fields on this second trip). At the same time, the period during which seals could be killed was extended to 5 May. (Apparently, the ban on second trips and on the killing of seals after 20 April did not suit the St. John's interests, who felt that such measures were unnecessary and costly; and as will be discussed below, St. John's businesses dominated the industry by the mid 1890s.) However, in 1898, the government prohibited second trips once again, adjusted the period during which seals could be killed to between 12 March and 1 May, inclusive. The time of departure was changed slightly to 8:00 a.m. from 2:00 p.m. And another Act in 1903 made it illegal to bring a cargo of seal pelts into a Newfoundland port if the ship involved had not cleared from the colony.

Another conservation issue which arose frequently was the practice of shooting seals. This became a more important issue in the early twentieth century because the older wooden wall steamers became more dependent on this method when they discovered that in the quest for whitecoats, they could not keep up with the steel steamers which been introduced in 1906. However, shooting was wasteful and many condemned it. In 1914, George Whiteley pointed out that only one out of every four seals shot was actually taken. He said that he and William Winsor had shot 450 cartridges one day and bagged 90 seals. He went on to describe how he had seen, on many occasions, thousands of old seals fleeing before gunners, with blood streaming out of bullet holes and "some with as many as three spurts of blood coming from parts of the body." He even mentioned the "awful cruelty of indiscriminate shooting of old seals at a time when every animal is supposed to need protection most."[77] However, the lobbying power of the firms with big investments in wooden walls was such that any attempt to ban shooting was doomed to failure. The legislative efforts to cope with the decline in the seal stocks were feeble and enjoyed very little success.[78]

By the end of the century, the seal fishery was virtually ignored by the officials who compiled the fishery reports which were published in the *Journals of the House of Assembly of Newfoundland*. This was a case of public perception finally falling in line with reality. One of the exceptions was a short report in 1905, as follows:

> The Newfoundland hair [sic] seal fishery is the largest of its kind in the world, and in point of seniority comes next to the cod fisheries. For many years its importance was second only to that of the codfishery, but since the introduction of steam, and the total disappearance of sailing vessels, with the exception of a few small schooners employed in what is termed "the Gulf Fishery" its value to the commonwealth has seriously declined ....
>
> It is noteworthy that this industry has been preserved from the total extinction with which it was threatened when second trips of steamers were permitted, whereby many thousands of breeding seals were destroyed, a large proportion escaping fatally wounded.[79]

Thus, by the beginning of the twentieth century, the seal fishery's contribution to the total Newfoundland export trade had declined to the extent that it had not, in fact or fiction, a major role to play in the colony's economy.

Summary: 1860-1914

The economy of Newfoundland after 1860 suffered greatly from the decline in the seal fishery as production and prices declined. Although the cod fishery generally performed well in terms of production, the decline in prices after c.1880 created considerable problems. Furthermore, efforts to diversify the economy through the exploitation of other resources on land and sea were not sufficiently successful to compensate for the problems brought about by the deepening failure of the seal fishery. Only by massive borrowing and rapidly increasing the public debt was the Newfoundland economy able to survive.

General Summary

In its contribution to the economy from the beginning of the nineteenth century, the seal fishery suited perfectly the situation in which the cod fishermen in Newfoundland found themselves in c.1815. It did not interfere with the cod fishery; in fact, it complimented and supported it—and allowed an abnormally large number of cod fishermen, especially employee fishermen, to live permanently in Newfoundland. It employed the same men and the same ships and used the same ports and infrastructure as the cod fishery—especially the Labrador branch of the industry. Everything changed, however, as the resource became depleted, and the situation which prevailed in the second half of the century, especially after the mid 1880s, was far different from that of the early years. In summary, the seal fishery had made an unparalleled contribution to Newfoundland and nothing—including, agriculture, mining, paper milling, railroads or free trade with the United States—could take its place. As will be seen in later chapters, its impact extended far beyond its vital role in the economy.

NOTES

1. *PP*, 1817, VI, 465-514. See Shannon Ryan, "The Newfoundland Cod Fishery in the Nineteenth Century" (MA thesis, MUN, 1972) for much of the information in this chapter.
2. *PP*, 1817, VI, 485. Contrary to what has been concluded by some writers, merchants in Newfoundland promoted and encouraged agriculture.
3. A letter from a Mr. Knight, n.d., Sir John T. Duckworth Papers, 1810-12. PANL.
4. Prowse, *Newfoundland*, p. 398.
5. *Mercantile Journal*, 14 February 1817.
6. Ryan, *Fish out of Water*, pp. 258-60, "Newfoundland's Saltfish Exports and Population: 1803-1914." This table also contains statistics on the per capita exports of saltfish. However, please note that beginning with the figures for 1827, "Per Capita" statistics have been listed in an incorrect column.
7. Ryan, *Fish out of Water*, pp. 258-60. The figure for summer fishermen seems abnormally high and probably includes summer residents who left the island for the mainland at the end of the fishing season. The present writer divided the number of residents recorded in the 1911 Newfoundland census (242,619) into the average number of cwts exported annually during 1905-14 (1,407,000) and obtained the figure 5.8.
8. See Ryan, *Fish out of Water*. The value of saltfish in the international market place varied during the century, but it never returned to the inflated high prices received during the Napoleonic War. In fact, there was a serious decline in fish prices beginning in the 1880s.
9. See Ryan, *Fish out of Water*, for a discussion of this subject.
10. Ryan, *Fish out of Water*, p. 39.
11. Ryan, *Fish out of Water*, pp. 261-2.
12. The governors reported Newfoundland's population annually until 1830. See Ryan, "Abstract." The local government ordered censuses to be taken in 1836, 1845 and 1857. See the *Census of Newfoundland*.
13. Richard Whitbourne, "A discourse and discovery of New-found-land" (London: 1620, 1622 and 1623), in Cell, *Newfoundland Discovered*: p. 121.
14. Ryan, "Abstract."
15. Head, *Newfoundland*, p. 14.
16. Head, *Newfoundland*, p. 199.
17. Head, *Newfoundland*, p. 201.
18. Prowse, *Newfoundland*, p. 398.
19. Prowse, *Newfoundland*, p. 396.
20. John Mannion, "Patrick Morris and Newfoundland Irish Immigration," *Talamh an eisc: Canadian and Irish Essays*, ed. Cyril J. Byrne and Margaret Harry (Halifax, 1986), pp. 180-202. Mannion quotes from the *JHA*, 1837, "Report of the Committee on Agriculture," Patrick Morris, Chairman.
21. Robert MacKinnon, "The Growth of Commercial Agriculture around St. John's, 1800-1935: A Study in Local Trade in Response to Urban Demand" (MA thesis, MUN, 1981).
22. MacKinnon, "Agriculture," p. 34.

23 MacKinnon, "Agriculture," p. 36. This description also applies to the farms the present writer observed in the Harbour Grace area during the 1940s and 1950s.
24 'Water Street' was the name given to the street that gradually developed along the shoreline of each harbour. Firms built wharves, warehouses and retail shops on 'Water Street'. The almost exclusive use of wood in local construction made fire a major hazard. After the fire of 1846, when most of St. John's was destroyed, the merchants involved rebuilt their highly flammable wooden seal oil vats on the south side of the harbour. However, this did not save the town from an even more disastrous fire in 1892.
25 *Newfoundland Blue Books*, 1826. The information on 1838, 1839 and 1840 is taken from the appropriate *Blue Books*.
26 *JHA*, 1857.
27 *Newfoundlander*, 3 May 1852.
28 *Newfoundlander*, 12 March 1829.
29 *Newfoundlander*, 8 July 1830.
30 *Public Ledger*, 15 February 1831.
31 *Royal Gazette*, 8 March 1831.
32 Henry Bliss, *Statistics of the Trade, Industry and Resources of Canada and the other Plantations in British America* (London, 1833), pp. 70-1. See also *Royal Gazette*, 2 July 1833.
33 Philip Tocque, *Wandering Thoughts or Solitary Hours* (London, 1846), p. 188.
34 *Newfoundlander*, 5 March 1846.
35 *Newfoundlander*, 11 May 1846.
36 *Newfoundlander*, 4 March 1847.
37 *Newfoundlander*, 18 March 1847; quoted from the *Conception Bay Herald*, 3 March 1847.
38 *Newfoundlander*, 20 April 1848.
39 *Newfoundlander*, 5 April 1849.
40 *Times*, 2 March 1853.
41 *Newfoundlander*, 22 December 1853. Quoted from the *New York Freeman's Journal*, "Report from the Exhibition at the Crystal Palace, New York."
42 Prowse, *Newfoundland*, p. 450.
43 *Evening Herald*, 3 March 1897.
44 The following information on the saltfish industry has been taken from Ryan, *Fish out of Water*.
45 Information on fish prices has been taken from Ryan, *Fish out of Water*, pp. 262-3.
46 Beginning in 1865, government financial records were recorded in dollars and cents.
47 *JHA*, 1863. For an excellent overview of the history of agriculture in Newfoundland, see Catherine F. Horan, "Agriculture," *ENL*. Much of the general information in the following discussion has been taken from this account. See also, Sean Cadigan, "The Staple Model Reconsidered: The Case of Agricultural Policy in Northeast Newfoundland, 1785-1855," *Acadiensis*, XXI, 2 (Spring 1992), 48-71.
48 *JHA*, 1863. "Proceedings of the Select Committee of the House of Assembly on Agriculture." This committee examined witnesses and received briefs throughout March 1863.

49 For an excellent account of the growth of small industry, see John Lawrence Joy, "The Growth and Development of Trades and Manufacturing in St. John's, 1870-1914" (MA Thesis, MUN, 1977).

50 See James K. Hiller, *The Newfoundland Railway: 1881-1949* (Pamphlet, St. John's, 1981).

51 See F. F. Thompson, *The French Shore Problem in Newfoundland* (Toronto, 1961).

52 See William Reeves, "The Fortune Bay Dispute: Newfoundland's Place in Imperial Treaty Relations under the Washington Treaty, 1871-1885," (MA thesis, MUN, 1971); William Reeves, "Our Yankee Cousins: Modernization and the Newfoundland-American Relationship, 1898-1910" (Ph.D. thesis, Maine, 1987); and David Davis, "The Bond-Blaine Negotiations: 1890-1891," (MA thesis, MUN, 1970).

53 *Royal Gazette*, 26 April 1853.

54 *Royal Gazette*, 25 December 1855.

55 *Royal Gazette*, 24 February 1858.

56 *Patriot*, 18 February 1861.

57 *Royal Gazette*, 1 August 1865.

58 *Royal Gazette*, 14 August 1866.

59 *Royal Gazette*, 31 May 1853.

60 *Royal Gazette*, 22 January 1867.

61 *Royal Gazette*, 12 December 1871.

62 *Royal Gazette*, 4 March 1873.

63 *Newfoundlander*, 28 February 1873.

64 *Times*, 28 March 1866.

65 For Acts relating to the seal fishery during this period, see *Statutes of Newfoundland* (St. John's, 1873-1916).

66 *Times*, 15 February 1873.

67 *Times*, 13 March 1867.

68 *Royal Gazette*, 11 August 1874.

69 *Royal Gazette*, 11 August 1874.

70 See *Royal Gazette*, 20 August 1878 and 21 August 1881 for examples.

71 *Royal Gazette*, 19 August 1879.

72 *Newfoundlander*, 10 May 1878.

73 *Evening Telegram*, 2 March 1885.

74 *Evening Telegram*, 6 May 1899.

75 *Evening Telegram*, 7 April 1891.

76 *Evening Telegram*, 5 April 1890.

77 *Evening Telegram*, 7 March 1914. A letter to the editor.

78 There are aspects of the legislation covered here which deal with other matters, particularly working conditions.

79 *JHA*, 1905, p. 155.

CHAPTER 3
Vessels and Ports

THE STORY OF the seal fishery's place in the overall economy of Newfoundland does not present the whole picture of the enterprise during the period up to 1914, because the seal fishery was an industry with definite organizational, regional and technological features, which had an ever-changing influence on the ship owners and ports directly and indirectly involved. The seal fishery was, first of all, established in only certain harbours which supplied processing facilities and shipping, and attracted further investments in labour and capital. This chapter discusses the types of involvement of the sealing centres in this industry, as well as the reasons for and the consequences of their participation. At the same time, the evolution of the seal fishery during 1814-1914 will be examined in terms of the changes that occurred in the size, shape and operation of this industry—physically and fiscally—and the subsequent effects upon Newfoundland. This examination will be organized around the three types of fleets to prosecute the seal fishery: the sailing vessels; the wooden wall steamers; and the steel fleet.

Age of Sail

Conception Bay shows the way

The sealing fleets during the early years of the spring seal fishery were made up of small vessels and shallops carrying small crews. As the Introduction noted, in 1803, for example, 19 vessels from St. John's averaging 39 tons and carrying an average crew of 10 men went 'to the ice' in the spring seal fishery. Conception Bay outports sent 21 vessels and 14 boats (shallops) with an average burthen of 46 tons and 31 tons, carrying an average crew of 11 and 9 men, respectively. Twenty-three other vessels sailed from Trinity and Bonavista. Conception Bay quickly dominated this industry.

Although, as the figures above show, St. John's was not late in becoming involved in the seal fishery, this new industry did not expand in that port as rapidly as it expanded in Conception Bay. From 19 ships carrying 188 men in 1803, the St. John's fleet grew to 32 ships (68 per cent increase) with 545 men by 1819.[1] Meanwhile, the Conception Bay fleet expanded far more dramatically, from 35 vessels and shallops carrying 369 men to 140 vessels (400 per cent increase) with 2,203 men during this same period. In 1803, the number of men from St. John's was half the number from Conception Bay, while the number of vessels from St. John's was a little over half that from Conception Bay. In 1819, the total number of men in the crews from St. John's was one quarter the number of men going to the ice from Conception Bay; and the number and total burthen of the ships of the latter was four times as large as that of the former. As Tables

3.1, 3.2 and 3.3 illustrate, this imbalance was most extreme in 1820, when St. John's sent 40 vessels with 666 men to the ice floes, while Conception Bay sent 170 vessels and 3,082 men. Again, the size of the St. John's branch of this industry was less than one quarter of that of Conception Bay. As Table 3.1 establishes, the number of pelts brought into port by each of these fleets shows an even greater imbalance in favour of Conception Bay. Although the figures on the St. John's sealing ships are incomplete during the period leading up to 1819, the figures on the annual catch of seals are complete, and these show that the capital's seal fishery stagnated, while that of Conception expanded during this period. For example, in 1816, the catch of the Conception Bay fleet amounted to 103,359 seals, while that of St. John's, to only 27,485; in 1818, the figures were 118,228 and 24,735, respectively; in 1819, as indicated in the tables, the figures were 179,051 and 39,052, respectively; and finally, in 1820, they were 152,502 and 24,132. The Conception Bay outports had taken an early lead in this industry.

The reason why St. John's was slower to take greater advantage of this new industry is probably to be found in the expanded commercial role that the capital had assumed during the Napoleonic and Anglo-American Wars. As already pointed out Governor Gower wrote in 1805 about the changing nature of St. John's as follows:

> This Harbour is no longer a mere fishing station, built round with temporary Flakes, Stages, and Huts of trifling value, but ... it is a port of extensive Commerce ... importing near two thirds of the supplies of the whole Island, and furnished with extensive Store-Houses and Wharfs for trade, containing a quantity of Provisions, Stores for the Fishery, British Manufactures and West Indian Produce, as well as Fish and Oil ready for exportation....[2]

The war turned St. John's into a commercial centre because it was well protected by the British navy, and many outport merchants felt safer importing and exporting through this port. While the defeat of the French and Spanish navy at Trafalgar removed that menace to shipping, the outbreak of war with America in 1812 created more serious concerns among shippers, and most continued to ship through St. John's. Not only did this activity create enormous local employment, but the presence of the navy, itself, produced economic spin offs. The Reverend Lewis Anspach, who resided in Harbour Grace during 1801-13, wrote of St. John's during this period:

> The increased circulation produced by the enlarged naval and military establishments, the numerous captures brought in by the several ships on that station, and the general practice of merchant vessels, freighted in the out-harbours, coming round to Saint John's for the purpose of joining convoy, gave to that town an unusual degree of prosperity and of consequence.[3]

Governor Keats, in 1813, also commented on the fact that "St. John's became the emporium of the Island in consequence of this extended war, with a population of nearly 10,000 inhabitants, seems to have grown out of its character from a Fishery to a large Commercial Town."[4] Thus, St. John's was not in a position to take immediate advantage of the new seal fishery and needed time to adjust to the new circumstances in which it found itself after the war.

The ports of Conception Bay, on the other hand, thrived as fishing ports, rather than commercial ports, during the wars and were much better placed to expand into the seal fishery when the opportunity arose. Governor Gower, in his report on the year 1804 (submitted in 1806), wrote of the seal fishery and Conception Bay as follows: "The inhabitants of Conception Bay who are particularly active and enterprising in it, have greatly improved their circumstances by it, and have this year fitted out double the number of vessels that they employed in it before."[5] The fact that Gower felt able to make so firm a judgement this early in the industry's history indicates the strong impression that the pursuit of seals by Conception Bay vessel owners and fishermen had made on him.

Not only did Conception Bay take the lead over St. John's in the seal fishery during this period, but it outstripped by far the other regions. Although the reported list of St. John's sealing ships is not complete for 1816, the figures for the rest of the colony seem to be thorough, and an examination of these reports is essential in order to understand what was happening at this time. In 1816, Conception Bay's combined fleet consisted of 93 ships while that of Trinity contained 6, Bonavista's contained 4 and Ferryland's contained 5. In 1818, Conception Bay sent 100 vessels to the ice, while Trinity sent 5 and Bonavista, 4; in 1819, the numbers were: Conception Bay, 140 vessels; Bonavista, 4; Trinity, 5; Ferryland, 1; and, as has been shown, St. John's was the headquarters to a fleet of 32 vessels. Furthermore, one can look at the number of men who went to the ice on these sealing vessels to see Conception Bay's dominance at this time. In 1816, Conception Bay sent 1,558; Trinity Bay, 91; Bonavista, 56; and Ferryland, 55.[6] In 1818, the number of men from Conception Bay was reported to be 1,524; from Bonavista, 58; and from Trinity, 84. In 1819, Conception Bay sent 2,203; St. John's, 545; Bonavista, 59; Trinity, 67; and Ferryland, 12. Looking at 1820, Conception Bay sent 3,082 sealers to the ice and St. John's sent 666 for a total of 3,748; Bonavista, Trinity and Ferryland together sent 270. Finally, Conception Bay's prominence is reflected in the percentages of the sealing industry in terms of pelts taken and the vessels and men involved in 1819 and 1820. In the former year, the Conception Bay ports produced 79 per cent of total pelts taken in the spring seal fishery; they owned 77 per cent of the entire Newfoundland fleet; and they employed 76 per cent of all fishermen-ice hunters. For St. John's, the figures were, respectively, 17, 18 and 19 per cent; and for all other ports together, 4, 5 and 5 per cent. In 1820, the Conception Bay ports' share amounted to 81 per cent of the total spring kill, 75 per cent of the fleet originated from there, and 77 per cent of all men involved were engaged by ship owners in these ports; for St. John's, the figures were, respectively, 13, 18 and 17 per cent; and for all other ports combined, 6, 7 and 6 per cent. From the statistical information alone, it is obvious that Conception Bay was the centre of the early Newfoundland seal fishery and, by 1820, controlled at least 75 per cent of the seal fishery (and a similar proportion of the Labrador cod fishery).

This perception of Conception Bay's paramount importance is apparent from the few contemporary reports relating to this issue. Anspach described it as follows:

> Conception Bay is undoubtedly the first district in the Island of Newfoundland, on account not only of the number of commodious bays, harbours, and

coves, which it contains, but also of the general ease and independence of the greatest proportion of its numerous planters, of the industry and intrepidity of the masters and crews which it sends yearly to the seal and cod fisheries, and of the very superior degree in which it contributes to the importance and value of Newfoundland.[7]

Later in the century, Joseph Hatton and Reverend Moses Harvey referred to Conception Bay as "the most populous and commercially important of all the seats of population...."[8]

Unfortunately for the student of the early history of Conception Bay, very few of the annual governors' reports (and none at all during this early period) distinguish the various harbours and settlements in this Bay. For statistical purposes in the reports, the Trinity Bay ports of Trinity, Old Perlican and New Perlican were often recorded separately; the harbours of Bonavista Bay were usually divided into two lots—'Bonavista and King's Cove', and 'Greenspond, Salvage, and Gooseberry Island', and the Fogo district was usually divided into five sub-districts—'Fogo', 'Twillingate and New World Island', 'Bard Isles [Barr'd Islands]', 'Joe Batt's Arm' and 'Tilting (sometimes incorrectly referred to as Tilton)'. It is rather inexplicable that these lightly populated harbours and islands to the 'Northward' warranted individual reports, whereas the more historically distinct and heavily populated ports on the north shore of Conception Bay were reported on as a unit. As a consequence, valuable information that could help researchers understand the history of the various ports in this bay has been lost. Nevertheless, the literary evidence provides a good, if non-quantitative, indication of the comparative importance of the different harbours in Conception Bay during this period.

It is obvious that Harbour Grace was the leading centre among the Conception Bay 'outports'. Anspach described this two-hundred-year-old community at the beginning of the nineteenth century as follows:

> Harbour Grace is the principal town of this district; it has several steep and barren rocks in its entrance, a bar which renders it dangerous at particular seasons to large ships, and an extensive beach, round which there is a capacious harbour where ships ride in the greatest safety.[9]

Furthermore, Harbour Grace became the mercantile, religious and judicial centre for much of the island north of St. John's, and its importance is quite apparent by the 1830s. Anspach described Carbonear as the "next town in importance," and then went on to describe the other chief outports in this bay:

> Farther up from Harbour Grace, as far as Holy-Rood, are several considerable settlements, formed on the borders of deep bays, which are separated by high perpendicular rocks of two or three leagues in length and scarcely more than a mile in breath.[10]

No doubt, he was describing Brigus, Bay Roberts, Port de Grave, Cupids, Harbour Main and Spaniard's Bay.

As indicated at the beginning of this chapter, the sealing fleet that originated in Conception Bay and St. John's consisted of small vessels carrying only a few crew members each. In the early days of the seal fishery, the burthen of the vessels employed was between 30 and 60 tons, and they averaged 10 to 16 men in their crews.[11] W. A. Munn wrote: "At first there were shallops, decked in fore and aft with movable deck boards in the centre of the boat. The shelter cuddies at each end gave the crews some protection when they remained out over night."[12] The early vessels were small with schooner rigging, but as Tables 3.2 and 3.3 illustrate, the size of the vessels increased during this period, and the average number of men carried increased accordingly. By the late 1820s, the average burthen of sealing vessels reached 60 tons, and the average size of each crew was about 20 men. Munn informs us that the style of construction of the early sealing vessels changed. He wrote:

> The small 'Western Boats' with their apple-cheek bows soon developed into the sharp stem with glancing bows that rose on the ice pans of its own accord, and helped to crush the floes with the weight of the heavier vessel [sic].[13]

It seems the problem lay in trying to compromise and build a vessel that was suitable for sealing which, at the same time, could be used in the Labrador cod fishery. Munn and Anspach agree that "the whole plan of the Seal Fishery was altered by the enterprise and industry of the inhabitants of Conception Bay, who contrived a method to conciliate the interest of both the seal and cod fishery...."[14] Meanwhile, building larger and stronger vessels became the trend, and William Munden of Brigus, who is referred to as the man "who taught them [the people of Brigus] all their business" was the first to build a sealing vessel of over 100 tons when he built the *Four Brothers*, a schooner of 104 tons, in 1819.[15] The second vessel of over 100 tons to be built for the seal fishery was the schooner *Experiment*, of 108 tons burthen, which was built in Carbonear for the firm of Gosse, Pack and Fryer. The *Experiment*, under Captain Heighton Taylor, brought in an unusually large cargo of over 5,000 pelts on its first voyage.[16] Munn credits another Conception Bay man, also from Carbonear, for solving several problems associated with the construction of a good sealing vessel. He says:

> We must mention particularly Richard Taylor of Carbonear, who went by the name of "The Thoughtful Man" who first thought of protecting [the] bows with 'Iron sheathing'. He was also the man who put 'false beams' in to strengthen the hull during the sealing voyage. It was soon found that the hold of the vessel had to be divided into pounds. When the voyage was prolonged, the seal fat began to run, and the cargo was liable to shift with the motion. Quite a number of vessels were lost in this way before this method of preventing the shifting of the seals was adopted.[17]

Another contemporary report points out that the construction of two vessels of 120 tons each in Conception Bay in 1825 marked the turning point in the general acceptance of bigger vessels, and the trend in this direction increased.[18]

The vessels were schooner rigged in the early years, but it was soon discovered that it was necessary to make the masts higher in order to catch as much wind as possible among the ice floes. Then, a temporary square top sail was installed on the foremast to make it easier to ram the ice and to reverse when necessary. Munn says that a schooner so rigged was referred to as a 'Beaver Hat Man'.[19] The rig of the 'Jack Ass Brig' was, according to Munn, a "square rig on the foremast, one square top sail, and one topgallant sail on the main-mast for temporary use." But it was the real brig with square sails on both masts that became the most common and the favourite.

These ships usually left port after 17 March, and sometimes nearer the end of March at the beginning of this industry's history. For example, in 1803, vessels and shallops from Conception Bay were engaged in the spring seal fishery during the period from 20 March to 15 May, and vessels from St. John's were engaged from 22 March to 16 May.[20] A report published in 1849 lists the dates of departure of the St. John's fleet for most years up to 1837[21] and indicates that up to and including 1828, the fleet from the capital usually left port after 17 March; in 1827, the date was 22 March, and in 1828, it was 1 April. One factor that could have contributed to these late departure dates was the continuing practice of chasing the old seals as they swam north among the loose ice in April; another explanation may relate to the fact that St. John's harbour (as well as the other harbours in Conception Bay and to the northward) froze over during the winter months and were late thawing. In fact, the local government passed an Ice Cutting Act in 1837 which provided for the establishment of an 'Ice Committee', and it was its responsibility to assemble a work crew and to oversee the cutting of a channel through the harbour ice from the docks to open water whenever ice prevented the sealing vessels from leaving port.[22] In any case, the fleet left St. John's on 7 March in 1829, and the first departures in 1835 and 1836 were on 1 March. By the late 1820s, everyone involved was aware that in order to fully utilize the seal populations, the vessels had to arrive among the ice floes before the white coats left the ice and entered the water, generally around the third week of March. Therefore, the fleets began leaving port in early March.

By 1820, the seal fishery was well established, with the Conception Bay fleets taking over 75 per cent of the total number of pelts harvested. When converted roughly into value, using the figures in Table 2.6, this means that in 1819 and 1820 combined, the seal oil itself was worth about £112,000 over the two-year period to the people of this bay—an average of £56,000 per year. Considering the fact that the population of Conception Bay was 13,393 in 1819 and 12,840 in 1820, the sum of £112,000 spread over two years was a substantial addition to the local economy. During these two years, this bay was home to 33 per cent and 30 per cent of the island's population, respectively. This means that less than one-third of the colony's population was producing over three-quarters of the total seal oil and leather exported. Meanwhile, the population of St. John's amounted to 9,400 people in 1819 and 9,251 in 1820, and thus contained 23 per cent and 22 per cent of the colony's total population, respectively. In all, St. John's and Conception Bay contained 55 per cent of the total population in 1819, and 52 per cent in 1820, and accommodated these people in a fairly small, concentrated area because the seal and Labrador cod fisheries allowed for population centralization.[23]

St. John's expands

The figures for the 1820s have many gaps, but as Table 3.1 illustrates, by 1831-33, the industry had grown considerably. The total yield averaged 515,782 pelts during this three-year period; the total number of vessels prosecuting this industry averaged 376 annually; and the total number of fishermen employed averaged 4,457.[24] Conception Bay and St. John's dominated the industry, as one can see, but the latter was growing more rapidly. During this period, Conception Bay, on average, provided 56 per cent of the total fleet, employed 54 per cent of the men and produced 56 per cent of the pelts—a considerable decline, proportionately, from the beginning of the decade. At the same time, the St. John's vessels made up 34 per cent of the total fleet, employed 33 per cent of the men and produced 33 per cent of the pelts—an enormous actual and proportionate increase. This left the remainder of the colony to supply 10 per cent of the fleet, employ 13 per cent of the fishermen and produce 11 per cent of the pelts. Therefore, there had been a dramatic expansion in the St. John's industry during the 1820s, at the expense of that of Conception Bay; to a lesser extent, the smaller ports had also increased their actual and proportionate share of the total fleet, the number of fishermen and the total yield—also at the expense of Conception Bay.

The real growth of the St. John's industry was even greater than these figures suggest, however, because many outport vessels had begun to bring their pelts to the capital, where they were sold and processed, enabling St. John's firms to make additional profits; furthermore, these businesses began to supply outport sealing vessels. The "Accounts of the Spring Seal Fishery" quoted above for the period beginning in 1815 show that in 1826, for the first time, it is evident that Ferryland sold its 2,000 pelts to St. John's.[25] In 1827, outport vessels brought 59,728 pelts to the capital, and 788 tuns of oil were produced from those. In this same year, St. John's produced 1,512 tuns of oil from its own pelts and 788 tuns from the outport pelts, for a total of 2,300 tuns, which was almost as much as Conception Bay's 2,747 tuns of oil.[26] It was likewise in 1828, but the records are incomplete for the next two years, so one can only surmise that this practice continued. In 1831, it was reported that 13,066 pelts from Trinity Bay, as well as 94,916 from other outports, were sold for processing in St. John's. This latter quantity was converted into 1,300 tuns of oil, which, when combined with the town's own production of 2,887 tuns, resulted in a total production of 4,187 tuns—close to Conception Bay's production of 4,273 tuns.[27] The following year, it was reported that Ferryland sold most of its catch[28] of 6,182 seals to St. John's and that Twillingate sold its catch of 13,652 there as well.[29] Finally, in 1833, the last year under discussion here, reports of the catch from Port de Grave, Trinity and Twillingate and Fogo stated that the "greater part of these seals [were] sent to St. John's and there manufactured." When it is seen that Port de Grave harvested 100,302 pelts and manufactured only 590 tuns of oil, whereas St. John's, with a catch of 125,874 seals, was able to manufacture 1,851 tuns, one can see that about 1,000 tuns of oil were produced in St. John's from Port de Grave's pelts alone. In any event, this "Account" indicates that St. John's produced 1,851 tuns of oil from its own pelts and 1,742 tuns from pelts brought from outports.[30] Therefore, in 1831, St. John's manufactured 3,493 tuns of seal oil in all, compared with 4,351 tuns manufactured in Conception Bay.

In 1831, consequently, St. John's and Conception Bay manufactured 7,841 tuns of oil, which made up about 93 per cent of all oil produced on the island;[31] calculating from Table 2.6, the total value came to about £200,000—£185,000 of which was divided between St. John's and Conception Bay. This share of production was vastly disproportionate to the percentage of population living in the two areas. By 1830 (the last year for which population figures are available during this period), the population of St. John's amounted to 15,265, which was 26 per cent of the island's total population, up 4 per cent since 1820; Conception

Bay's population had reached 17,859 people, or 31 per cent of the total number of inhabitants, up 1 per cent on 1820's figures.

Meanwhile, other ports in Newfoundland became involved, in varying degrees, in the spring seal fishery, but it is not always easy to differentiate the produce of the traditional winter seal fishery from that of the new industry. However, enough information is available to give an indication of what was happening. In 1815, according to the "Accounts," Trinity Bay was reported as having sent 5 vessels with a total of 359 tons burthen and carrying in all 76 men to the ice fields.[32] The following year, 6 ships with a total burthen of 408 tons and carrying a total of 91 men were involved; the figures for Bonavista were recorded as 4 ships, 250 tons and 56 men; and Ferryland's fleet consisted of 5 ships, 161 tons and 55 men. Production was as follows: the Trinity Bay fleet—6,793 pelts and 51 tuns of oil; Bonavista—4,306 pelts and 64 tuns; and Ferryland, 13,110 pelts and 13 tuns.[33] A tun of oil generally required between 70 and 80 pelts, which means, that Trinity sold part of its catch elsewhere and Ferryland, nearly all. The records for 1819 are somewhat more complete and give a better indication of the seal fishery north of Conception Bay. In that year, the seal fishery in that area produced as follows: Trinity—43,749 pelts and 609 tuns of oil; Bonavista, King's Cove, Greenspond, Salvage and Gooseberry Island—14,878 pelts and 198 tuns of oil; and Twillingate and New World Island—2,000 pelts and 20 tuns of oil.[34] It is obvious that Trinity did not process all its pelts; but at the same time, one must conclude that these figures contain information on both the winter and spring seal fisheries because the same "Account" gives a separate report on the spring seal fishery, which was prosecuted as follows: Trinity—5 ships with a total burthen of 165 tons and total crews of 84 men who harvested 4,133 pelts that produced 45 tuns of oil; Bonavista—4 ships (tonnage not given) which carried 58 men, caught 3,495 seals and produced 41 tuns of oil. The spring seal fishery did not account for all of the 14,878 pelts produced in the Bonavista area mentioned above, and therefore, there was obviously a strong winter seal fishery in that area in 1819, and likewise in Trinity.

To summarize the 1820s briefly, it is sufficient to note the number of vessels and men engaged in the spring seal fishery. In 1821, Trinity had a fleet of 10 vessels which employed 174 men; and Bonavista, 6 vessels and 97 men. In 1822, Trinity possessed 9 vessels with 146 men; and Bonavista, 5 vessels and 72 men.[35] The situation remained unchanged during the following years, and Ferryland, also, continued to send several vessels to the seal fishery each year. In the meantime, a small but erratic winter seal fishery continued to be prosecuted by the landsmen. In 1827, Trinity, Bonavista and Ferryland sent 12, 5 and 3 vessels to the ice, respectively;[36] in 1828, Trinity, Ferryland, Burin and Placentia sent 15, 4, 3 and 1, respectively;[37] in 1829, Trinity and Bonavista sent 13 and 1, respectively;[38] in 1830, Trinity sent 16, Great Placentia sent 3, Bay Bulls sent 2, and Little Placentia, Merasheen and St. Lawrence sent 1 each.[39] Finally, during the three-year period 1831-33, the Trinity fleet averaged 24 vessels per year, while in 1832, Twillingate sent 10, Ferryland sent 5, and Placentia, 3; and in 1833, Twillingate/Fogo sent 7, Ferryland and Placentia sent 4 each, Bay Bulls sent 3, and Burin, 1.[40] The population of Trinity had also increased and amounted to 5,161 people by 1830; however, its proportion of the total population of the island remained the same as in 1820, at 9 per cent. While the records are incomplete and certain ports only participated in the seal fishery on an intermittent basis, it can be seen that Trinity slowly expanded its fleet of sealing vessels during this period and gradually increased its catch.

By the beginning of the 1830s, it is apparent that there had been a considerable increase in the seal fishery, and almost all of this increase, with one exception—Trinity—had been confined to Conception Bay and St. John's. In fact, the two major centres controlled about 90 per cent of the industry, and with St. John's beginning to process much of the catch from the smaller ports, this percentage figure is probably a conservative estimate. While the industry in Conception Bay expanded in real terms, it is obvious also that this district had lost the formidable lead it had held over St. John's at the beginning of the 1820s. However, given the infrastructure of the capital and its overall importance, it is not surprising that it turned its attention toward the seal fishery with such success; nor is it surprising that it began to become the centre for the manufacturing of pelts from the smaller producers, who found it both impractical and financially unsound to invest in facilities when the vagarious nature of the seal fishery was so well known.

Seal oil processing

The manufacture of seal oil was a relatively simple procedure. Anspach described the operation in the early 1800s as follows:

> When the seals [i.e., pelts] have been landed, the fat is separated from the skin, and cut up into small pieces, which are put into puncheons or into vats, and there left to melt by the heat of the sun and weather. These vats, some of which are of sufficient capacity to contain from fifteen to twenty tuns, are square vessels constructed of studs and thick planks dove-tailed, and tarred all over on the outside. At each of the corners and at the bottom are fixed chemps [sic] of iron for strength; the inside is lined round with a grating composed of rods slanting from the rim of the side to the bottom, and at the distance of about six inches from the bottom is a fauset [sic], or hole stopped up with a plug; this is intended to let out the water which is *rendered* (a sealer's term) by the fat of the seals, or thrown in by the rain or snow, and which naturally sinks to the lowest parts. At two-thirds of the height from the bottom, and at equal distance from the top and bottom, are similar openings of which the lower part is furnished with a piece of thick leather to let out the oil which is thus rendered by the fat: this is called *virgin*, or white oil, is considered as the best, and obtains the highest price, the finest being that which runs from the upper opening. After all the oil that could be extracted by this process has been obtained, the *blubber* [which is left] ... is boiled over a large fire in copper cauldrons. This last operation which, on account of the intolerable stench which attends it, is generally carried on in places at some distance from the towns, produces ... common seal-oil of an inferior quality.[41]

The oil which was rendered out first, because of pressure rather than decomposition, was the valuable 'pale seal' or 'cold drawn' in greatest demand (see Chapter 1). In his excellent pamphlet published in 1852, Samuel George Archibald described the process as it existed 40 years after Anspach observed it, and little had changed. Archibald went into more detail, though, and pointed out that one could expect the first and best oil to be running by 10 May and that it took two to three months for this quality to begin to turn dark. As discussed in Chapter 1, this allowed Newfoundland's best seal oil to appear on the markets in early to mid-summer. He also described the stench that accompanied the production of seal oil during the course of the hot summer months and indicated that every seal fishing harbour with its own vats was subject to this nuisance.[42] However, there is some indication that there were early efforts to speed up the process, and Chafe thinks Oliver St. John, of Harbour Grace, was successful in boiling the fat. This does not seem likely because of the continued references to the high quality of cold drawn oil, although, as pointed out, the last of the fat was boiled to produce the lowest quality dark oil.[43]

However, the production of oil was speeded up considerably sometime shortly before 1871, according to a contemporary report:

> The fat, after being weighed, is thrown into huge vats, where the oil is extracted by heat and pressure, then drawn off and barrelled for exportation. This is a tedious process; and of late years, the great innovator, steam, has been called on to quicken the extraction of the oil. By steam-driven machinery the fat is rapidly cut into minute pieces, then steamed, stewed, pressed, and the oil passed into stout casks. By this process the work of two months is completed in a fortnight. Not only so, but the disagreeable smell of the oil is removed, the quality improved and the quantity increased.[44]

Samuel George Archibald was involved in the introduction of steam into the oil manufacturing process, and it seems that John Munn perfected it. Munn's brand of 'Steam Refined Pale Seal Oil' became, according to Chafe, the standard on all foreign markets. Chafe reported in 1923: "Old Mr. Munn used to say that one of the greatest compliments paid him was when Hon. Stephen Rendell of Job Bros & Co., wrote and asked him if he would tell how he got his oil so pale and free from smell."[45] Later, the practice of allowing the oil to age under glass in the sun improved the product further. The price of these improvements meant that the production of good quality seal oil became a much more expensive proposition—beyond the resources of smaller producers and outports.

Conception Bay and the growth of Harbour Grace

Beginning in the 1830s, it becomes possible to differentiate the major harbours of Conception Bay and to examine the developing seal fishery in each. Unfortunately, it was at this time that the local governors ceased to report to the Colonial Office on the state of the inhabitants and fishery of Newfoundland because the fishing station/colony had been granted representative government in 1832 and had thus accepted the responsibility for monitoring its own local situation. Additionally, the Newfoundland migratory cod fishery declined dramatically after 1815 and consequently, there was less reason for Britain to remain apprised of what was happening there. Thus, while the information on this area becomes more specific geographically, there is considerably less detail on developments surrounding the seal fishery. Nevertheless, an effort must be made to examine the major harbours of Conception Bay—Harbour Grace, Carbonear and Brigus—and note the participation of a number of smaller harbours in this bay as well.

Although the information on the relative importance of the three harbours referred to above is spotty and somewhat contradictory, it can be assumed that

Harbour Grace was the most important.[46] As mentioned before, by the 1830s, Harbour Grace had become a judicial and diocesan centre as well as a commercial centre of considerable worth. Jukes observed and commented on this outport in 1839-1840 as follows:

> [Harbour Grace] is a pretty-looking little town, consisting of one long straggling street along the north side of the inlet or harbour, the houses being mostly painted white, and standing on a narrow flat with a rocky ridge just behind them. Its population is about 3000. It has altogether a more English and neat appearance than most places in Newfoundland. It contains, moreover, a very decent inn, which at this time even St. John's is destitute of.[47]

Although it is almost impossible to demonstrate the relative importance of these three principal Conception Bay harbours during the 1830s, the information for the 1840s is much more illuminating. In 1844, the Conception Bay fleet was composed as follows:[48]

	Vessels	Tons	Men
Harbor Grace	61	6,190	1,753
Carbonear	41	3,447	1,143
Brigus	31	3,012	1,034
Port-de-Grave	10	860	279
Cupids	10	777	283
Bay Roberts	11	944	302
Colliers	2	178	50
Chapel Cove	1	81	26
Spaniard's Bay	5	520	153
Musquitto	1	72	24
Total	173	16,081	5,047
Average	(Tons - 93)		(Crews - 29)

Harbour Grace possessed a strong lead in terms of these figures, and this is in line with other indicators. However, the statistics for Conception Bay in 1847, which are also available, show the following:[49]

	Vessels	Tons	Men
Harbor Grace	48	5,379	1,585
Brigus	66	5,991	2,141
Carbonear	37	3,894	1,157
Spaniard's Bay	4	354	131
Mosquito	1	72	28
Total	152	15,690	5,042
Average	(Tons - 103)		(Crews - 33)

The record for Brigus needs some explanation because it was probably responsible for Munn's comment: "The town of Brigus was a source of jealousy to the whole of Conception Bay; the wealth amassed there was wonderful. Statistics show Brigus, 1830 to 1850, as next to St. John's in importance."[50] Nevertheless, it must be noted that these figures differ to some extent from the figures reported by Governor Le Marchant for 1847, which are as follows:[51]

	Vessels	Tons	Men
St. John's	95	9,353	3,215
Brigus	66	5,010	2,111
Carbonear	54	4,634	1,672
Hr. Grace	51	5,084	1,684
Ports to the Northward	74	5,803	2,123
Total	340	29,884	10,805

While there is no doubt that Brigus was among the three most substantial ports in Conception Bay, most indicators suggest that it was in third place behind Harbour Grace and Carbonear as a fishing and sealing centre throughout this period. It is, of course, obvious that the reported number of 66 vessels represents all vessels clearing from the custom house in Brigus and would include vessels from smaller harbours in the area that did not have custom houses from which the ships could clear; indeed, one of the sources specifically mentions that this is a "list of vessels cleared from the Custom House at Brigus." Therefore, the fleet reported for Brigus in 1847 no doubt includes the ships belonging to Cupids, Colliers, Port-de-Grave, Bay Roberts and Chapel Cove—all of which were recorded individually in the 1844 report (above). Geographically, Brigus is well-situated to serve the harbours on that part of the coast—from Harbour Main on one end to Bay Roberts on the other. Similarly, Table 3.4 illustrates that the increase in the Brigus 'fleet' was accompanied by the disappearance from the records of most other fleets in the area. Nevertheless, this does not explain the discrepancy that exists in the sources concerning Carbonear (and, to a lesser extent, Harbour Grace), unless Le Marchant was aware of and included ships from Carbonear and Harbour Grace that cleared from other northern ports.

Neither the history of Carbonear nor that of its seal history has been well documented during this period. Anspach wrote in his book published in 1819: "Carbonier, formerly Carboniero, or Collier's Harbour [sic], the next town in importance, has likewise a spacious harbour, but by no means so safe [as Harbour Grace], on account of its greater exposure to the easterly winds."[52] The scattered available statistics support the view that Carbonear was next to Harbour Grace in importance, and one long piece of correspondence to the Conception Bay *Weekly Herald*, which published in Harbour Grace, complained bitterly about the more favourable position that the latter enjoyed over the former:

> I am not, Mr. Editor, in the habit of writing for newspapers, so I must beg you and the public to excuse my style, while I narrate a few instances, in which it will clearly appear that Carbonear has been neglected and insulted, while her rival, Harbor Grace, has been petted and pampered in a manner and degree perfectly disgusting. I must tell you plainly, once and for all, that

> several of the leading intelligents [sic] of Carbonear, require exactly the same things for our town as are established in Harbor Grace. We think ourselves of as much importance as the Harbor Gracians; *we have not the same things*; this is the point on which we feel we have a right to complain....[53]

The writer went on to list in great detail Harbour Grace's advantages: a far superior legal infrastructure; a military garrison; a lighthouse; free port status; a stone Episcopal church; a Roman Catholic church and priest's residence; a 'neat' Wesleyan Meeting House; a larger Grammar school; a regularly published weekly newspaper, the *Weekly Herald*; an appointed Councillor in the colonial government; two watchmakers; and one tinker. The writer then concluded:

> To these standing grievances may be added the casual slights and insults we are subject to, for instance, on Wednesday last [20 August 1845], the Governor and Prince Henry of the Netherlands, during their trip to the Bay, in the steamer "Unicorn", landed at Harbor Grace and walked about the town for half an hour; whereas at Carbonear, the steamer only just looked into the harbor, turned up her nose, and passed on to St. John's, as if we had been infected with the Small Pox.
>
> From the foregoing statement it must be evident to persons of the smallest understanding and the most obtuse sensibilities, that Carbonear has deep and great reason to complain of the Executive Government; the consideration of our wrongs is exciting; it may become so to a dangerous degree:- Who can account for the consequences of disappointed agitation in such a cause? Had not Switzerland her William Tell? And in more modern times, have we not seen the revolutions of France and St. Domingo?

This letter demonstrates not only the paramount importance of Harbour Grace at this time, but also the rivalry that existed among the Conception Bay outports—a rivalry that prevented the 'Bay' from taking a united front on issues involving their common rivalry with St. John's.

Meanwhile, although it has been shown that Brigus was, during the first half century, the third most important sealing centre in Conception Bay, it is possible that at times it's importance exceeded that of Carbonear as Table 3.4 suggests. However, as noted, the number of clearances is not primarily an indication of the extent of economic activity, although it does illustrate a port's intrinsic importance.

Finally, as indicated by the above figures, there were a number of other harbours in the Bay where planters and merchants owned sealing vessels and participated in this industry as well as in the Labrador fishery. Although information is scarce, it is most likely that the smallest ports—such as Mosquito, Spaniard's Bay and Harbour Main—ceased to own sealing ships during the 1850s. Probably, the increasingly bigger ships being demanded by the industry forced the smallest operators out of the business; others moved their operations to bigger centres. For example, William Donnelly supplied 4 sealing vessels in Spaniard's Bay in 1847, and 5 in 1852. However, from that point on, this firm operated out of Harbour Grace. Similarly, Arthur Thomey, who owned, supplied and operated the sealing vessel *Eliza* out of Mosquito in 1847, had 2 vessels in 1852 (one of which was captained by his son Henry). By 1862, Henry Thomey was captain of a sealing vessel belonging to Ridley and Sons, Harbour Grace— the *Isabella Ridley*, which, at 154 tons burthen and employing a crew of 66 men, was the largest vessel the company owned at that point;[54] thus the Thomeys were no longer ship owners and Captain Henry had become a resident of Harbour Grace. Similarly, with its larger harbour, Bay Roberts began to expand during the 1850s and may have absorbed some of the investment that was available in the region. These harbours, but especially Brigus, and to a lesser extent Bay Roberts, benefitted from the expansion of the seal fishery during the first half of the century, but by 1862, Harbour Grace seems to have been best poised to dominate the industry. In that year, the sealing fleet that cleared from this port numbered 60 vessels with a total burthen of 7,633 tons and 2,919 men, and exceeded that of St. John's, which consisted of 48 vessels with a burthen of 6,173 tons and 2,513 men—ignoring, for the moment, vessels belonging to these ports but clearing from others.

Changing role of St. John's

Meanwhile, with its array of newspapers, it is easier to follow the fortunes of the sealing fleet of St. John's. As Table 3.2 illustrates, its fleet followed closely on the heels of that of Conception Bay throughout the 1830s and 1840s. The figures are incomplete and somewhat misleading in the late 1830s and early 1840s because there are insufficient records to enable one to get a complete picture of Conception Bay, and the ships were switching from 'Old Measurement' to 'New Measurement' in calculating burthen. 'New Measurement' makes ships look smaller on paper, so that a ship registered at 140 tons, 'Old Measurement', would be registered for 115 tons, 'New Measurement'.[55] Although the complete figures are available to show the development of the St. John's sealing fleet during the late 1830s, they can be confusing, as ships were reclassified over a period of several years. Thus, during the late 1830s and early 1840s, it is wise to ignore burthen and concentrate on the number of vessels and the number of men employed. In 1837, 121 sealing vessels left St. John's (and the fleet was valued at £130,000, or an average of £1,074 per vessel).[56] By 1839, the number of vessels had declined to 76 and remained in the low 70s until the fleet began to grow again in 1843. At the same time, the number of men declined from 2,940 in 1837 to 2,029 in 1839, remained just over 2,000 in 1842 and then climbed quickly to over 3,000 men in 1843.

However, by the 1840s, St. John's was less dependent than ever on its own catch of seals for the manufacture of seal oil. In 1847, for example, 334,270 pelts were processed in St. John's, while 110,910 were processed in Conception and Trinity Bays; in fact, by then, only Harbour Grace and Carbonear were consistently processing their own pelts. In 1847, over 4,500 tuns of seal oil were produced in St. John's, which, at about £28 on the London market (according to Table 1.8), was worth over £125,000. Using the same calculation with the 110,910 pelts manufactured in Trinity and Conception Bays, it can be seen that about 1,500 tuns of oil were produced, worth about £41,000. This practice continued, and it was reported that in 1857, 1858 and 1859, there were manufactured in Newfoundland a total of 530,733, 398,166, and 396,303 pelts, respectively; in St. John's 379,533, 331,666, and 277,303 pelts, respectively, were processed; Conception Bay—105,000, 36,000, and 99,900; and totals of 46,200, 30,500, and 19,100, respectively, were processed in Trinity Bay, Bonavista Bay, Fogo, Twillingate, Green Bay and other ports.[57] This same report points out that there were 15 businesses engaged in processing seal oil in St. John's in 1859. They ranged, in descending order of activity, as follows: McBride and Kerr—40,234 pelts which contained 18,051 cwts of fat;[58] Bowrings—34,220 pelts and 14,792 cwts; Baine Johnstons—33,892 pelts and 14,973 cwts; J. and W. Stewart—26,158 pelts and 11,239 cwts; Jobs—25,933 pelts and 11,707 cwts; W. and H. Thomas and Co.—23,832 pelts and 10,763 cwts; and R. O'Dwyer—20,130 pelts and 8,654 cwts. These firms were followed, in descending order, by: Stabb, Row and Holmwood with about 14,000 pelts; William Hounsell and Co.; K. McLea and Sons; P. and L. Tessier; and finally, by L. O'Brien with less than 10,000. Lastly,, the smallest operators were: Archibald and Bartlett; R. Alsop and Co.; and Brooking, Son and Co. The seal catch was down considerably in 1859, so this gives a rather distorted view of the manufacturing component in St. John's. However, it does indicate that not all sealing suppliers and ship owners in St. John's were engaged in processing, at least not in 1859, when there were 30 suppliers of sealing ships listed for this port. It should also be noted at this point that the firms to watch during the second half of the century are those that were most heavily involved in the manufacture of seal oil and not those who were simply prominent in the supply trade.

St. John's versus Conception Bay

Both St. John's and Conception Bay were the chief beneficiaries, naturally, of the several large seal catches in the 1840s, and were again so in the 1850s. However, in terms of its own sealing fleet employing its own inhabitants and in terms of the catch produced by this fleet, the capital trailed its competitor during the 1850s. Combining the figures for the St. John's and Conception Bay fleets from Tables 3.2 and 3.3, one can calculate that in 1834, 36 per cent of the combined fleets cleared from St. John's and 64 per cent from Conception Bay. In terms of total tonnage, the figures were 38 per cent and 62 per cent, respectively; in terms of men employed, the proportions were 37 per cent and 63 per cent. Although it is impossible to ascertain whether and to what extent ships from these areas cleared from northern ports, or to what extent St. John's and Conception Bay employed vessels belonging to the northern ports, the evidence suggests that these northern ports still retained much of their traditional infrastructure and operated independently of St. John's to a considerable degree (a situation which had changed by the 1860s). When one combines the two fleets in 1854, St. John's share of total clearances had dropped slightly to 32 per cent of the fleet, 31 per cent of the tonnage and 31 per cent of the men employed. In 1862 (the last year during which sailing vessels were exclusively used), St. John's had slipped further behind and accounted for only about 26 per cent of the combined St. John's-Conception Bay fleet clearances, 26 per cent of the combined tonnage and 26 per cent of the combined crews. By 1862, St. John's had become, primarily, the processing centre for the seal fishery, while the northern outports and some of the Conception Bay outports were engaged increasingly in providing the pelts. It appears that Conception Bay, with its strong Labrador cod fishery, a fairly large shore fishery and extensive subsistence agriculture, was better suited to engage in the spring seal fishery. On the other hand St. John's, had become more urbanized and commercial; for example, in 1869, over 80 per cent of the larger factories in the colony were located there.[59] As well, it had become the political and bureaucratic centre of the colony. It is very likely that the St. John's sealing operation, employing vessels and men indigenous to the area, would have continued to decline if other factors had not intruded in the 1860s.

Ports to the Northward

In the meantime, a spring seal fishery had continued to operate in Trinity Bay, Bonavista Bay and points north. The records are fragmentary, however, and give the impression, at first glance, of an industry which was waxing and waning. Such, however, was not the case. During this period, vessels went annually to the ice from throughout this area and in increasing numbers, although references are few. In 1837, it was reported that the "schooner *Endeavour*, of Twillingate, with about 250 [pelts], arrived here from the ice, having received some damage."[60] The following year, it was reported on 27 March that the schooner *Lark*, belonging to Greenspond and commanded by Captain Blackmore, arrived in St. John's with about 1,000 pelts. In the same newspaper, it was also reported that the *Lark* had spoken to the *North Star* of Bonavista, which also had 1,000 pelts, and that two vessels belonging to Trinity, the *Jane* and *Sally*, owned by J. B. Garland and Co. and Slade and Kelson, respectively, had been driven on shore by the ice and wrecked.[61] In 1847, 13 vessels from Trinity with a total burthen of 1,227 tons and 435 men were engaged in the seal fishery; 19 sealing vessels sailed from Catalina with a total burthen of 1,435 tons and 566 men; 10 vessels went to the ice from Hant's Harbour with a total burthen of 846 tons and 315 men; and 14 vessels sailed from New Perlican with a total burthen of 1,324 tons and 498 men.[62] Besides the old firm of R. Slade and Co. and a number of individual owner-operators, most of the suppliers were St. John's firms, and in one case, Harbour Grace (Munns). In 1852, Trinity Bay South (which included Hant's Harbour and New Perlican mentioned above) sent 15 sealing vessels with a total burthen of 1,343 tons and carrying in all 562 men to the ice; once again, St. John's firms were among the chief suppliers.[63] In 1853, St. John's supplied three Bonavista Bay vessels: Baine Johnstons supplied the *Huntress* from King's Cove; Hunters and Co. supplied the *Jasper* from Salvage; and Lawrence O'Brien and Co. supplied the *St. Margaret* from Flat Island.[64] That same year, 17 vessels from the port of Trinity were reported as engaged in sealing, and at least several were supplied from St. John's. Catalina, also in Trinity Bay, had a fleet of 28 sealing vessels in 1854, and, while no firms dominated, a fair number were supplied by St. John's firms, with such names as Warren, Stewart, Stabb, Rogerson and O'Brien, while Ridleys, from Harbour Grace, supplied three.[65]

And the trend continued. In 1855, 22 vessels from Trinity were engaged in sealing, 23 more from Catalina and 29 from Greenspond (10 supplied by Brooking, Son and Co., and 11 by J.and W. Stewart—both of St. John's; only 2 by William Cox and Co., the local English firm);[66] 4 came from Hant's Harbour, 11 from Heart's Content, 6 from New Perlican, and 30 from Greenspond.[67] In 1857, 28 sailed from Trinity Bay South (Hant's Harbour, Heart's Content and New Perlican) with 26 supplied by St. John's firms and 1 by John Rorke, Carbonear; 20 vessels came from the port of Trinity in 1857, only 4 of which were supplied by Slade from the area, and the rest by St. John's;[68] the same year, there were 22 from Catalina and 30 from Greenspond, most of which were supplied by St. John's.[69] In 1858, 28 sealing vessels sailed from Greenspond and 22 from Catalina, with St. John's supplying nearly all of Greenspond, while Slade supplied 3 in Catalina and Harbour Grace supplied 7.[70] In 1859, Trinity sent 10 vessels to the ice and Bonavista, 12.[71] In 1860, the reports from outside Conception Bay and St. John's were as follows: Trinity—15 vessels; Trinity Bay South—12; St. Mary's/Ferryland—6; Greenspond—20; Fogo Island—6.[72] Fogo, Greenspond and St. Mary's were entirely supplied by St. John's, while Slade supplied 4 in Trinity and 3 in Catalina, leaving most of the remainder to St. John's. Catalina sent 14 vessels, Trinity—10, and Greenspond—19 in 1861.[73] The trend was unmistakeable: a strong northern spring seal fishery dominated entirely by St. John's suppliers in the case of Greenspond and almost entirely in the case of Trinity and Catalina. (Trinity South was an exception and seems to have had a fair number of individual owners and suppliers.)

St. John's had ceased to grow as the base for a sealing fleet, in large part, no doubt, because it did not have a sufficient supply of manpower at its ready disposal; more than likely, the fleet depended on the fishermen from the neighbouring harbours, such as Pouch Cove and Torbay, for its sealers. Furthermore, most of these were shore fishermen who fished in their own communities which, meant that St. John's did not have the same outlet for a large fleet of vessels such as Harbour Grace possessed, with its extensive Labrador cod fishery. Consequently, it made economic sense for the St. John's merchants to supply vessel owners in the northern ports, who had a ready labour supply on hand and a considerable Labrador floater fishery; St. John's vessels could also be chartered to planters on the northeast coast to be used in the Labrador cod fishery. In addition, St. John's probably assisted with the financing of vessels on the northeast coast and certainly purchased their pelts for processing.

The reason for the domination by St John's is not hard to find. Most of the long-established saltfish firms on the northeast coast to the north of Conception Bay had disappeared by the 1840s because of market problems in the saltfish industry. During the Napoleonic and Anglo-American Wars, they had prospered from the cod fishery, but had not become established in the spring seal fishery. Therefore, while St. John's and Conception Bay were able to manage throughout the depressions in the saltfish trade during the 1820s and 1830s because of the lucrative seal fishery, the other firms could not make the change quickly enough. Only the Slades, under a variety of company titles, and William Cox and Co. survived, and then, only into the 1860s. James Simms, the Attorney General, summed up the situation in 1849:

> ... in the Northern District beyond Conception Bay, very few of the once great number of such Merchantine houses are now remaining, for St. John's has become, through the process of thirty years, the emporium of the Island, and absorbed to a great extent the supplying of planters and fishermen, where of course they resort spring and fall for supplies and for settlement of accounts.[74]

While Simms was interested in the cod fishermen, it is obvious that the same conclusion can be drawn with regard to the seal fishery.

Vessels

As already discussed, the average size of each ship and the complement of sealers increased during the period. These ships were owned and supplied and operated in a fashion that changed as well. At the beginning, ownership was fairly widespread, with planters building ships which could be employed in both the seal and Labrador cod fisheries. Governor Gower reported on the situation in 1805 as follows:

> The means by which the seal fishery has got to the present extent in this District, are, that the Merchants finding the employing so many craft on their own account would take up more of their attention than they could properly bestow, encouraged the people to build craft for themselves, advancing all the necessary supplies on credit for three years, which was a good spur to industry, while the Merchants reserved to themselves the benefit of supplying these men both for the Seal and Cod fisheries. Many of the successful Adventurers cleared their Vessels in two years, and became real Owners, and at the end of three years those who could not accomplish that desirable object, were obliged to have their Vessels sold or give them up at a Valuation.[75]

However, the practice of planters building and owning their sealing vessels began to change. In the first place, those who were unusually successful, avoiding the bankruptcies which plagued all fishing operations during periods of low seal catches and poor cod fish markets, evolved into small mercantile enterprises for at least awhile. Those who could not survive the depressions were unable to re-enter the seal fishery because of the trend toward larger, more expensive vessels.

Ownership varied from the individual with a few shares in a vessel to the firm with a number of craft. For example, Table 3.5 shows the broad spectrum of firms, partnerships and individuals involved in the ownership and supply of vessels in 1853 in St. John's—no particular type of ownership predominated here. Harbour Grace, in that same year, was a different scene. There, Munns and Ridleys dominated the seal fishery (as well as the Labrador cod Fishery). The former supplied 27 vessels, and the latter, 24; and these figures do not take into account their vessels which cleared for the seal fishery from northern ports. At the same time, in this port, there was still a number of individual vessels owners in the seal fishery. In addition to Arthur Thomey, who has already been mentioned, it can be seen that the *Elizabeth and William* was owned by John Stevenson and sent to the ice under command of a Stevenson; the *Eliza* was owned by Patrick Strapp and sent to the ice under a Strapp; the Gordons owned and operated vessels; and William Parsons, Thomas Power, Isreal [sic] Gosse and Robert Walsh were individual vessel owners. In Carbonear, the major firms were Pack, Gosse and Fryer and John Rorke, while here, too, there were several owner-operators. Brigus never seemed to develop a large vessel-owning firm, and most reports that mention Brigus suppliers simply refer to them as "Sundry" suppliers; indeed, ships clearing from that port were often supplied by outside firms, unlike the situation in Harbour Grace or Carbonear during this earlier period. As Table 3.5 indicates, in 1853, Baine Johnstons outfitted one Brigus vessel, while Jobs outfitted three. Bay Roberts was the site of one major sealing operation—James Cormack—and Pack, Gosse and Fryer supplied four vessels in that port, while the remainder depended on sundry suppliers. This is an over-simplified description of the situation because many participants owned shares in various vessels. Ownership of sealing vessels continued to be spread among companies and wealthier individuals, with the former becoming more common during the latter stages of the seal fishery.[76]

These early sealing fleets were usually insured through mutual societies. In 1829, it was reported: "The scheme of the Mutual Insurance Society of Conception Bay, consists of 106 vessels in the seal fishery, nearly all First Class, valued at £51,050, and insured for £49,800."[77] Obviously, not all vessels were insured, or at least not insured under this arrangement, because there were about 175 sealing vessels in Conception Bay at that time.[78] The following year, it was advertised in St. John's that the "underwriters of St. John's" would present a silver watch, worth five guineas, to the man who would kill and haul on board his vessel the greatest number of seals." However, to qualify, the vessel had to be "insured by the underwriters—whether sailing from St. John's or an outport."[79] This would appear to be an attempt by an insurance company from outside the colony (the advertisement was signed, J. Boyd, Agent) to attract Newfoundland business. By 1844, there appears to have been two mutual insurance companies in Conception Bay, and it was reported that year that 112 vessels, out of the total Conception Bay fleet of 173 vessels, were insured in the 'Harbour Grace Club' for a combined value of £83,000.[80] By 1847, there were certainly two companies so involved, as Governor Le Marchant pointed out:

> The vessels in Conception Bay are insured, in mutual societies, that is a certain number of owners enter into an agreement with each other, that they will pay all losses that may occur to each other's vessels during the season. There are two of this description now in Conception Bay, one at Harbour Grace, the other at Brigus. Each one has a secretary who keeps the records of the Society for which he is paid 15/ [shillings] for each vessel insured. There are also three surveyors to inspect the vessels previous to proceeding on the voyage and to see that they are properly equipped to encounter all dangers. They are paid a small sum for their services. The insurance in the Brigus Society has been very light indeed only five vessels having been lost since the year 1833, whereas the Harbour Grace Society has been very unfortunate lately, the losses being very heavy. The vessels of St. John's are insured in a Society and a certain premium is charged each vessel according to her class.[81]

In 1852, a list of vessels insured in the St. John's Mutual Insurance Association was published and included 68 vessels insured for the total amount of £50,375.[82] Thus, it would seem that there were three local mutual insurance companies involved in this business (probably, in addition to London underwriters).

The seal fishery brought to light new problems and interpretations in the laws surrounding marine insurance and salvage. The most notorious case was the one which involved the brig *Kingaloch* and the brigantine *Dash* (see also Chapters 4 and 5). Both sailers were well-known. Chafe writes, for example: "1846—'Spring of the Great Fire', first arrival April 18th. *Dash*, Capt. Barron [master], landed 9,646 seals"; "1847—First arrival *Kingaloch*, Burke [master], 6,400, March 30th, to L. O'Brien."[83] The *Kingaloch* was owned by Lawrence O'Brien and Co. and commanded by Capt. John Burke, while the *Dash* was owned by the master Capt. John Barron when the famous incident occurred. Whenever it appeared that a vessel was about to be dragged by 'running ice' through dangerous rocks and shoals, the crew would leave their vessel and follow along as quickly as possible over the moving ice. Depending on how tight the ice was and the actions of the sea and wind, this was often a dangerous and slow process. However, if the vessel was carried along safely past the danger, the men would board the vessel and carry on the voyage. On the evening of 13 April, the *William, Kingaloch, Hound, Rake, Caledonia* and *Dash* were carried by the ice through the run between Peckford Island and White Island. The crews left them, but attempted to keep up with their vessels. The *Kingaloch* was the second to go through the run safely, and some of the crew of that vessel were back on board before nightfall. The *Dash* did not get through before the following morning, was seen by the *Kingaloch*'s crew to be deserted or abandoned, and was boarded by part of the latter's crew and brought to St. John's. The resulting dispute ended up in the supreme court, with Barron and his witnesses arguing that the *Dash* had not been abandoned and that it was common practice to leave a vessel under these conditions. It was an intricate case because some vessels in similar situations were not recovered by their crews; in other cases, crews did not even make an effort because their vessels were inaccessible. On the other hand, the salvors who boarded the *Dash* raised sails and left for St. John's immediately; obviously they were not anxious to remain in the vicinity. The court stated that in this case, as in a case involving the salvage of the *Caledonia*:

> Salvage is a reward offered and given to stimulate exertion and encourage enterprise in those who have no interest in property exposed to imminent peril for the preservation of that property, for the general interest of trade and commerce, and for the general benefit of owners and underwriters.[84]

However, given the unique features of the seal fishery and the circumstances of this case, the amount of salvage awarded to the salvors was set at one-sixth the value of the property. Thus, Barron paid O'Brien £470 for the return of his vessel.

There were many vessels abandoned during the season of 1852, but most were salvaged; this was not the pattern in 1862. In that year, when ships became jammed in the ice and the crews were forced to abandon them, most of them set fire to their vessels, reporting later that they were a hazard to shipping. However, it seems that the captains had instructions to burn their vessels under these circumstances so that the owners could collect insurance rather than leave them for salvage. One case, that of the *Caroline*, was reported more widely than the others. All the men escaped to shore, but not before setting fire to their vessel. They explained that they considered it a menace to shipping and therefore burned it and all their provisions as well. One newspaper, along with reporting the incident, wondered why the men could not have saved any provisions without endangering their lives—"not securing even a few biscuits for their own sustenance."[85] The same newspaper reported later in more detail:

> We have already noted the lamentable loss of shipping engaged in the seal fishery this spring. Some of them are reported to have been fired from prudential motives—the protection of life and property at sea and which might at any moment be imperilled by collision. This is all very well as long

as it can be shown that prudence was the prevailing impulse. It is feared that, however, that [sic] certain parties are guilty of having, in a more wanton and unjustifiable manner, violated the strongest sections of the maritime law and it remains for the underwriters to institute the strictest investigation touching the loss of vessels this spring. It is thought that the loss to underwriters will be little short of £25,000 and that the aggregate loss to the owners of the sealing fleet will scarcely be short of £100,000....[96]

There is no record of any investigation into the circumstances surrounding these fires; however, it seems evident that owners were more interested in collecting some insurance than having others claim their abandoned ships as salvage. In 1852, there had been many vessels abandoned, no reports of fires, but many reports of vessels successfully salvaged. In 1862, out of about 25 vessels abandoned, only one was salvaged while all the others burned. This disastrous year was followed by another in 1864 and contributed to the reduction in the number of sailing vessels available for the spring seal fishery (see Chapter 7).

In the meantime, it is apparent that bigger and bigger sealing vessels, mostly brigs by this time, had become the norm. The average burthen of the Conception Bay fleet reached 104 tons in 1853, 112 tons in 1857 and 125 tons in 1862. When the *Thomas Ridley* was built by Michael Kearney in Carbonear for John Rorke and launched on 25 February 1852, it was reported to be the largest sealing vessel in the island—260 tons (old measurement) and 170 tons (new measurement),[87] although it was recorded by the custom office as having a burthen of 164 tons. As larger sealing ships were acquired, the crews they employed became bigger and increased from an average of 31 men each in 1846 to over 50

Figure 9 - St. John's sailing fleet departing for the ice fields c.1860. There was always a great deal of activity and an air of excitement when the fleet prepared to leave—magnified when the harbour was frozen and the sealers were required to cut channels to tow the vessels to open water.

men each in the 1860s. By this time, St. John's had slightly bigger ships, which averaged 129 tons burthen in 1862 as compared with 125 tons average in Conception Bay. This was probably due to the fact that a number of small outports in Conception Bay continued to employ small vessels, thus reducing the overall average of vessel size in that region; and one must remember that small vessels continued to be utilized in the Conception Bay-Labrador cod fishery.

Instead of continuing to build their vessels, the ship owners began to import them in increasing numbers, and this trend seems to have been related to the fact that larger ships were increasingly preferred. In 1845, one observer wrote:

> We believe that over forty vessels of an improved class and of substantial build, expressly intended for this employment [i.e. sealing], have been constructed in Pictou, P.E.I., etc., and sold in this market within a few months past; and these will replace those lost, or which are two inferior a description.[88]

In March, another report described the launching of a number of sealing vessels—brigs and schooners—in Harbour Grace and Carbonear in the late winter of 1845, including three built for Ridley, Harrison and Company. It added that Punton and Munn were not having vessels built in Harbour Grace but pointed out that "extensive concessions by purchase, have been made to the numerous fleet already belonging to that enterprising establishment."[89] Nicholas Smith, who was born in Brigus, wrote that his father came from Dartmouth, England, in 1839 as a ship builder for a St. John's firm and that in 1843, "my father was asked by the leading seal-killers of Brigus, Capt. William Munden, Capt. A. Bartlett, Capt. N. Norman, and many others, to come to Brigus and supervise the building of vessels for the Seal-fishery."[90] (It is interesting to note here that Smith makes no reference to his father being invited by a firm, but rather by captains; Brigus does not seem to have developed supplying and processing enterprises such as were found in Harbour Grace and Carbonear.) In 1851, a St. John's newspaper reported that Ridleys, Harbour Grace, had recently launched two new vessels—the *Brothers*, 108 tons, and the *Greyhound*, 153 tons—and that these were "beautiful specimens of native talent in shipbuilding...." Furthermore, it pointed out, they were the equal of anything "brought from the neighbouring colonies..." and that government action to support local shipbuilding was needed.[91] In fact, the newspaper was so impressed by this launching that four days later, it again praised the vessels and the builders, Kearney and Stevenson.[92] It would appear from these expressions of praise that local shipbuilding had declined considerably within a few years. Similarly, in 1853, another newspaper drew its readers' attention to the importance of sealing vessels when it pointed out that the Conception Bay fleet had been "augmented by a score of very superior vessels, built for the most part (we regret to say) in the neighbouring provinces."[93] The folly of this practice was summarized by another local newspaper in that same year:

> The entire [sealing] outfit, we should suppose, amounts to about 360 vessels averaging over 100 tons each, with an aggregate little short of 15,000 men. Pricing the vessels at £1000 a piece we have £360,000, (irrespective of provisions) of floating capital now upon the waters. In the brief space of 8 years the whole of this capital disappears for the fleet wears out within that

period and must needs be replaced. Hence £40,000 a year (besides the cost of annual repairs) is expended for a portion of the decked vessels with which we prosecute the fisheries of this island! They should be all built in this colony.[94]

W. A. Munn wrote at the beginning of the twentieth century:

> Ship-building was a great industry during the first half of the last century. Prior to 1840, nearly every vessel prosecuting the Seal Fishery was native-built. Every Harbour of importance on the Eastern Front of our Island built its own vessels; Twillingate, Fogo, Greenspond, Bonavista, King's Cove, Trinity and Hant's Harbor as well as every Harbour in Conception Bay and several ports South of St. John's, where they all carried on with great spirit. The crews belonged to the place, and the fat was nearly all manufactured into oil in the same Harbor.[95]

It is possible that Newfoundland ship builders and the available timber were inadequate for the task of providing the ever larger vessels which were required. For whatever reasons, it had obviously become cheaper to purchase shipping from outside Newfoundland rather than to build locally. The change from local shipbuilding to the importation of shipping had serious implications for the economy of the colony as well as for the individual economies of the outports because of the considerable work and capital involved.

Summary of the age of sail

The age of sail in the seal fishery reached its peak in the early 1840s in terms of total production and per capita production. For example, in 1844, Newfoundland exported over 685,000 seal skins and over 10,000 tuns of oil (including cod) valued at over £300,000 (see Tables 2.5 and 2.6). That year, as Tables 2.2 and 2.3 demonstrate, about 8,800 men from Conception Bay and St. John's were employed in the industry, and sailed in nearly 300 vessels valued at about £221,000.[96] However, the decline in production, the decrease in local shipbuilding and the centralization of seal oil manufacturing in St. John's and, to a lesser degree, Harbour Grace, were evident throughout the late 1840s and the 1850s.

Thus, 1862 was the end of an era, although this is apparent only in retrospect. The introduction of steamers into the seal fishery in 1863 was the beginning of a new period which brought many changes coinciding with many other developments. Newfoundland, in the first half of the century, was shaped by the seal fishery in just as marked a manner as it had been shaped by the cod fishery centuries earlier. Prowse wrote:

> The seal fishery, besides altering social customs, largely increased the importance of the out-ports; the statistics of the early part of the century show us that there was a considerable foreign trade in the Island outside of St John's.... The effect of the seal fishery was to add materially to the wealth of the various settlements where it was carried on. First of all there was the building of the ships, as prior to about the Forties, when the Conception Bay merchants began buying the slop-built vessels from the Provinces, nearly every vessel in the seal fishery was native built; the crews belonged to the place; in many cases the seals also were manufactured into oil in the same harbour—Twillingate, Fogo, Greenspond, Trinity, besides the Conception Bay ports had vats. All this ... brought strength and wealth to the out-har-

bours, and nourished the growth of a great middle class—the traders and sealing skippers; the two occupations were often combined.[97]

By the beginning of the 1860s, Newfoundland could reflect on a largely successful half century. In spite of horrendous difficulties in the saltfish trade, the small population had maximized the cod and seal resources and had changed Newfoundland society. As pointed out earlier, the demand for oil in Britain and the discovery of the seal resource off the island's coast were crucial in the development of the economy during this period. Nowhere was this more critical than in the outports of Conception Bay, especially Harbour Grace, followed by Carbonear, Brigus and Bay Roberts. The people of Trinity Bay and Bonavista Bay were also heavily dependent on this resource. In John's, after a slow start, the saltfish firms adopted the seal fishery and the inherent advantages of the colony's capital helped it become the major supplying and processing centre for the seal oil industry.

The population of Newfoundland had continued to grow. While fishermen spread out into the vacant harbours and coves of the south and northeast coasts, creating more and more small settlements, the population growth in the Conception Bay communities was dense, compact and centralized because men earned their livelihood away from their harbours in two enterprises mutually dependent on each other. By 1857, out of a total population of 124,228 people, Harbour Grace contained 10,067, Carbonear—5,233, Port de Grave (including Brigus)—6,489, Bay de Verde—6,221, and Harbour Main—5,386, for a total of 33,396 in Conception Bay—excluding part of the south shore of the bay, which was counted as part of St. John's West. The population of St. John's, meanwhile, had grown to 30,476, and had become more urbanized, commercialized and industrialized. Despite the competition from the availability of new fishing rooms along the south and northeast coasts, Conception Bay and St. John's continued to grow and, by 1857, contained over 51 per cent of the colony's population within its boundaries.

Age of the Wooden Walls

Major companies

Prior to discussing the place of steamers in the Newfoundland seal fishery, it is necessary to provide an overview of the firms that became the leading owners and operators of the sealing steamers. As one can see from Table 3.5, approximately 30 St. John's firms and/or individuals were engaged in sealing in 1853. The composition in Harbour Grace consisted of 2 giant firms, a smaller one and about 10 individuals for a total of 15 establishments; at Carbonear, with 13 owner/operators, there was a mixture of firms and individuals involved. There were at least 6 firms, mostly from St. John's, operating from Brigus and a couple of St. John's firms in Port de Grave and Harbour Main. (The northern ports are discussed below in a somewhat different context.) By 1914, however, 8 firms controlled the whole of the sealing industry, with 4 operating 16 steamers, and the remaining 4, one each. The principal firms from among this group and the major ones not surviving to 1914 are those deserving consideration here. To simplify the examination of the introduction of steam, this summary discusses the companies[98] in the order in which they became involved in the steamer seal fishery.[99]

The first two companies to purchase steamers for this purpose were Walter Grieve and Company (hereafter, Grieves) and Baine Johnstons—both closely connected with each other. Walter Grieve came from Scotland to Newfoundland in the 1820s and was employed by Baine Johnstons, where his uncle, William Johnston, was a partner. He became manager of the firm in 1837, when his uncle died, and in 1851, when the other partner, Walter Baine, died, Grieve and his brother, James Johnston Grieve, became the owners of Baine Johnstons. James spent most of his time in Britain, usually on company business, but later as an M.P. Their nephew, Robert Grieve, became manager of Baine Johnstons when Walter established Walter Grieve and Company in St. John's and became a partner in Grieve and Bremner in Trinity. Later, Walter Baine Grieve, son of James Johnston Grieve, became the manager of Baine Johnstons in 1872, when his cousin Robert returned to Britain; he remained in charge until his death in 1921.

Bowrings was the third firm to invest in sealing steamers. This firm was founded by Benjamin Bowring, a watchmaker from Exeter, who set up a small business in St. John's in 1811. In 1816, he brought out his family and established a permanent residence in St. John's, expanding into the dry goods business and, later, into the saltfish trade and the seal fishery. By the 1850s, his four sons (Charles, Henry, Edward and John) operated as Bowring Brothers in St. John's and as C. T. Bowring and Company in England, and when they left the business, their sons and grandsons succeeded them.

Another St. John's firm, Jobs, was the next to follow the trend towards steam. John Job was born in Haccombe, Devon, orphaned as a child and became a ward of a Devonshire merchant, Samuel Bulley, who operated a business in St. John's. Job came to St. John's c.1780 to work for Bulley, married Bulley's daughter and, eventually, became a partner in the firm Bulley, Job and Company. In 1809, John Job settled permanently in Liverpool and transferred the English branch of the business to that city. In 1839, he retired and his four sons became the sole proprietors of the firm that became Job Brothers and Company in St. John's and Job Brothers in Liverpool. John Job's son, Thomas Bulley Job, managed the company in St. John's during mid century and was succeeded by his sons and other members of his family.

Two individuals who became involved in the steamer seal fishery at an early date were Alexander M. MacKay and Capt. Philip Cleary. The former came from Nova Scotia and was superintendent of the Anglo-American Telegraph Company and founder of the St. John's Electric Light Company. The latter was an experienced ship owner, born in St. John's, and the owner of the S.S. *Ariel*, a small steamer used in the Newfoundland coastal mail service.[100] Both invested in the operation of a steamer in the seal fishery, either individually or jointly with each other, for a number of years.

J. and W. Stewart and Company (hereafter, 'Stewarts') was a large saltfish and sealing firm which also recognized that the use of steam was the best approach to the sealing industry. Not much is known about this firm, except that James and William Stewart came from Greenock, Scotland, and were established in the fish trade at the beginning of the century. James was a partner in the firm of Rennie and Stewart and withdrew from that firm in 1819 to go into business with his brother, William. (Prowse credits them—probably in partnership with Rennie—with being the first to export saltfish directly from Newfoundland to Brazil.) James and William died in the 1830s, and their business was thereafter controlled by British shareholders and managed by agents. The most significant of these agents was Robert Alexander, who assumed control in 1861 and ran the business successfully until his health began to fail c.1878.[101]

Harvey and Company (hereafter, 'Harveys') originated as the Bermuda Trading Company, with headquarters in Bermuda and branches in the Caribbean and various ports in North America. When this company was dissolved in 1767, the St. John's branch became identified with the saltfish trade to the British West Indies. The name 'Harvey' became associated with the firm following the arrival in St. John's of Eugenius Harvey from Bermuda in 1820. The firm operated under a variety of names during the following years, but when Augustus W. Harvey, nephew of Eugenius, joined the firm in 1860, the name was changed to Harvey, Tucker and Company. Joseph Outerbridge, another nephew of Eugenius, arrived from Bermuda to join the firm in 1861; the firm became Harvey and Company in 1862, and Outerbridge was heavily involved in its development during the remainder of the century.

Nicholas Stabb, born in St. John's in 1803, became the principal owner of Stabb, Row and Hammond, a firm engaged in sealing and cod fishing. He and his brother Ewen established Ewen and Nicholas Stabb; however, the business was destroyed in the St. John's fire of 1846. Nicholas Stabb then launched his own firm of Nicholas Stabb and Sons (hereafter, 'Stabbs').

Stephen March ventured into the commercial life of the colony as a merchant in Old Perlican, his birthplace, by investing in the cod and seal fisheries. He moved his operation to St. John's, entered politics and finally retired in the 1870s. He left his business, Stephen March and Sons, to his two sons—Nathaniel and Stephen R.

Harbour Grace was the headquarters of the large firm of Ridleys. The origins of this business go back to Thomas Ridley, who was born in Ireland and came to Newfoundland in the 1820s to work with his uncle, William Bennett, in his establishment in Carbonear. After Bennett died in 1826, Ridley moved to Harbour Grace and started his own business, Ridley and Company, which later became Ridley, Harrison and Company, and still later, Ridley and Sons. Beginning in about 1850, Thomas Ridley's son, Thomas Harrison Ridley, managed the business, which was heavily involved in the Labrador cod fishery as well as the seal fishery.

Ridleys' principal competitor, Punton and Munn,[102] also had its headquarters in Harbour Grace. John Munn came from Scotland to Newfoundland in the 1820s and worked as a bookkeeper at Baine Johnstons in St. John's. Shortly after the fire of 1832 in Harbour Grace, Munn and Capt. William Punton purchased the property belonging to the recently bankrupt firm of William Danson and established Punton and Munn. The firm became engaged in the shore and Labrador cod fisheries and also in the seal fishery. Punton died in 1845 and Munn's relatives, apparently, bought out his share, but the name remained the same until there was a reorganization in 1872, when it was changed to John

Munn and Company (hereafter, 'Munns'). John's son, William Punton Munn, and his nephew, Robert Stewart Munn, became partners at that time. John died in 1879 and William Punton Munn was lost at sea in 1882, leaving Robert Stewart to operate the business, now the largest in the Labrador fishery, until his death in 1894.

There were two other companies involved in the Newfoundland seal fishery during the latter part of the century which owed their origins, not to the old Newfoundland saltfish industry as many others did, but to the Scottish whale and seal fisheries. They were the Dundee Seal and Whale Fishing Company (hereafter, 'Dundee Company') and Alexander Stephen and Sons (hereafter, 'Stephens'), both actively engaged in the Greenland seal fishery and the Arctic whale fishery, which had begun to fail.[103] The former attempted to prosecute the seal fishery off Newfoundland in 1862 with the steamers *Camperdown* and *Polynia*, and although the venture was a complete failure, with the *Camperdown* requiring repairs in St. John's, it demonstrated to the locals the advantages of steam. (Grieves and Baine Johnstons purchased steamers for the 1863 season.) The first venture of the Dundee Company was a disappointment, and with the recovery of the Greenland seal fishery, this company lost interest in Newfoundland until there was a later serious downturn in the Greenland fishery. By then, 1877, the Scottish firms had become convinced that experienced Newfoundlanders would have to be hired and processing facilities set up in St. John's. Stephens invested in a new major enterprise in 1877, followed by the Dundee Company in 1878.

Transitional 1860s

By the 1860s, when the above fishing establishments (as well as many others, excepting the Scottish firms) were generally in place, several events caused Newfoundland to reach a crossroads. There was political unrest, stemming partially from depressed economic conditions and partially from demographic changes which had taken place in recent years, leaving the Protestant population numerically greater than the Roman Catholics and anxious to translate their numbers into political power. There was a major depression in the saltfish trade, which was experiencing low catches, declining demand and low prices. And there were the developments on the mainland, where the other colonies were planning a union; joining the rest of British North America, many Newfoundlanders thought, would resolve their social, religious and economic problems. Rowe writes: "The decade from 1860-1870 began with [political] violence and turbulence; the years of unredeemed adversity and disaster continued right up to the end of 1869."[104] Although the whole colony suffered, the outports dependent on the traditional spring seal fishery were forced to make the biggest adjustments—adjustments that did not always result in solutions.

Wooden walls

The introduction of steamers into the sealing industry was a very significant development. It was inevitable that it should happen because Scottish whalers and sealers in Greenland waters had been demonstrating for years that in certain circumstances, steam power was indispensable, for it enabled vessels to manoeuvre quickly and effectively under confined conditions. The whaling vessels were required to follow the whales and to follow the whale boats when they were in the process of capturing these sea mammals; and they, too, were engaged in an Arctic spring seal fishery by mid century. In addition, they often operated along the outer edges of the Arctic ice floes, where manoeuvrability was also essential. The necessity to be able to manoeuvre quickly is what separated Arctic whaling and sealing from other maritime activities. So, it was not long after steam had proved itself in the Arctic that steamers were introduced to the Newfoundland seal fishery.

As early as 1855, it was reported that Samuel George Archibald was to invest in a sealing steamer. One local newspaper ridiculed the idea as follows:

> This project, at first sight, would seem to be feasible enough; but we doubt very much its success.—Its extra expense, we apprehend, would be an insuperable obstacle to the pressing into the sealing business vessels of such a description. Certainly their superior power would be a great advantage amid ice; but we question whether the snorting and groaning and puffing and whizzing of a steamer would be very attractive and engaging to young seals just about to enjoy a snooze in the sun! And to alarm the old ladies and gentlemen,—Harps, Bedlamers and Hoods—nothing, we guess, would do the business in a more workmanlike manner than the everlasting pumping, bumping and thumping of the screw! We have no more faith in this contemplated enterprise than we have in the other [sealing in Greenland waters]. We think the old mode of prosecuting the seal-fishery, with superior class vessels, is the best and only mode of doing it successfully. In this way fortunes have been made, and may be made again with the same kind of

Figure 10 - Ridleys' S.S. *Retriever*, a fully-rigged, low-powered steamer of 237.95 net tons and 50 nominal horse power (nhp), was the first sealing steamer employed by the Harbour Grace firm. As this illustration demonstrates, men on early steamers had to haul the pelts to their vessel because early steamers did not possess sufficient power to force channels through the ice.

perseverance and industry which marked the days of our fathers. Steam will not supply the absence of these pre-eminent qualities—and these can dispense with steam. Business carried on in a fast manner is very likely to end in the same way as the career of fast young men—in ruin. We, however, should have no objection whatever to the trial of a steamer's qualities for the prosecution of the S.F.—if it were only for the purpose of putting an end to the idea altogether. For allowing that this description of craft were all that could be desired for the purpose, what use would it be to the hardy and intrepid sealing master of Newfoundland? Could he, or any of his class, ever hope to become the owner of a steamship? Alas! the answer is at hand. No, no![105]

Although the writer was certainly mistaken in his general views on the subject, as will be seen below, he was accurate in pointing out that the average "hard and intrepid sealing master of Newfoundland ...[could never] hope to become the owner of a steamship." However, steam was introduced in 1863.

Most of the wooden-wall steamers were built in Dundee by Stephens. The earliest ones were under-powered, by later standards, and were fully rigged for sail as well. For example, the S.S. *Retriever*[106] had three masts and carried square

Figure 11 - The S.S. *Terra Nova*, 450.28 net tons, 120nhp, was a wooden whaler/sealer built in 1884 in Dundee by Stephens. This steamer began sealing in 1885 and continued annually until 1903, when it was bought by the British Admiralty and sent to relieve Scott's S.S. *Discovery* in the Antarctic. It then returned to Newfoundland's sealing fleet and, aside from another interval with Scott's second Antarctic expedition, continued sealing until 1942. It was lost of Greenland in 1943, thus ending a career that included 51 springs to the seal fishery with a total catch of over 850,000 seals—the third largest total in the history of the industry.

sails as well as jibs, and a primitive 50 nominal horse power (nhp) engine. The *Lion*, another steamer from the 1860s period, had a 75 nhp engine but was fully rigged as well.[107] As the century progressed, the wooden walls were improved, and larger ones were built with more powerful engines. The 210-ton *Wolf* was the most powerful steamer up to the end of the 1866 season. The following year, the *Lion*, at 292 tons, the *Nimrod*, at 226, and the *Panther*, at 246, were all larger; and the *Esquimaux*, which was sent from Scotland to Newfoundland for that year only, had a burthen of 465 tons. For a couple of years, beginning in 1870, the *Montecello* from Boston prosecuted this fishery as well. The *Eagle* was introduced to the industry by Bowrings in 1871 and, at 343 tons, was the largest of the early regular vessels. Grieves brought the *Wolf II* to Newfoundland, and Stewarts, the *Ranger*, both registered at 353 tons, in 1872. Baine Johnstons bought the 376-ton *Bloodhound II* in 1873, and Jobs bought the 465-ton *Neptune* that same year. The *Micmac*, at 463 tons, was acquired by Baine Johnstons, and the *Bear*, at 468 tons, was bought by Grieves in 1874. The *Proteus*, with a 467-ton burthen, was added to the fleet by Stewarts in 1875. In 1877, Stephens, established a branch in St. John's, and brought out the 522-ton *Arctic* to prosecute the seal fishery. The *Terra Nova* (one of those steamers that was to become synonymous with Newfoundland), at 450 tons burthen, was brought to Stephens' St. John's branch in 1885. Most of the fleet registered in the 300- to 500-ton range during this period, and those that survived the hazards of the industry, for any length of time, were re-fitted and were often re-classified in terms of tonnage. The *Walrus*, at 183 tons and owned by Stewarts, and the *Kite*, at 190 tons and owned by Bowrings, were the two smallest sealing steamers during this period. Capt. Farquhar's *Newfoundland*, at 568 tons, was the largest to be added to the fleet and, during the 1890s and up until 1906, was the only sealing steamer to exceed 500 tons. While most wooden walls were built in Dundee by Stephens, there are a number about which the records are silent, and a few that were definitely built elsewhere, including Norway, Aberdeen and the Canadian mainland.[108]

The early engines were not only of low power but inefficient, in that the nominal horse power generated did not actually translate into the power that reached the propeller. Therefore, the use of sail power was important to them on their journeys to and from the ice fields, and the engines were essential for following open leads in the ice. It was only as engines became more powerful and efficient that wooden walls could actually make progress by battering or ramming the ice. While the early steamers could depend on only 50 to 60 or even 80 nhp, the later ones, like the *Newfoundland* at 130 nhp and the *Southern Cross* at 100 nph, were a considerable improvement. As engines improved, sails became less and less important, until finally, in the twentieth century, most of the old surviving wooden walls had their masts shortened—to be used for hoisting the pelts on board and for use by the barrel men and scunners.[109] Those vessels that lasted to the turn of the century were generally refitted with larger and more efficient engines. One account from Dundee summarized these developments and pointed out that:

> ... a gradual but steady development has taken place in this type of craft. The earlier vessels of the Alexander type gave place to the large full-powered

vessels such as the *Arctic, Bear, Neptune, Proteus, Resolute, Thetis,* and *Terra Nova*, with the similarly built craft for Arctic exploration....[110]

Given its apparent advantages, steam power was enormously attractive to the sealing firms. Munn says it "revolutionized this business" and Prowse says that it "completely changed the whole aspect of affairs."[111] They were both correct, but there is no evidence to suggest that any far-sighted individual ever came even close to foreseeing what a tremendous change steam would make to the industry.

Companies and wooden walls

Grieves and Baine Johnstons introduced steamers to the Newfoundland seal fishery. In 1863, the former purchased the *Wolf*, with a burthen of 210 tons, and sent it to the ice with 110 men under the command of Capt. William Kean. That same year, Baine Johnstons purchased the *Bloodhound*, with a burthen of 153 tons, and sent it out with 100 men under Capt. Alexander Graham. The *Wolf* returned with 1,340 pelts, and the *Bloodhound*, with 3,000—both very poor performances. This did not discourage investors, and the following year, Baine Johnstons added the *Osprey* (sometimes called the *Ospray*) to its fleet. In that year, the fleet of three steamers, with 311 men, brought in a total of only 1,059 pelts—a complete failure. (However, this was a disastrous year for the seal fishery in general.) In 1865, these same three steamers brought in a total of 19,086 pelts, which were manufactured into about 250 tuns of oil worth about $58,000, or about £12,000 stg. (calculating from Table 2.8).[112]

The following year, Bowrings purchased the *Hawk*, with a burthen of 172 tons, and sent it to the ice with 120 sealers under the command of Capt. Edward White, Sr., who had been a successful master of sailing vessels at the seal fishery since 1836. In two trips, the *Hawk* brought in a total 10,700 pelts, the *Bloodhound* brought in 12,700, while the two other vessels had only meagre catches.[113] In

Figure 12 - The St. John's fleet of wooden-wall steamers preparing to depart c. 1890. Photograph by James Vey.

1867, the fleet increased by four as Grieves bought the *Lion*, with a burthen of 252 tons; Baine Johnstons, in partnership with Capt. Abram Bartlett, purchased the *Panther*, of 246 tons; Prowse and Sons (a new enterprise which participated only during this year) purchased the 465-ton *Esquimaux*; and Jobs bought the *Nimrod*, at 226 tons, and hired Capt. White away from Bowrings. (Baine Johnstons employed the *Panther* in the seal fishery every year until 1902—with the exception of 1895, when the company was undergoing restructuring because of the colony's financial crisis—and Jobs employed the *Nimrod* each year until 1907.) However, with the exception of Capt. William Ryan in the *Bloodhound*, who brought in 9,200 pelts, and Capt. Robert Dawe in the *Lion*, whose catch amounted to 5,300, the season was a failure for the steamer fleet, with the eight vessels bringing in only 28,050 pelts in all. In fact, the mighty *Esquimaux*, with an experienced captain and a crew of 165, brought in a total of only 150 pelts, and it seems that the owners, Prowse and Sons, went out of business and returned the steamer to Scottish interests. (The *Esquimaux* returned to the Newfoundland seal fishery in 1878 under the ownership of the Dundee Company.)

An incident occurred in 1867 which seems to have reflected on the high expectations associated with the entry of the *Esquimaux* into the seal fishery. At 465 tons burthen, it was a giant of a steamer for that period in the seal fishery, with the *Lion*, the next largest, registered at only 292 tons, and the *Panther*, next, at 246. Moreover, the *Esquimaux* was placed under the command of Capt. Terry Halleran, who had been previously master of the biggest sailing vessel in the seal fishery—the *Kate*, which, at 222 tons, was larger than some of the steamers and was owned by Grieves. Halleran had an excellent reputation and is mentioned on three occasions in Chafe's records:

> 1859—March 7th—Vessels sailed for the seal fishery. Capt. Terry Halleran took the lead, 30 others followed him during the day. First arrival from the seal fishery March 27th, brigt. *Zambesi*, Capt. Terry Halleran, with a splendid trip of 9,500 seals; 1860 ... and on April 21st the brigt. *Livingston[e]*, Capt. Terry Halleran, arrived with 4,890 seals; and 1861 ... on March 26th the *Livingston[e]*, Capt. Terry Halleran, arrived with 8,600 prime seals.[114]

Capt. Halleran was inexperienced as a steamer master and unlucky as well, and he seems to have jammed the *Esquimaux* in the ice. In any event, the incident resulted in a short song being composed, ridiculing Halleran and illustrating the competitiveness which steam was bringing to the seal fishery. It was entitled "Captain Bill Ryan left Terry Behind":

> Terry is a fine young man
> But he has a lot of 'chaw',
> He thought to do the devil and all
> When he got the *Esquimaux*.
>
> The *Mary Joyce* is stuck in the ice,
> And so is the *Osprey* too,
> Captain Bill Ryan left Terry behind
> To paddle his own canoe.

> Though Terry is a fine young man,
> He couldn't cut no shine.
> Although he had the *Esquimaux*,
> And Dawe has got the *Lion*.
>
> The *Wolf* and the *Lion* kept very good time,
> And so did the *Panther* too,
> But Captain Bill Ryan left Terry behind
> To paddle his own canoe.[115]

Although the overall performance of the steamer fleet was far from impressive during this five-year period from 1863 to 1867, the merchants (with the one exception, Prowse, mentioned above) were convinced that investment in steam power was going to pay dividends eventually; their belief did not falter during these experimental years. One can only assume that the captains were persuasive in arguing that out among the Arctic ice floes, steamers were intrinsically more efficient than sailers. Certainly the *Esquimaux* was the only steamer to be withdrawn from the seal fishery after the 1867 season.

The remaining seven steamers were much more successful in 1868 and brought in a total of 57,800 pelts. Capt. Alexander Graham returned to the seal fishery this year. As noted above, Capt. Graham had been hired by Baine Johnstons to command the *Bloodhound* in 1863 and had spent three years with that firm and the same steamer. He left the seal fishery after 1865, only to return in 1868, in charge of the *Lion* for Grieves. The *Lion* was the high liner[116] that year, with 15,500 pelts, and Graham's nearest rivals were Captains White and Ryan in the *Nimrod* and *Bloodhound*, with 10,500 and 10,200 pelts, respectively. Captain Graham was consistently one of the most successful sealing captains and commanded Grieves' steamers until he retired after the 1874 season.

The following year, 1869, this fleet was joined by the 248-ton *Merlin* owned by A. M. Mackay, who continued to own and supply it himself or, on a few occasions, in partnership with Capt. Philip Cleary. (The *Merlin* was lost at the ice in 1882.) In addition, the *Ariel*, of only 78 tons, was sent out by its owner, Capt. Cleary, and brought in only 300 pelts. Meanwhile, the total catch of the steamer fleet increased to 76,620 pelts. This was the year that two of the most prominent captains—Capt. Edward White and Capt. William Jackman[117]—each made three trips to the ice in one season—the former in the *Nimrod* and the latter in the *Hawk*. In 1870, Stewarts finally invested in steam and bought the *Walrus*, which the firm was to use for 21 years. And an American firm, Ludlow and Company, sent the *Montecello* and created a problem for the Newfoundland Government because its products could enter the United States of America duty-free. Its first voyage was a failure, but the next, in 1871, was quite successful; ownership was transferred to Harveys, but the steamer was lost in the ice during the 1872 season and direct American involvement ended. The following year, 1871, Bowrings introduced the 343-ton *Eagle* to the seal fishery and sent it annually to the ice until 1893. At the same time, Jobs bought the *Hector*, which they had rebuilt and renamed the *Diana*, in 1892 and which continued to prosecute the seal fishery until 1922. The *Eagle* and the *Hector* were placed under the command of Captains Jackman and White, Sr., respectively, and not only were they the high liners for 1871, but each again made three trips. (Chafe records that Capt. White set a record that year, for he brought in a total 31,644 pelts valued at £19,800; he received £949 as his share, while his crew of 170 men received £38 14s each.)[118] That year, the 12 steamers were quite successful and brought in 178,769 pelts. One contemporary reported that Jobs' two steamers brought in pelts worth a total of $246,000 in a fishery that lasted less than two months.[119] This was also the year that Capt. Arthur Jackman, William's brother, was appointed master of his first steamer, the *Hawk*. 'Viking Arthur' was to go to the ice for 36 years in steamers, in most cases, belonging to Bowrings.[120] And finally, the *Wolf*, the first of many steamers to be lost while prosecuting the seal fishery, sank on its second voyage of the season; the crew was brought home safely by Capt. Graham in the *Lion*.

During the following year, 1872, the tenth in the history of the steamer seal fishery, 4 more steamers were added to the fleet to bring it up to 16. The 259-ton *Greenland* was bought by Stabbs and used by this firm until 1880, after which it was sold to Munns in Harbour Grace.[121] Another steamer whose name is synonymous with the Newfoundland seal fishery, the *Ranger*, of 353 tons burthen, was added to the fleet this year. (This steamer was operated by Stewarts until the end of 1893, and then it had other owners for several years; it was taken over by Bowrings in 1903 and completed the 1941 season for that firm. This steamer spent a record 68 springs to the seal fishery, missing only 1915, and brought in 934,007 pelts, the second largest number for any sealing steamer.) During the year 1872, its first at the seal fishery, the *Ranger*, under the command of Capt. P. Mullowney, from Witless Bay, was the high liner, bringing in 16,345 pelts out of a total catch for the 16-steamer fleet of only 76,261 pelts. Stabbs also added the *Iceland*, of 287 tons burthen, to the St. John's fleet this year, and, like the *Greenland*, it became the property of Munns in 1881, transferring to Baine Johnstons in 1895. A small operation, Cleary and Company, brought back the 78-ton *Ariel* this year, but its 48 men only managed to take 900 seals.

This was a peculiar year for the seal fishery; the catch was low for the size of the fleet and the season was late. The *Hector*, under Capt. White Sr., returned to St. John's first, and it did not arrive until 10 April and carried only 1,220 pelts; Capt. White returned to the seal fishery again, presumably after re-coaling, managed to secure 4,000 old seals, and arrived back in port 20 May. Capt. William Jackman, in the *Eagle*, was also unfortunate, and although he stayed out until 11 May, he and his 180 men brought in only 2,500 pelts, while his brother Arthur in the *Hawk* had only 1,000. (This was also a tragic year for the people of Bay Roberts and for the Dawe family, because the sailer *Huntsman* sank, taking the lives of Capt. Robert Dawe, his son and 40 crewmen; and it was an unfortunate one in general, with the loss of both the *Retriever* and the *Bloodhound*, although no lives were lost in the latter cases—see Chapter 5.)

Meanwhile, of all the outports, only Harbour Grace decided to invest in steamers for its seal fishery, and Ridleys was the first outport firm to do so. In 1866, this company purchased the 237-ton *Retriever*, which it placed under the command of James Murphy, a well-known sealing master from Catalina. On its first voyage, this steamer, with only 105 men, brought in 17,450 pelts and returned to the ice for a second trip, during which it took another 5,950. (The men made the enormous share of $303 each.) The following year, this firm added the 245-ton *Mastiff* to its fleet, placed it under the command of Capt. James Murphy and hired Capt. Isaac Bartlett of Bay Roberts to take over the *Retriever*. (The *Mastiff* must hold the record for the number of captains who commanded it—18—during its career, and it outlasted all its owners: the Ridleys, Paterson and Foster and, finally, the Munns; with the rest of Munns' fleet, its ownership passed to Baine Johnstons in 1895, and it sank in 1898 while at the ice.)

Finally, Munns followed the lead of the other principal firms and arranged for the 290-ton *Commodore* to be built for the 1871 season. This steamer was placed under the command of Capt. Azariah Munden of Brigus, who was recognized as the 'Commodore' of the Harbour Grace fleet—hence the steamer's name. (Capt. Azariah was the son of Capt. William, who has already been mentioned as the captain who broke the 100-ton ceiling on sailing ships when he constructed the 104-ton *Four Brothers* in 1819; that vessel was rebuilt by Azariah in 1851 and rechristened the *Three Sisters*.) Azariah had been pressing Munns for several years to invest in steamers, but the company had a great deal of capital tied up in sailers and, presumably, was reluctant to expand further. In 1873, Munns added the *Vanguard*, 322 tons, and placed Capt. Munden in charge, while appointing Capt. N. Hanrahan from Harbour Grace to take over the *Commodore*. Capt. Munden supervised the construction of the *Vanguard* and had been told by John Munn to christen it the *Admiral* in his own honour; this honour Capt. Munden refused and chose instead the name *Vanguard*. During the remainder of the 1870s, Munns' steamer fleet consisted of these two vessels.

Mercantile changes and the decline of outport firms

By the mid 1860s, it is obvious that whatever was left of the old West of England mercantile presence that had existed on the northeast coast was gone. A list of Newfoundland's exporters and exports was compiled for the "Chamber of Commerce Minute Book," and this collection of data shows the situation as it existed in 1865.[122] The major saltfish exporters in St. John's, in descending order of importance, were as follows: Grieves, with a total of 97,000 cwts of saltfish exported; P. and L. Tessier—59,000; Bowrings—58,000; Baine Johnstons—57,000; Jobs—45,000; Stewarts—36,000; McBride and Company—29,000; Harveys—25,000; Brooking and Company—23,000; Muir and Duder—20,000; Lawrence O'Brien and Company—16,000; C. F. Ancell—16,000; Stabb, Row and Holmwood—15,000; Kenneth McLea and Sons—14,000; and C. F. Bennett and Company—9,000. A number of other firms exported small amounts, usually less than 3,000 cwts, and these included Ridleys, Munns, Charles Cowan, H. C. B. Thomas, A. Goodridge and Sons, M. H. Warren and Company, and Hunt and Henley. In all, these firms exported from St. John's about 541,000 cwts in that year. Fewer of them were involved in the exportation of seal skins and seal oil or in the catch itself. Again, Grieves led with 468 tuns of "seal and whale oil" (which, given the absence of any whaling, means seal oil) exported, followed by: Baine Johnstons, with 443 tuns; Bowrings—407 tuns; Tessier, Stewarts and

McBride, with over 200 tuns each; Jobs and Harveys—100 tuns each; and Brooking, Muir and Duder, O'Brien, and William Hounsell and Company, with smaller amounts each—for a grand total of 2,477 tuns of oil. The same companies exported the seal skins, with Baine Johnstons, Bowrings and Grieves dominating.

In Harbour Grace, from where a total of 111,000 cwts of saltfish was exported, Ridleys exported 72,000 cwts while Munns exported 37,000—almost the entire catch for that town. This harbour also exported 737 tuns of seal oil as follows: Ridleys—367 tuns; Munns—288; and W. J. S. Donnelly—82 tuns. From Carbonear, which exported 11,000 cwts of saltfish, Ridleys sent 4,000 cwts abroad, and John Rorke, 7,000, but no seal skins or seal oil were exported. Looking briefly at the other outports that have been mentioned in this study, one finds the following. Jobs exported about 2,000 cwts of saltfish from Hant's Harbour. From Trinity there were 18,000 cwts exported: Grieve and Bremner—5,000; and Brooking—13,000. Fourteen thousand cwts were exported from Catalina: Ridleys—11,000; and Bowrings—3,000. William Cox and Company and Brooking exported 9,000 cwts of saltfish from Greenspond. Cox exported 5,000 cwts from Fogo, and Muir and Duder—2,000, for a total of 7,000 cwts. And in Twillingate, Cox exported 6,000 cwts, Muir and Duder 5,000 and the executors of J. Slade and Co., 2,000. A significant point was the fact that outside St. John's and Harbour Grace, no seal oil was shipped abroad, except for a little exported by the old West of England interests: Cox in Greenspond exported 60 tuns of seal oil and about 4,000 seal skins; Cox in Fogo—95 tuns and 6,000 seal skins; and the executors of Slade in Twillingate—5 tuns and 400 seal skins. Thus, the entire export trade in seal oil and skins was now in the hands of St. John's and Harbour Grace, but the latter was confined to its own fishery, while the former had become the market for the pelts of the rest of the northeast coast.

The exportation of saltfish from the Labrador coast was reported separately, and a brief examination of the situation there will complete the picture. Munns and Ridleys dominated the saltfish trade there in 1865, with the former exporting 58,000 cwts of fish, and the latter, 50,000 cwts, out of a total exportation of 196,000. Donnelly exported another 10,000 to increase Harbour Grace's share to 118,000, and another 10,000 was sold abroad by John Rorke from Carbonear. The 70,000 cwts that remained were exported by Tessier, McBride, Grieves, Baine Johnstons, Warren, Stewarts, Jobs, Bowrings, Stabbs and O'Brien. This relatively small involvement in the Labrador cod fishery by St. John's firms has already been mentioned as a possible explanation for the fact that much of the St. John's sealing fleet was sailing from the outports at this time. It does not seem that the capital had a population of potential Labrador fishermen close at hand, and without them, it was not possible for St. John's to continue to expand its sealing operations in its own harbour; it had to make use of the men and, often, the ships in the outports. At this point, Ridleys was the largest exporter of Labrador-cured saltfish in the Newfoundland trade and, in fact, the biggest saltfish exporter in the entire colony. At the same time, Munns, with its combined exports of Labrador- and shore-cured saltfish, was about equal to the largest exporter of shore fish—Grieves.

By the middle of the 1860s, the centralization of the seal fishery and the centralization of both the trade in saltfish and the trade in seal oil and skins was almost complete. William Cox and Company, a Poole firm, retained a small operation on the northeast coast, but the Slades were winding up their business. The latter company, under a variety of names, had establishments in Fogo, Trinity Bay and Carbonear. In addition, it had been involved in the Labrador fishery, especially at Battle Harbour, since the 1770s. The Carbonear business had been sold earlier to John Rorke.[123] In the spring of 1852, Robert Slade, the senior partner of the Fogo-Twillingate premises, began to curtail his business and stopped all credit to all customers—doing business in cash or barter only.[124] In the fall of that year, he tidied up his books by forgiving one-half the amount of each old debt if the other half was paid immediately.[125] In 1861, the Slades in Trinity went bankrupt, and that business was taken over by Walter Grieve and Alexander Bremner.[126] It seems that the Fogo establishment was disposed of around that time because there is no mention of it in the "Chamber of Commerce Minute Book" quoted above. And they sold the Battle Harbour business to Baine Johnstons in 1871.[127] In the meantime, William Cox and Company closed its Fogo and Greenspond branches in 1867 and the Twillingate premises in 1868, thereby bringing to an end Poole's 250 years of participation in the cod fishery and in the history of Newfoundland on that part of the island's coast.[128] By this time, outside of St. John's, only Harbour Grace was in control of its cod and seal fisheries—but this control was tenuous.

This centralization of the seal and cod fisheries in St. John's was not an uneventful and unrecognized development. The old-style firms could see what was happening to them. For example, Newman and Company, with premises on the island's south coast, complained about the competition from the St. John's merchants and the saltfish market gluts they caused.[129] Similarly, Cox and Company was usually upset about the competition from St. John's and found that "few of them are to be trusted."[130] The adaptation of the more modern St. John's firms to the changing conditions widened the chasm between them and the old style firms.

There were a couple of reasons why the St. John's firms, the Harbour Grace firms and John Rorke in Carbonear were able to survive when the others were going under. These firms, which had grown up on the island, were operated directly by their owners, while the old English and Jersey firms were owned in the British Isles and operated by agents (i.e., managers). Thomas R. Job stated the feelings of the local firms about agents when he wrote: "The business under one's own supervision is hazardous enough nowadays without leaving it to servants."[131] Related to this was the withdrawal of many of the local firms, especially in St. John's, from the operation of retail establishments in the outports. Rather than continue the older traditional merchant-fishermen operation, the St. John's firms dealt with small traders on a wholesale basis. This made it easier for them to control credit. The other reason for the relative success of the local firms was their heavy involvement in the seal fishery. For example, a rare document shows the operation of Job Brothers during 1861-71. The loss of £9,000 in 1864 is explained by the note: "Seal Voyage very bad"; the profit of £15,000 shown in 1869 is followed by the explanation: "*Nimrod* (sealing vessel) very successful"; and in 1871, which was the most favourable year of the period and showed a profit of £18,000, another note reads: "*Hector* and *Nimrod* both very successful."[132] Thus, the St. John's firms entered the 1870s with a very strong advantage.

At the same time, economic changes also adversely affected companies outside St. John's that were locally run and had strong seal fisheries. In 1858-59, Pack, Gosse and Fryer in Carbonear went out of business, leaving Rorke as the only substantial mercantile operation in that harbour. Much more serious, from the point of view of the outport firms versus the firms in St. John's, were the financial problems of Ridleys which arose in 1870 and which caused the company to go bankrupt in 1873. The size and importance of this enterprise has already been intimated, and when it went bankrupt with liabilities amounting to £250,000, it was not only a major blow to Harbour Grace but also to the Labrador fishery, to the fisheries of the northern ports and to the whole economy. It provided the unwelcome evidence that a large business could go under even while engaged in what was (at least on the surface) a well-balanced operation involving both the cod and seal fisheries. It was another casualty of the depression of the 1860s.

When placed within this commercial framework, the competition between steam and sail in the crucial seal fishery was to produce significant results. It was this competition and its results which dominated the industry, including ports and ship owners, during the 1860s and 1870s.

Decline of sail

In the meantime, the sailing fleet continued to play a very important role in the prosecution of the seal fishery. As pointed out, however, this fleet was an amorphous entity, and the vicissitudes of its various components are difficult to trace. The introduction of steam was a gradual development which took place during the 1860s and 1870s, so that the effects of the new technology on the traditional practices can be discerned over a specific and compact period of time.

In 1863, the first year that steam was introduced to the Newfoundland seal fishery, St. John's sent a fleet of 38 sailing vessels, with an average burthen of 120 tons and carrying 2,004 men, to the ice, but there are no statistics to indicate how many vessels supplied by this port cleared from the northern ports. However, all the regular major firms were represented among this fleet, with the following sending between one and six vessels each: P. Rogerson and Son; Baine Johnstons; McBride and Kerr; Bowrings; Stabb, Row and Company; Stephen March; E. White; J. McLaughlin; Goodridge and Company; A. Goodridge; McLea and Sons; Hounsell and Company; R. Alsop and Company; Thomas and Company; and R. O'Dwyer.

Conception Bay had a larger sailing fleet in 1863. From the Harbour Grace customs house, 41 sailing vessels, with a total burthen of 5,166 tons and employing 2,073 men, cleared for the ice. They were supplied, and many actually owned, by the following firms: Ridleys—17; Munns—16; Donnelly—5; Rutherford Brothers—2; and Daniel Green—1. The fleet which cleared from Carbonear was made up of 27 sailing vessels, having a total burthen of 3,460 tons and carrying 1,381 men, and it was supplied as follows: J. Rorke—8; Munns—3; Moses Wilshear—3; and a number of small suppliers. Clearing from Brigus were 24 vessels, with a total tonnage of 2,757 and 1,213 men in all, while

9 vessels, with a total burthen of 1,135 tons and a total of 497 men, left Cupids; in neither case, Brigus nor Cupids, is it possible to ascertain the suppliers. Bay Roberts reported that 12 sailing vessels cleared from there, with a total burthen of 1,555 tons and 672 men. With supplier names like Dawe, Mercer, Bartlett and Parsons, it is obvious that most were supplied locally.

The 'ports to the northward' continued to support the sailing component of the seal fishery. Seven sailers cleared from Trinity for the ice that year: 3 were supplied by Baine Johnstons; 2, by Brooking and Company; 1, by Bowrings; and 1, by Grieves. Farther north, 10 sailers cleared from Catalina: 4 were supplied by Ridleys; 2, by K. McLea and Sons, and 1 each, by Stewarts, Jobs, and P. Rogerson—thus 4 were supplied by Harbour Grace firms and 6, by firms from St. John's. And finally, 16 sailing vessels cleared from Greenspond: 14 were supplied Stewarts, Brookings, Baine Johnstons, K. McLea, L. O'Brien, and McBride and Kerr—all of St. John's; and 2, by William Cox and Company.

While this list does not cover the whole island because very few of these annual records are complete (and especially missed are reports from Twillingate and Trinity Bay south), it is apparent that the seal fishery had come under the control of the St. John's merchants by 1863, with a large share of the business going to two specific Harbour Grace firms.

The 1860s were difficult years for the sailing vessels. The weather and ice conditions prevalent on a couple of occasions resulted in heavy losses—as in 1864, when between 26 and 30 vessels were lost (see Chapter 5)—and production was the lowest since the industry became established (see Chapter 2). In addition, there was an increase in competition, which has already been noted, and as a result, the weaknesses of the sailers vis-a-vis the steamers were highlighted.

Problems with research sources make it difficult to quantify the trends in the sailing vessel seal fishery during this period. Two difficulties arise at this point: firstly, the growing emphasis on steamers and their peculiarities, their novelty, the competition among the hundreds of masters to be appointed to them, their effectiveness and speed all attracted attention; the sailers, on the other hand, were considered second rate, and records of their sailing to the ice fields were not as conscientiously reported in the local papers, particularly since many of them cleared from northern ports.

However, during the latter years of the 1860s, it seems that the number of sailers employed in the seal fishery remained fairly firm. For example, as Table 3.6 illustrates, Harbour Grace averaged 53 clearances, including steamers (except the *Mastiff*, which Ridleys always sent to Catalina for a crew and which cleared from there), during 1867-70. In addition, isolated reports make it clear that other Harbour Grace vessels, as well, were clearing from a number of ports, which means that the size of this fleet was much larger than has generally been recognized.[133] For example, both the *Royal Gazette* and the *Public Ledger* indicate that Harbour Grace supplied 90 sealing vessels in 1868, distributed as follows: 40 vessels supplied by Ridleys; 40 supplied by Munns; 7 by Donnelly; and 1 each by Green, Devereaux and McBride. The latter newspaper describes the fleet as the "List of Vessels Supplied for the Seal Fishery at the Port of Harbor Grace" and sub-divides it according to the port from which each ship sailed. Thus, Ridleys' 40 vessels sailed as follows: 20 from Harbour Grace; 4 from

Greenspond; 6 from Catalina; 2 from Heart's Content; 3 from Port de Grave; and 5 from Carbonear. Munns' 40 vessels followed a similar pattern: 27 from Harbour Grace; 5 from Brigus; 6 from Carbonear; and 1 each from Leading Tickle and Greenspond. The remaining 10 sailed from Harbour Grace.[134]

Another report in the *Royal Gazette*, this time in 1870, states that the number of the vessels in the sealing fleet were as follows: St. John's—27; Harbour Grace—84; Carbonear—19; Brigus—19; Bay Roberts—13; Trinity—13; and Hant's Harbour—7—for a total of 182. However, an accompanying note points out that the number for St. John's "does not include the many vessels belonging to St. John's fitted out and sailing from northern ports."[135] It is obvious that the sailing fleet from St. John's was not being fully reported. The important point is the fact that the sailing fleet managed to compete with the steamers during the 1860s when, as has been shown, the latter were slow to demonstrate their superiority.

A new development was taking place by this time which may help explain why the numbers of sailers in the seal fishery remained reasonably large during the 1860s and that was the involvement of smaller sailing vessels in this industry. The Chamber of Commerce first remarked on this phenomenon at its general meeting in 1867, when it praised the fact that small vessels had begun to go to the ice, especially vessels that had never gone before.[136] The following year, the Chamber remarked on their use again and was pleased, in spite of the poor luck that they had in 1868:

> The seal fishery of the past spring proved to be in some respects disastrous, on account of a much larger number than usual of small craft having been fitted out for sealing in the Gulf of St. Lawrence, as well as on the northern parts of the Island. Many of these owing to boisterous stormy weather were wrecked and lost, but notwithstanding these casualties, opinion seems to be gaining ground that the substitution of small vessels, in place of those of the largest class now employed would be a beneficial change, as they can be built in this island, are less expensive in outfit and would be more generally and diversifiedly scattered through the ice fields surrounding our coasts....[137]

The growing use of small vessels during this period is evident when one examines the tonnage of sailing vessels clearing port, and fortunately, the statistics are nearly complete for 1868. Ridleys sent out vessels of 40 tons burthen that year, Munns sent some as small as 30 tons, and one cleared from St. John's with only 22 tons burthen and carrying eight men. The most likely explanation for this phenomenon is the fact that the firms stopped investing in sailers for the seal fishery because of the decline in the price of seal oil beginning in 1865 (see Table 2.8) and the growing interest in steamers. On the other hand, very little expense was involved in supplying small Labrador schooners or south coast schooners built for the winter fishery, for a voyage to the ice.

Meanwhile, the steamers called at the outports to take on crews for the trip to the ice and this also softened the challenge of the new technology that the men now faced; some steamers cleared customs from the outports. In 1868, for example, the *Mastiff*, as always, cleared from Catalina with 130 men; the *Panther* cleared from Brigus with 120 men; and the *Wolf*, from Trinity with 105. In addition, it is evident that other steamers, after clearing customs in Harbour

Grace or St. John's, stopped en route to the ice to pick up crew members. Thus, fishermen from the major outports who wished to go sealing were not dependent on the dwindling supply of older sailers, nor on the smaller vessels that were now in use, but could look forward to a berth on a steamer. Granted, the terms under which the catch was shared were not as favourable on steamers, where the men shared one-third the value of the catch, instead of the one-half shared by men on the sailers. However, the work involved on a steamer was less strenuous (see Chapter 4), and the possibility of success, so much greater. In any event, the fact that fishermen were picked up by the steamers and given the opportunity to go to the ice obscured the fact that the outports' share of the trade was declining, at least as far as the fishermen were concerned. Thus, although the battle between the two technologies was joined, its significance was not always obvious.

The sailing fleet remained a force to be reckoned with for a number of years. Furthermore, Munns and Ridleys may have even increased their fleets of these vessels in the late 1860s (depending on the actual extent of their involvement, as yet undetermined, in the northern outports). In any event, these firms were not reluctant to maintain strong fleets of sailers at this time. The same applies to St. John's, which supplied 20 sailers in the Brigus seal fishery alone in 1868.

While it is impossible to quantify the sailing craft engaged in sealing, the record of vessels clearing from Harbour Grace is available, and an examination of that port can illustrate the overall decline that occurred during the latter part of the nineteenth century. As can be seen in Table 3.6, the number of clearances from Harbour Grace decreased dramatically between 1867 and 1886. In the former year, 50 vessels, carrying over 2,500 men, left that port for the ice fields; in 1886, only 3 steamers cleared with 575 men. (However, this table does not record the fact that Munns supplied 4 steamers during this period.) In fact, the decline was even more precipitous than 50 down to 3 because, as indicated, Harbour Grace supplied a much larger fleet than 50-plus in the 1860s, when vessels from that port cleared from other ports on the northeast coast and from other Conception Bay harbours.

In every respect, the battle for survival waged by the sailing fleet was being lost; and it was a one-sided battle from the beginning. In 1872, it was reported:

> The outfit for the seal fishery this year will be smaller than that of last season, both here [St. John's] and in Conception Bay. Owing to the number of steamships engaged in this business, the owners of sailing vessels consider the chance of obtaining a fair catch of seals very small; hence the falling off in the sealing fleet.[138]

Another report, in 1874, shows just how rapidly the balance was shifting in favour of steam:

> The seal fishery this year will be chiefly prosecuted by steamers—no less than 23 being about to leave or have left, St. John's on that voyage. The few sailing vessels of the sealing fleet left port last week.... We wish them success.[139]

The following year, the remaining vessels in the sailing fleet demonstrated, once again (amidst a continuing decline in the price of oil—see Table 2.8), that they were not able to compete with the steamers. On 24 August, the Chamber of Commerce regretted

> to have to report that with few exceptions the sailing vessels were entirely unsuccessful caused by the enormous quantity of ice packed on the coast and the unusually severe weather. The steam fleet was generally successful and on the whole the catch by steamers was fully up to an average.[140]

Steam was, in every sense, pulling ahead of sail, and in 1877, when conditions looked good for the few sailers at the ice, one of the capital's newspapers once again lamented the loss of sail while, at the same time, accepting the inevitability of steam:

> One cannot help regretting at such a time as this the great decrease in the number of sailing craft as compared with old time outfits. They attract a larger degree of what may be distinctively called popular interest than do the steamers, and with fair reason; for there can be no question that when the former do succeed, the results are more advantageous to the men and to the general public than are those that come of the good trips of steamers.
>
> We have, however, to remember that ... as a rule we must trust much more to steamers for successful voyages. It is very much a question of wind and weather; and as steam gives comparative independence of these conditions, it takes high vantage ground in our calculations of probable issues. Steam power can prosecute this fishery amid surroundings which render helpless and impotent the best manned and best equipped ship that has to trust her movements to the winds; and for this reason steamers will generally bring into port thousands of seals which would never be taken by vessels less able to cope with the risks of the adventure.[141]

Another writer reminded his readers that the decline in sail was leading to an increase in unemployment—a sad result and one "for which we are yet without a remedy."[142]

By 1881, it is most probable that the vast majority of brigs had disappeared from the seal fishery. It had been reported in 1848 that eight years was the average life span of one of these vessels; therefore a decline in investment in wooden sealing vessels after 1863 would be quickly felt in the industry. Consequently, it is to be expected that the wooden fleet had declined considerably by 1881 when it was reported to be as follows:[143]

	Vessels	Tons	Men
St. John's	15	5,107	3,294
Harbour Grace	11	1,884	1,175
Catalina	4	1,007	670
Carbonear	2	299	146
Brigus	3	321	165
Total	35	8,618	5,450

Cross checking with the clearances from Harbour Grace, one can see that the 11 vessels recorded above consisted of 8 sailers—including the famous 154-ton *Isabella Ridley* under the command of the equally famous Capt. Arthur Thomey—and three steamers—the *Commodore, Mastiff* and *Iceland*. However, the S.S. *Greenland* had been cleared from Harbour Grace for Greenspond on 3 the S.S. *Greenland* had been cleared from Harbour Grace for Greenspond on 3 February;[144] therefore, Table 3.6 does not show the complete picture, since the *Greenland* is excluded from the figures shown. In the meantime, when one examines Chafe's records, it is clear that the sealing steamers cleared as follows: St. John's—13; Greenspond—2 belonging to St. John's and 4 belonging to Harbour Grace; and Catalina—4, Channel—2, and Trinity—1, all belonging to St. John's. It seems most likely that 2 sailers were among the 15 vessels clearing from St. John's, as reported in the *Public Ledger*. In the case of Harbour Grace, 3 steamers were recorded twice; once as having left their home port and again as having left Greenspond. But, of course, the problem of making an accurate count of the sealing fleets has arisen before, and allowances must be made for this. Therefore, it seems more than likely that the old sailing fleet had declined to as few as 15 vessels—2 from St. John's, 8 from Harbour Grace, 2 from Carbonear, and 3 from Brigus—by 1881. The evidence suggests that Harbour Grace was fighting a rearguard action at this point—that the old sailing fleet had disappeared everywhere else, with a very few exceptions, and thus, the sailing fleet clearing from this town was the end of the line for this old industry.

On 29 April 1882, when the brig *Isabella Ridley*, under the command of Capt. Henry Thomey, Arthur's son, arrived from the ice fields, Capt. Henry reported that he had seen thousands of seals but that the ice was too thick to penetrate. He was confident, however, that the steamers would be successful.[145] The *Evening Telegram* wrote at the beginning of the seal fishery that year that it was essential that the spring's endeavour be successful. Then it continued and drew a picture of sealing in days gone by:

> It is said and no doubt truthfully too, that many remarkable changes have taken place since then, but none of them for the better. Not a few of us can still remember the busy aspect presented by the various mercantile establishments of this and the Bay metropolis [Harbour Grace] during the winter season when hundreds of staunch little vessels were repaired and made ready for the voyage—thus affording remunerative employment to large numbers of our people. These were lively times in Newfoundland and we long for their return in vain. The "festive" season, too, was far more festive than now. On St. Stephen's Day thousands of the outport "toilers of the sea" crowded into our great commercial towns to tender to their "old skippers" the compliments of the season and make sure that their berths were kept in readiness for them, and these annual visits from the "baymen" as we used to call them were beneficial in more than one respect. Besides amusing us by exhibitions of many-coloured costumes and general jollity, they circulated no small amount of hard cash during their brief stay, for, be it remembered, their circumstances were much better than the majority of our "seal hunters" of today ... and later in the season, when they came with their guns, bats, clothes bags and boxes to get their vessels ready for sea, their presence infused new life into every branch of trade and made the whole city throb with business excitement.[146]

This story became a familiar one during the following years. In 1883, it was reported:

> The brig *William*, Capt. Stephen Whelan arrived at this port from the Northern icefield about noon yesterday with between six and seven hundred old seals. Capt. Whelan's news seems to be in the main very similar to

what we have received from the commanders of other sailing vessels. The steamers took the lead and kept it all spring, and as they passed through the different patches of hoods and harps everything in the shape of a seal was picked up. Verily, the days of our sailing fleet are numbered.[147]

By now, the era of the sailing seal vessel had truly ended.

Meanwhile, by the mid 1880s, fishermen were beginning to use their small schooners in the seal fishery in a manner reminiscent of their forefathers at the beginning of the century. These schooners had been built for other purposes: on the northeast coast, they were primarily used in the Labrador cod fishery and for coasting (carrying freight and fish to and from major centres, especially St. John's); on the south and west coasts, they were built for the winter cod fishery just off shore and also for coasting. In 1886, a letter to the editor of a local newspaper pointed out that the number of schooners in the seal fishery was increasing, and the writer was relieved because he was convinced that the costs were lower and the rewards to the fishermen were higher.[148] The *Candid*, for example, a schooner with ten men, brought in 1,019 pelts that spring, which were sold for £443 15s 1d (currency) with the men receiving £21 8s 9d each, or about $85. In 1888, five schooners (with a burthen of 27 to 50 tons each) from Channel went sealing, as did five from Sandy Point (15 to 38 tons burthen) and seven from Trinity (ranging from 41 to 68 tons burthen).[149] Another example involved the firm of James Baird, St. John's, which supplied the following schooners for the Gulf seal fishery in 1893: "The *Alice May*, Capt. Buffet, 56 tons, about 14 men, from Channel; The *Albert F.*, Capt. Keeping, 32 tons, about 10 men, from Channel; and the *Dorothy*, Capt. Gallop, 48 tons, about 18 men, from Codroy."[150] (The *Dorothy* was crushed in the ice and lost, but all lives were saved.)[151] The fact that a small schooner, the *Island Gem*, Capt. Rodger commanding, with about 2,700 pelts taken from about 100 miles east of the Funks, was the first to arrive in St. John's in 1893 with a 'load' of pelts did not go unnoticed by observers.[152] There were many reports, this year, of schooners arriving at various ports around the island with 'small' quantities of pelts. While these little schooners did not contribute much to the total economy, they did not cost much and could make a considerable difference to a family's income. They continued to operate throughout this period and were reported carrying landsmen's pelts into St. John's as well as the results of their own catches.

It is apparent that the schooner seal fishery was an important source of income to small groups of people in particular areas. Often this source of income was quite valuable: In April 1895, the schooner *Rise Over*, commanded by Capt. William Eastman, landed 1,406 pelts in St. John's, and the crew shared $39.88 each; the *Amazon*, commanded by Capt. John Eastman, landed 1,191 pelts and paid each man $44.32 each; and the *Charming Lass*, commanded by Capt. Garcia, unloaded 1,103 pelts, and each man received $54.43. These shares were comparable to the shares earned by the steamer crews during this period, and all the men made a profitable voyage. However, the three vessels employed 21, 18 and 16 men, respectively, which was far removed from the 200 to 300 men employed by each of the large steamers. However, not all reported schooner arrivals were successes; a couple of days after the arrival of the three above, the *Portland* arrived in St. John's with a crew of 35 men and a cargo of 30 pelts.[153]

While it is impossible to ascertain the ports of origin of these schooners, it is probable that they were from the south and west coasts, particularly the latter, where there was less Arctic ice. A contemporary commented the following year:

> The schooner *Annie*, Capt. Phillip Smith, arrived here last evening, before the storm came on. She is from Hant's Hr., and is fitting out for prosecution of the Gulf seal fishery. Her crew have been selected at Hant's Harbour and Scilly Cove. It is to be hoped that the venture will prove a paying one. Sailing vessels are so comparatively scarce going to the ice these years that they almost excite wonder. Of course, there are a few schooners sailing from the West Coast.[154]

This point was substantiated by other accounts. For example, in 1896, a report (one of many) from the southwest coast stated that the following schooners were sealing: *Harvest Home*, Capt. Solomon Gillam, from Burnt Island (wrecked in the ice near Prince Edward Island); *Sisters*, Capt. Collier, from Channel; *Comet*, Capt. Bragg, from Channel; *Beatrice*, Capt. Barter, from Channel; *May Queen*, Capt. John Gillam, from Channel; and *Florence B*, Capt. John Baggs, from Rose Blanche. The same report pointed out that there were several others, and all of them were being supplied by James Baird, St. John's.[155] It is obvious that at least one St. John's merchant was adopting the practice that had flourished in the cod fishery at various times; by supplying planters and vessel owners, merchants could reduce their own risks, as they were doing in the saltfish trade at this very time. However, it is not clear that St. John's merchants were supplying all the schooners from the southwest coast that spring, and there were, at least 26, ranging in burthen from 20 tons to 106 tons and averaging 52 tons.[156] Also, two other schooners from Bonne Bay were lost in the Straits of Belle Isle that spring, which suggests that sealing craft were operating all along the west coast—despite the disadvantages experienced by Newfoundlanders living and working on the French Shore prior to the Anglo-French agreement of 1904.

By 1897, it seems that the schooner seal fishery had become, primarily, a southwest coast industry which concentrated on the Gulf seals. One correspondent wrote in February of that year that the schooner masters from Bonne Bay to Burgeo were "earnestly praying that the owners of steamers will send their ships on the front, so as not to interfere with them...."[157] However, their wishes were not fulfilled, and steamers did go into the Gulf in March, but according to reports, the schooners managed to get their share.[158] As one commentator wrote: "Capt. Young [of the *Winnie L* from Channel] has done remarkably well this season, and almost all the Codroy fleet have followed in his footsteps."[159]

Meanwhile, in 1898, the local government passed legislation to encourage the prosecution of the seal fishery by offering financial assistance to vessel owners. The legislation granted to the outfitter of every registered vessel that prosecuted the seal fishery from a port in Newfoundland the sum of four dollars per registered ton of each vessel (with some variations). The conditions under which this bounty could be claimed were published under the title "Regulations for outfitting for the Seal Fishery of sailing Vessels upon which it is intended to claim bounty."[160] These included the following requirements for each vessel: to be well sheathed with one-and-one-half to two-inch sheathing; to be well pounded; to carry a spare rudder; to carry hawsers of a specified length and size; to carry one boat for every four men; to carry for each man one bag of bread, one-quarter barrel of flour, ten pounds of pork, ten pounds of butter, one pound of tea and three gallons of molasses; to remain at the seal fishery from 1 March to 25 April; and to undergo the appropriate surveys and inspections. This was particularly applicable to the schooner owners, because any vessels that charged for berths were excluded. Apparently, the amendment which excluded vessel owners from both charging for berths and receiving a bounty was forced on the government by the opposition under Robert Bond.[161] However, one editorial writer was not impressed by this initiative:

> It is generally recognized that with the exception of Channel and a few other ports on the Western portion of the island, the proposed four dollars a ton bounty for sailing vessels will not be taken advantage of. The risk to schooners will not justify the few middle-men who own them in sending to the ice. In order to insure, say a fifty ton schooner at five per cent, it would take $50, valuing the schooner at $1000. Deduct this $50 from $200, the amount of the bounty money, and $150 are left to outfit the schooner. No man who has a good schooner is going to risk her on such poor inducements. It would be better to take this money and distribute it among the poor.[162]

The wisdom of this measure was questioned further, in great detail, by an anonymous writer from Placentia Bay on the island's south coast:

> The bounty of four dollars per ton to sailing vessels is of great importance to our friends of the northern ports, who have large vessels to engage in this important industry; but as is generally the case with bounties, it reaches the pockets of those who are least deserving of such assistance. The schooner holders are generally merchants or those so closely identified with them, that money for this purpose drifts into the pockets of the latter and is of no material advantage to the men who engage in this hazardous pursuit. The sealers are deserving of every encouragement; but this hoodwinking legislation is too apparent and the effort to build up business men at the expense of the taxpayers will not meet with general approval.... But what of those 50,000 men who yearly engage in the codfishery? No assistance for those; no encouragement; but they are to be taxed to death to keep up a fishery which turns them in nothing.... I think the Government should look to our fishermen and encourage the shore and Bank fishing as well as the sealing, giving a bounty of four dollars a ton to every craft engaged in it.[163]

One has to agree that the regulations under which the bounty was to be paid seemed to require a fair amount of expense, but the pounds, sheathing, spare rudder, hawsers, boats and food were basic requirements and, no doubt, had already been adopted by serious schooner owners. Consequently, it is difficult to accept the fact that the conditions under which the bounty was granted were onerous.

Nevertheless, it is impossible to ascertain any increase in the productivity of the sealing schooners as a result of this bill because of the unpredictability of the catch; but from the financial records, it is apparent that grants ranging from $1,000 to $2,500 were paid for shipbuilding in the years immediately following the passage of this legislation; and this sum had risen to $16,000 annually by the end of the period.[164] In the meantime, the fleet of sealing schooners in Channel alone numbered an impressive 27 in 1898, and there is no doubt that there were more sailing out of other southwest coast ports.[165]

Nevertheless, it seems that the legislation that had been passed to encourage sailing craft to engage in the seal fishery had not made much difference. Many felt that the "good schooners that fit out for the Gulf fishery would prosecute the voyage anyway ... [and] the bounty otherwise only tempts the fitting out of unsafe vessels which get no seals and only wreck themselves."[166] This point continued to be reiterated for some time to come.

The reports on the schooners engaged in the seal fishery continued to appear in the local papers, and it is obvious that a considerable seasonal activity had grown up on the southwest coast by this time. The schooners in this area had a distinct advantage over their counterparts on the northeast coast. On the southwest coast, the winds were usually offshore, which meant that it was very unusual for sailing vessels to become jammed in port—a common occurrence on the northeast coast. However, in 1899, the results of the Channel schooner seal fishery were disappointing, as only two made a profitable voyage while the rest did not average 100 pelts each.[167]

The following year, 1900, was much better, and although the total voyage was not recorded, overall satisfaction with the fishery was expressed. It was pointed out that the 50-ton *May Queen*, under Capt. John Gillam, had brought in a cargo of pelts to St. John's numbering almost 1,700 and that his crew of 12 received $83.54 each. At the same time, the crew of the *Jubilee* received $66.47 each for their load of 1,376 pelts.[168]

Nonetheless, by 1901, there had been a decline in the Channel sealing fleet of schooners, because only 6 cleared that port for the ice in that year.[169] These schooners and others from the area, however, did well, and during the month of May that year, there were reports of several arrivals in St. John's: the *Maggie A*, under Capt. Dicks, with 1,100 pelts; the *Swan*, under Capt. Adam Power, with 640; the *Goodship Jubilee*, under Capt. W. Carter, with 930; the *Orlando* from Codroy, under Capt. P. Young, with 1,200; the *Eldora*, under Capt. Seeley with 1,703; and the *May Queen*, under Capt. John Gillam, with 804. Also, the schooner *Gardiner* from Greenspond, under Capt. Alph Barbour, arrived with 1,000 pelts, and another craft, also from Greenspond, with 700. The *Orlando*'s crew of 17 received $39.62 each; the *Eldora*'s crew of 19 received $60.70 each; and the *Maggie A* 's crew, $42 each.[170]

The next year, 1902, was also a rewarding one for the Gulf schooners, and the following were reported: the *May Queen*, under Capt. John Gillam, with 2,500 pelts; the *Winged Arrow*, under Capt. J. Davis, with 2,500; the *Escort*, under Capt. S. Hall, with 1,100; the *Enthusia*, under Capt. Bond, with 530; the *Maggie A*, under Capt. Dicks, with 975; the *Eldora*, under Capt. Seeley, with 1,381; and Capt. Carter, who lost his schooner the *Goodship Jubilee* earlier, arrived in another schooner with 8,000 pelts. In addition, the *Escort* brought 385 pelts for Paul Hall, whose 10 men received $21.76 each, and 670 pelts for Stephen Hall, whose 8 men received $28.42 each. At the same time, the crew of the *Maggie A* received $53.71 each, while the crew of the *Eldora* received $63.46. Jobs purchased all the pelts from these Gulf schooners.[171]

The schooners from the southwest coast continued to prosecute the seal fishery, but like the early sailing fleet on the northeast coast, they found it increasingly difficult to compete with the steamers, and ice conditions caused problems on occasion. In 1903, because of the ice, the schooners could not get out of Channel until 14 March, and then, at least one crew found that "the seals were badly cut up by the steamers and Capt. Gillam's crew were forced to travel six miles over the ice with every seal brought on board."[172] Other schooners had similar experiences, and some failed to even meet expenses. It seems that the peak of the schooner seal fishery had passed by this time, because in 1905, it was reported: "Only one sailing vessel, the *Jubilee*, Capt. W. Carter, will prosecute the sealing voyage in the Gulf this season, leaving Channel this morning. Owing to the severity of the winter and the Gulf being packed with ice the past ten weeks the other schooners would not venture."[173]

However, one must keep the different branches of the sealing industry in perspective. The steamers were the giants of this fishery—the large factories of the industrial revolution—while the schooners were the little cottage industries. In 1900-02, these two types of vessels produced as follows:[174]

	Steamers		Schooners	
	Pelts	Value	Pelts	Value
1900	353,099	$483,528	4,739	$6,081
1901	344,936	388,501	5,680	6,438
1902	274,219	402,435	16,998	22,820

Nevertheless, these small sailing vessels were also used for fishing and trading during the appropriate seasons of the year, and their sealing voyages were vital to the small economies of the isolated outports of the south and west coasts.

Larger firms and centralization

Other implications of the change from the older technology to the newer concerned the disappearance of smaller investors, especially in the outports but also in St. John's, and the concentration of the proceeds from this industry into fewer and fewer hands. Responding to the general disappointment with the economy, despite the biggest and most valuable seal fishery since the 1850s, the *Newfoundlander* recognized and elaborated upon these problems in 1871:

> The state of things we have lately witnessed and of which we hear so much, as being consequent upon the Seal Fishery, is peculiar and suggestive. That Fishery, as we all know was the most productive within memory. From this fact we should naturally expect to flow an unusually large circulation of money throughout the various departments of business. Marketable commodities in general should change hands actively, and shop-keepers should be able to report a brisk money trade and exhausted stocks. Instead of these cheering signs and sounds, we have heard all through the spring the frequent complaint from business people that money was seldom or never so scarce; that sales were slow, and the shop trade little better than depressed, even in those articles which one would expect to see in lively demand. So that, notwithstanding the good beginning of the year with a bumper seal fishery, the result does not appear, as far as surrounding indications go, to have told in any marked or appreciable degree on the condition of the large

majority of our people. We do not believe this statement would hold good of the people of any other country after a season of like prosperity, and so remarkable a peculiarity deserves consideration.

That so much wealth has been so speedily realized is undoubtedly due to the employment of Steamers in the Seal Fishery; and however, gratifying the fact of this success, it is at the same time to be observed that as these steamers are all the property of capitalists, the proceeds of the voyage necessarily become more centred in their hands and proportionately withdrawn from the hands of the seal catchers and from general distribution. The crews of steamers share but one-third of the catch, two-thirds belonging to the owners; and as a general rule the crews are so circumstanced that they find it either a matter of interest or necessity to expend a large part of their earnings in supplying their wants from the stores and shops of the merchants whose steamers have given them employment. In this way it is clear that the bulk of the voyage remains with the proprietors of the steamers; and hence we are enabled to account for the lack of money and for the dullness in various channels of business which would have been enlivened by a more equable division of the profits of the enterprise.

But the superior efficiency of steam for the Seal fishery being established, we have to look to this agency for its future prosecution, and the great likelihood is that with the stimulus of recent success, the steam sealing fleet will increase. What may be the effect of this increase on the fishery itself, is a speculative question which we need not now discuss; but the employment of steamers, advantageous as it is in some respects, will not only have the effect of apportioning the greater part of the voyage to the capitalist, but will also throw out of employment half the men formerly engaged in sailing craft. We take, for instance, either of the largest trips of last spring—say 28,000 seals, this would give a fair paying voyage to nine sailing vessels carrying four hundred men; while the steamer did the work with about 180 men. Is this an argument against steam? Certainly not. We should no more think of so using it than we should try to show the benefits of stage coaches as compared with railways. Steam has shown what it can do in Sealing as well as in its application to other industries, and we all hope and believe it will hold its place. But the facts still warn us that in those prosperous results which we thankfully acknowledge, the conditions of abiding good to the general population are wanting. The drawbacks are such as must arrest the attention of thoughtful men in their calculations of our future fortunes. The point inevitably and most forcibly suggested, is the necessity of new employments for those who will be cast adrift by the abandonment of sailing vessels in the pursuit of which we speak; and this problem will demand solution at no distant day.[175]

The writer of the above, while failing to recognize the problems of over-exploitation of the resource and declining prices in the market place—the general dilemma—does identify the crucial local problems: the inevitable disappearance of the sailing fleet; the subsequent decrease in employment and the decline of the outport economies; and the concentration of the wealth of the seal fishery in the hands of a select few. This writer demonstrated considerable prescience. That same year, the St. John's Chamber of Commerce pointed out that the large catch of seals in general (including 178,769 by St. John's 12 steamers and 74,938 by Harbour Grace's three steamers) would "lead to a material increase in their [steamers'] number the ensuing season."[176] The Chamber's forecast was correct, and in 1872, the number of steamers from St. John's increased to 16, setting the stage for the conquest of the industry by steam.

Between 1865 and 1870, as Table 3.7 illustrates, the steamers' share of the pelts produced increased from 8 per cent to over 50 per cent; and this does not indicate the whole picture, because the companies that owned the steamers were also processing the oil from the pelts brought in by the sailers, except in the case of Harbour Grace. During the 1870s, the steamers continued to increase their share of the investment in manpower and ships, as well as their share of the catch. Seventeen steamers went to the ice from St. John's in 1873, supplied by the firms now established in the industry and commanded by the masters already experienced in the steamer fishery or from established sealing families. The fleet increased to 19 in 1874, dropped to 17 in 1876, rose to 19 in 1877 and to 22 in 1878, and increased further to 24 in 1880; but the catch was low that year, and the number of St. John's steamers declined to 20 the following year.

The seal fishery declined in the 1880s and the number of steamers decreased also, dropping from 24 vessels in 1880 down to 14 in 1885. The number of steamers in the capital increased to 16 during 1886 and 1887 and then declined to 15 for the next four years. The fleet increased again to 16 in 1892 and to 18 the following year, which experienced the worst seal fishery since the 1860s. The next year, 1894, was a terrible year for the steamers, which numbered 16 (but an unusually good year for the landsmen sealers, especially in the Twillingate area). Of course, it was in December 1894 that Newfoundland's two banks 'crashed', throwing firms out of business and bringing the fisheries to a near standstill.

Robert Stewart Munn died a few days after the 'Bank Crash', and the firm of John Munn and Company went out of business. The company had been struggling to stay afloat in the midst of a crisis in the markets for Labrador-cured saltfish. Munns was the biggest supplier and exporter involved in the production and sale of this product, and it was hard hit from the early 1880s, when French-cured fish, supported by French bounties, began underselling. It would be simplistic to suggest that the decline in the seal fishery caused the bankruptcy of this firm, but it is significant that the firm's four steamers—the *Greenland*, *Iceland*, *Mastiff* and *Vanguard*, with about 900 men in all—brought in only about 20,000 pelts in 1894 and a similar number in 1893. Furthermore, there had been a number of times when one or more of these steamers had not even paid expenses during the previous decade.[177]

At the same time, it was becoming apparent that the competition was no longer only between the outports and St. John's, or between sail and steam; now that market problems were making the seal fishery a greater investment risk, there was serious competition among the St. John's steamers as well. A local newspaper described in 1883 what was to become more common than most participants were prepared to admit:

> The steamer *Falcon*, belonging to the enterprising firm of Messrs. Bowring Bros., and commanded by Capt. William Knee, arrived at this port about nine o'clock today with 21,000 prime seals. Capt. Knee tells a "strange and eventful story" in relating what he has passed through since the commencement of the voyage. The *Falcon*, it seems, was one of the first vessels to strike the seals this spring, and having a "Tip-top crew", in a very short time they

managed to kill and pan, or haul together, more than enough to load the steamer. But there were pilferers around, it is said, and some of these—like the free-booters of the Whiteway-Shea Ring—plundered everything in the shape of a seal they could lay their hands upon, whether dead or alive, marked or unmarked. Why, they didn't even respect the flags hoisted over the different pans, but carried them off also, compelling Capt. Knee and his crew to kill a great many more seals before they got their good ship loaded; and almost heart-rending to relate, when they steamed into port this morning, the *Falcon* hadn't a solitary yard of bunting to display—all had been stolen! In fact, she moored up to Messrs. Bowring's wharf more like a funeral barge than a steamer with a full cargo of well-developed young harps. Truly the demoralizing influence of the present Government must be something dreadful to contemplate, when its "tentacles of corruption"—if we may be permitted to use the simile—even extend to our naturally honest and straightforward sealing masters.[178]

Leaving aside the farfetched connection between this incident and the government's reputation, it was a sad indication of the type of situation that was developing. In the very next edition of this newspaper, Capt. Samuel Blandford, master of the S.S. *Neptune*, having discovered that he and his crew were being accused of theft, responded in a letter to the editor, which throws further light on the problems associated with panning. He wrote:

> I arrived from the seal-fishery at 11 o'clock A.M. yesterday and had not gone ashore when I heard from more than one of my friends that I had been accused of stealing panned seals whilst in the prosecution of the voyage, and that my name had been used by many as having taken seals panned by the crew of the S.S. *Falcon*.
>
> I deny having taken one seal other than my own, and defy Capt. Knee to prove that I have.
>
> My friends reported that flags were taken away and seals stolen by my men. Now I will lay the facts bare. About seven o'clock on the morning of Friday, the 30th March, I steamed up to 'panned' seals which were 'flagged'. I was myself at that time in the barrel; my men, seeing many seals about, jumped on the ice and some of them knocked down four of the flags from one pan and replaced them with my flags. Seeing this, I immediately left the barrel and, going on the ice, ordered my flags to be taken away and same flags as knocked down replaced. This was done with three, the fourth (a small red handkerchief about 12 in. square) I could not find; but at once took my own handkerchief and placing it on the pole, put it up; then, joining my own ship, hoisted the ensign and called my crew on board. One man before getting over the vessel said: 'I have hauled 14 seals to the pan with the flags on.' To which I replied: 'Let them remain there.' Not one seal from them was put on board the *Neptune*. When this occurred two vessels were within 50 yards of my vessel, portions of their crews were on the ice panning seals, and some of the men were not many yards from me whilst I was replacing the flags. To this, no doubt, the Captains of these two steamers with many of the crew, can testify, if necessary. Having all my crew on board, I steamed away about six miles and picked up my seals (some 7000) killed and panned....
>
> I much regret that Capt. Knee has made use of my name in this transaction so freely, and, as I have reputation and character to sustain, I challenge him to a proof of his assertions; and further, I will add that, had Capt. Knee placed

men to watch his panned seals, as I do, he would be better enabled to say justly who it was that took his seals, if he has lost any.[179]

Despite Capt. Blandford's plea of innocence, he does describe how his men removed four flags from Capt. Knee's pans (all flags were identifiable) and replaced them with the flags of their vessel, while about 50 yards away, the officers and men of two other steamers looked on. One is left wondering whether this was the first time the men behaved in this way and also whether Capt. Blandford had ever intervened before.

However, the growing competition was not confined only to sail versus steam or to steamer versus steamer; it was also found in the relationship between crews and landsmen—both after the same dwindling resource. In 1886, for example, Capt. White, master of the S.S. *Hector*, complained to his employer, Jobs, that after his crew had panned 5,000 pelts just off Twillingate, about 2,000 local people showed up and "took them all." The *Mercury* reported this injustice and was immediately attacked by the *Evening Telegram*, whose editor sided with the local inhabitants. He wrote:

> Of course they "took them all," and they were perfectly justified in doing so. What legal or divine right, we should like to know, have sealing steamers to "kill, pan and claim," indiscriminately, right under the eyes of shoremen, without giving the latter a chance at all? We submit that crews of steamers possess no such right, and that it is a piece of gross injustice on their part to run in close to land and take the bread out of the mouths of poor people who live along the shore and depend upon the harvests of the sea for the support of themselves and families. Let crews of steamers kill their tows and haul them on board. This privilege they are entitled to. But they must not suppose that they hold a "patent right" to kill and pan every seal on the icefields, to the deprivation of their fellow-countrymen who may not have the advantages of prosecuting the voyage in powerful steamers.[180]

Leaving aside the feud between the *Mercury* and the *Evening Telegram*, this incident illustrates the increasing competition for the declining seal resource. In a similar vein the people on the south and west coasts complained in 1891 that the St. John's steamers had killed the adult hoods while the young were too immature to be worth taking but, at the same time, were too immature to survive on their own.[181] These are examples of the growing tensions in Newfoundland that were fuelled by the declining seal and Labrador cod fisheries—tensions which had already led, in 1883, to a riot in Harbour Grace resulting in five deaths, and to the collapse of the government in 1885.

The old sealing outports were dying. Trinity, for example—not a large port but one that had had a well-balanced economy, based on the cod and seal fisheries—was severely affected by the changes that had occurred in the seal fishery as well as by the general decline in that industry. A correspondent from there summarized the situation as it existed in March 1886:

> There is not a day's work of any description for the people to do. I have not heard of the capture of one seal in the bight this winter. Very few customers are to be seen in the shops. Indeed I do not think that shopkeepers here now make enough profit to pay for the coals they burn. When you look back, some 20 or 30 years, to the days of the sailing vessels, what a contrast! Then the harbour of Trinity was alive with industry. Just about this time of year

there would be a couple of fine brigs launched, and the fitting out of these would afford plenty of employment to the people. Some would be building punts, others sawing sheathing for the decks, and others again preparing pound boards and beams, all being employed right down to the poor crippled man in the corner making 'noggins' and buckets, and the women and children making birch brooms. Then the shops would be filled with customers to the number of two or three thousand people, buying up their 'crops' for the ice. Now, however, things are changed and all this is merely a 'memory of our departed greatness.' Well may we say 'the steamers are the curse of the country!' Every man in Newfoundland should protest against the further employment of steam in the prosecution of the seal fishery; we should start an agitation in every district and keep at it until the old profitable method has been revived. It seems to me such a glaring piece of injustice that a few merchants should reap the entire profits of this fishery and make fortunes in the same, while the great bulk of the people are in a semi-starving condition.[182]

This was also the time when several districts in Conception Bay submitted petitions to the House of Assembly, asking that legislation be passed to prohibit the use of steamers in the seal fishery. Apparently, the *Harbour Grace Standard* played down the issue, noting that one of the petitions contained only 200 names, while neglecting to explain that others contained over 700. (The *Standard* was owned by a member of the family that owned Munns.) A correspondent to the *Evening Telegram* asked:

Have not these steamers had the monopoly of our seal fishery long enough? Rights of property! Such Trash! Have no one else rights but the merchants? Who remunerated the owners of the sailing vessels, who were brought to begging by the introduction of steamers, for this loss? Who remunerates craft now lying idle on account of steamers? ... I may answer no one! Why? Because they are not merchants, therefore their interests are to be neglected. Knowing how they were represented, the Hr. Grace petitioners did not expect much support to their petition from their own members, one of them being absent, master of a 'sealing steamer,' one other holding his seat by order of the firm owning the same steamer and solely under that firm's control, and the other member, I am told acting as solicitor for the Dundee steamers. Looking at the above facts the petitioners could not expect much more other than they got; but could they not reasonably hope for support from other members, say of Trinity, Bonavista and Green Bays? But no, not one voice was raised in its support. — If the House of Assembly is only convened to look after the interests of the merchants, what need of the farce of an election?[183]

The success of the steamers in eliminating the old sailing sealers was complete by the early 1880s, and the chances of reviving the earlier shipping technology was out of the question (although, as pointed out, smaller Gulf schooners continued to operate). However, it is obvious that fishermen from the northeast coast were learning to adapt to the new circumstances as best they could. They began to commute to St. John's in order to acquire berths to the ice. As early as 1875, for example, when 21 vessels cleared from Harbour Grace to the ice fields from that port, at least 500 men from the Harbour Grace area travelled to St. John's on the regular steam ferry across Conception Bay to take up their berths on the St. John's vessels. (The fact that the ferry had a narrow escape is what impressed the local newspapers, and consequently, the event was reported.)[184] There are many examples of this happening. The *Evening Telegram* reported in 1880: "The time is fast approaching when we may expect to see our principal streets and thoroughfares thronged, as heretofore, with vast numbers of the 'bone and sinew of the country,' viz: our sealing friends from the outports."[185] The newspaper went on to repeat its often stated view that the decline of the outports and the old sealing fleet was a great misfortune and that this opportunity for the outport fishermen to find employment on St. John's vessels was a poor substitute.

In his autobiography, Nicholas Smith, of Brigus, writes about his second trip to the seal fishery, c.1880, on the S.S. *Panther*, under the command of Capt. Abram Bartlett, also of Brigus. Smith was given his berth by Capt. Bartlett, and he and the other men had to find their own way to St. John's, a distance of about 60 kilometres. Smith describes what happened as follows:

> Conception Bay was frozen over that spring, and there was no way of getting to St. John's except to walk around Conception Bay, while a team of horses hauled the men's clothing, etc. About fifty men of us left Brigus for the journey, and some eight or ten horses, at dawn on March 1st, and at 8 p.m. we arrived in the city, every man feeling fine. However, the horses did not arrive until midnight; feeding them and giving them a rest was the cause of the delay.[186]

Smith does not even mention the possibility of going on one of the sailing craft that left Brigus and other nearby harbours that year, although he discusses, in a fair amount of detail, the process involved in his and his friend's obtaining berths.

By the mid-1880s, with the completion of the railway to Harbour Grace, sealers from Conception Bay used this form of transportation to reach St. John's, and the trend extended to the other outports as the rail line passed the heads of Trinity, Bonavista and Notre Dame Bays. In fact, special trains were put in service to bring sealers from Conception Bay to St. John's, and special cars for the purpose were attached to the scheduled trains running to and from central Newfoundland (see Chapter 4). However, the adjustment that the fishermen were making in order to participate in the annual seal fishery was not contributing much to the economies of their towns and outports, as commentators were constantly pointing out. Of course, some steamers continued to clear from outports where they also hired sealers. However, it seems that Trinity and Catalina were no longer ports of call after the rail line passed Clarenville because of the number of men who were prepared to walk to the Clarenville station and go to St. John's by train. However, as Chafe reports, Greenspond remained a popular port of clearance and, no doubt, a place to hire experienced sealers as well; and, of course, for those steamers sealing in the Gulf of St. Lawrence, the port of call and clearance throughout this period was Channel (now Channel-Port aux Basques).

Competition from outside Newfoundland

While Newfoundland steamers and sailers had competed with each other—and the latter had lost—the former were forced to contend with competition of another sort—competition from outside steamers that encroached on what Newfoundlanders considered as their resource. Two steamers from Dundee, the *Polynia* and the *Camperdown*, were the first of such vessels to participate in the Newfoundland seal fishery, in 1862—the year before Grieves and Baine Johnstons introduced steamers into the industry; in fact, it has been pointed out that these steamers demonstrated the advantages of steam, even though they were completely unsuccessful in this first attempt. The *Montecello* from Boston, supplied by Ludlow and Company of that city, tried to take advantage of the resource during 1870, 1871 and 1872. That steamer had one successful year and was lost at the ice in its third year; the Americans did not try again.

In the late 1870s, two Dundee firms that were engaged in the whale and seal fisheries in the Arctic returned to participate in the Newfoundland seal fishery. It was reported by the Chamber of Commerce in 1876 that a Dundee steamer, the *Arctic*, was engaged in the seal fishery on the front "in close proximity to our local Steamers."[187] The Chamber went on to say that the appearance of the vessel on Newfoundland's coast had created considerable uneasiness, but that "this vessel's catch, (yielding about 40 tons oil) is a result not likely to stimulate a renewal of that enterprize." However, a second firm, Stephens, sent out the *Aurora* and *Arctic* in 1877 and, at the same time, ordered that local processing facilities be established. By now, Newfoundland public opinion was much more positive. A local paper reported:

> The arrival of the steamers *Arctic* and *Aurora* a few days since from Scotland, to proceed hence to the Seal-fishery, gives an important addition to our Steam Sealing fleet. These two vessels, manned almost wholly here, will take off over four hundred men who would otherwise have been idle and adrift about our streets. Since these boats arrived there has been a continual "rush" upon them by applicants for berths. Many of our best and ablest young men seem to have been totally dependent upon them for means of subsistence for the coming month, and are right glad of the chance thus offered.[188]

The next year, the Dundee Company did likewise, sending out the *Esquimaux* and *Narwhal*, and again, the press responded positively:

> Within the last few days four steamers *Arctic*, *Aurora*, *Esquimaux* and *Narwhal*, belonging to the Dundee Company and intended for the Seal fishery have arrived here. We need hardly say how welcome to our people are these visitors, which will take off for their crews about one thousand of our best men, who would otherwise probably have been condemned to spend their spring in idleness.[189]

By 1878, the two companies had overcome initial local hostilities and were employing about 1,000 Newfoundland sealers at the ice, and an indeterminate number ashore; the Dundee interests had become indistinguishable from the other firms. They added the *Resolute* and *Thetis* to their fleets in the 1880s and, in all, owned 12 sealers and made 93 sealing trips to the ice from Newfoundland. However, their participation declined in the 1890s, not because it was uneconomical per se, apparently, but because the Newfoundland seal fishery was not lucrative enough to offset the losses being sustained in their summer whaling and sealing endeavours off Greenland.[190]

A more significant challenge to the monopoly of the Newfoundland sealing firms was that mounted by Nova Scotia. In January 1893, it was reported that the inhabitants of Cape Breton were beginning to express an interest in the seal resources in the Gulf of St. Lawrence, which were being harvested each year by St. John's steamers and by west and south coast schooners.[191] However, when the invasion of Newfoundland's resource—and that was how it was viewed locally—occurred, it emanated from Halifax, not from Cape Breton. Capt. J. A. Farquhar purchased the steamer *Newfoundland*, which had been built in Quebec in 1872 and used on the winter mail service between Halifax and St. John's for a number of years, and was considered by observers to be well-suited for the Newfoundland seal fishery. The steamer was reported by the *Halifax Herald* as being well built, machinery in "splendid order," boiler and donkey boiler "almost new" and "was thoroughly overhauled, caulked and sheathing replaced."[192] This paper went on to point out that Newfoundland merchants and even vessels from the north of Scotland were taking "many hundreds of thousands of dollars" from this resource every year. Then, it was discovered in late February that Capt. Henry Thomey from Harbour Grace had been hired to go to Halifax and take the *Newfoundland* to the Gulf seal fishery.[193] Capt. Thomey was accompanied to Halifax by about 60 Newfoundland sealers, and on 4 March, it was reported that all was ready for the sealing voyage, with the added advantage of the fact that the vessel did not have to abide by Newfoundland laws regarding date of departure or any other aspect of the operation.[194]

The trip got under way in a very inauspicious manner—that is, if one can believe the Newfoundland media, which followed events with a keen, if rather biased, eye. On 9 March, it was reported that the *Newfoundland* had broken down several times and had finally limped into Cow Bay, Cape Breton, where Capt. Thomey and the Newfoundlanders were stranded. According to reports, Capt. Farquhar was not living up to his part of the deal, and the men had to send a message to the Newfoundland Prime Minister, Sir William Whiteway, asking for assistance; apparently, the Newfoundland Government replied promptly that the men be "amply provided for while there, and that they be sent on here by the earliest opportunity." The *Evening Telegram* concluded by stating:

> To our mind, it does not reflect any great credit upon Canadian speculators to endeavour to induce men from this colony to undertake such a hazardous voyage in an unseaworthy vessel.
>
> We are very pleased to know that our wise and humane Government are attending to this matter, and that they most heartily discountenance this attempt on the part of the proprietors and agent of the S.S. *Newfoundland* to speculate at the risk of the lives of our people.[195]

The headlines over the above commentary give some idea of the outlook of this newspaper, which had been lambasting the Newfoundland steamer operators for years—an attitude of enormous disapproval of the Halifax expedition: "Heartless Outrage; Capt. Henry Thomey and 60 of our Hardy Sealers sent away by Hon. M. Monroe, the Agent [Farquhar's] here, and left to Starve on the Streets of Cow Bay"; "*Newfoundland* condemned as unseaworthy and unfit to proceed to the Seal Fishery." (The picture becomes somewhat unclear at this point because it is not absolutely certain that Capt. Thomey left the vessel, although the indirect evidence suggests that he did.) On 16 March, the *Evening Telegram* reported:

> From reliable members of the steamer *Newfoundland*'s crew, who have just arrived here by the S.S. *Portia*, we get all the corroborative evidence necessary respecting the wretched condition of that ship and her utter unfitness for the seal fishery in which she is now engaged.[196]

However, later reports mentioned the *Newfoundland* on a number of occasions, and by 3 April, this steamer was reported in the Gulf and engaged in sealing.

Despite gloomy projections, the *Newfoundland* was quite successful in the Gulf and killed a cargo of 10,000 young and old seals which was the equivalent of 17,000 white coats. Not only was this a respectable load, but in addition, the vessel was the first steamer from the seal fishery to enter St. John's harbour, which was a double shock to the local establishment.

The *Newfoundland* was reported as under the command of Capt. J. A. Farquhar, but that, in itself, is no proof that Capt. Thomey was not on board as the 'ice captain'. It was quite common for sealing steamers to have two captains: one with a foreign master's certificate because the sealing steamers were leaving Newfoundland; and one with a lifetime's experience in the seal fishery.[197] Thus, Capt. Farquhar could have been in charge of clearing the *Newfoundland* out of Halifax and taking it to the ice fields, while it would have been Capt. Thomey's duty to carry out the sealing operation. However, the lack of reports about Capt. Thomey suggest that he left the steamer in Cow Bay and returned to Newfoundland with at least some of the Newfoundland sealers who had been with him. As evidence of this, when the steamer reached St. John's, Capt. Farquhar reported that they would have acquired a full load if he had had a complete crew.[198]

Farquhar's success caused some consternation in St. John's, which can be seen in the pages of the *Evening Telegram*. This newspaper was so convinced that the project would fail right from the beginning that it took some time for it to regain its equilibrium. At the same time, Farquhar was not able to sell his pelts immediately. Although the steamer arrived on 4 April, it was not until 10 April that Baine Johnstons bought the cargo.[199] By then, this company was aware that its two steamers had not done well, because a published report on 6 April stated that the most recent news from the *Hope* indicates that this steamer had only two pelts.[200] It arrived on 11 April with 2,687 pelts and paid its crew only $6.65 per man—not enough to pay the price of their berths and crops. The other Baine Johnstons steamer, *Panther*, arrived on 24 April with only 3,893 pelts and paid its men a little more—$17.27 for two months' work—and this amount was needed to pay for the berth and crop.[201] On the other hand, the crew of the *Newfoundland* made $68 per man, a share over 50 per cent higher than that paid on any 'Newfoundland' steamer that year; the next highest share was that received by the crew of the *Eagle* —$43.36.[202] Added to this was the fact that 1893 was an unusually unproductive year for the seal fishery, and no matter how much the St. John's merchants disapproved of Farquhar's encroachment, his pelts were a welcome addition to the limited stocks available to the oil plants that year.

Captain Farquhar was not a person to pass up an opportunity, and having done well in 1893, he prepared for the next season, deciding, as in 1893, to operate from Halifax. The *Evening Telegram*, in a much more conciliatory tone, wrote:

> The genial captain will again be in command of the S.S. *Newfoundland* on her sealing voyage the coming season. The steamer has for some time past been undergoing extensive repairs and alterations at Halifax, so as to fit her for the seal fishery. Capt. Farquhar will have no trouble in getting seal killers to go with him to the ice this year. Already many have applied for berths in the *Newfoundland*.[203]

Reports at the beginning of 1894 indicate that Capt. Farquhar took a great deal of care in preparing for the spring seal fishery. He engaged men from Halifax and the Cape Breton outports as well as a few from Newfoundland, and arranged to have coal available at Cow Bay.[204] He, again, did not concern himself with Newfoundland sealing legislation, leaving port when it suited and killing seals before the Newfoundland steamers were allowed to do so. On this occasion, he went to the front and managed to acquire about 5,000 pelts before the steamer's propeller was severely damaged. It arrived in St. John's on 2 April, before any local steamer, dropped off the few Newfoundland sealers and then left port, not being heard from again until it was reported stuck in the Gulf ice, where it had tried to take more seals on the way back to Halifax. In any case, the catch for that year was sold in Halifax, and five Newfoundland seal skinners were brought over from Newfoundland to remove the fat from the skins. How and where the rest of the processing was carried out has not been discovered to date.[205] With the sealing front a long way from Halifax, this must have proved to be too expensive and involved too much risk, so in 1895, Capt. Charles Dawe, Bay Roberts, was hired as ice captain, and the *Newfoundland* cleared from Bay Roberts with, presumably, at least some sealers from that area in the crew. Capt. Farquhar took command again in 1896 and during subsequent years until the ownership of the *Newfoundland* passed to Harveys in 1904. After 1895, he operated according to the legislation of Newfoundland.

Problems and changes

Although it is obvious that the St. John's fleet of sealing steamers and the firms and captains that owned and commanded them fared better than the owners and masters of the many sailing vessels that had lost the battle of sail versus steam of the 1870s and 1880s, the steamers, too, had their problems. As indicated earlier, the seal fishery was only one component of the economy by the late 1880s—and a component whose value was shrinking. When one combines this with the depressed state of the cod fishery and the growing expense of the steamers themselves, one can understand the financial crunch that the St. John's (and Dundee-St. John's) sealing industry was in by the beginning of the last decade of the century.

Grieves stopped operating in 1887 and the managing director, Robert Thorburn, went on to found Thorburn and Tessier, which did not become involved in the steamer seal fishery. Grieves' steamers were taken over by a new combination of investors who operated as the Newfoundland Steam Sealing and Whale Fishing company.[206] Stewarts went bankrupt in 1893;[207] Munns, in 1894. Baine Johnstons was not able to participate in the seal fishery in the spring of 1895 but had recovered sufficiently to return to the industry by 1896. With the exception of Bowrings, which came through the 1894 crash unharmed, the companies that survived the financial crisis were forced to compromise on their debts and reorganize. With the financial squeeze of the late 1880s and early 1890s, all companies were striving desperately to stay in business. It is no wonder that Capt. Samuel Blandford, having killed only 6,000 seals in the *Neptune* with a 310-man crew in 1893, was heard to remark, while returning to St. John's the following year, with only 5,000 pelts: "I have only a few more days to live; when I get ashore Mr. Job will hang me."[208] However, it appears that the St. John's firms that were engaged in the seal fishery weathered the depression of the late 1880s and the financial crises of 1894-95 better than the firms that were dependent solely on the saltfish trade. Among the latter were: Thorburn and Tessier, which went out of business owing the Union Bank $458,000; Edward Duder and Company, which owed the Commercial Bank $668,676; Goodfellow and Company, which owed the Commercial Bank $169,326; and John Steer, whose debt has not been recorded. On the other hand, Baine Johnstons owed the Union bank $618,000 and managed to survive; Jobs closed temporarily but recovered quickly; and Bowrings was relatively unscathed.[209] As a result, the steamer owners, scared of compounding any future losses in the seal fishery, re-organized their investments during this period by creating limited liability companies for each ship.

The steam seal fishery recovered dramatically in 1895 from its disappointing performances in 1893 and 1894; in the latter year, 133,000 pelts were taken, while in 1895, the total catch amounted to 235,000 pelts. This, of course, was a normal fluctuation for the seal fishery, whose annual catch was dependent on the men and vessels which made it to the seal herds. The industry also benefitted, in the short term, from the fact that second trips were permitted once more, beginning in 1895; this state of affairs lasted until new legislation banned second trips once again in 1898.[210] In the meantime, the industry continued to fluctuate widely.

After the bank crash in 1894, the seal fishery was monopolized by the St. John's steamers (although there remained the small schooner fishery—discussed above—and the landsmen fishery—discussed in the next chapter). Harbour Grace's departure from the prosecution of the industry marked the end of an era; from this point on, fishermen from the outports who wished to pursue the seal fishery were forced to look to St. John's companies, their agents and captains for berths. (However, Munns' old oil plant in Harbour Grace continued to be used by St. John's, so that there was some local secondary employment derived from the industry.) The railway became the favourite form of transportation to St. John's from Conception Bay and, to a large extent, from the more northerly regions, although some steamers continued to pick up their crews, or at least a portion of them, on the way to the ice. For example, in 1896, the *Labrador, Greenland, Leopard, Walrus, Kite, Algerine, Ranger, Panther* and *Diana* cleared from Pool's Island (Bonavista Bay north, in the vicinity of Greenspond, Badger's Quay, Newtown, Wesleyville and Cape Freels); the *Vanguard* cleared from Carbonear; the *Mastiff* cleared from Bonavista; the *Hope* cleared from Fogo; the *Harlaw, Nimrod* and *Iceland* cleared from Channel; the *Neptune, Esquimaux, Terra Nova* and *Aurora* cleared from St. John's; and the *Newfoundland*, under Captain Farquhar, who also supplied the *Harlaw*, cleared from Bay Roberts.[211] The majority of steamers hired most, if not all, of their men before leaving St. John's, but depending on various circumstances, including weather, there were generally commitments to local merchants to hire a few in other ports. In the case of the *Newfoundland* leaving Bay Roberts, Capt. Farquhar made it a practice to hire men from that harbour; and the *Vanguard* leaving Carbonear also hired locally, probably to make it more convenient for some of Munns' former sealers to join this vessel, which had belonged to Munns. In 1898, 5 steamers cleared from Channel; 11 from St. John's; and 1 each from Greenspond and Bay Roberts (Farquhar).

This became the standard pattern. Steamers often left St. John's and entered a port, usually in Bonavista Bay north if the steamer was proceeding to the front, and then on the legal day for clearing port, these steamers had an advantage over their competitors who cleared from St. John's. In 1896, Capt. Farquhar (who was forced to obey local legislation in order to use a Newfoundland port as his point of departure) was charged with clearing port before the lawful time. On this occasion, the *Newfoundland* left Bay Roberts early, having cleared for Seldom-Come-By, but did not enter that port. Capt. Farquhar was charged but argued that he could not enter Seldom-Come-By because of the ice, and when he found himself and his steamer still at sea when the date of leaving port arrived (10 March), he decided that it was in order for him to proceed directly to the ice floes. He produced his log and witnesses to give evidence to that effect and to also testify that the *Newfoundland* was among seals on 13 March but did not kill any because the season had not opened. He was also able to show that on Sunday, 15 March, he restrained his men from going on the ice because it was illegal for Newfoundland steamers to allow the men to kill seals on Sunday. He convinced the judge of his innocence and avoided the heavy fine which a conviction would have entailed.[212] (It is worth noting that Farquhar was not intending to take on any men in Seldom-Come-By but that he did expect to receive news of the whereabouts of the 'whelping' ice.) Another report concerning this procedure of sailing from St. John's to a northern outport involved the *Hope* in 1897. This steamer returned to St. John's near the end of February after an unsuccessful attempt to reach Greenspond—"an experience that seldom falls to the lot of our sealers (i.e., steamers) en route to the northern bays for crews." Even "under full steam and canvas," the *Hope*, under the command of Capt. Abram Kean, could not make it past Baccalieu because of ice.[213] Again, it is noteworthy that reference was made in this case to the fact that the St. John's steamers hired crews in northern ports.

In addition, only a few participants in the seal fishery were involved in seal oil production—and, once again, these firms were concentrated in the capital. The manufacture of seal oil had become more sophisticated; instead of letting the fat render naturally, the process was speeded up considerably with the application of steam heating. The final product was further improved by the new system of allowing the oil to mature in shallow vats covered with glass. Some idea of the increased speed of processing can be gained from the report on 5 May 1898 to the effect that Jobs had 1,000 tuns of seal oil, manufactured from 80,000 pelts, ready for export. Not only was this an early date to have such a quantity processed, but as the same report went on to point out, this was "the product of the trips of the firm's own steamers, the *Neptune, Nimrod* and *Diana*—and the purchased cargoes of the steamers *Ranger* and *Walrus*, and of some schooners."[214] While sealing had become almost completely centralized and monopolized by St. John's merchants, the manufacture of seal oil was even more centralized and in the hands of a small cartel.

The competition between St. John's and the outports may have ended with survival of the latter's fleet of steamers, but there was increasing competition among the St. John's steamer owners and captains and crews all during this period. While competition for berths to the ice was very evident, here the main focus is on the fleet, the owners' demands and the captains' efforts to meet these demands.

Throughout this period, competition was best illustrated by the increase in the thefts of pelts. The first real publicity given to this topic occurred in 1898, when the survivors of the *Greenland* disaster (see Chapter 5) complained that they were forced to continue sealing because other crews had stolen their pelts (reportedly, the crew of the *Aurora*, under Capt. Abram Kean); as a consequence of the theft, the *Greenland*'s men were forced to seek out more seals and were caught in a storm in the attempt.[215] Another incident that spring involved the theft of some of the *Iceland*'s pelts—allegedly, also by Kean's men. Apparently, after the *Iceland*'s crew had panned about 20,000 pelts and had them flagged (i.e., marked with their ship's flag), the *Aurora*'s crew came by, knocked down the flags and took some pelts by force.[216] The following year, 1899, a report stated that:

> ... the most bare-faced robbery has been perpetrated by the crew of the S.S. *Diana* upon the crew of the *Newfoundland*, according to reports more than 7,000 seals were stolen, and the ship's initial cut out of the pelts. It is further said that Capt. Farquhar has unquestionable evidence of this glaring plunder, and that the matter will be placed before the court.[217]

Although no further action seems to have been taken, there was considerable debate again over the pros and cons of panning pelts.[218] In 1900, a local correspondent wrote:

> The amount of thieving that has taken place at the seal fishery this year is unprecedented, if all reports are to be believed. By the *Aurora* letters were received [in St. John's] from three captains [still at the ice], each accusing the other of taking seals, the largest number being placed at five thousand.[219]

Capt. Samuel Blandford is reported to have said about his missing pelts: "These people were afraid that we would be too deeply laden and would never reach home, so are bringing the fat in for us."[220] On this same occasion, Capt. Arthur Jackman in the *Terra Nova* complained that two steamers had stolen thousands of his pelts. Apparently, almost 60,000 pelts were panned over a 15-mile distance, and none of the crews were too discriminating when there were no more seals left to kill and it was time to hoist the pelts aboard.

Killing and panning without any attempt to load had long been criticized as wasteful by observers of the industry, but the high cost of the steamer operation and the many factors that could force a vessel to return to port 'clean' all combined to encourage killing and panning to the exclusion of loading while there were seals available. The presence of other steamers in the vicinity only increased the pace, and even the fear of another steamer appearing over the horizon drove the men to kill and pan. They had a strong legal claim to pelts they panned; they had no claim, based on first discovery, to live seals. Nevertheless, theft of panned pelts had become a serious problem and even the practice of carving the ship's initials in the fat of the pelt did not prevent it.

By the latter part of April of that year, 1900, the *Terra Nova* and *Neptune* were both suing the *Esquimaux*.[221] Capt. MacKay of the *Esquimaux* settled out of court with Capt. Jackman of the *Terra Nova* by giving the latter 1,000 pelts. However, the other case went to court; Capt. MacKay lost and was forced to give Capt. Blandford of the *Neptune* 2,000 pelts, although the jury was unanimous in its opinion that Capt. MacKay was personally blameless.[222] In 1902, it was reported that the theft of pelts was worst than ever. One captain had a pan of several thousand pelts with his firm's flag over them and returned to find that the flag had been removed from the pole and another installed; however, the thieves had failed to notice that the original firms initials had been branded on the base of the pole.[223] The problem of theft continued to plague the industry.

As part of the competition to gain an advantage over the other owners, Capt. Farquhar, in 1903, decided to clear from a Nova Scotian port in order to reach the Gulf seal herd earlier than the others. However, in order to prevent Newfoundland steamers from doing this, the local legislature passed a law prohibiting the importation of seal pelts into Newfoundland unless the steamer involved had cleared from a local port under the regulations in effect. This had the effect of plugging this gap in the legislation and also kept the fleet bound to St. John's, where infrastructure, expertise and commercial connections were all centred. Capt. Farquhar came on to St. John's with his load of pelts in the *Newfoundland*, and was required to post a bond of $4,000 before unloading them. In the meantime, he acquired documentation from the Canadian Department of Marine testifying to the fact that he had been required to leave port early in March to go to the assistance of two Canadian steamers, the *Minto* and *Stanley*.[224] He was thus able to satisfy the Newfoundland authorities that the *Newfoundland* should not have been made subject to the legislation and the $4,000 was refunded to him.

In 1904, Capt. Farquhar left the industry briefly. At the same time, the Canadian press complained about the unfairness of Newfoundland legislation governing the industry. A local newspaper, convinced that Farquhar was behind the complaints responded:

> We have proved that we are not "too green to burn." In 1900 Capt. Farquhar was compelled to pay duty on his herring outfits. In 1901 he was compelled to pay duty on his sealing stores. In 1902 he was compelled to transport his crew to their homes. In 1903 he was penalized for breaking our sealing laws. Four times in succession he tried to get ahead of us and was beaten every time. No wonder he got tired and quit![225]

Capt. Farquhar had made the greatest effort, as a neighbouring Nova Scotian, to break into the sealing industry, and he had been a successful captain.

However, he concluded that he could not operate from Nova Scotia for geographical and legal reasons, and shortly before the 1904 season, he sold his steamer, the *Newfoundland*, to Harveys and gave up this fishery. (Nevertheless, his departure would not be permanent, and in 1906, he would return to the Newfoundland seal fishery in the *Havana*, in 1910, in the *Harlaw*, and in 1912 and 1913, in the *Seal*.[226])

The sealers' strike in 1902 (see Chapter 6) brought matters to a head and focused attention on, among other things, the somewhat precarious financial state of the steamer industry. One of the owners, James Baird, spoke of the difficulty of making a profit from this industry and the impossibility of increasing the shares to the sealers. He pointed out that the industry was much more expensive to operate than it had been in the days of the sailing ships; it now took $20,000 to outfit a steamer and two-thirds of a load to "square the yards." He said that of the 400,000 seal skins from 1901, only 100,000 had been sold, 70,000 were still in storage in St. John's, while the rest were in storage in England and the United States. He said that the risks were such that no new vessels were entering the trade, that the two large Dundee firms had abandoned it, and that the *Labrador* and *Leopard* had not made a cent for three years and had sunk twice the capital of the company.[227] However, in actual fact, these steamers had not been quite the failures that Baird maintained. The *Labrador*'s voyage had been a complete failure in 1897, but it had managed to pay its crew $19.10 per man in 1898, $26.77 in 1899, $56.17 in 1900 and $38.46 in 1901. The *Leopard* had also failed to take any seals in 1897, but it had paid $27.50, $41.07, $47.44 and $29.08 per man during the following four years, respectively. However, while these shares were average (for example, the average share per man in 1900 was $40), it is true that the shareholders were forced to reorganize their investment in these steamers a couple of times during this period.[228] Walter Baine Grieve, speaking on behalf of Baine Johnstons, pointed out that the cost of having the steamers available for the seal fishery meant that they were sometimes lying up for 11 months without employment and that this was a real cost that had to be taken into account as well. While the strike was settled with a compromise, it was symptomatic of the financial squeeze that the steamer industry was finding itself in by this time. Nevertheless, Baird and Grieve were not portraying the whole picture. It was true that the limited companies that owned the steamers may have lost money, but this was at least partly because the price of fat was kept artificially low by the companies that purchased the fat, and it was on the manufacturing end of the business that the shareholders made their profits (see below).

Nonetheless, despite the problems involving theft, dissatisfaction among the sealers, declining prices and the continual threat from Nova Scotia-based steamers in the Gulf, the St. John's firms managed to hold on to the steamer seal fishery during this period and to increase their control.

Expansion, rising prices and mercantile control

The decline of the Norwegian cod fishery and the decrease in the quantity of cod oil in the British markets improved the demand for seal oil in the early years of the twentieth century. Furthermore, as already explained, by 1903, the St.

John's sealing industry had become profitable too, and firmly under the control of the oil plant owners rather than the steamer owners, per se. It was reported in May of that year that the profits of the seal fishery had become enormous and that the price had risen from £18 to £28 stg. (or $140) per tun on the London markets. In the case of the skins, the only markets had previously been in England, but a new tanning process developed in the United States had created a new market at good prices. It was said that the three seal oil plants in St. John's had made a profit of $150,000 the previous year and were about to realize a profit of $200,000 in 1903. Furthermore, the three refineries—Bowrings, Jobs and Murray and Sons (Baine Johnstons)[229]—had leased a fourth refinery, belonging to Prowse, and were keeping it closed so that they would have no competition; thus, they were the "only three on this side of the water that are available for this work."[230] The report goes on to point out that there should be 5,431 tuns of oil, worth $760,340, produced from the 8,146 tons of fat brought in. Since the fat had been valued at $3.75 per cwt and purchased from all sources at that price, the refineries had spent $305,475 for the fat and skins. This was paid, in many cases, into their own accounts because they owned 13 out of 22 steamers in 1903. Also, given the fact that the demand for skins had risen recently, with the price at 90 cents, the 317,760 skins alone would bring the exporters $285,984—a sum almost equal to the entire cost to them of the whole seal catch—skin and fat.

The monopoly of the oil manufacturers did not go unnoticed. One writer reported as follows:

> With the introduction of limited liability companies, the ownership of all, or nearly all, of the ships changed hands, with the result that many interested as shareholders in a ship, or ships, do not now participate in the manufacture, and as in the crew's case, have no voice in fixing the price of seals. When the shipowner is also the manufacturer it is immaterial to him what the price of his two-thirds of the catch is; it simply becomes a case of robbing Peter to pay Paul, or withdrawing money from one pocket to put it into the other.[231]

Given the fact that the manufacturers paid only $3.25 per cwt for young harp pelts in 1900, $3.50 in 1902, and $3.75 in 1903, while at the same time taking advantage of the rapid rise in oil prices in London and the United States, one can see why commentators and sealers complained about the injustice of the system. As it was pointed out, "all competition in the price of seals has been effectively cut off." In fact, the newspaper in question maintained that a profit of between 50 per cent and 60 per cent had been made by the three manufacturers of seal oil who owned 13 steamers; neither the crews that supposedly owned one-third of the pelts nor the 9 steamers outside the combine participated in this profit. The writer concluded:

> No one begrudges the steamer owner a fair profit on what has always been a hazardous venture, and more or less a lottery, but the manufacture of the produce is not characterized by these elements, and should not be allowed to absorb an undue amount of the sealers' earnings, and the capital of others not interested in the manufacture.[232]

What had happened by 1903, in effect, was that competition and the continual elimination of weaker producers through a pure system of 'survival of the fittest' had resulted in a small combine of three manufacturers in St. John's controlling the processing and export of the oil and skins at considerable profits while, at the same time, setting the price for pelts to be paid to the other steamer owners and to the sealers. The concentration of the industry could not be carried much further.

The results of the seal fishery were not quite as rewarding in 1904 as they had been the previous year. Bowrings processed the cargoes of 9 steamers with a total of 117,000 pelts and produced 2,110 tuns of oil; Jobs handled 7 cargoes with 96,024 pelts and produced 1,784 tuns; and Murray and Sons, 5 steamers, 71,346 pelts and 1,292 tuns of oil; a total of 284,370 pelts and 5,186 tuns. These figures were below those of 1903, when 317,562 pelts were manufactured into 6,274 tuns of oil,[233] but, given the evidence of profits from this industry, the three firms enjoyed a lucrative monopoly.

The firms tightened their operating rules in 1905, in order to eliminate disputes and to facilitate the management of their enterprises. Principal representatives of each firm met at Bowrings' office, under the chairmanship of the Hon. E. R. Bowring, and drew up a set of rules which all parties eventually signed. It was no accident that Bowrings took the lead in this initiative and that the firm was recognized as the premier enterprise in the business. The new agreement arrived at by the sealing firms was strictly a business agreement, without any input by representatives of the sealers. In this, it was reminiscent of the practice initiated by St. John's saltfish merchants at the beginning of the nineteenth century, when they began meeting regularly each August to decide on the price to be paid fishermen for their fish and cod oil.[234] One of the major inferences to be drawn from the signatures to this agreement is the extent to which the sealing firms had created limited liability companies. For example, James Baird signed on behalf of the Newfoundland Sealing Company Limited, and one of the Job partners signed on behalf of The Erik S.S. Co. Ltd. and The Avalon S.S. Co. Ltd.[235]

The main point of this agreement was the clause which required all agreements between sealers and steamer owners to be similar so that sealers would not find it to their advantage to leave one vessel for another. The crop—meaning the few necessities that almost every sealer was forced to purchase—would be set at goods to the value of nine dollars, but the sealer would be required to pay twelve dollars for his crop at the end of the voyage. This was intended to compensate the suppliers for the occasions when the vessels returned from the front clean and the sealers could not pay for their crops. In actual fact, this did not happen often. In only 5 out of the total of 65 sealing voyages in 1903, 1904 and 1905 were the men's shares insufficient to pay for their 'nine dollar' crops; and in these cases, the crops were partly paid because the supplier withheld the money coming to the sealer to a total of twelve dollars. For example, the following men's shares were retained by the steamer owners in those cases where the men had taken a crop: in 1903, the *Windward*—$5.51 per share; in 1904, the *Kite*—$3.96; in 1905, the *Leopard*—$6.62, *Bloodhound*—$3.94, and *Southern Cross*—$2.74. Therefore, the companies would have received partial payment on goods that were sold at high retail prices in any case. In the meantime, crews of other steamers that paid over $12 per share did not always earn much. For example, the *Panther* paid $16.53 per share in 1903 after a voyage lasting till 6 May—almost two months; those men who had taken a crop cleared

$4.53; and there were other cases too numerous to mention. The companies were very skilled at protecting their investments.

Other clauses in this agreement included free berths, permission for sealers to live on board the vessels while waiting to leave port, and a one-third share of the catch. Crews were to be encouraged to come to St. John's, no doubt as a stimulus to local business, but if necessary, owners could send their vessels to the outports for men and sail from there. The *Kite*, for example, signed on 30 men in St. John's in 1906, and then cleared for Channel, where it signed on 60 men.[236] Where feasible, a medical man was to be provided on the larger ships; and the captain was to be paid 4 per cent of the gross value of the catch. In an attempt, no doubt, to increase the men's shares without decreasing their own, the firms agreed that they would carry 25 per cent fewer men than the 1898 legislation allowed. Thus, instead of hiring a maximum of 270 men on the largest ships, they set their own maximum at 203. (Smaller vessels could legally carry three men for every seven tons of gross registered tonnage.) The owners were taking a firm grip on overseeing the operation, so that even the sealing voyage itself was beginning to resemble a monopoly, with all conditions of employment being made universal throughout the fleet.

Summary of the age of the wooden walls

The Newfoundland sealing fleet was locally built until the 1840s, when shipping was imported for that purpose from the British colonies on the mainland. Beginning in the 1860s, Newfoundland firms turned to steamers in their prosecution of the seal fishery, and Dundee shipbuilders, with their experience in the construction of Arctic whalers and sealers for their local needs, became the main suppliers of these steam-powered vessels. This process, whereby the Newfoundland seal fishery employed ever larger craft, had implications for the overall development of the industry and of the economy itself. Not only did the trend towards the use of bigger vessels lead to the purchase of ships from abroad, which meant the decline in the local shipbuilding industry, but also to the elimination of all smaller and/or weaker firms as the amount of capital required to participate in the industry became the deciding factor. This led to the concentration of the industry in the hands of a few St. John's firms by the early years of the twentieth century. Prowse summed it up well:

> When Mr. Walter Grieve sent the first sealing steamer to the ice it was a poor day for Newfoundland. The only consolation we can lay to our hearts is that steam was inevitable; it was sure to come, sooner or later, the pity of it is that it did not come later. Politics and steam have done more than any other to ruin the middle class, the well-to-do dealers that once abounded in the out-ports.[237]

The Newfoundland seal fishery during the period from 1860 to 1906 was dominated by the wooden steamers. As the sailing craft left the industry, they were not replaced (with certain minor exceptions), and only the firms in St. John's—and, for a while, the firms in Harbour Grace—were able to afford the expensive steamers. Over 90 per cent of all pelts brought to the oil plants during the ten-year period ending in 1905 (see Table 3.7) were produced by these steamers. The manufacture of seal oil became concentrated in the hands of three

Figure 13 - The S.S. *Adventure*, 895.94 net tons, 213 nhp, was the first steel steamer to prosecute the Newfoundland seal fishery. Harveys' introduction of this steamer into the industry prompted other sealing companies to acquire steel steamers as well, and "from 1906 to 1914 Newfoundland had the finest fleet of Sealers and Ice-Breakers in the world." (Mosdell, *Chafe*, p. 26)

St. John's firms, which formed a monopoly and controlled the price paid for fat. The fleet of sealing ships was considerably reduced and, by 1905, consisted of only 22 steamers owned by firms in St. John's, which had become, by now, the only sealing centre of the colony.

Age of the Iron Clads

Iron clad S.S. Adventure

Harveys, in 1906, with the introduction of the 829-ton iron-clad[238] steamer *Adventure* to the seal fishery, raised the stakes exponentially. The *Adventure* set a style, not only for a new sealing fleet, but for a class of ice breaker—a vessel with a stem that was designed to ride up on the ice and break it with its own weight instead of ramming it to crack open a lead. This was to be the final phase in the evolution of the sealing fleet. The *Adventure* was special and a contemporary report makes this clear:

> Yesterday a steamer was launched by the Dundee Shipbuilders Co., Ltd., which marks a distinct advance on any of these [earlier steamers], and built as it is for the sealing trade of St. John's, Newfoundland, seems not unlikely to be destined to be the forerunner of many vessels of a similar type. The S.S. *Adventure* which name has been given the boat, is especially constructed for ice navigation, but, unlike previous vessels for that trade, she is built completely of steel to Lloyd's highest class, and far in excess of their requirements, her stem, stern post, rudder and steam steering gear being fully twice the weight and power required for vessels of her size. Her frames and beams are much above usual size and more closely placed, while her plating is correspondingly heavy, the plates at forward and along the side

> being up to one and one-half inches thick. She is also fitted for further strength during the sealing season with tiers of heavy portable wood beams, these being removed when the vessel is engaged during the rest of the year as an ordinary trading steamer.
>
> She is ... 265 feet by 88 feet by 21 feet moulded depth, fitted with all the latest appliances for the rapid handling of cargo, and is lighted by electricity, including a powerful searchlight for use in the quest of seals. She will be fitted with engines ...[and] three boilers of sufficient power to obtain a speed of twelve knots....[239]

Thus, the *Adventure* was built to cope with ice that no previous sealing steamer, or even Arctic exploration steamer, had ever been designed to handle. When the new vessel arrived in St. John's from Newcastle after an 11-day trip in bad weather, it attracted a large crowd of spectators. A local reporter described the event in part:

> As soon as the steamer was berthed, hundreds of people visited the wharf, curious to see the new vessel. She represents a departure from the other sealing vessels now in use, being built entirely of steel, no wooden sheathing whatever. Instead of this a belt of steel eleven feet six inches wide, and one-half inch at its thickest surrounds the vessel, and it is the opinion of the constructors much superior to wood.[240]

The description goes on to repeat many of the points in the earlier report and states that portable beams, which were laid across the vessel inside to strengthen the sides against the ice, were twelve inches square and laid in a tier seven feet apart. Another unique feature was the presence of seven air and water-tight bulkheads, with even the engine and boiler room separated. Although the decks were made of steel, decks on the bridge were wooden, and it was planned to install a temporary wooden flooring over the main deck before proceeding to the ice in order to protect the pelts from damage from the frozen iron. Finally, the uniqueness of the stem was described, especially "the way in which it slopes back on reaching the water, allowing her to run up on the ice and break it." This last feature has remained part of the design of modern ice-breakers.

The 1906 sealing season began auspiciously, with reports in January to the effect that:

> Many thousands of old seals are now in White Bay, beating their way North to meet the northern ice floe. Numerous lots were seen in the water and on the slob ice.... An old sealer said he never before saw them in such numbers at this early date.[241]

As the fleet departed for the ice, all eyes were on the new steamer, *Adventure*. One is reminded of the interest aroused by Prowse and Sons in 1867, when that firm had brought out the *Esquimaux* from Scotland and placed it under the command of the best sailing captain available in the seal fishery, Capt. Terry Halleran. It had been a disastrous season; eight steamers, five of which made two trips, brought in a total of just over 28,000 pelts. The industry had remained on hold, and it was not until the early 1870s that improved results had increased investment. The experiment with the *Adventure*, on the other hand, was far different.

The *Bloodhound*, under Capt. Darius Blandford, was the first to return to port in 1906, reaching Harbour Grace, where Baine Johnstons was still doing some of its processing, on 27 March with 18,756 pelts. Capt. Blandford described how 15 of the steamers, in one day, cleaned up one huge patch of seals, each vessel getting 7,000 to 10,000 whitecoats. He was impressed by the *Adventure* and reported: "She struck the seals on the 11th and cut a track for the *Vanguard* and *Neptune*. Her speed was terrific through the ice. While other steamers butted she made a cut through the solid ice."[242]

The next to arrive was the *Diana*, under Capt. Alpheus Barbour, which entered St. John's on 30 March with 17,459 pelts. Capt. Barbour also described how the *Adventure* "proved herself to be a wonderful ice boat" and that the "track cut by Captain Harry [Henry Dawe] made it easy for the *Neptune* and the *Vanguard* to steam into the patch also."[243] When the *Adventure* arrived, there was a clear interest in gathering as much information about this vessel as possible. One report stated that:

> throngs of people lined the docks to see her entrance, and Capt. H. Dawe was cross questioned by the various merchants who met him at the dock, inquiring about their own vessels. Among those present were, Hon. John Harvey, Mr. E. F. Harvey, Hon. E. R. Bowring, Mr. W. C. Job, Mr. J. Munn, Mr. F. Halfyard, Mr. R. Grieve, Mr. M. Winter....[244]

The early indications of a prosperous season proved to be true, and the performance of all vessels ranged from moderately good to very good. The lowest share was paid by the *Viking*, which Bowrings had sent to the Gulf under Capt. William Bartlett; that vessel paid $25.04 per man, still comparing very favourably with the previous year, 1905, when 11 vessels paid less than $25 per man. The *Adventure*, under Capt. Henry Dawe, brought in 30,193 pelts and paid its men $81.65 each—the largest cargo and the highest shares for the voyage, and over 6,000 pelts and $10 more than the nearest-placed wooden steamers.

Meanwhile, the price paid for whitecoat pelts by the companies had risen steadily, from $3.25 per cwt in 1900 to $4 by 1906. The total net value of the largest cargo was almost $46,000 in 1904 for the *Aurora*'s load of 34,849 pelts, and almost $48,000 in 1905 for the *Eagle*'s cargo of 32,064. In 1906, the *Adventure*'s load of 30,193 pelts was worth almost $50,000 and the 203 men's share of $81.65 each was the highest in over ten years. In addition, the 1906 season was the best since 1901 and the total catch of 341,836 seals was never to be equalled by the fleet again. Thus, not only did the result of the 1906 voyage convince the investors that the resource was viable, but the performance of the *Adventure* persuaded them that steel ships were the ultimate answer to harvesting problems; prospects looked increasingly good.

Iron clads versus wooden walls

This 1906 season demonstrated to all that competition had entered a new phase, and to see the *Neptune* and *Vanguard* following in its wake through the icefields and thus accessing the centre of a big patch must have been galling to the captain and crew of the *Adventure*. And in spite of the number of seals available, accusations of theft were made later that year. In addition, there had been legislation recently passed which prohibited killing seals on Sundays, and captains brought charges against each other over this action. Capt. Abram Kean in Bowring's *Terra Nova* accused Capt. Samuel Winsor in the *Walrus*, owned by G. Browning and Sons, of bringing into St. John's pelts from seals killed on a Sunday. Mr. J. LeMessurier, Department of Customs, brought a similar charge against Capt. Arthur Jackman in the *Eagle*. Because all the steamers were near each other, the captains were forced to testify against each other, with Kean testifying against Jackman, although both were employed by the same firm. Jackman was convicted and forced to pay the standard fine of $2,000.

Winsor argued that his men had disobeyed his orders and had gone overboard to kill seals when they saw the *Eagle*'s crew killing seals under the bow of the *Walrus*. Winsor said that the men had come to see him after breakfast on Sunday and that his conversation with them went as follows:

> Men—"Captain! What are we going to do?" Winsor—"Boys, I can't send you on the ice today, it is against the law." Men—"Well Captain, 'tis pretty hard to see another crew taking the bread out of our mouths. Flesh and blood can't stand it Captain, and we'll have to do something." Winsor—"I don't intend to let a man go on the ice today."

According to Winsor, he then went below and did not come on deck until the following morning; he did not hear until later that some of his men had killed between 9,000 and 10,000 seals on Sunday, of which about 3,000 were stolen. The judge pointed out that it was strange that the law allowed hoisting pelts aboard, stowing them away, and doing other things, including stealing pelts from each other, and yet prohibited killing; he was sympathetic to Capt. Winsor's plight and fined him only $1,000.[245] The wooden walls were about to experience new levels of competition, where theft would be a minor problem.

In 1906, when the first steel steamer was introduced, 24 sealing steamers of varying sizes (excluding the *Havana* from Halifax, under Capt. Farquhar) were engaged in the St. John's seal fishery.[246] Five of these had been built in the 1860s: the *Walrus* (219 tons), *Erik* (412), *Labrador* (256), *Nimrod* (277) and *Panther* (247)—all between 200 and 300 tons, except the *Erik* at 412 tons. Eleven had been built in the 1870s, as follows: the *Diana* (275 tons), *Ranger* (354), *Greenland* (259), *Iceland* (287), *Newfoundland* (568), *Vanguard* (323), *Bloodhound* (314), *Kite* (190), *Leopard* (217), *Neptune* (465) and *Aurora* (386), and all except the last were built before 1873. One vessel, the *Kite* was less than 200 tons, four were between 200 and 300 tons, four were between 300 and 400 tons, while the *Neptune* had a burthen of 465 tons, and the *Newfoundland*, 568 tons. The *Algerine* (233 tons), *Viking* (276), *Terra Nova* (450), *Southern Cross* (325) and *Virginia Lake* (440) were built in the 1880s. Only one of the fleet was built in the 1890s—the *Grand Lake* (463 tons), in 1892. The *Eagle* (418) was built in 1902, and, of course, the

Adventure (826) entered the industry in 1906—it was the arrival of the latter that signalled a turning point in the history of the St. John's fleet.

The voyage the following year, 1907, was not as productive, and the total value of pelts brought in declined from $608,000 in the previous year to $455,000. The *Neptune*, under Capt. George Barbour, was the high liner, with 31,000 pelts worth about $59,000, and his 203 men received $96.37 each. However, the *Adventure* was next, with nearly 25,000 pelts worth about $46,000, and the men each received $74.84. The *Virginia Lake*, under Capt. Jacob Kean, was next with an excellent trip. (Kean was Capt. Samuel Blandford's former second hand; Blandford had just retired because of illness.) Although the catch was down in 1907, the price was up to $4.20 per cwt for whitecoat pelts, and the future market for seal oil and skins looked bright. In 1908, the catch declined further, but once again, there was an increase in price—this time to $4.50 per cwt for whitecoat pelts. On this occasion, Capt. Henry Dawe in the *Adventure* led the fleet, with over 27,000 pelts worth almost $49,000, and he paid his men $79.79 each. He was followed by Capt. George Barbour in the *Neptune*, who brought in about 24,000 pelts worth $43,000, and each of his 203 men received $70.16. These were followed by Capt. William Bartlett in the *Viking*, with a cargo worth $36,000, and Capt. Jacob Kean in the *Virginia Lake*, whose cargo was worth $34,000.

However, it was a 'hard' spring, in that the ice was thick and tight, and the *Adventure* demonstrated its superiority in hard ice conditions by returning first to port. Many of the other vessels in the fleet were in a battered and damaged condition; three, the *Panther*, *Walrus* and *Grand Lake*, sank—the first two with their bows stove in and the *Grand Lake* from the pressure of the ice while jammed.[247] The large voyages of the successful steamers, nevertheless, encouraged the industry to overlook the vessels that did not pay shares large enough to cover the men's crops (i.e., $12): the *Algerine, Aurora, Kite* and *Nimrod* in 1907 (with the *Southern Cross* paying only $12.63); and the *Southern Cross, Eagle* and *Labrador* in 1908. In other words, the failures were ignored. Instead, the firms had come to the conclusion that as the seals were there and the price was rising, the vessels that reached the herds would make fortunes for their owners, while those that could not do so would be left behind, literally and figuratively.[248] Therefore, Harveys, the managers of the sealing company that had owned the *Grand Lake* and the *Panther*, "opened a share list for the building of two new sealing steamers" for the following spring, similar to the *Adventure*, but smaller.[249] All the shares were bought up quickly, and Jobs opened a share list to build a similar steamer; these, too, were all purchased.[250] Each company was prepared to get its share of whatever seals were on the ice, and if steel was the answer, steel would be used.

Three of the major firms, Bowrings, Jobs and Harveys, adopted the new technology wholeheartedly in 1909. Bowrings weighed in with the largest steamer to that point, the steel *Florizel* at 1,980 tons burthen; Jobs brought in the steel steamer *Beothic* at 471 tons burthen; and Harveys added two steel ships, the *Bellaventure* and *Bonaventure*, at 466 and 446 tons, respectively. Only Baine Johnstons stuck entirely to the old wooden walls. Bowrings was the most innovative in applying new technology by installing wireless on the *Florizel* for the first time in the history of the seal fishery.

While the local press was relatively silent on the day-to-day activities of the sealing fleet during this season, because of the political crisis that was gripping the country,[251] one newspaper pointed out that the sealing ships had to cope with the "heaviest ice for years."[252] Also, both the *Vanguard* and the *Virginia Lake* broke their main shafts and sank, without loss of life. (The latter was the last of the "Lake" line of coastal steamers that were also used in the spring seal fishery, and all three were lost while pursuing this industry; the *Windsor Lake* sank in 1896 and the *Grand Lake*, in 1908.) The *Bloodhound* narrowly escaped a similar fate and was towed into St. John's, also with a broken shaft, by the *Diana*.

The stars of the fleet did well and the two top voyages were almost identical. Bowrings' *Florizel*, under Capt. Abram Kean, was the high liner; it brought in 30,488 pelts (including over 10,000 young hood or 'blue back' pelts) for a total value of $54,000 and paid its crew of 203 men $88.33 each. Jobs' *Beothic*, under Capt. George Barbour, brought in 34,837 pelts worth $53,610 and paid its men $87.67. Harveys' *Bonaventure*, under Capt. John Parsons, was third, with 31,188 pelts worth $46,000, and paid its men $75 each; that company's *Bellaventure*, under Capt. Job Knee, was close behind, with 27,000 pelts worth $43,000, and paid out $70.34 per share. Bowrings' wooden wall, the *Eagle*, under Capt. Joseph Kean, was a close fifth, with pelts valued at $36,000; and Harveys' original steel steamer, the *Adventure*, still under the command of Capt. Henry Dawe, followed next, with $33,000 worth of pelts. The wooden steamers *Viking*, *Terra Nova* and *Aurora* all had respectable loads, but the *Newfoundland*, *Neptune* and *Southern Cross* earned too little to cover the men's crops, and the *Labrador* paid only $13.47 per man. Finally, as pointed out, both the *Vanguard* and *Virginia Lake* went to the bottom. To summarize, the 5 steel steamers, carrying 1,015 men, brought in about $230,000 worth of pelts, or 50 per cent of the total value, while the 15 wooden walls, employing 2,565 men, brought in the remaining 50 per cent. It was very obvious where the profits lay.

It was also apparent that the older wooden walls could not keep up with the new steel steamers, and in 1910, the owners got together and agreed that the former would be given a head start on their way to the ice and be permitted to clear port after 8:00 a.m. on 12 March, while the latter were required to wait until 6:00 p.m.[253] This was a fairly productive season with no vessel failing to pay shares to its crew, although the shares paid to the men of nine vessels were only in the $20 and $30 range. The *Florizel*, under Capt. Abram Kean, set a record, with over 49,000 pelts worth a total of $90,800, and paid $148.36 per man. The *Bellaventure*, under Capt. Job Knee, brought in a cargo worth over $68,000 and paid $112 per share, while the *Bonaventure*, under Capt. John Parsons, did not do so well, with a cargo of pelts worth $35,000, and paid its men only $57 each. The *Adventure*, under Capt. Dawe, did even worse, with only $21,000 worth of pelts. However, despite the relatively poor performance of the two steel ships belonging to Harveys, the five steel steamers, with 1,015 men, brought in pelts worth a total of $277,000, while the fourteen wooden walls, with 2,349 men, brought in only $351,000 worth.

Technology took an additional step forward this year with the installation of wireless on the *Florizel* and *Eagle*, both belonging to Bowrings. While the company had already tried wireless on the *Florizel* in 1909, it was its presence on two vessels, in 1910, and the resulting capability of communicating conditions and results, which was probably instrumental in the success of the *Eagle*, under Capt. Joseph Kean (son of Captain Abram), with 30,000 pelts worth $54,000 and shares of $89. News by wireless was relayed to St. John's during the voyage, and it was noted that the *Florizel*, when reporting on its catch, did not publicize its position. The public suspected that the "*Florizel* took the 'outside cut' on her way north, with the *Eagle* on the inside run, and that both communicated results daily."[254]

The *Labrador* changed hands this year when, for $11,750, it became the property of James Baird, whose representative outbid Mr. Tasker Cook at the auction sale. Baird had supplied it in 1909, when it belonged to the Newfoundland Sealing Company, and it was reported after the sale that it was well worth the money paid for it; if it brought in 15,000 pelts, it would pay for its cost and make a profit in its first year.[255] Actually, it brought in only 3,700 adult hood pelts, but, since these were the equivalent of about 10,000 whitecoat pelts, it was not a total loss. Jobs experienced a loss as well, when its wooden wall, the *Erik*, broke its shaft and needed to be towed into port by another of their vessels, the *Beothic*.[256] More unfortunately, Baine Johnstons' *Iceland*, in the Newfoundland seal fishery since 1872, went to the bottom after being crushed in the ice—fortunately without loss of life.[257] A contemporary summed up what many people were becoming concerned with—the losses:

> The report of the abandonment of the sealer *Iceland* and the critical condition of the *Newfoundland* [possibly he means the *Erik*], draws attention to the disasters which have attended the Newfoundland seal fishery in the last four years. For a long period this venture was attended with success to both underwriters and owners, but since 1907 the former have had very heavy losses. Last year the *Vanguard* and *Virginia Lake* were abandoned with broken shafts; In 1908 the *Walrus*, *Panther* and *Grand Lake* were lost; while in 1907 the *Leopard* and *Walrus* [means *Greenland*] were both wrecked. The [insurance] rate paid this year, viz., 50s. per cent for the trip, compares with 30s. per cent, which was the rate current four years ago, but the [insurance] business continues to be a loss; and ... will make it difficult for Newfoundland owners and shippers to place their risks in the future.[258]

However, although the losses and damages to sealing vessels were alarming, with prices at $4.50 per cwt for whitecoat pelts, the industry continued to pay profits to the successful steamers and the men who crewed them.

There were very few changes in the fleet in 1911, with one exception—the extension of the application of wireless communication to the industry. Up to now the firms had become accustomed to receiving information from their captains through the nearest Marconi stations at Channel, Belle Isle and elsewhere; however, only Bowrings had actually employed the technology on board its vessels the *Florizel* and *Eagle*. Jobs recognized the value of this new application and equipped the *Beothic* and *Neptune* with it in 1911.[259] Harveys installed it in the *Adventure* and *Bellaventure* also, and it was reported that the *Adventure*, *Bellaventure* and *Beothic* corresponded with each other and practised with the new equipment en route to and from the northeast coast transporting sealers to St. John's that spring. They found the system quite satisfactory, were able to signal each other from up to 60 miles apart, and were sure they could signal farther than that.[260] In all, four steel steamers (the *Florizel, Adventure, Bellaventure* and *Beothic*) and two wooden walls (the *Eagle* and *Neptune*) had wireless during the 1911 season. Now, the firms could be kept informed of their vessels' activities on a regular and continuous basis, with prearranged code words for sensitive information. The airwaves were full of messages from the fleet during the spring of 1911—to such an extent, in fact, that a local newspaper occasionally ran a headline "No Sealing News."[261] In addition, now that captains from the same firm could communicate with each other and receive coded messages regarding the whereabouts of seals, the application of this communications technology took a dramatic leap forward because these masters could direct each other to the seal patches.

In addition to this technological change, the St. John's sealing firms made certain changes to their standing agreement. In the first place, they cancelled the agreement to employ 25 per cent fewer men than the maximum set by legislation. This legislation of 1898 stipulated that the steamers could not carry more than three men per seven gross tons up to a maximum of 270 men. The owners, in their own agreement, had reduced this in 1903 by 25 per cent for a maximum of 203 men. However, in 1911, the owners decided to return to the government regulations, allowing each vessel to increase the number of men it carried. For example, the five steel ships carried 203 sealers each in 1910, as did the wooden walls *Newfoundland, Eagle* and *Neptune*. In 1911 the *Beothic* and *Florizel* carried 270 men; the *Bonaventure, Bellaventure, Neptune* and *Eagle* each carried 269; and the *Newfoundland* and *Adventure* employed 267 each. Over all, employment increased from 3,364 men in 19 vessels in 1910 to 3,973 men in 18 vessels in 1911—an increase of over 600 jobs. It seems obvious that all the big firms agreed that the large steel ships could carry more pelts than 203 men could be expected to take, and therefore, their agreement had to be changed in order to obtain as much profit as possible from the annual fishery. Finally, in an effort to allow the whitecoats to grow a little larger, the owners of vessels operating at the front agreed not to kill any seals in 1911 before Wednesday, 16 March.[262] This was the context in which the steamer seal fishery was carried out in 1911.

It was a comparatively good year, and it was obvious from the beginning that the steel ships would again out-perform the wooden walls. One of the first wireless messages dealing specifically with the catch of seals came from the *Beothic* at the icefields and pointed out that the "steel ships ... are in the main patch and will take the lion's share from it, as the ice being in heavy sheets will make it very difficult, if not impossible, for most of the wooden ships to force through."[263] Another incident highlighted the differences, which were beginning to surface regularly, by this time, between the wooden and steel steamers. A headline in a local newspaper read: "Steel Ships Slowly Loading—No Definite News of the Wooden Fleet Yet."[264]

There were no huge voyages this year, and only one steamer, the *Beothic*, brought in a cargo worth $50,000; additionally, with the larger crews, the men's shares were even smaller than would normally have been the case since 1903. The *Bellaventure* paid the lowest share of the steel ships, at $40 per man, while the others all paid in the $50 and $60 range. Even so, these five steamers, with

1,345 men, brought in about $218,000 worth of pelts, while the fourteen wooden walls, with 2,628 men, brought in only about $276,000 worth, and the wooden walls *Algerine, Neptune* and *Eagle* paid only $16, $17 and $19 per man, respectively. As well, five wooden walls did not return to port until the first week of May. The price of pelts remained at $4.50 per cwt. The firms believed in making big profits from better ships; thus, in May, Bowrings sold the wooden wall *Aurora* in England for Antarctic exploration work, and all companies began to plan their strategy for the following year.[265]

The sealing industry decided to gamble a large investment on the 1912 season. Six new steel steamers were brought in to participate, and new owners entered the trade: the *Fogota*, of 238 tons, and *Sagona*, of 420, owned by Crosbie and Company; the *Nascopie*, 1,004 tons, owned by Jobs; the *Stephano*, of 2,143 tons, owned by Bowrings; the *Seal*, of 277 tons, owned by Capt. J. Farquhar of Halifax but sailing from St. John's; and the *Lloydsen*, of 247 tons, owned by Tasker Cook and Company, St. John's. As one can see, Bowrings and Jobs each invested heavily, aiming at size and capacity as well as quality. At last, Baine Johnstons decided to invest in a steel steamer and had one built in Glasgow—the *Erna*; but it was lost on the trip to St. John's, taking with it 33 crew members and three passengers (see Chapter 5). The new steel additions to the fleet increased the number of this type of steamer from 5 to 11; the wooden walls, with the sale of the *Aurora* the previous year, now numbered 13. Thus the total sealing fleet numbered 24 vessels.

On the whole, 1912 was a terrible season. Capt. William (Billy) Winsor, in the *Beothic*, was the high liner; he brought in just over $60,000 worth of pelts and paid his men $73.82 each. The *Nascopie*, under Capt. George Barbour, and the *Adventure*, under Capt. Jacob Kean, each brought in $35,500 worth of pelts, and the *Stephano*, under Capt. Abram Kean, brought in pelts worth $25,000. The older wooden walls, the *Ranger* and the *Eagle*, brought in respectable cargoes valued at $22,000 and $21,000, respectively. The *Fogota*, with an unusually small crew of 85 men (probably because, as sometimes happened, it was unable to complete its complement of men in a northern port due to heavy ice), brought in about $17,000 worth of pelts and paid the second highest men's shares at $65.30 each. The *Ranger* with 125 men, paid a fairly high share as well of $59.02 per man. The *Adventure* and *Nascopie* paid $41.22 and $43.71, respectively. Meanwhile, neither the *Erik, Viking, Bloodhound, Labrador, Sagona, Florizel, Bellaventure* nor *Bonaventure* paid enough to cover the men's crops, and the *Seal* paid only $13.76; the *Viking*, $12.76; and the *Newfoundland*, $14.11. Twenty-four vessels with 4,179 men brought in only 175,000 pelts worth a total of $329,000, which made 1912, considering the capital invested, one of the worst seasons in the history of the spring seal fishery to that point.

Nineteen vessels participated in the seal fishery in 1913 (although twenty-one actually left port for the ice fields), and it was an average year. There was, however, an obvious increase in the level of stress, anxiety and competition among the companies and the crews. Bowrings' *Algerine* had been lost the previous year while on charter in the Arctic,[266] which left that company short one vessel. Crosbie's *Fogota* had been put on the Halifax, Sydney and St. Pierre mail run, so that company sent only the *Sagona* to the ice fields.[267] The *Labrador*, while en route to Channel, sprang a leak in stormy weather near St. Pierre and barely managed to make it ashore.[268] Meanwhile, the *Beothic* and *Bonaventure* collided while racing each other out of St. John's harbour. The former cut across the latter's bow, and the latter, in order to avoid the rocks, was forced to ram the *Beothic* in the port quarter. This caused so much damage to the *Beothic* that it was forced to return to St. John's and go on dry dock, where it was discovered that extensive repairs would have to be completed in New York.[269] The *Lloydsen* was badly damaged at Channel while preparing to clear for the Gulf seal fishery and was forced to return to St. John's. It was repaired and left 27 March to try its luck with the old seals in the Gulf. It was successful and returned to St. John's 6 May with $16,000 worth of bedlamer and adult harp pelts and paid its crew $40 per man. The *Kite* also had an unusual experience that spring. It proceeded to Groais Island (off the coast of the Northern Peninsula) to pick up its crew, taking only 11 men from St. John's. It could not reach its destination because of heavy ice and managed to hire only 6 men from Seldom Come By, Trinity and Catalina, making a total of 17 men. Then at the ice fields, Capt. Yetman signed on 21 stowaways from other vessels to make a total of 38, the smallest crew of any steamer ever.[270] During this season, also, Harveys did not repeat their wireless experiments because they had been unremunerative. However, Jobs' *Nascopie* and Bowrings' *Stephano*, *Florizel* and *Eagle* were so equipped, although its advantages were not obvious. So in several respects, it was an unusual year.

The fishery proceeded as usual, otherwise. The 19 vessels hired their men in St. John's from among the hundreds who arrived in the city with their 'tickets' or 'on spec'. The *Bellaventure* arrived in St. John's from Pool's Island, King's Cove, Trinity and Catalina, with 500 sealers and Captains Job Knee, George Barbour and Jacob Kean on board. Most Conception Bay sealers arrived by train. However, the total number of men involved was down this season, partially due to the loss of the *Labrador*, the severe damage to the *Beothic*, the long delay which resulted in a smaller crew on board the *Lloydsen* and the inability of the *Kite* to reach Groais Island.

The men of the *Stephano*, under Capt. Abram Kean, received the largest share, or 'bill', at $85.56 for their load of 37,900 pelts valued at $70,000. The *Nascopie*, under Capt. George Barbour, brought in 32,000 pelts worth $55,000 and paid $67.04 per share. The wooden walls *Neptune*, under the famous Capt. Bob Bartlett, and *Viking*, under Bob's father, Capt. William Bartlett, paid even higher shares than the *Nascopie*, at $75.64 and $72.13, respectively, but they carried fewer men and brought in smaller loads. It was a moderate year in terms of production; no man made less than the price of a crop, although on the *Ranger*, the share was only $12.53, while the next lowest shares were paid by the mighty *Adventure* and *Bellaventure*, whose men received only $15.50 and $18.37 each, respectively. In terms of the total value of the cargoes, the five top vessels, in descending order, were the *Stephano*, *Nascopie*, *Neptune*, *Viking* and *Florizel*, ranging from $70,000 down to $36,000. Thus, it was a mixed year, with all firms having to take losses. It was disappointing to Harveys that its three steel steamers did so badly (the *Bonaventure* paid only $25.25 per man), while their wooden wall, the *Newfoundland*, was relatively successful and paid $50.11

a share. Jobs lost the whole voyage of the *Beothic* and had costs besides, but did well from the *Nascopie* and *Neptune* and not too badly from the *Diana* and *Erik*. The old firm of Baine Johnstons sent only one steamer to the ice this season, the wooden wall *Bloodhound*, which brought in a respectable $31,000 worth of pelts and paid its crew $60.04 per man. Bowrings did best of all, with the *Stephano* as high liner and the *Viking* well loaded. The *Florizel* was somewhat of a disappointment because of its capacity, but it got a good share of pelts. The *Eagle* also did well, but the *Ranger* and *Kite* did poorly. Crosbie, Farquhar and Tasker Cook all had reason to be satisfied with the performance of their vessels.

The first seven steamers to arrive back at St. John's from the ice fields were steel and arrived during the period from 30 March to 11 April, while the last nine were wooden walls, which arrived between 15 April and 6 May. The longer work period for sealers on the wooden walls had become a recognizable problem. This difference had become evident almost immediately with the introduction of steel steamers to the seal fishery in 1906; in that year, the *Adventure* was fourth back from the ice, arriving on 1 April. There was a distinct pattern in all arrivals during the next several years; the steel ships generally returned first, usually with larger cargoes consisting almost entirely of whitecoat pelts. This was happening because the steel ships were able to break through the ice to the rich whelping patches, covered with whitecoats, when wooden walls often could not; even when the latter could break through, the steel ships could do so faster, so that when the wooden walls arrived, the patches were 'cut up'. Thus, the wooden walls were often forced to chase after bedlamers and old seals in the water, especially as the seals began to leave the ice and take to the water in early April for the long swim back to the Arctic. In 1910 and 1911, this trend was not quite as evident because the early and late returning vessels were a mixture of steel and wooden steamers. In 1912, however, it was indisputable; the first nine steamers to return were the steel ships, beginning on 28 March, and the last twelve were wooden walls, some of which did not return until the first week of May.

It was early in 1913 that the Fishermen's Protective Union took up the cause of the sealers on the wooden walls. William Coaker wrote the owners of the wooden walls, asking if they would consider dividing their crews into two parts: one-third would be hired as gunners with their assistants or 'dogs', and would be committed to the whole trip; two-thirds would be hired as batsmen for the whitecoat season and would be guaranteed their return to nearby ports on or before 5 April. Coaker argued, correctly, that only a few gunners could operate in a given area and that carrying a full complement of sealers around for an extra month was expensive to the owners and delayed many men from cod fishing preparations.[271] Although the problem could be easily recognized, it is understandable why a solution was not easy to implement. The vessels could find themselves anywhere in all kinds of weather, and if a vessel came upon a patch of seals, old or young, it might be able to acquire a load fairly quickly with a full complement of men. Sometimes, old seals and bedlamers got caught out of the water on tight ice and were killed and sculped by batsmen in situations where a large crew working quickly made all the difference. Also, there were no convenient ports where two-thirds of a crew could be dropped off quickly and with any expectation of getting home conveniently while carrying boxes and bags and quantities of seal meat. Finally, it would make it extremely difficult to divide up the catch fairly, and many sealers would have to wait for their money. Thus, while it was recognized that this was a labour problem, it demonstrated clearly another advantage of steel over wooden ships.

The 1914 sealing fleet was almost evenly divided; there were eleven wooden and ten steel steamers. The fleet of wooden walls was made up of rather old vessels, whose numbers had been in decline since 1880. The steel fleet, on the other hand, was composed of new vessels and at its peak, although the absence of the *Lloydsen*, on charter in Europe,[272] meant that this segment of the fleet was one vessel below its top complement of eleven.

There were a few changes from 1913. Capt. Samuel Wilcox succeeded Capt. Bob Bartlett on the *Neptune*; Capt. Isaac R. Randell succeeded Capt. Job Knee on the *Bellaventure*; and in a move that was to have disastrous results, Harveys removed the wireless set and operator from the *Newfoundland*, under the command of the 29-year-old Capt. Westbury Kean, in order to cut expenses. Most of the wooden walls cleared from ports near the icefields: The *Newfoundland* and *Ranger* from Pool's Island; the *Bloodhound*, *Diana* and *Eagle* from Wesleyville; and the *Erik*, *Terra Nova*, *Viking*, *Southern Cross* and *Neptune* from Channel. One exception, the *Kite*, cleared from St. John's on 13 March for the front. Capt. Farquhar's steamer, the *Seal*, was the only steel steamer to engage in the Gulf fishery, and it cleared from Channel. Crosbie's two steel ships, the *Sagona* and *Fogota*, steamed to Wesleyville and cleared from there. The seven remaining steel vessels, the largest of the fleet, all cleared St. John's beginning at 8:00 a.m. on 13 March. James Murphy, a local songwriter, composed a ballad to mark their departure and noted Capt. Billy Winsor's caution in taking the *Beothic* out at 8:30 in number five position, behind the *Adventure*, *Bellaventure*, *Bonaventure*, and *Nascopie* in that order, and followed by the *Stephano* and *Florizel*. It was remembered that Capt. Winsor had ruined the previous year's voyage completely by colliding with the *Bonaventure* in his haste to reach open water. Murphy wrote:

> Bill Winsor sailed at half-past eight,
> He came in a half hour late,
> A burnt child it dreads the fire,
> That's why maybe that Bill did tire.[273]

While this voyage produced an average return, all activities were overshadowed by tragedy. The *Newfoundland* found itself still jammed in the ice on Monday, 30 March, but the captain, Wes Kean, could see his father Capt. Abram give a pre-arranged signal with the derrick on the *Stephano*, about five miles away, indicating that there were seals in that neighbourhood. The next morning, Capt. Wes sent his men, under the command of the second hand (first mate), to the *Stephano* to get directions from Capt. Abram and begin killing seals. A storm commenced that afternoon Tuesday, and the men were unable to get back to their vessel; Wes Kean, concluding that they had stayed on the *Stephano*, did not continue to blow the vessel's whistle during the storm, which raged until Thursday morning. Capt. Randell in the *Bellaventure* picked up most of the living and dead on Thursday and returned to St. John's. He was overtaken by

Capt. Winsor in the *Beothic*, anxious to get the recognition of high liner and first arrival for the year; his insensitivity was remarked by all. The *Beothic* brought in the biggest cargo of over 28,000 pelts, worth almost $62,000, with shares per man of $75.81. However, two wooden walls from the Gulf fishery, the *Erik* and *Terra Nova*, brought in comparatively bigger loads, worth $40,000 and $51,000 and paid their men $77.79 and $83.90, respectively. The *Ranger* failed to pay the men's crop with its shares of only $7.18 each. But the attention of all Newfoundland, as well as that of eastern Canada, the northeastern United States and Great Britain, was on the tragedies of that spring.

Out of 132 men from the *Newfoundland*, who spent over 48 hours on the ice, 78 died, and many of the rest were crippled and maimed for life from frost bite(see Chapter 5).[274] Meanwhile, after days of awaiting its arrival, Newfoundlanders were forced to accept another unthinkable tragedy; that the *Southern Cross*, returning from the Gulf, had sunk in a storm, taking 173 men to their deaths. Out of a total population of about 250,000, Newfoundland had lost 251 men in the prime of life; confidence in the industry never recovered.

Prior to the terrible tragedies of 1914, there was a general feeling that the wooden steamers could not survive very long in competition with the steel ships. The following wooden walls were lost at the seal fishery during the era of the steel ships: the *Greenland* and *Leopard* in 1907; the *Grand Lake, Panther* and *Walrus* in 1908; The *Vanguard* and *Virginia Lake* in 1909; the *Iceland* in 1910; the *Labrador* in 1913; and the *Kite* and *Southern Cross* in 1914. In addition to the loss of these 11 wooden walls, the *Algerine* sank in Hudson Bay in the summer of 1912 while on charter, and the *Nimrod* was sold to England in 1907. Not only were the steel vessels more productive in general, but they were also more efficient in reaching the whelping ice and taking the younger, more valuable whitecoats. For example, 83 per cent of the total number of pelts brought in by the *Adventure* over the nine-year period from 1906 to 1914 consisted of whitecoat pelts, while only 71 per cent of the total number of pelts brought in by the large wooden wall *Newfoundland* during this same period were whitecoats.

Although it looked like the old wooden walls could not survive, other factors were coming into play. The steel steamers were specially built for sealing and for working in ice and were the best in the world in their class, but they could not make a profit from work which involved only one or two months' activity a year; consequently, they were used for other things. Unfortunately, these ice breakers were not always the best or most convenient or economical vessels for other purposes, especially when the wooden walls were available and perfectly sufficient for summer work, and much cheaper. The Hudson's Bay Company chartered both types to supply their northern stores; the Canadian Government chartered them for northern work; and some, like the *Fogota* and *Sagona*, were used in the coastal freight and passenger service, while the *Florizel* and *Stephano* were used in the St. John's-New York passenger trade. Nevertheless, the sealing fleet had evolved to such a point technologically that it could not be supported economically, and while the study of the post-1914 period is outside the scope of this book it must be pointed out that the steel fleet collapsed during the First World War. W.A. Munn summed it up best in 1923:

From 1906 to 1914 Newfoundland had the finest fleet of Sealers and Ice-Breakers in the World. We had a fleet that any Country might be proud of, and which judging from all human expectations, placed assured success within our grasp.

These powerful iron clads set such a pace for the wooden steamers beyond the limit of what they could possibly stand, and the result was that they broke them up in their attempts to force them through the ice to keep up with their more powerful rivals.

The records show the loss of some of the wooden walls every year, and in a very short time there would have been none of them left.

The iron clad steamers were built at a time when the cost of ship-building was exceedingly low but there is such a thing as paying even too much for success, and while nothing could compete with these steamers in ice floes, still, that work is only for one or two months of the year, and during the other ten months the steamers found it impossible to compete with larger steamers built more economically for carrying freight.

When the great war started in 1914 the owners of these Iron-clads took the first opportunity to dispose of this splendid fleet, so that within two or three years most of these iron-clads were sold to foreigners.[275]

The *Beothic, Adventure, Bellaventure* and *Bonaventure* were sold to Russia in 1915-16 to be used on the sea route to Murmansk. The *Stephano* was sunk by a German submarine off Nantucket in October 1916, without war-risk insurance, and the *Florizel* was totally wrecked near Cape Race, with the loss of 99 lives, while en route to New York in February 1918.

The *Seal* continued to prosecute the seal fishery until it was lost in 1922. The *Fogota* went to the front on a few occasions, and the *Sagona*, more frequently, but each was employed primarily in the coastal trade. But the expectation that the steel ships would displace the wooden walls was not met, and this antiquated industry remained heavily dependent on the few old surviving wooden walls until its demise in the early 1940s.[276]

Although St. John's monopolized the industry by 1914 the victory was a hollow one, with sealing playing a much reduced role in the economy. In addition, the capital's fleet of steel sealers/ice breakers was not destined to remain a permanent part of the shipping of that port.

Summary of the age of the iron clads

The iron clads were impressive steamers and contributed to the centralization and monopolization of the seal fishery by the most powerful firms in the capital. There was no denying their effectiveness but they were too expensive for this industry. While they could successfully compete with the older wooden walls in the sealing industry the latter were generally quite adequate for other summer and fall work and much cheaper to operate. Consequently, the "age of the iron clads" was glorious but brief.

General Summary

The seal fishery began as a cottage industry in that it could be prosecuted by planters with small shallops and schooners. This was one of the reasons why

Newfoundland sealing competed so effectively with British-Greenland whaling at the beginning of the nineteenth century. However, even though early sealing was relatively inexpensive in comparison to whaling the vessels in use were considerably larger than the small 3- to 5-man boats used in the shore cod fishery. The two industries also differed in that the former needed a more complex infrastructure—seal skinning facilities, oil rendering containers, primary leather processing equipment—than did the latter. The traditional cod fishery involved the production of saltfish by salting and drying, and the production of cod oil by rendering and could be carried out in any of the thousands of coves and harbours situated along the coastline. The spring seal fishery, on the other hand, gravitated towards those harbours that were able to accommodate vessels, and bypassed the many coves where the cod fishery had always managed to flourish. In fact, as the industry acquired larger vessels, only the more commodious harbours were suitable. Similarly, as the manufacture of seal oil became more complex, and the qualities more definitely recognized and labelled, the facilities needed to process this product became more sophisticated and expensive. Thus fishermen from smaller coves and harbours such as Bishop's Cove, Upper Island Cove, Spoon Cove, Bryants Cove, Crockers Cove, Freshwater, Salmon Cove, Perry's Cove, Spout Cove, Small Point, Broad Cove, Blackhead, Western Bay, Ochre Pit Cove, Northern Bay, and Gull Island, for example, sought spring employment on the sealing vessels of the neighbouring ports of Harbour Grace and Carbonear.

Not only were shelter, space and facilities important for the establishment of a sealing industry, but considerable specialized expertise was necessary as well. Only a relatively large and wealthy outport could provide the craftsmen needed to maintain the ships and their rigging and equipment. For example, in 1857, there were in the outport of Harbour Grace 100 'mechanics' in a population of 5,095 people, while in the district of Fogo, there were 9,717 people living in at least 72 communities and only 31 men were recorded as 'mechanics'.[277] As time went on, the sealing investors needed more and more specialized craftsmen to maintain and operate the vessels and the oil plants.

The sealing operation found a natural home in Conception Bay, with its big fleets of North Shore and Labrador fishing vessels available for spring sealing. Consequently, the industry flourished in this area at the beginning of the nineteenth century, with Harbour Grace leading the way, followed by Carbonear and Brigus. St. John's took some time to enter this fishery but, by the late 1820s, had become a major player. During this early period, vessels went to the ice from a variety of harbours on the east coast, but by mid century, it was almost exclusively confined to the bigger centres which could handle the larger vessels and the more sophisticated manufacturing.

As the same time, the growth of the Labrador fishery became ever more important because the increasing population needed summer employment as well as sealing in order to prosper. Because neither Harbour Grace, Carbonear nor Brigus—to name the most important outports—could absorb more fishermen into their local shore fisheries, growing numbers of men (and women) went to the Labrador coast each summer: most as sharemen and 'shipped' men and women employed by planters; others as planters themselves; and all described as 'stationers'. Another group, who were vessels owners (or leased vessels from the merchants) and known as 'floaters', transported the stationers and sometimes fished from shore but, more often, sailed from harbour to harbour along the coast, looking for the best fishing grounds. It was the combination of the seal and Labrador cod fisheries that made Harbour Grace the leading sealing centre by the 1860s.

With the introduction of steamers into the seal fishery in 1863, St. John's found itself better situated to dominate the industry. The two major sealing firms in the colony—Ridleys and Munns—had invested so heavily in sailing craft that they could not shift to steam very easily. In St. John's, on the other hand, merchants were not so hampered, and being greater in number, it was not long before a fleet of steamers was firmly established there.

Sail could not compete with steam, and it was not long before the sailing fleet had practically disappeared—first from the smaller outports, such as Trinity and Catalina; next from Carbonear and Brigus; and finally, from Harbour Grace. The latter outport's two mercantile houses, Ridleys and Munns, eventually acquired four steamers, but Ridleys went out of business in 1870—destroyed by the depression of the 1860s. Munns continued to operate a large business in Labrador saltfish, cod oil, seal oil and seal skins, but the 1880s were terribly depressed years, and that firm closed in 1894. The St. John's firms and steamers now dominated the industry.

Competition within the capital city continued, and the firms drove their captains, who in turn, drove their men to produce. The seal herds declined (along with the steep decline in the price of oil), and steel steamers were introduced in 1906 in an effort to make the most of this dwindling resource. Many wooden walls were destroyed, trying to keep up with the steel ships, and it became increasingly difficult to operate the latter economically. However, by forming a cartel, the three firms that owned the oil plants were able to dictate prices and make considerable profits. By 1914, the sealing fleet was firmly in the hands of St. John's, with all the attending industries as well. From nearly 400 sailing vessels and 14,000 men operating out of prosperous outports in the 1850s, the industry had declined to 21 steamers and 3,959 men centred in St. John's in 1914. However, by now, a truly efficient seal fishery had become too expensive to operate and most of the steel ships were soon disposed of and not replaced.

NOTES

1. Shannon Ryan, "The Origin and Early Growth of Newfoundland's Seal Fishery." (A paper presented to the annual meeting of the Canadian Historical Association, Guelph, June 1984), p. 15. The statistics in the above paper were taken from the CO 194/23-87.
2. CO 194/44, fols. 115-7. Gower to CO, 18 July 1805.
3. Anspach, *History*, pp. 260-1.
4. CO 194/54, fols. 159-74. Keats to CO, 18 December 1813.
5. CO 194/45, fols. 17-47. "Gower's Report for 1804."
6. Please note that the records do not always differentiate between the outport of Trinity and Trinity Bay; the same applies to the outport of Bonavista and Bonavista Bay. While most of the fleet of each bay was owned by firms and individuals in each of these principal outports, certainly during this early period, this was not

entirely true at all times. The present writer will continue to use the name of the outports and the bays interchangeably as they appear in the records.

7 Anspach, *History*, p. 298.
8 Hatton and Harvey, *Newfoundland*, pp. 166-7.
9 Anspach, *History*, p. 299.
10 Anspach, *History*, pp. 299-300.
11 CO 194/45, fols. 17-47. "Gower's Report for 1804," and CO 194/129, fol. 147. Le Marchant to CO, 4 May 1848. See also Tables 3.2 and 3.3 for an indication of the average size of vessels and crews during this early period.
12 Mosdell, *Chafe*, p. 20.
13 Mosdell, *Chafe*, p. 16.
14 Mosdell, *Chafe*, p. 20.
15 Mosdell, *Chafe*, p. 22.
16 Mosdell, *Chafe*, p. 39.
17 Mosdell, *Chafe*, p. 16.
18 CO 194/129, fols. 128-53. Le Marchant to CO, 4 May 1848.
19 Mosdell, *Chafe*, p. 17. Most of the information on rigging has been taken from this source.
20 CO 194/23, fol. 517. "Account of the Seal Fishery for 1803."
21 *Royal Gazette*, 24 April 1849.
22 *Newfoundlander*, 2 March 1837. It should be pointed out that in recent years, St. John's harbour has not frozen, and the other former sealing harbours seldom freeze; no doubt, the raw sewage that is being dumped there in modern times acts as an efficient anti-freeze.
23 See the governor's annual reports for 1819 in CO 194/62, fol. 150; and CO 194/64, fol. 23.
24 In order to find the total number of men, it is necessary to use a different source in the CO 194 correspondence. Each year during 1831-33 (and for some years previous), the governors sent to the Colonial Office, along with the usual report on the state of the inhabitants and the fisheries, an "Account of the Spring Seal Fishery." The three that have been used here can be found as follows: CO 194/83, fol. 35; CO 194/85, fol. 186; and CO 194/87, fol. 18. These reports show a very slight variation in the number of ships clearing from St. John's and Conception Bay, as expressed in Tables 3.1, 3.2 and 3.3, but this does not affect the conclusions being drawn.
25 CO 194/72, fol. 390.
26 CO 194/74, fol. 355.
27 CO 194/83, fol. 25.
28 The word 'catch' was used widely to describe the result of the season's voyage. See *DNE*, p. 90.
29 CO 194/85, fol. 186.
30 CO 194/87, fol. 18.
31 CO 194/87, fol. 18. According to this report, Newfoundland produced 8,428 tuns of seal oil in 1833, while the report used for drawing up Table 2.6 stated that 8,639 tuns were exported. For the sake of consistency, the present writer has used the former "Account."
32 CO 194/57, fol. 12.

33 CO 194/59, fol. 27.
34 CO 194/62, fol. 150
35 CO 194/64, fol. 143; and CO 194/65, fol. 121.
36 CO 194/74, fol. 355.
37 CO 194/78, fol. 70.
38 CO 194/80, fol. 84.
39 CO 194/81, fol. 45.
40 CO 194/83, fol. 25; CO 194/85, fol. 186; and CO 194/87, fol. 18.
41 Anspach, *History*, pp. 418-25.
42 Samuel George Archibald, *Some Account of the Seal Fishery of Newfoundland* (St. John's, 1852).
43 Mosdell, *Chafe*, p. 33.
44 *Royal Gazette*, 18 April 1871. The author goes on to say that the oil "is used largely in lighthouses, for machinery, and in the manufacture of the finer kinds of soap."
45 Mosdell, *Chafe*, p. 33.
46 See Linda Little, "Plebian Collective Action in Harbour Grace and Carbonear, Newfoundland" (M.A. Thesis, MUN, 1984). Little examined the conflicting reports and concluded that Harbour Grace was the most important outport in Conception Bay.
47 Jukes, *Excursions*, Vol. 1., p. 35.
48 *Newfoundlander*, 28 March 1844. Note that Harbour Grace is often spelled without the 'u' and that the spelling for Mosquito varies.
49 *Newfoundlander*, 25 March 1847.
50 Mosdell, *Chafe*, p. 21.
51 CO 194/129, fols. 128-53. Le Marchant to CO, 4 May 1848. Note that "Ports to the Northward" refers to all ports north of Conception Bay and includes Trinity, Bonavista, Catalina, Fogo and so on.
52 Anspach, *History*, p. 229.
53 *Weekly Herald*, 27 August 1845.
54 The *Isabella Ridley* was built in Pugwash, Nova Scotia, under the supervision of Capt. Henry Thomey, who commanded this vessel at the ice for 25 years, averaging 3,800 pelts annually over that time. At the end of its career, the vessel was tied up in Harbour Grace for a couple of years until the moorings broke in a gale and it drifted across the harbour and was smashed to pieces on Feather Point. *Daily News*, 27 February 1896.
55 CO 194/129, fols. 128-53. Le Marchant to CO, 4 May 1848.
56 *Newfoundlander*, 16 March 1837.
57 *Express*, 31 May 1859. This is a rare report; very little is known about the seal oil processing plants. It is equally rare to find references to the amount of fat obtained from the pelts.
58 In this case, 50.2 lbs. of fat were obtained from each pelt; other averages from the figures below range from 48.1 lbs. to 50.6 lbs.
59 Joy, "Trades and Manufacturing," p. 3.
60 *Royal Gazette*, 4 April 1837.
61 *Royal Gazette*, 27 March 1838,
62 *Newfoundlander*, 1 April 1847.

63 *Public Ledger*, 23 March 1852.
64 *Courier*, 16 March 1853.
65 *Public Ledger*, 11 April 1854.
66 *Public Ledger*, 27 March 1855; and *Courier*, 11 April 1855.
67 *Public Ledger*, 21 March 1856; and *Courier*, 9 April 1856.
68 *Newfoundlander*, 19 March 1857.
69 *Public Ledger*, 14 April 1857.
70 *Public Ledger*, 30 March 1858; and *Newfoundlander*, 5 April 1858.
71 *Newfoundlander*, 24 March 1849; and *Express*, 5 May 1859. It seems that previously, Bonavista ships had cleared from Catalina.
72 *Public Ledger*, 23 March 1860; and *Courier*, 4 April 1860.
73 *Courier*, 23 March 1861.
74 CO 194/131, fol. 269. James Simms to Le Marchant, 23 June 1849.
75 CO 194/45, fol. 199-205. "Gower's Report for 1805."
76 Thanks to Keith Hewitt, who checked on vessel ownership and provided the present writer with his data.
77 *Newfoundlander*, 16 April 1829.
78 *Royal Gazette*, 21 April 1829.
79 *Public Ledger*, 9 March 1830.
80 *Royal Gazette*, 26 March 1844.
81 CO 194/129, fols. 128-53. Le Marchant to CO, 4 May 1848.
82 *Pilot*, 13 March 1852.
83 Mosdell, *Chafe*, p. 40.
84 "The *Dash*" in Brian Dunfield (ed.), *Newfoundland Law Reports, 1846—1853: Decisions of the Supreme Court of Newfoundland* (St. John's, 1915), pp. 375-84. Judgement rendered in September 1853. See also the *Newfoundlander*, 1 September 1853. "The decision of the Vice-Admiralty Court."
85 *Times*, 12 April 1862.
86 *Times*, 16 April 1862.
87 *Express*, 6 March 1852.
88 *Royal Gazette*, 7 January 1845.
89 *Courier*, 5 March 1845.
90 Smith, *Labrador Fishery*, pp. 10-1.
91 *Times*, 8 March 1851.
92 *Times*, 12 March 1851.
93 *Express*, 6 January 1853.
94 *Weekly Herald*, 23 March 1853.
95 Mosdell, *Chafe*, p. 16.
96 *Royal Gazette*, 26 March 1844. This same account also reported that, in addition, 50 to 60 vessels were fitted out in Trinity Bay this year.
97 Prowse, *Newfoundland*, pp. 451-2.
98 See Mosdell, *Chafe*, p. 81, for the names of the sealing firms in operation in 1914. The important and well-known companies will be given full titles when first introduced but thereafter, for reasons of brevity and in order to avoid cumbersome

repetition, where possible, they will be referred to by abbreviated names. For example, Baine, Johnston and Company will become 'Baine Johnstons'; Bowring Brothers will become 'Bowrings'; and Job Brothers—'Jobs'. Locally, companies were known by the abbreviated plural, as in, "He worked at Jobs" or "She shopped at Bowrings." More well-known examples of this practice can be found in the use of 'Eatons', 'Simpsons' and 'Lloyds' in place of these firms' longer names. Incidentally, Job is pronounced like the name of the long-suffering but patient biblical character.

99   Unless otherwise indicated, the following information has been taken from *DNLB*, *ENL* and Keith Matthews, "Profiles of Water Street Merchants" (MUN, CNS: Typescript, 1980).

100 Wendy Martin in *Once Upon a Mine: Pre-Confederation Mines in Newfoundland* (Montreal, 1983), p. 24, wrote: "No character in the story of early Newfoundland mines appears as consistently and in more divergent enterprises as Captain Philip Cleary. Between 1865 and his death in 1907, his name surfaced across the Island in ventures associated with marble, coal, asbestos, oil, pyrite and copper. It is said that none of them ever made respectable profits while in his grasp."

101 Prowse, *Newfoundland*, p. 403; and Melvin Baker, "Alexander, Robert," in *DCB*, Vol. XI.

102 See also Donald K. Regular, "The Commercial History of Munn and Company, Harbour Grace" (student paper, CNS, MUN, 1973?).

103 For the best accounts of these operations, see Chesley A. Sanger, "The 19th Century Newfoundland Seal Fishery and the Influence of Scottish Whalemen," *Polar Record*, XX, no. 126 (1980), 231-52; Sanger, "The Newfoundland Seal Fishery and Scottish Whalemen in the Nineteenth Century," a paper presented to the 15th General Assembly of the International Geographical Union, Tokyo, Japan, 2 September 1980; Sanger, "Dundee Steam-Powered Whalers and the Newfoundland Harp Seal Fishery," a paper presented to the Annual Conference, British Association for Canadian Studies, Southampton, 23-25 March 1988; and Sanger, "The Dundee-St. John's Connection: Nineteenth Century Interlinkages Between Scottish Arctic Whaling and the Newfoundland Seal Fishery," *Newfoundland Studies*, IV, no. 1 (1988), 1-26.

104 Frederick W. Rowe, *A History of Newfoundland and Labrador* (Toronto, 1980), p. 293.

105 *Patriot*, 24 December 1855.

106 When it is obvious that steamers are being discussed, the prefix S.S. will be omitted.

107 See Ryan, *Seals and Sealers*, pp. 12-3, for copies of paintings of these two steamers under full sail.

108 See Ryan, *Seals and Sealers*, for the origins and vital statistics of most of the important wooden walls.

109 The barrel man stood in a barrel high on a mast, searching for seals, usually with a telescope (spyglass) or binoculars. The scunner stood in a barrel on the foremast and directed the vessel through the ice by shouting instructions to the bridge master, who relayed them on to the man at the wheel. See *DNE*.

110 *Evening Telegram*, 2 February 1906. Quoted from the *Dundee Advertiser*.

111 Mosdell, *Chafe*, p. 25; and Prowse, *Newfoundland*, p. 452.

112 £1 stg. was equal to about $4.86 (Newfoundland) while £1 local currency was equal to $4. The dollar became the official currency in 1865, but it was quite some time before its use became widespread.

113 For statistics on the steamer seal fishery, see Mosdell, *Chafe*; and Ryan, *Chafe*. For biographical information on the most important steamer captains, see Mosdell, *Chafe*; Ryan, *Seals*; *DNLB*; and *ENL*.

114 Mosdell, *Chafe*, pp. 41-2.

115 Gerald S. Doyle, *The Old Time Songs and Poetry of Newfoundland* (St. John's, 1927), p. 65. Reprinted in Ryan and Small, *Haulin' Rope and Gaff*, p. 32. The *Mary Joyce* was a small St. John's sailer of 58 tons, which was the only vessel (clearing from St. John's, at least) owned by E. Smith and Co. Its name might have been inserted into the song to emphasize the contrast with the *Esquimaux*, or it may actually have been part of the incident which involved becoming jammed in the ice. 'Chaw' here means 'talks too much' see *DNE*. The unfortunate Halleran did not receive another steamer command. Robert Daw[e] commanded the *Lion*, and although he brought in 5,300 pelts—the second largest load—for whatever reason, he never commanded another steamer. In 1872, he, his son and many of his crew were lost in the sinking of his schooner, the *Huntsman*, while at the seal fishery (see Chapter 5). In 1867, the *Wolf* was commanded by Capt. John Bartlett, Sr., and the *Panther* by Capt. Abram Bartlett. Fragments and versions of this song can be found in several sources.

116 A 'high liner' in this context refers to the captain or the vessel with the biggest and, usually, most valuable cargo of pelts in any given spring or in any given trip. See W. J. Kirwin and G. M. Story, "The Etymology of *High Liner*: Problems of Inclusion in *The Dictionary of Newfoundland English*" in *American Speech*, LXI, no. 3 (1986), 281-4.

117 In 1867, Capt. William Jackman saved 27 lives during an autumn storm on the Labrador coast. He swam to a wrecked fishing vessel 27 times, bringing one person ashore on his back each time.

118 Mosdell, *Chafe*, p. 42.

119 *Royal Gazette*, 18 July 1871. "Correspondent of the Boston Traveller," 24 May 1871. He also pointed out that one merchant in Harbour Grace cleared $180,000 on 100,000 seal pelts. This was most likely the firm of Munns.

120 Capt. Arthur Jackman commanded Commodore Perry's vessel on the latter's first polar venture in 1886.

121 After Munns' bankruptcy, ownership was transferred to Baine Johnstons, and in 1898, while under the command of Capt. George Barbour, 48 of the crew were lost on the ice in what became known as the *Greenland* Disaster.

122 "Chamber of Commerce Minute Book" (1866-1875, Vol. V), pp. 46-52. This is a rare and valuable source. All statistics have been rounded off.

123 Nimshi Crewe, "A Descriptive Monograph on the Slades," William White Collection.

124 William Cox and Company Letter Book, 1858. A letter from Charles Edmonds, agent at Twillingate, to Messrs. William Cox and Company, Poole, 30 April 1858.

125 Cox Letter Book, 30 April 1858.

126 Crewe, "Slades."

127 Crewe, "Slades." Although they had no direct role to play in the history of the seal fishery and its development on the northeast coast, it is important to point out that the same decline in the West of England firms that can be seen happening on that coast was also happening on the south coast. For example, the long-standing Jersey firm of P. W. Nicolle and Co., with its Newfoundland headquarters in Jersey Harbour, Fortune Bay, went bankrupt in 1863 with liabilities amounting to £54,000. (See Newman and Company Letter Book, No. 64, pp. 279-80.) In 1864, Newman and Co., with its headquarters in London (but originally in the West of England),

closed its premises in Burgeo. (No. 64, pp. 300 and 337-39.) Meanwhile, Nicolle's premises in Jersey Harbour were taken over by Nicolle de Quitteville and Co., which, in turn, sold out to Degroucy and Co. in 1872 and sold the Labrador coast premises at Forteau and Blanc Sablon to Philip Simon in 1876. (John and William Boyd Letter Books, 1875-1878, 16 May 1877.) In 1882, Newman and Co. pointed out that it and Degroucy were the only two old-style firms left in Newfoundland. (Newman, Book No. 67, pp. 233-6.) In 1886, when the Jersey bank suspended business, Degroucy and Co. was forced to declare insolvency. (No. 68, pp. 117-8.) Newman sold its Gaultois establishment in 1900 and the premises in Harbour Breton in 1907, bringing to an end, after 300 years, its involvement in the Newfoundland saltfish trade and in Newfoundland's development in general.

128 Cox Letter Book, 1865-67. Letters dated 12 September 1866, 9 March 1867, 19 March 1867 and 29 March 1867.
129 Newman, Book No. 67, pp. 195-6; and No. 70, pp. 216-7.
130 Cox Letter Book, 1865-67, 19 March 1867.
131 Job Family Papers, Thomas R. Job to his father, Thomas Bulley Job, 16 December 1864.
132 Job Business Papers, 1810-85.
133 Ryan, "The Newfoundland Cod Fishery in the Nineteenth Century," p. 24.
134 *Royal Gazette*, 17 March 1868; and *Public Ledger*, 18 March 1868.
135 *Royal Gazette*, 15 March 1870.
136 *Royal Gazette*, 13 August 1867.
137 *Royal Gazette*, 11 August 1868.
138 *Royal Gazette*, 5 March 1872.
139 *Royal Gazette*, 10 March 1874.
140 *Royal Gazette*, 24 August 1875,
141 *Newfoundlander*, 13 April 1877.
142 *Newfoundlander*, 6 March 1877.
143 *Public Ledger*, 25 March 1881. See the *Weekly Herald*, 23 March 1853 for details on the average lifespan of a wooden vessel.
144 Book of Coasting and Fishing Ships. [Vessels clearing from Harbour Grace.]
145 *Evening Telegram*, 29 April 1882. Capt. Thomey was one of the most experienced sailing seal captains and one of the most successful as well. His failure this year was symptomatic of the decline of the sailing fleet and of the outports.
146 *Evening Telegram*, 7 March 1882.
147 *Evening Telegram*, 30 April 1883.
148 *Evening Telegram*, 4 February 1886.
149 *Evening Telegram*, 11 March 1888.
150 *Evening Telegram*, 6 March 1893.
151 *Evening Telegram*, 21 March 1893.
152 *Evening Telegram*, 28 and 29 March 1893.
153 *Daily News*, 27 and 29 April 1895.
154 *Evening Telegram*, 27 February 1896.
155 *Evening Telegram*, 27 February 1896. Another report, on 26 March, states that the captain of the *Florence B* was William Buffett, not John Baggs. See report of 14 May for an account of the total loss of the *Harvest Home*.

156 *Evening Telegram*, 26 March 1896.
157 *Evening Herald*, 20 February 1897.
158 *Evening Herald*, 16, 18 and 22 March 1897.
159 *Evening Herald*, 22 May 1897.
160 For example, see *Evening Herald*, 28 January 1898.
161 *Evening Telegram*, 28 January and 4 February 1898. See also *Statutes of Newfoundland*, 61 Vic. Cap. II.
162 *Evening Telegram*, 28 January 1898.
163 *Evening Telegram*, 18 February 1898.
164 *JHA* (1898-1914).
165 *Evening Telegram*, 15 March 1898.
166 *Evening Telegram*, 31 March 1898.
167 *Evening Telegram*, 8 May 1899.
168 *Evening Herald*, 19 April 1900.
169 *Evening Herald*, 13 March 1901.
170 *Evening Herald*, 3, 4, 10 and 14 May 1901. As one can see, the information on each schooner is not always complete.
171 *Evening Herald*, 18, 19, 22, 23 and 24 April 1902.
172 *Evening Herald*, 12 May 1903.
173 *Evening Herald*, 13 March 1905.
174 *Evening Herald*, 13 May 1902.
175 *Newfoundlander*, 17 June 1871.
176 *Newfoundlander*, 18 August 1871.
177 Ryan, *Chafe*, pp. 185-201.
178 *Evening Telegram*, 7 April 1883.
179 *Evening Telegram*, 9 April 1883.
180 *Evening Telegram*, 1 April 1886.
181 *Evening Telegram*, 15 May 1891.
182 *Evening Telegram*, 19 March 1886. 'Noggins' were small wooden casks, especially such casks sawn in half. In this case, they would be used on sealing ships to bring the food from the galley to the bunk, where the men ate in groups of three or four. (*DNE*, p. 350.) The 'crop' consisted of the personal supplies and equipment that each sealer purchased before leaving port. It could include a knife, steel, special boots, waterproof clothing and food. (*DNE*, p. 122.)
183 *Evening Telegram*, 25 March 1886.
184 *Newfoundlander*, 5 March 1875.
185 *Evening Telegram*, 14 February 1880.
186 Smith, *Labrador Fishery*, p. 22.
187 *Newfoundlander*, 18 August 1876.
188 *Newfoundlander*, 6 March 1877.
189 *Newfoundlander*, 22 February 1878.
190 See sources cited in endnote 103, above.
191 *Evening Telegram*, 12 January 1893.
192 *Evening Telegram*, 4 February 1893.

193 *Evening Telegram*, 25 February 1893.
194 *Evening Telegram*, 1 and 4 March 1893.
195 *Evening Telegram*, 9 March 1893.
196 *Evening Telegram*, 16 March 1893.
197 Rare were the sealing captains who had a foreign master's certificate; Capt. Bob Bartlett and Capt. Abram Kean were two such men.
198 *Evening Telegram*, 4 April 1893.
199 *Evening Telegram*, 4 and 10 April 1893.
200 *Evening Telegram*, 6 April 1893.
201 Mosdell, *Chafe*, p. 60. 'Crop' and 'berth' will be discussed in greater detail in Chapter 4.
202 *Evening Telegram*, 3-6 April 1893; and Mosdell, *Chafe*, p. 60.
203 *Evening Telegram*, 30 January 1894.
204 *Evening Telegram*, 14 and 28 February 1894.
205 *Evening Telegram*, 2, 11 and 23 April 1894.
206 Mosdell, *Chafe*, pp. 59-62; *ENL*, Vol.2, p. 748; and *DNLB*, pp. 338-90.
207 Newman, Book No. 69, pp. 307 and 314-6.
208 *Evening Telegram*, 16 April 1894.
209 *Times*, 30 January 1895; *The Crown vs the Directors and Manager of the Commercial Bank of Newfoundland* (St. John's, 1895); *The Crown vs the Directors of the Union Bank of Newfoundland* (St. John's, 1895); and Patrick Joseph Hickey, "The Immediate Impact of the 1894 Bank Crash" (BA Honours Dissertation, MUN, 1980).
210 *Statutes of Newfoundland*, 55 Vic. Cap. X and 61 Vic. Cap. IV.
211 Unless indicated otherwise, basic data on the steamer fleet has been taken from Mosdell, *Chafe*.
212 *Evening Telegram*, 8 and 11 April 1896.
213 *Evening Herald*, 1 March 1897.
214 *Evening Telegram*, 5 May 1898.
215 *Evening Telegram*, 28 March and 5 April 1898.
216 *Evening Telegram*, 6 April 1898.
217 *Evening Telegram*, 4 April 1899.
218 *Evening Telegram*, 11 April 1899.
219 *Evening Herald*, 29 March 1900.
220 *Evening Herald*, 2 April 1900.
221 *Evening Herald*, 20 April 1900.
222 *Evening Herald*, 8 and 21 May 1900.
223 *Evening Herald*, 1 April 1902.
224 *Evening Herald*, 16 April 1903.
225 *Evening Herald*, 28 January 1904.
226 *Evening Herald*, 8 February 1905. Harveys had dropped out of the seal fishery in 1896, when it lost the *Windsor Lake*, and re-entered it in 1903 under the name A. J. Harvey and Company (hereafter, Harveys). The firm bought the *Grand Lake* and *Panther* that year, and in 1904, added Farquhar's *Newfoundland* to its fleet. However, it is apparent that each one of these steamers constituted a separate

company, at least by 1905—i.e., New Panther S.S. Co. Ltd., Grand Lake S.S. Co. Ltd. and S.S. Newfoundland Sealing Co. Ltd.
227 *Evening Herald*, 10, 11 and 12 March 1902.
228 Mosdell, *Chafe*, pp. 59-68.
229 Murray and Sons, an oil manufacturing company, was a subsidiary of Baine Johnstons. In February 1905, at a meeting of sealing operations held in St. John's and discussed below, Baine Johnstons signed an agreement on behalf of Murray and Crawford. Also, in another report in the *Evening Herald*, 9 May 1905, Baine Johnstons was listed as one of the seal oil manufacturers along with Jobs and Bowrings.
230 *Evening Herald*, 27 May 1903.
231 *Evening Herald*, 18 June 1903.
232 *Evening Herald*, 24 June 1903.
233 *Evening Herald*, 9 May 1904.
234 CO 194/45, fols. 17-47. "Gower's Report for 1804."
235 Fortunately for the modern researcher, the relentless compiler and researcher Levi Chafe, an observer with personal and professional contacts throughout the business community at this time, has left an account of the principals involved in the industry. See the short biographical sketch of Levi Chafe in the "Introduction," Ryan, *Chafe*, pp. 7-13.
236 *Evening Telegram*, 3 March 1906.
237 Prowse, *Newfoundland*, p. 453.
238 The terms 'iron clad' and 'ironclad' were first used to describe the early warships that were built of wood and sheathed with iron cladding. By 1906, the so-called 'iron clads' were built entirely of steel and were thus truly steel ships. See Peter Kemp, ed., *The Oxford Companion to Ships & the Sea* (Oxford, 1976), p. 420. Chafe used the term 'iron clad' to describe the steel steamers introduced into Newfoundland at this time, and this term has become more commonly used than the term 'steel'. In the interest of historical accuracy, the present writer uses the term 'steel', except in the headings and sub-headings, where the traditional term 'iron clad' is used.
239 *Evening Telegram*, 2 February 1906. Quoted from the *Dundee Advertiser*.
240 *Evening Telegram*, 6 March 1906.
241 *Evening Telegram*, 20 January 1906.
242 *Evening Telegram*, 28 March 1906.
243 *Evening Telegram*, 30 March 1906.
244 *Evening Telegram*, 2 April 1906.
245 *Evening Telegram*, 21 to 30 April, 1906. This was Capt. Jackman's last voyage to the ice; he died in 1907, having first served as captain of a sailing vessel at the age of 22 and taking command of his first steamer in 1871. From 1871 to 1906, Capt. Jackman never missed a spring as captain of a steamer and, during his last years, was recognized as "commodore" of the fleet.
246 See Mosdell, *Chafe*; and *Evening Telegram*, 22 May 1906. The tonnage of each vessel is included in brackets after its name. See also Table 3.8 for the class of each of Newfoundland's sealing steamers.
247 *Evening Telegram*, 7 April 1908.
248 See Mosdell, *Chafe*, for the information on catches, prices and other facts used here.
249 *Evening Telegram*, 18 April 1908.

250 *Evening Telegram*, 21 April 1908.
251 The election of November 1908 produced a tie of eighteen members each for the ruling Liberal Party and the opposition Peoples' Party. The governor refused Prime Minister Bond's request to dissolve the House and the latter was forced to tender his resignation in late February 1909. Peoples' Party leader Morris assured the governor that he could form a government and was invited to do so. When the House met on 30 March it failed to elect a speaker and Morris was granted a dissolution. In a hard-fought election on 8 May, Morris' party won a majority. The campaigns of each leader during the period from November 1908 to May 1909 fully occupied their supporters, including the media.
252 *Evening Telegram*, 12 April 1909.
253 *Evening Telegram*, 25 February 1910.
254 *Evening Telegram*, 21 March 1910.
255 *Evening Telegram*, 14 January 1910.
256 *Evening Telegram*, 25 April 1910.
257 *Evening Telegram*, 2 April 1910.
258 *Evening Telegram*, 25 April 1910.
259 *Evening Telegram*, 16 January and 17 February 1911.
260 *Evening Telegram*, 6 March 1911.
261 *Evening Telegram*, 28 March 1911.
262 *Evening Telegram*, 2 March 1911.
263 *Evening Telegram*, 20 March 1911.
264 *Evening Telegram*, 21 March 1911.
265 *Evening Telegram*, 26 May 1911. On 30 May, both Jobs and Bowrings let it be known that their firms would have new steamers for the following year.
266 *Evening Telegram*, 8 January 1913.
267 *Evening Telegram*, 12 February 1913.
268 *Evening Telegram*, 6 March 1913. See Chapter 5 for the details of this voyage.
269 *Evening Telegram*, 13 and 29 March 1913.
270 *Evening Telegram*, 26 April 1913.
271 *Evening Telegram*, 4 February 1913. Letter to the editor from William F. Coaker, 3 February 1913.
272 *Evening Telegram*, 21 February 1914.
273 *Evening Telegram*, 13 March 1914.
274 Cassie Brown, *Death on the Ice* (Toronto and New York, 1972).
275 Mosdell, *Chafe*, pp. 26-7.
276 Ryan, *Chafe*.
277 *Census of Newfoundland* (St. John's, 1857).

CHAPTER 4
Fishermen—Ice Hunters

THE MEN WHO engaged in the seal fishery were, first and foremost, cod fishermen who had discovered a new resource and learned how to harvest it. Because of the nature of this new fishery, the men who pursued the seals among the ice floes were referred to as 'ice hunters', at least by 1833, when a song was published that began with: "Come all you jolly ice-hunters and listen to my song."[1] A slightly later song, first published in 1842, began with the line: "We'll sound the hardy sealers praise, a wild and cheerful strain."[2] Like the vessels that carried them, the cod fishermen who engaged in the quest for seals came to be referred to increasingly as 'ice hunters' or 'sealers', and the nineteenth century was their age.

As earlier chapters have indicated, during the early years of the nineteenth century, Newfoundland cod fishermen became adept at hunting seals among the ice floes. In 1803, the records show that 77 vessels with a combined burthen of over 2,100 tons sailed 'to the ice' carrying 998 men: 306 from Bonavista; 135 from Trinity Bay; 369 from Conception Bay; and 188 from St. John's.[3] In 1804, the numbers rose significantly to 1,639 men, of whom 120 came from Bonavista and 152 from Trinity while 430 and 729 were from St. John's and Conception Bay, respectively. Others were reported as coming from Ferryland, Greenspond and Fogo.[4] Conception Bay and St. John's gave early indications that they would be leading centres in this new industry.

Ice hunting in its earliest period was a democratic activity, with ten or twelve men in an open boat together learning how to operate in dangerous fields of ice and how and where to find their prey. Munn described the first ice hunting craft as being simple shallops, decked in both fore and aft with movable boards in the centre for stowing the pelts. He points out that the "shelter cuddies at each end gave the crews some protection when they remained out over night,"[5] but according to him, at first, these shallops seldom went out of sight of land. However, in 1800, Captain William Bartlett of Brigus was out after seals and, not finding any in the vicinity of Baccalieu, decided to 'follow on', first to Cape Bonavista, then to the Funks, and eventually to the Labrador coast, where he finally found the ice and the seals and returned with a load of pelts.[6] The enterprise expanded rapidly, as already described, and within a short few years, there were thousands of ice hunters in hundreds of vessels engaged in the spring seal 'fishery'.

Manpower

Sail and manpower

The identities of the men who built the seal fishery remain tantalizingly indistinct. As Tables 4.1 to 4.5 illustrate, there were hundreds of captains employed in this industry by the 1830s—some bearing the old surnames of the original West of England planters, such as Pinn, Pynn and Pyke in St. John's and Pike, Taylor, Howell, Pynn, Parsons, Sheppard and Davis in Conception Bay. Others were obviously from among the newly arrived Irish settlers who had emigrated during the Napoleonic War period, such as Shea, Murphy, Ryan, Power, Dalton, Cummins, Brennan, Callahan, Mulcahy, Burke, Hennessey, Dooley, Mackay, Casey, Fitzgerald, Walsh, Kavanagh, Sullivan, Barron, Feehan, Kelly, Kennedy and McCarthy in St. John's and Conception Bay. Some names were to remain definitely associated with certain centres. Chafe records that such surnames as White, Jackman, Ryan, Graham, Duff, Rhodes and Cummings were associated with St. John's, while Chafes and Mullowneys came from the Southern Shore.[7] Also, as Table 4.3 indicates, Howell, Davis, Taylor, Pike, Parsons, Penny, Oats, Butt, Forward, M'Carthy, Hanrahan, Finn and Ash were noted Carbonear families. Similarly, Keefe, Pike, Power, Dwyre [or Dwyer], Davis, Murphy, Kelly, Hogan, Hearne, Sheppard, Stevenson, Pynn, Cooney and Parsons (six captains in 1836) were and remained prominent Harbour Grace names. As Table 4.5 shows, one finds captains in Bay Roberts by the name of Snow, Mercer, Davis, Russell, Delany (or Delaney); Dawe, Richards, Batten, Mugford and Andrews (four sealing captains in 1838) are among the most prominent in Port de Grave; the names of Norman, Munden, Whelan, Percy, Wilcocks (or Wilcox), Spracklin, Walsh and Burke appear in the Brigus records; the names of Whelan, Norman and Spracklin also appear among the captains from Cupids, along with three Ledroes (LeDrews). These examples are intended to illustrate the number and diversity of the surnames of the captains in the seal fishery during the early decades of the industry. Chafe was correct when he wrote:

> Conception Bay captains are too numerous to mention—The Mundens, Normans, Bartletts, Antles, Wilcox's, Smith, Clarkes, Dawes, Noel, Parsons, Thomey, Howells, Penneys, Pikes, Pynns, Gosse, Taylors, Spracklins, St. John, Fury, O'Brien, Flynns, Pumphreys and Murphys.[8]

The surnames demonstrate that by the 1830s, the Irish half of the population in Conception Bay and St. John's had become heavily involved in the seal fishery as captains and/or ship owners. This suggests that there had been considerable upward mobility within the industry over the previous 20 years, considering the fact that so many Irish had arrived to seek employment at that time as fishing servants in Conception Bay and St. John's.

Information on some of the early captains has survived. Captain Arthur Thomey, originally from Mosquito, was a successful sealing captain on his own account and, later, commanded sailing ships belonging to Ridleys in Harbour Grace. His son, Captain Henry, born c.1822, became master of a sealing vessel at the age of 20 and continued in command until he was almost 80 years of age, and "rarely failed to bring in a good trip."[9] William Munden, born in Brigus in 1776, is reported to have "taught them [the Conception Bay fishermen] all their business [about sealing]."[10] He was convinced that bigger vessels were more efficient and effective for sealing and, despite being ridiculed by other skippers, had the *Four Brothers*, of 104 tons burthen, built and launched for the 1819 season. Chafe concludes that William Munden was one of the most successful

Brigus captains, along with John Norman and William Whelan.[11] Linda Little, in her Master of Arts thesis, writes:

> The sealing masters were likewise not a homogeneous group, although they were predominantly men of a respectable social standing in the communities. Some masters had little or no financial investment in sealing vessels and others had extensive investments. An examination of a sample of 76 masters revealed that 43 percent never owned a share in a vessel over a 25 year period, while 16 percent owned from 129 to 275 shares over the same quarter century. Socially, masters could have worked up from fisherman status through the ranks or could have been born into a prosperous planter or merchant family.[12]

Except for the above fascinating minutiae and Little's examination, the information on the early sealing masters is most inadequate. However, it is fair to conclude with Little that some, like the Thomeys and Mundens, were born into relatively important planter families, while others worked their way into their commands through experience and intelligence within the context of a rapidly growing industry.

This is not to suggest that all captains were uniformly successful. One of the earliest extant ballads describes a sealing voyage on board the sailing vessel *St. Patrick* under the command of Captain Tom Casey. Captain Casey took the vessel into White Bay (always a risky decision, given its long v-shape which could become packed solid with ice in a northeast wind), and it became jammed and missed the seals. The men walked ashore in disgust and found their own way home. The incident was recorded in a ballad entitled, "We will not go to White Bay with Casey anymore:"

> Tom Casey being commander
> Of the *St. Patrick* called by name.
> With 28 as brave a boys
> As ever ploughed the main.
>
> While sailing down by Baccalieu
> Five whitecoats we took in.
> Cheer up, my boys, Tom Casey cries,
> Five thousand we'll bring in.
>
> The poor man was mistaken
> Upon that very day
> Which left us jammed in White Bay
> Until the last of May.
>
> And now we're clear of White Bay
> And landed safe on shore,
> We will not go to White Bay
> With Casey any more.[13]

While little is known about the early captains, even less is known about the men who served on the ships for a share of the voyage. Their numbers increased rapidly in the early years. Conception Bay experienced an increase in the number of fishermen engaged in the spring seal fishery from an average of 3,101 annually during the five-year period from 1824 to 1828 to an average of 4,874 men exactly ten years later in 1834-38—an overall increase of 57 per cent. The numbers in St. John's, although not as high as those of Conception Bay, were quite substantial and increased from 1,485 men to 2,911 over this same period—a much higher overall increase of 96 per cent.[14] The combined population of St. John's and Conception Bay in 1828 was reported to be 33,024 people, and in that year, 5,333 men, or 16 per cent of the population, went to the seal fishery. In 1836, the total population of the two areas amounted to 42,041 people, and the average number of sealers totalled 7,785 annually during the 1834-38 period,[15] or about 19 per cent of the population. Consequently, by the 1830s, one must assume that every able-bodied cod fisherman in Conception Bay and St. John's went to the ice every spring. This suggests that men from all social groups participated.

Little summarizes the descriptions of the social classes in Harbour Grace and Carbonear that have appeared in the contemporary literature.[16] The group at the lowest social position were the servant shoremen, who "were basically unskilled labourers who were unable to find work in the fishery and so worked on shore fetching and carrying and running errands." The next group were the fishing servants who "were better off, at least having regular employment for the season, either in the inshore fishery or on the Labrador coast." Little does not discuss "shoremen" thoroughly, but leaves the reader with the impression that this was a very small group of people. However, that was not the case, as one can see from the governors' annual reports. For example, during the 1821-26 period, the reported number of "shoremen" in the cod fishery averaged 5,137 and included such skilled shoremen as splitters, salters, headers and the masters of the voyage.[17] These shoremen were always considered on a par with the regular fishing servants employed in the island's shore fishery and on the Labrador coast. Better off than these were the independent fishermen who owned fishing rooms on the island or on the Labrador coast. They usually depended on family members to assist with the fishery: fathers and sons, fishing and splitting and salting; the elderly, women and children, spreading and turning the fish on the flakes or on the 'bawn',[18] and taking care of the other chores, such as gardening and tending the animals. Above this large body of men were the more skilled and specialized craftsmen, publicans, smaller shop keepers, tradesmen and captains and planters. The skilled tradesmen included the highly skilled splitters and salters (who were often considered shoremen) in the cod fishery and the seal skinners and gunners in the seal fishery. Many people could be classified in more than one category: it was not unusual to find a publican who was also a shop keeper; a butcher would often work as a seal skinner for the several weeks that it took to separate the fat from the skins after the pelts were landed every spring; a planter might also be a substantial farmer, often leaving part of his family to run the farm while he proceeded to the Labrador coast for the summer's fishery; an independent fisherman might work slaughtering animals in the fall and as an animal 'doctor' in the spring. The planters owned their fishing rooms; the smaller ones were basically independent fishermen while the larger planters owned more than one room or a large room and employed a number of men and a woman or two; the wealthier of these owned schooners and blended in with the smaller merchants. The important merchants, clergy, judges, senior civil servants and, later, politicians made up the top social classes; and, of course, the merchants were often elected to public office or served as magistrates.

However, Little concluded that there was a remarkable degree of homogeneity among the sealers:

> The seal fishery held a unique position in the social and economic fabric of Conception Bay. Socially, the seal fishery provided a more rigid and evident class structure than existed at any other time of the year. While the complex social structure of the communities persisted, work relationships for the three months of the seal fishery were comparatively straight forward. Each sealing vessel required a supplier, a master, and a crew of sealers. Labour and management were fairly distinctive, with the only grey area being the position of the masters.
>
> Fishermen, shoremen, servants, and tradesmen of different socio-economic status went sealing together. All united as sealers where demand for labour was great and opportunities were good.... No one could be barred from participation in the hunt through poverty as the few supplies needed for the journey could be obtained on credit from the local merchants. Working age men who could walk to Carbonear or Harbour Grace and find a berth could go sealing. The relative [sic] lucrative nature of the hunt prevented fishermen of a respectable status from snubbing the hunt and all its hardships and hazards as a menial task fit only for servants. Thus, men who prosecuted the cod fishery under a wide variety of relations to the means of production, were in the same 'class' in the seal fishery.[19]

While Little is undoubtedly correct in her descriptions of the seal fishery of the 1830s, and the society in which it operated, one could not go so far as to imply that any working-age man could go sealing. The key phrase here is "find a berth." As will be shown later, sealing was tremendously demanding on the men, and the exertions that it necessitated were beyond the capabilities of many fishermen—especially among the older, the arthritic and the injured. Nevertheless, it was reported retrospectively in 1916 that in the 1830s, "men were scarcer then and a boy of 15 could get a berth as easy as a grown man can now."[20] Therefore, one must return to the fact that, given the high proportion of the men in the seal fishery during the 1820s and 1830s, practically every able-bodied fisherman was sealing during these years.

The expansion in employment slowed dramatically after the 1830s. During the five-year period from 1854 to 1858, St. John's sent an average of 3,254 sealers to the ice annually—about 11 per cent of its population of 30,476 (as recorded in the 1857 census). Conception Bay—pulling ahead—sent an average of 7,585 sealers to the ice during this period, or about 23 per cent of its 1857 population of 33,396. In the case of St. John's, the number of men so employed had increased only 12 per cent during the 20-year period since 1834-38, while there had been an increase of Conception Bay sealers of 56 per cent during this same interval. However, these statistics do not tell the whole story; as has already been pointed out in the previous chapter, the records are very imprecise when one tries to determine which ships and men belong to each port. It is impossible to determine if all ships clearing from Brigus or Harbour Grace, for example, belonged to these ports; vessels from neighbouring ports without customs houses cleared from the nearest port where a customs house was located. Similarly, it is likewise impossible to distinguish ships belonging to northern ports from those that simply cleared from there. Problems of identifying the home ports of vessels makes the identification of home ports of crews, already difficult, now practically impossible. Nevertheless, it is apparent that nearly all able-bodied fishermen north of Ferryland district, but particularly those in St. John's and Conception Bay, found employment in the spring seal fishery. This fact was to change in the 1860s.

Steam and manpower

Both the introduction of steam and the decline in the seal fishery in the 1860s affected personnel. As Table 4.6 indicates, 1,418 men left St. John's on vessels clearing from that port for the ice fields in 1869, with the steamers carrying the largest crews—140 men on board the S.S. *Nimrod*, for example. While it is impossible to determine the exact number of outport men who went to the ice on St. John's vessels, there must have been some at least, because existing evidence indicates that vessels from both Conception Bay and St. John's had begun to take on fishermen from other outports and bays. The S.S. *Mastiff*, under Captain James Murphy and owned by Ridleys, Harbour Grace, cleared from Catalina. Murphy, who was a native of Catalina, was always ordered by his employers to pick up a large proportion of the crew in that outport, but some Harbour Grace men always made up part of the contingent. Nevertheless, Harbour Grace, with over 2,500 men on 51 vessels, supplied and owned primarily by Munns and Ridleys, was the main centre of employment in the seal fishery that year. Conception Bay ports, as a whole, with a total of 5,327 men at the ice, provided nearly 60 per cent of the manpower, according to the information in Table 4.6. However, the situation changed rapidly with the introduction of steam and the centralization of the steamer fleet in St. John's; in 1881, less than 1,200 men were on board the vessels clearing from Harbour Grace, and by 1885, this figure had declined to less than 600, as Table 3.6 illustrates. By this time, the sealing industry as prosecuted in sailing ships from Conception Bay had virtually collapsed, with only 80 men on one sailing vessel from Bay Roberts in 1884, 65 men on one from Brigus, and 295 on five from Harbour Grace (which also had a fleet of four steamers).[21] In 1885, as far as can be ascertained, St. John's sent out 14 steamers with 3,523 men; Harbour Grace sent 3 steamers with 705 men; and Conception Bay and St. John's together may have employed 4 or 5 sailing vessels as well.

With this trend came dramatic changes in the nature of employment at the ice, two of which were immediately apparent. Firstly, there was a steep reduction in the number of jobs for masters and second hands (i.e., first mates). This meant that most of these people were thrown back into the ranks of the ordinary sealers, and prospects for intelligent, hard-working young ice hunters to acquire command of their own vessels—without the proper connections—ceased to exist. The practical monopolization of captaincies by several prominent families was the end result. The second change was the fact that, with the exception of a few small vessels that continued to operate on the southwest coast, all fishermen were required to seek employment from the small collection of steamer owners in St. John's (and in Harbour Grace, in a small way, until 1894). Now the sealers were forced to compete in a climate of the declining supply of berths to the ice, controlled, ultimately, by merchants in St. John's. Those who were fortunate enough to overcome the obstacles and obtain berths to the ice annually were envied by their neighbours, many of whom were forced to resort to emigration to the United States, especially to Boston. Conception Bay was most seriously affected by this development, and the economy in that area practically collapsed because it was dependent on the seal fishery and on the cod fishery on the Labrador coast, which also declined severely, beginning in the mid-1880s.

Steamer captains

The employment situation in respect to captains' positions took an interesting turn, beginning in the 1860s. As already seen, there were hundreds of experienced sailing captains available to guide the sealing vessels on their annual quest for pelts. Yet, these experienced men quickly passed from the business. The depression and competition resulted in the loss of their jobs and sometimes the loss of their vessels; a new generation of captains took over.

In 1863, the first year that steamers prosecuted the seal fishery from Newfoundland, the two vessels involved, the *Bloodhound* and the *Wolf*, were commanded respectively by Capt. Alexander Graham, a Scot, and Capt. William Kean, from Pool's Island. Both captains did poorly, and Kean did not receive another steamer command (whether by his choice or not is unknown) until he was given the *Micmac* for two seasons in 1874 and 1875.

The following year, 1864, a third steamer, the *Osprey*, was added, and Graham continued with his command of the *Bloodhound*, while the *Osprey* and the *Wolf* were commanded by Capt. James Gulliford and Capt. P. Feehan, respectively. Gulliford does not appear in any of the newspaper accounts of sailing ship clearances during the several years prior to 1864, which suggests that he was brought in from Scotland. A Capt. Feehan was reported clearing St. John's port for the seal fishery in the sailing vessel *Triumph* in 1859, 1860, 1861 and 1862, and in the *Afton* in 1863. However, in 1864, besides the P. Feehan mentioned above in the *Wolf*, another Capt. Feehan cleared from St. John's in the *Afton*. The explanation is most likely the fact that the *Afton* was commanded by P. Feehan's son, because captains often apprenticed their sons and nephews into the business. All three steamers had abysmal seasons, however, and Feehan, whose steamer brought in only nine pelts, did not get another command; he was replaced by Capt. Patrick Skinner, who commanded the *Wolf* during 1865 and 1866. Skinner was another whose name does not appear among the many sailing captains that were recorded in the previous years. He did well enough, comparatively speaking, with over 7,000 pelts in 1865, but badly in 1866, and did not receive another command. Meanwhile Graham, who had a considerable reputation and had brought in a respectable 6,000 pelts in 1865, left his command, and probably Newfoundland, and did not return as captain for three years.

In time for the 1866 season, another steamer, the *Hawk*, was added to the fleet. Capt. Edward White from Tickle Harbour, Bonavista Bay, was given command of this steamer, and Capt. William Ryan was put in charge of the *Bloodhound*. Unlike Feehan, Ryan's origins are known. His father had been Capt. Charles Ryan of Water Street, St. John's, and "among the most renowned sealing masters in the days of the *Nine Suns*, the *Duck*, the *Gull*, the *Goose* and the *Drake*."[22] William was born in 1825 and went to the seal fishery in 1836 at the age of eleven. He was most likely the Capt. Ryan who cleared St. John's for the ice fields in the sailing vessel *Isabella*, supplied by R. Alsop and Company, each year from 1857 to 1864, inclusive. In 1865, he went to the ice as second hand with Graham in the S.S. *Bloodhound*, while another Ryan, supplied by Bowrings, commanded the *Isabella* in 1865 and 1866. It is impossible to ascertain whether Feehan had had a similar induction into the steamer fishery, but it appears that White, a very successful sailing captain, was hired without any experience on steamers. Ryan and White were both successful, bringing in about 13,300 and 10,700 pelts, respectively.

In 1867, the fleet was doubled to eight steamers, and Gulliford, who had had little success for three seasons, was replaced as captain of the *Osprey* by James Winsor, who was a local sealing master and was left in charge of that steamer for two unsuccessful seasons. White was given command of the newly constructed *Nimrod*, and the *Hawk* was given to Capt. Benjamin Snelgrove. Snelgrove was an experienced sailing vessel captain from Catalina and had been captain of the *Kitty Clyde* from that port since at least 1858. He brought in only 400 pelts in 1867 in the *Hawk* and, during the next several years, commanded the sailing vessel *Young Prince* out of Catalina. One of the most famous St. John's captains, Terry Halleran, was given command of the largest steamer in the seal fishery to date—the *Esquimaux*—and, as already pointed out in the previous chapter, was also a disappointment, with only 150 pelts to show for a month's work by 165 men. That seems to have ended his career—certainly as a steamer captain. Robert Dawe, from Bay Roberts, was given command of the S.S. *Lion* that year and brought in 5,300 pelts—not a poor result in a very unproductive year, with even the experienced Captain White taking only 2,600 in two trips. However, Dawe owned a business in Bay Roberts and, as far as it can be ascertained from the records, was involved with his family in sealing, cod fishing and the coastal trade. Perhaps he gave up command of the *Lion* after this unproductive, if not disastrous, voyage. In any event, Dawe resumed his voyages out of Bay Roberts in his 120-ton schooner, the *Huntsman*, which carried crews of 50 to 60 men; as described in Chapter 5 below, the *Huntsman* sank in 1872 while prosecuting the seal fishery off the Labrador coast, taking Dawe, his son and over 40 other lives. Meanwhile, in 1867, Abram Bartlett of Brigus entered the steamer seal fishery as captain and part owner (with Baine Johnstons) of the *Panther*. He commanded this steamer until his retirement 17 years later, whereupon it was commanded by his son, William, until it passed to new owners in 1896. As already mentioned, 1867 was not a successful year for the spring seal fishery, and only just over 28,000 pelts were taken—9,600 of these by Ryan in the *Bloodhound*, which was the only steamer to show any degree of success.

Up to this point and for some years to come, the sealing operators were learning, by trial and error, to prosecute the seal fishery with steamers instead of sailing vessels. Local captains with extensive and successful experience in the sailing vessel operation and Scottish captains with similar experience in Arctic sealing and whaling were both hired. Their comparative success rates were similar. Skinner and Gulliford seem to have been Scottish, as do Gent, who was unsuccessful in the *Wolf* in 1868 and 1869, and Hagen, who failed badly in the

Ariel in 1869 and 1872. On the other hand, Alexander Graham from Scotland was successful and was followed later in the century by other successful Scottish masters. Newfoundlanders who do not appear to have been able to make the successful change from sail to steam included: William Kean who commanded the *Wolf* in 1863; Snelgrove and Halleran, who each served one year in 1867; and James Winsor, who had two unsuccessful seasons in 1867 and 1868. Newfoundlanders Ryan and White made the transition to steam successfully, as did William Jackman in 1868 (in 1867, he took the *Lion* out on its second trip), Pierce Mullowney in 1870, Arthur Jackman in 1871,[23] and Azariah Munden in 1871.

The transition to steam seems to have been more traumatic than most observers have realized. Prowse, who observed this transition, recognized the significance of the event from a technological angle and wrote: "Politics and steam have done more than any other cause to ruin the middle class, the well-to-do dealers that once abounded in the outports."[24] However, Prowse did not seem to realize that the change from wind to steam power required a change in skills of command at least as great as those of the technology itself. Although the records are not available to fully substantiate this conclusion, it appears evident that the successful steamer captains were (with the exceptions already noted) those who acquired experience as second hands on board other steamers before taking their own commands. In other words, a second hand on a steamer for one season was better qualified to command a steamer than an experienced and successful sailing master. Consequently, by the 1870s, it is very likely that the steamer captains were in a position to train their replacements and to train those needed to fill the positions that were opening up with the expansion of the steamer fleet. Edward White trained his two sons, Edward, Jr., and Richard; Abram Bartlett trained his son, William; Arthur Jackman's son Thomas served as second hand with his father;[25] Abram Kean received his training under Job Barbour and went on to train his four sons, Job, Joseph, Nathan and Westbury;[26] and there were many other examples.

The development in Harbour Grace was similar to that in St. John's, although there does not seem to have been any effort to hire abroad. Both Munns and Ridleys invested in steamers—Ridleys in 1866 and Munns in 1871. Despite the considerable number of captains in the Harbour Grace area, Ridleys hired James Murphy[27] from Catalina as master of the S.S. *Retriever* in 1866. He was transferred to the S.S. *Mastiff* in 1867, and Isaac Bartlett[28] from Brigus took his place on the *Retriever*. Both must have had very favourable reputations, although one must question why Ridleys' most prominent sealing captain, Henry Thomey,[29] did not receive a steamer command until Ridleys closed and he was offered the *Commodore* by Ridleys' competitor, John Munn. Captain Richard Pike, Harbour Grace, who took over the S.S. *Retriever* after Isaac Bartlett transferred to a St. John's ship, also proved himself with Ridleys and joined the St. John's fleet after his employers went bankrupt. When the Munns invested in the first of their steamers, the *Commodore*, they asked Captain Azariah Munden of Brigus to take command. Here, they had also made a good choice (besides which, Munden was a relation through marriage). In any event, the Munns ceased to operate in 1894 (although the trustees of the firm sent the fleet out in 1895), and the only steamer fleet outside St. John's was absorbed into the capital.

In all, 29 captains had served on the Harbour Grace fleet between 1866 and 1895, and all had been sailing captains and mostly local men. Captains Pike, Jeffers, Hanrahan,[30] Keefe and Kennedy were Harbour Grace (or, in some cases, Carbonear) natives. Jeffers, from Freshwater, Carbonear had been captain of the *Elfrida* for three springs—1867-69. The *Elfrida* had a burthen of 126 tons and carried 55 men, making it one of the larger vessels in the seal fishery. Jeffers lost that vessel on "Old Sow" (i.e., Mosquito Point) while entering Harbour Grace in 1869, and seven unfortunate men also lost their lives in this wreck. Jeffers moved to Ridleys and, for a while, commanded their sailing vessel the *Cabot*, which had a burthen of 126 tons and carried 66 men. In 1874, he was given command of the *Commodore* by Munns, which he commanded with moderate success for four years. In 1882 and 1883, he commanded the S.S. *Greenland* for the same firm, but without success.[31] Capt. Thomey, formerly of Mosquito, was then living in Harbour Grace; Captains Dawe and Gosse were from Bay Roberts and Spaniard's Bay, respectively; and of course, Munden hailed from Brigus. Others, mostly from within the bay, were also employed in the industry, even if very briefly, like Captains H. Curtis, T. Fitzgerald, John F. Noel, Thomas Green, Samuel Dawe, N. Hanrahan, Perry, Edward Murphy,[32] Jacob Winsor and John Winsor, all of whom served only one season (in the case of Fitzgerald and Perry, the former commanded the *Mastiff* on its first trip and the latter commanded it on the second, both in 1877). As in numerous St. John's cases, many Conception Bay captains could not make the move from sail to steam.

The owners demanded performance, and the seal fishery was, to a large extent, a lottery, with winners and losers every year. When the firms had their risks spread over a larger number of smaller sailing vessels, the losses were usually well covered by the profits, especially in the decades when the industry was nearing and at its peak. However, with the move towards steam by the 1860s and 1870s, the risks became consolidated in a smaller number of vessels, and the success or failure of its one or two steamers could mean large profits or bankruptcy for a firm. In addition, there was a very limited number of captains' positions in the steamer fleet and a strong pool of experienced, highly-regarded second hands. Thus, steamer owners could, and did, demand the best and refused to put up with second-rate captains. Abram Kean described his first steamer command as follows: "It was beyond doubt that any man in charge of a ship the size of the *Wolf*, if he were fortunate, had the best paying business. If he were not fortunate, he would not hold the job long."[33] In 1894 Captain Samuel Blandford, as already quoted, was reported to have remarked to one of his officers, while enroute to St. John's from the ice after a disappointing voyage: "I have only a few more days to live; when I get ashore Mr. Job will hang me."[34] Between 1863 and 1914, there were 35 sealing captains who served only one season in command of a sealing steamer, and another 22 were in command for only two years.[35] With a few rare exceptions, these captains were unsuccessful, and in some cases, for reasons beyond their control.

Obviously, only the most highly recommended second hands and the most successful sailing captains received appointments as steamer captains in the first place. For example, When Job Knee was appointed captain of the *Falcon* in

1889, he had had considerable experience at the front. He had first gone to the ice at the age of eight in 1860, and in all had served 15 years on a 'square rigger', sometimes as an officer under his father. He was also quarter master on the S.S. *Micmac*, and served on the S.S. *Kite*, which his father, William Knee, commanded in 1877. He then followed his father to the S.S. *Eagle*, where he was promoted to second hand. In 1880, his father was given command of the S.S. *Falcon*, and with Job as second hand, father and son took that steamer to the ice together for nine years. When his father was transferred to the command of the S.S. *Kite*, Job was promoted to captain of the *Falcon*.[36] Similarly, when John Clarke was given the appointment as captain of the S.S. *Southern Cross* in 1910, he had already served as second hand for 18 years in the Gulf of St. Lawrence, 6 of them with Capt. William Bartlett in the S.S. *Viking*.[37] With backgrounds such as those, one can see that many of the captains who lasted only one, two or three years were simply unlucky, and therefore unwanted, regardless of their family connections. However, sealing captains had always had wide latitude in choosing their second hands and master watches and, not unreasonably, chose brothers, sons, nephews and in-laws when appropriate. When one combines this with the fact that a long apprenticeship in all aspects of the operation was needed to turn an intelligent lad into a successful sealing captain, one can understand why these positions became concentrated in the hands of a small number of families by the beginning of the twentieth century (see Tables 4.7 and 4.9).

Changes in manpower

The labour force involved in the seal fishery was forced to adapt to the changes wrought by the introduction of steam, production and market demand. The number of men who could find employment declined from a peak of about 14,000 in the 1850s (with the total population in 1857 at 124,000) down to 4,000 on steamers in 1914 (and a few hundred on the small schooners of the southwest coast and others parts of the island). However, by the latter year, the total population of the colony was in excess of 240,000. Meanwhile, the landsmen (and women and children) continued to take seals whenever the opportunity arose.

Unemployment was always a problem in Newfoundland during the winter months as well as late fall and early spring—not in the sense that there was not work to be done, from gathering firewood to caring for livestock, but rather in the sense that there was no immediate remuneration to pay for necessities that the previous summer voyage of cod fish could not cover. The 1820s were noted for their deprivation, as were the latter 1840s. In May 1847, for example, it was reported:

> The distress of the North Shore of this [Conception] Bay, and of various settlements of Trinity Bay, is still on the increase. Another fruitless sealing voyage has sent hundreds to their homes without a morsel of food or a penny in their pockets...[38]

However, there was always an optimistic outlook that the next season would be a prosperous one. Indeed, this optimism was generally rewarded, and depressions were followed by even greater bursts of investment and prosperity.

In April 1858, for example, it was reported that "the *Adelaide* belonging to Mr. J. Rorke [Carbonear], could not proceed upon the voyage for want of men."[39] However, after this sealing season proved to be below average, the same newspaper reported that there was gloom and depression in St. John's, mostly "amongst the shopkeepers and sellers generally who miss the supply of cash which a good sealing voyage brings greedily to help sales."[40] Only by the 1870s, with the combination of developments already described, did a feeling develop that permanent and drastic changes had arrived.

The decline in the outport sealing fleets reduced the number of jobs available to the outport fishermen at a dramatic rate, until finally, the men were all forced to depend on the steamers in the capital city. Instead of going on board a sailing vessel in his own community among his relatives and neighbours, the sealer was now required to find transportation to St. John's, and only a select number managed to obtain a 'berth' on a steamer. In 1873, after the fleets had departed St. John's and Conception Bay, a correspondent wrote:

> It is much to be regretted that so many able men are left ashore, unable to procure berths—it is estimated that not more than a third of the available labor finds employment at the ice this season. The Steam fleet this year numbers twenty vessels, and with its increase is proportionately diminished the area [sic.- need?] for the industry of our laboring men. No one can find fault with the prosecution of this fishery by steamers as the means most likely to succeed—capital here as elsewhere possessing an undoubted right to adopt such appliances as it may deem best suited to the object for which it is invested; but this unfortunate effect upon labour is nevertheless to be deplored. In countries more advanced and with resources more developed, such untoward issues find correctives in new directions of industry opening up as former ones are closed. Here, so far, we are without those compensations; and it ought to be with us all a subject of serious consideration, how far means may be devised to meet in some degree this rapidly increasing demand for employment, which arises from the cause already referred to. The subject does not hitherto seem to have commanded due attention or inquiry.[41]

Even in 1875, when Harbour Grace had its own fleet of steamers and there were sailing vessels operating from Conception Bay, it was reported that 500 men had arrived in St. John's on board the passenger steamer *Hercules* from Harbour Grace on their way to the seal fishery. This event was reported only because the captain lost his way in a snow storm and, thinking that they had rounded Cape St. Francis, ran the steamer ashore. Fortunately, the shore ice absorbed most of the impact, and the engines were reversed and the steamer brought safely into Portugal Cove without loss of life. The reporter of this incident reminded his readers that the men had experienced a "providential escape" and that such overcrowding should not be allowed.[42]

By the 1880s, observers had begun, for the first time, to refer to the "good old days," when "hundreds of staunch little vessels were repaired and made ready for the voyage—thus affording remunerative employment to large numbers of our people."[43] It was also remarked that "our great commercial towns" could expect visits from thousands of fishermen during the "festive season," who would spend "no small amount of hard cash during their brief stay for, be it remembered, their circumstances were much better than the majority of our 'seal hunters' of today."[44]

Although men continued to travel by sail and steam from their homes in the northern outports to St. John's every winter, the Conception Bay men came to depend more and more on the railway after the line to Harbour Grace was completed in 1884. Extra engines and cars would be put in service to handle this increase in traffic, and 200 or more men would arrive at a time. For example, on 7 March 1892, two trains arrived from Harbour Grace: one with one engine, three cars and 150 sealers; and the other—the regular train—with two engines, five passenger cars, two freight cars, one lumber car and 250 sealers.[45] An interesting scene was described as follows:

> Quite a lot of sealers came out by the train this evening and divers methods were employed by them to reach their boarding places. In one batch twelve men formed in two lines of six each and, with a brand new canvas bag on each left shoulder, marched boldly for the Seaman's Home. This was followed by another batch of eight in Indian file, with their gear on left shoulders. Another crowd hired a slide and each man sitting on his bag drove in state to their destination.
>
> As was the case every year, hundreds of sealers flocked into St. John's, signing on board the various steamers, looking for lodgings, consuming a little too much alcohol in some cases, attending meetings, etc.[46]

By now, however, the railway was being extended westward, and sealers from farther afield were using it as well. In 1893, it was reported that sealers from Trinity, Catalina and Bonavista had walked to Clarenville to join the train—a trek of up to 80 miles (130 km.) under very difficult walking conditions.[47] When the railway reached Gambo, many of the sealers of Bonavista north would walk the long distance from Wesleyville, Newtown and neighbouring outports to take the train at that station. And, of course, many from there and elsewhere who travelled to St. John's without a confirmed berth, or 'on spec', as they called it, were disappointed and often forced to return home at the government's expense—like the 200 men in 1897 who were given passes for their return journey.[48]

The mixing of sealers from the outports and from St. John's and vicinity did not go smoothly in the beginning. There were rows, usually involving alcohol, both before and after the voyage. The most serious disturbance of this nature occurred in 1875 and was reported in a local newspaper:

> A row of a very serious character took place in Water Street about midday last Wednesday, between a number of Bay Roberts men composing the crews of the steamers *Iceland* and *Greenland* on one side and some St. John's men on the other. It originated as often happens, between two of those opposing parties, but soon became general. The opening scene was on Messrs. Stabb's premises, but the prompt presence and active exertions of Inspector Carty, and a body of police served to repress the riot there. It quickly broke out, however, in other places and was continued on the ice, sticks, gaffs and knives being in speedy requisition, and cuts and blows freely exchanged between the combatants who were in a state of fury little short of madness. For an hour or two the whole town was in a state of alarm, as it was impossible to say to what an extent the disturbance might reach; fortunately, however, though severe bodily injuries were inflicted in several instances,

nothing worse resulted. The steamer *Greenland* sustained much damage from the violence of some men who forced their way on board.[49]

The governor, chief justice, priests, sealing captains, Inspector Carty and the mounted police were all present and involved in trying to maintain order. The correspondent concluded with praise for the latter (who were newly appointed to the recently reorganized police force): "Those amongst us who may have underrated the mounted police could have seen on Wednesday what ought to be enough to satisfy them of the efficient service this force can render in breaking up knots of rioters."

Although many men travelled to St. John's for their berths, others were taken on board at major centres, from which the steamers then cleared. Catalina, Trinity, Greenspond, Pool's Island, Wesleyville, Bay Roberts, Brigus and, of course, Harbour Grace were often used as points of departure. Beginning in the early twentieth century, a greater proportion of the fleet prosecuted the seal fishery in the Gulf of St. Lawrence and usually cleared from Channel. However, when the iron-clad or steel steamers were introduced into the industry, these were required to depart St. John's on a specified date, leaving the less efficient wooden walls the small advantage of clearing from a port nearer the ice floes. Above and beyond the considerations of efficiency, the St. John's steamer owners had mixed feelings about the wisdom of picking up their crews in the outports or requiring the men to sign on in St. John's. The latter meant increased business for the city, as the men made the necessary purchases there for the voyage, but at the same time, it gave the men a unique opportunity to gather in one group and become more aware of their numbers and strength. The strike of 1902 (see below) was the most famous occasion during which the outport sealers in St. John's acted as a united force, but there were lesser occasions as well, and the steamer owners came to respect their power as body.

Unusual ice hunters

There were a few unusual individuals reported among the many thousands of ice hunters who went to the ice—and, no doubt, many who were unreported. In 1893, the youngest 'man' to sign on was nine-year old Moses Gushue of Brigus, who went to the ice on board the S.S. *Nimrod*, under Capt. Moses Bartlett. Young Gushue was signed on at one-half a man's share and received $20.30 at the end of the voyage.[50] At the other extreme, it was reported in 1900 that the oldest man who went to the ice that year as a sealer was William Maddigan, Water Street West, St. John's, in his eighty-sixth year, who signed on the S.S. *Esquimaux*, commanded by Capt. Henry McKay. It was stated that Mr. Maddigan had been "going to the seal fishery since he was a young man, and believes himself still able to handle a bat or knife with anyone."[51] His share at the end of this trip was $28.56. Finally, another milestone was recorded in 1901, when the S.S. *Newfoundland*, under Capt. Farquhar, arrived in St. John's after a profitable trip, and it was reported, rather casually under the circumstances: "Her crew enjoyed good health and the lady doctor, Mrs. Levache, who made the trip in her, had easy work."[52] This report led to a discussion of other women who went to the seal fishery, and it was concluded that the first woman to do so was a Kate O'Driscoll, who was in the employ of the Clarke family of Brigus and who had gone to the ice as cook for two or three seasons on board the schooner *Goorkha*, under the command of Capt. William Clarke.[53] It was also reported that a Mrs. Warrington, who kept a boarding house in St. John's, made a trip in the 1870s. Finally, it was revealed that Mrs. Levache was not a lady doctor but the well-travelled wife of the steward, and had gone to the ice to observe the hunt.[54]

Stowaways

Other unusual, and sometimes newsworthy, participants in the seal fishery, were the stowaways who, by the 1890s, were being reported in the local papers. There is no indication when this became a phenomenon, but it seems likely that the introduction of large steamers made it easier to become a stowaway. In the early days of the sailing vessels, it would not be easy to hide away on board a small schooner or even on a larger brig, and it probably was not necessary because there was usually a demand for labour. Furthermore, boys of a very young age went seal hunting with their fathers, uncles and older brothers. For example, Capt. Samuel Blandford went to the ice for the first time at the age of thirteen in the company of his father,[55] and Capt. Job Knee, at the age of eight, accompanied his father (see above). Also, the labour involved on the sailing vessels (see below) was very demanding, and stowaways could expect gruelling work for little or, in most cases, no pay. Because the stowaways were forced to work and did not receive a share, the captains of the early period probably did not even bother to record their presence on the large brigs. On the steamers, they were usually put to work fetching and carrying for the officers and the firemen, but particularly the latter, and were generally well fed, often receiving a few coins at the end of the voyage. However, as their numbers grew, it became a constant battle of wits for the officers to seek them out and send them ashore before the ship cleared port. Stowaways were usually boys who were too young to sign on as sealers (despite the exception made for nine-year-old Moses Gushue, as noted above), but occasionally, adult men, unable to obtain berths, went to the ice as stowaways as well.

In 1893, a telegram from Whitbourne announced that ten stowaways had passed through there on their way to St. John's. Apparently, Capt. Arthur Jackman, in the *Eagle*, landed them at Greenspond, and from there, the local magistrate sent them 20 miles by boat to a point from which they walked 12 miles in snow up to their knees to Gambo, where they boarded the train to St. John's. The local St. John's correspondent wrote:

> I asked the leader, 'Con,' fraternally, how they escaped detection; if they got below deck. He looked at me with contempt, was disgusted at my ignorance, and, with one of his well-known smiles, replied, 'Not such fools. We mixed with the crew, and were not noticed until the roll was called at Greenspond.'[56]

These stowaways were obviously adults, as were the "twelve sad and disconsolate looking men" who were seen on their way to the city from Torbay one day in March 1895:

> They were not tramps, but ardent souls who had thought they could get to the icefields and perhaps make a good 'bill' without going through the, to

> them, unnecessary formula of signing articles, etc. Capt. Brett, of the S.S. *Hope*, differed with them on this little point, however, and the result was that this morning they were flung ashore in Torbay and brought to town by a carman named Patrick Fleming. The men, some of whom were very young looking, complained loudly of the captain for giving them no food.[57]

On another occasion, when Capt. William Winsor, enroute to the seal fishery in the S.S. *Panther*, called into Greenspond in order to deliver the mail, he landed four stowaways at the same time.[58] It seems that on the steamers, stowaways had a fair chance to escape discovery for awhile, but if detected while the vessel was near a port, they were immediately put ashore. However, if they could stay hidden until the steamer reached the ice floes, then they stayed out for the season. This is what happened with a boy Neal who disappeared from his home in St. John's on 8 March 1901. It was reported a month later: "The parents ... received word last evening that he was on the *Iceland* at the seal fishery and will return home when her seals are discharged."[59]

Captains did not like stowaways as a rule because they were of the opinion that if a particular steamer became known as a haven for stowaways, it would be overrun by them. Arthur Jackman informed the *Evening Telegram* on one occasion that:

> he makes "small pieces" of any stowaways found on board his ship, and that in order to get a living they must shovel coal in the stokehold. Furthermore, they are not allowed to go on the ice in quest of seals. They may get grub enough to keep them alive, but no share of sealers' wages, and no day's pay.[60]

The news report concluded with, "Woe to the unfortunate stowaway who is caught out with Captain Arthur."

Close watch was kept on the steamers by owners and officers in an effort to prevent stowaways, but with only limited success. It was noted in 1906, after the steamers had sailed: "Several stowaways are already known to have gone from here in the sealing steamers this morning, although a strict watch had been kept on the gangways the past twenty-four hours."[61] Some of these were put ashore at Greenspond or Pool's Island and made their way to St. John's via the train from Gambo, causing their parents and families "considerable anxiety" until they showed up in St. John's.[62] However, a month later, it was discovered that there had been seven stowaways on the *Adventure* that spring: "One of them returned by the *Diana* and the others came in last night [1 April] on the *Adventure*. While among the seals, they all earned their board and passage."[63] Similarly, it was reported the following year that two stowaways named Kane and Keneally stowed away on the *Neptune* and "earned their board on the steamer by helping the firemen."[64]

Not all stowaways were so lucky. Some were treated harshly and meanly, as Arthur Jackman had threatened—rather out of character in his case because of his well-known generous nature. In 1907, it was reported:

> The two lads, Leone Hickey and Leo Dempsy, who stowed away on the *Neptune*, fared off very badly, as far as getting their 'share' is concerned. They worked like Trojans, firing, strapping on, etc., and made themselves general favorites with all, and expected, if a big bill were made, to get

something. The crew made the largest bill this year—$97.37—and neither the Captain nor a member of the crew had the manliness to pass around the hat or even give a five cent collection, which surely wouldn't be missed from their big bill. The boys speak strongly of the niggardliness of the crew, and both say they aren't likely to stowaway with Capt. [George] Barbour, anyway, next year.[65]

One cannot generalize. The following year, 1908, one reads about a "baby sealer," Lidstone, who was a stowaway on the S.S. *Grand Lake* the year this vessel sank. He was "one of the last to leave the sinking ship...[and] he behaved so well that the captain promised to take him as a cabin boy with him next spring."[66]

One stowaway in 1911 used an ingenious ploy to carry out his scheme:

> Before the *Labrador* left here, one of the masters of watch went through the ship looking for stowaways, and while doing so he was helped by a young man who procured a lantern and went through all the dark places on the ship with the officer, but none could be found. Later it was discovered that the man who had assisted the other in his search was himself a stowaway, but being a fine husky fellow could not be distinguished from any other member of the crew. He went out in the ship and worked with the rest of the men.[67]

He was fortunate in that he seems to have been accepted and maybe received a share, but another adult stowaway, Sweeney, on board the S.S. *Ranger* complained that he had worked hard and had not received any reward.[68]

Stowaways continued to be a concern for the owners and captains, but the practice was not eliminated, and sealing steamers continued to have "the usual trouble with stowaways."[69] However, it was in 1913 that the most unusual incident occurred concerning stowaways. That year, as already described, the little, old wooden-wall *Kite* left with 11 men for the Groais Islands to pick up its crew. It was stopped by ice off Bonavista and decided to add to its crew by recruiting sealers from the nearby ports. However, after entering Seldom Come By, Trinity and Catalina, only 6 more men were found, and the *Kite* proceeded to the ice with a crew of 17, the smallest of any sealing steamer in the history of the colony. The voyage was only a partial success, but "when she arrived [in St. John's on 26 April] the *Kite* had 38 sealers, her former crew [of 17] being augmented by boys who had stowed away in the steel ships and were transferred to the *Kite* and signed on at the icefields."[70] The crew's share of the voyage was $19.85 per 'man'. While it was unusual for stowaways to play such a prominent role in the seal fishery, this incident indicates that a fair number of stowaways managed to make the trip each year. Even in this way, the industry introduced boys and young men to sealing, albeit grudgingly, and provided the basic apprenticeship that prepared many for a life at sea.

Ice hunters and St. John's

By the end of the period under study, St. John's had taken over the shipping, supplying, processing, financing and exportation of the sealing industry; only the men on the ice and the captains who commanded them originated in the outports. While it is impossible to quantify the proportion of sealers who were baymen, a glance at the list of those who died in the *Greenland* disaster (see Chapter 5) will show that out of the 48 victims, only 2 came from St. John's, 2 from Quidi Vidi, and 1 from Torbay. The remaining 43 came from points north—6 from Harbour Grace and 37 from harbours north of Harbour Grace. Moreover, the captains were under pressure not only to make profits but to commit themselves to their employers, who had other types of employment for their sealing steamers during the other ten months of the year. Therefore, with the exception of those who had their own businesses, like the Dawes, sealing captains were forced to move to St. John's with their households. They became further removed from their former neighbours and friends, the outport ice hunters.

At the same time, the St. John's men monopolized all the sealing jobs outside the actual killing and pelting. The seal skinners who had operated in every outport, slicing the fat from the skins during the week or more following the voyage, were now residents of St. John's (with a few remaining in Harbour Grace, where the oil plant continued to operate, even after the bankruptcy of the Munns). The engineers were, at first, hired from Scotland, but eventually, St. John's men acquired these skills, and the Scots became permanent residents. Furthermore, these men were needed as permanent employees of the steamers in the various off-season trades. Similarly, firemen and coal trimmers were needed to assist the engineers, and these jobs became confined to St. John's residents.

The point is that in most cases, the ice hunters were separated from the other trades on board ship and in the company's employ. Later, when William Coaker was able to harness the northern discontent against the St. John's merchants, he was not able to translate this discontent into a labour movement based on worker solidarity; the outport ice hunters, who became the backbone of the Fishermen's Protective Union, did not identify with the St. John's working class.

Berths

To go 'to the ice' during the spring seal fishery, the ordinary ice hunter needed a 'berth' on a vessel. At its very inception, when small schooners and shallops were employed in the industry, there is no doubt that families and relatives made up the crews of these vessels. Nevertheless, it is fairly obvious that boat owners from the very beginning did hire men to go on these sealing expeditions, and unlike the situation in the cod fishery, the men were required to pay in order to participate. It seems that this practice was justified by requiring the fishermen to contribute to the cost of the supplies needed for the four- to six-week voyage. In any event, by the time Anspach observed the operation during 1803-12, each man paid 40 shillings (£2) for his berth or "for their proportion of the provisions during the voyage."[71] No doubt, this was the standard charge at this early period in Harbour Grace, but to what extent the same charge was applied elsewhere is unknown. However, there is no reason to believe that there was any great deviation from this amount. However, the berth charge applied only to the regular ice hunters or batsmen because, as Anspach pointed out, the gunners, who were required to bring their own guns at this stage, received free berths. Much later in the study period, it was reported that in the 1820s, regular sealers in st. John's had to pay from £2 to £3 for berths, while in Conception Bay and points north, the charge was from 10 to 30 shillings.[72] However, in 1840, when he went to the ice as an observer on the brigantine *Topaz*, Jukes reported that most sealers were charged £4 currency for their berths, which would be about £3 4s sterling.[73] Other reports from this period state that in the early 1840s, berths cost £2 for batsmen, £1 10s for after gunners and were free for the bow or chief gunners.[74] By 1871, the prices charged for berths had become standard on sailing vessels, as far as one can see from examining two sealers' agreements, and were as follows: second hand—free; bow gunners—15s; after gunners—25s; batsmen—35s; batsmen with guns—25s; and cook—20s in one case, but free in the other. In addition, each man agreed, according to these 1871 documents, to cut ice for a channel if necessary when the vessel was leaving port and to go on the second trip if the ship owner so wished. And every gunner was required to provide himself with a good gun and two locks.[75] By 1877, the price of a berth seems to have declined to between 10 and 30 shillings.[76] In the meantime, however, 'berth charges', as they were known, continued to be imposed only on the crews who manned the sailing vessels and who shared one-half the catch. The steamers shared only one-third of the catch among the men and eliminated berth charges, which they replaced with a 'coaling charge' of $3 (see below).[77]

In the second half of the century, berths became increasingly difficult to obtain. In 1913, a sealer who had been to the ice on board the S.S. *Commodore* in 1872 reported how they had cleared from Harbour Grace on 21 April to make, what was common at the time, a second trip for old seals. The steamer ran aground in Bear's Cove and the voyage had to be called off, "which caused us great rejoicing as we did not want to go out so late in the season, but the man who refused to go the second trip need not apply for a berth the next spring."[78] Also, when the Dundee vessels began to prosecute the seal fishery, using St. John's as a base, one reporter wrote as follows:

> The arrival of the steamers *Arctic* and *Aurora* a few days since from Scotland, to proceed hence to the Seal-fishery, gives an important addition to our Steam Sealing fleet. These two vessels, manned almost wholly here, will take off over four hundred men who would otherwise have been idle and adrift about our streets. Since these boats arrived there has been a continual 'rush' upon them by applicants for berths. Many of our best and ablest young men seem to have been totally dependent upon them for means of subsistence for the coming month, and are right glad of the chance thus offered. They have now secured for awhile,—what, we fear, several of them have not lately had—an abundance of good food, while they and theirs are cheered with the hope of a successful voyage.[79]

Similarly, in 1881, it was reported that Conception Bay sealers were complaining that they were being overlooked and that the "northern nimrods" were getting all the berths to the ice.[80] Another newspaper report, published in 1882, regretted the passing of the "good old days," when on St. Stephen's Day, (December 26) "thousands of the outport 'toilers of the sea' crowded into our great commercial towns to tender to their 'old skippers' the complement of the season and to make sure that their berths were kept in readiness for them."[81]

As already pointed out, a long and major depression began in the 1880s, first affecting the seal fishery, then the Labrador cod fishery and, later, the shore fishery around the island. This depression was marked by the scarcity of berths to the ice, and many men who gambled their small savings on a fare to St. John's often were forced to rely on government relief for their return passage after failing to obtain berths. One report in 1892 gives some indication of the desperation associated with the effort to obtain a berth. On 5 March, a local paper reported:

> We have often heard of the feat of walking on people's shoulders, but never saw it really accomplished until this morning, when a sealer who had been fortunate enough to secure a berth in the *Esquimaux*, had to scramble his way out of Baird's shop, along the shoulders and heads of the crowd outside, the pressure inwards being too great for him to make his exit otherwise.[82]

Similarly, in 1895, when the S.S. *Newfoundland*, a Halifax steamer, arrived in St. John's to sign on a crew, a report stated that it "was hardly moored to the wharf yesterday when the cabin was besieged with men looking for berths."[83] It eventually completed its complement of sealers in Bay Roberts, where 150 men, out of between 1,700 and 2,000 applicants, received berths.[84] In 1901, this steamer was in the news again respecting berths, when Capt. Farquhar, the owner and skipper, asked one of his regular Newfoundland sealers to sign on 80 men for the coming voyage. However, it was reported that the request was taken advantage of and each man was charged one dollar for his 'ticket', such that, "Some of the women worked like slaves to get the cash demanded by this Shylock, so that their husbands might get to the ice."[85]

The scarcity of berths continued to present a problem to the fishermen. In 1905, it was reported as follows:

> It is utterly useless for Northern sealers to come on here for berths this season unless provided with tickets beforehand. The owners of steamers have not any to dispose of, nor can they get berths for some of their dealers. Several men from Trinity and Conception Bays are now on the way here hoping to secure a place on some ship, but it would be better for them to remain at home than spend their money coming here on a useless errand. The crews are reduced and the captains have their own men selected.[86]

Again, in 1907, complaints were published about the lack of berths available for Conception Bay and Southern Shore sealers,[87] and this continued to be the case for the remainder of the period. The cost of the berths had been an irritant throughout the whole of the industry's history, most especially during depressions, but simultaneously, other problems also arose for the sealers (see Chapter 6).

One final point in dealing with the subject of berths is the fact that they became part of the patronage available to captains, merchants, clergymen and politicians. For example, Peter Cashin wrote about the relationship between his father, the politician Sir Michael, and Capt. Arthur Jackman:

> Captain [Arthur] Jackman was a great personal friend of my father's and for several years before his death [in 1907] had given Mike Cashin, as he called him, some thirty berths to the ice in the *Eagle* for fishermen from the district of Ferryland.[88]

One local, well-informed poet, writing in the anti-government *Evening Telegram* in 1910, told of his (supposedly) own personal experience in trying to collect the berth promised him by a member of Prime Minister Morris' government in return for his vote. Several stanzas of his ballad, "Tricked," are as follows:

> He came round last year for Morris
>   And the people put him in;
> Like his master, sir, he promised
>   Right and left—through thick and thin—
> Not a man in our harbor,
>   Eli's, Harry's, Jack's, and Con's,
> But he'd get a berth this season
>   If they'd see him in St. John's.
>
> Bad luck be his!—but seeing him
>   In any decent place
> Reminds me of that ancient saw
>   About "the wild goose chase."
> But I will find that paltry pawn,
>   That man of little worth,
> That shoddy representative
>   Who promised me a berth.
>
> He said he knew each captain out,
>   Each merchantman knew him;
> From day to day throughout the bay
>   He told it with a vim.
> But now there's not in St. John's town
>   An alley-way or lane,
> But that same member courts; but, sir,
>   His dodgin' is in vain.
>
> This disappearing promiser,
>   This smiling sycophant,
> Who packs his purse with public gold
>   And recks not if we want;
> This man who lately on us fawned
>   Now spurns us as the dirt;
> We men of wives and families,
>   We men who need a berth.
>
> I've looked for him in Parliament,
>   I've sought him in the Court,
> His office I have haunted
>   From the 1st until the 4th;
> I've hung around his residence
>   From dinner-time till tea,
> But cannot catch the cad who said
>   He'd have a berth for me.
>
> But I'm waiting, and although I may
>   Not tread the 'bloody pans,'
> There'll be other business doin'

> When on him I lay my hands.
> Yes, he'll think the comet's tail has struck
> With shivering shock the earth'
> That lyin' politician, sir,
> Who promised me a berth.[89]

Illness complicated the process of obtaining berths beginning in 1911, when an outbreak of smallpox occurred on the S.S. *Newfoundland*. The 146 men who had been vaccinated previously were allowed to leave the steamer once their effects were disinfected and fumigated, and then the vessel proceeded to discharge the cargo of pelts. The 16 men who were infected were sent to hospital at Signal Hill; and the remaining 68 were quarantined for 21 days in a shed on the south side of the harbour.[90] Beginning in 1912, the firms demanded that sealers produce proof of vaccination before receiving their berths, and this resolved the problem.

Given the declining employment in the Newfoundland seal fishery during the latter part of the nineteenth century, it is obvious that the 'berth' to the ice became increasingly difficult to obtain as the industry itself, became less and less important.

Ice Hunters' Incomes

Fishermen

The income that could be earned from the seal fishery was the great attraction to the early Newfoundland cod fishermen. Income was based on the share system, with one-half the value of the total catch divided among the crew and the remaining half going to the vessel owner, who, in the early years, was often the skipper/planter.[91]

The earliest available information from which wages in the spring seal fishery can be calculated refers to 1803. In that year, it was reported that seals valued at £23,152 were taken in shallops and small vessels by 998 men from Bonavista, Trinity and Conception Bays as well as St. John's. This averages out at £11 12s per man overall, but ranges widely from a low of £5 8s in Bonavista Bay to £7 9s in Trinity Bay, £15 2s in Conception Bay and £17 17s in St. John's.[92] The season of employment was also recorded for the seal fishery during 1803, as follows: Bonavista Bay—20 March to 15 May; Trinity Bay—30 March to 20 May; Conception Bay—30 March to 24 May; and St. John's—22 March to 16 May. The season averaged about seven weeks in total time spent at the ice, but some vessels would have spent less time, and many would have made two trips in the seven-week period. In this respect, Anspach wrote: "They will, in general, make one trip within from four to six weeks, and when ice and seals are abundant on the coast, they make two trips before the latter end of May, each trip averaging from nine to twelve pounds sterling per man." Also, he concluded that men in Harbour Grace were clearing £9 to £12 per man per trip and that two trips were not unusual. This would mean that each man making two trips cleared from £18 to £24. Consequently, the average income of £15 2s per Conception Bay sealer described above included those who made two trips as well as those who made only one.

However, it would not be correct to assume that the spring seal fishery was limited to the time the fishermen spent on board their vessels. As Anspach pointed out, "Soon after Candlemas-day [2 February], they begin their preparations for the *seal-fishery*, fixing their craft and afterwards laying their stock of provisions." In Newfoundland terminology, the men 'went in collar' after 2 February and 'went out of collar' or 'broke collar'[93] after the sealing vessels were unloaded and cleaned for the cod fishery—14 or 15 weeks in all.

Comparable figures for the cod fishery for 1803 are not available. However, figures are available for the following year, 1804, and circumstances were not significantly different during the latter year. As Table Intro.5 indicates, wages for a fisherman employed as a midshipman were recorded as £28 in Trinity, £40 in Conception Bay and £30 to £40 in St. John's.[94] The midshipman was an experienced fisherman who was paid more than the foreshipman, but less than the boat master, who in that year was paid £30 in Trinity, £50 in Conception Bay and £40 to £50 in St. John's. Among the shoremen (on the bigger fishing plantations, because on the smaller operations, fishermen and shoremen often performed the same work), the splitter was the most highly valued next to the master of the voyage, who supervised the curing. In 1804, splitters were paid £28 in Trinity and Conception Bay, and £30 in St. John's. The salter was paid a couple of pounds less. However, the fishermen and shoremen were required to work in the fishery from various dates in early May until late October. Some men stayed in collar to his sealing captain or planter and accompanied him to the Labrador and would not go out of collar until late fall, when they returned home to the island. In all, the men would be employed—in collar—for nine months or more.

Although it is difficult to make accurate comparisons, fishermen who received £20 for the sealing voyage and £40 for the cod fishing voyage would not be out of the ordinary during these early years. Other things being equal, the £20 from sealing would, and did, make it worthwhile for many cod fishermen to remain in Newfoundland all year round. To appreciate the amount of basic food supplies that such an income could purchase, one can examine the prices of provisions in 1804, as presented in Table Intro.4. There it can be seen that in St. John's, prices were as follows: bread per cwt—20s to 28s; flour per barrel—44s to 50s; beef per barrel—£4; pork per barrel—£4 10s; butter per lb.—1s; molasses per gallon—6s 6d; and rum per gallon—6s to 7s. In Conception Bay, the prices per similar units were listed as: bread—25s; flour—46s to 60s; beef—£4 15s; pork—£6 10s; butter—1s 6d; molasses—7s 9d; and rum—7s. In St. John's, a consignment of supplies, consisting of one barrel of pork, one barrel of beef, five barrels of flour, five cwts of bread, twenty lbs. of butter, ten gallons of molasses and one gallon of rum, cost about £30 6s 6d, while the same supplies in Conception Bay cost about £37 2s 6d. (The spread was probably less than this, because there seems to be rather big differences between the prices of flour, pork and butter, in particular.) In any case, this is an indication of what could be purchased within the incomes of experienced fishermen at this time, with those from St. John's being better off than their co-workers in Conception Bay.

Saltfish prices, wages and the cost of living all increased dramatically during the latter stages of the Napoleonic War. This brought about a real increase in the standard of living and a dramatic increase in settlement. It is not clear to what extent the seal fishery contributed to fishermen's incomes during the prosperous Napoleonic War period, but it was most likely not seen as having an essential role in the economy because the huge increase in saltfish prices had made any other economic activity unnecessary. It was during the post-war depression that began in 1815 that income from the spring seal fishery became crucial to the very survival of a substantial settled fishery in Newfoundland. Prowse wrote that the worst of "this period of calamity came to an end" in the spring of 1818, when for the first time in several years, "in less than a fortnight scores of little vessels returned, loaded to the scuppers with fat."[95]

References to individual shares for fishermen on sealing ships are non-existent during most of the early period. There are figures that give an indication of the total value of the industry to the economy, which have already been examined, and the number of men engaged in the annual ice hunt was occasionally recorded. These records give an indication of the growing importance of the seal fishery. Finally, as has been noted earlier in this chapter, the writers of the 1880s looked back on the 'good old days', when the industry was booming and providing direct and indirect employment for numerous fishermen and tradesmen.

An examination of the number of men in the spring seal fishery, the quantity of oil exported and the average prices the oil fetched, as reported in the governors' annual records, can be used to show income in general terms. In 1830, the value of seal oil exported amounted to about £160,000, and 5,735 men were reported to be engaged in this fishery. If one assumes that one-half this amount was shared among the men, that would mean that the men earned about £14 each. Meanwhile, the skins, at one or two shillings each, would have given each man £2 to £5. In that same year, it was reported that the three classes of fishermen, made up of masters, men servants and dieters, totalled about 20,000 men (including the elderly and retired, apparently) on the English shore. They produced (with the help of women and children) about 906,000 cwts of saltfish for export at between 8s and 12s per cwt. Assuming that the fish was sold at 10s per cwt, on average, then the total value of saltfish exports came to about £453,000, or about £23 per man, when divided by the 20,000 above. When one considers the bare survival of the cod fishery during these years, it must be concluded that the £16—£19 per fishermen earned in the seal fishery was crucial to the survival of many families—and this refers to the direct contribution in income from the seal fishery only.[96] Meanwhile, the quantity and value of seal oil exports increased in 1831 to 8,761 tuns, worth about £206,000, while the total value of saltfish exports declined to about £360,000, making the proportion of each fisherman's income that was dependent on the seal fishery even larger.[97] Given the fact that the fishermen of St. John's and Conception Bay were engaged in the seal fishery in greater numbers, absolutely and proportionately, than those from elsewhere in the colony, it would not be unreasonable to conclude that during the 1820s, the income, of fishermen in these two areas was drawn equally from the cod and seal fisheries.

Unfortunately, averages do not go near explaining the personal economics of the seal fishery. Lives and ships were lost in the same season that saw other fishermen make a year's pay in three weeks and merchants making a year's profit during the same period. No other industry in the history of Newfoundland has ever been, to such an extent, a lottery as has been the case of sealing. In 1827, a local newspaper reported that the cargoes of 71 sealing vessels arriving from the ice amounted to 123,150 pelts in all.[98] According to Tables 3.2 and 3.3, sealing vessels from St. John's and Conception Bay carried an average of 19 men each in 1827, which means that about 1,350 men were employed on these vessels and shared in the proceeds. At 70 pelts per tun, the cargo produced about 1,760 tuns of oil, which was valued at £23 per tun, according to Table 2.6, for a total of approximately £40,460; one half of this was shared among the men, which means that each fisherman received about £15 as his share of the voyage. In addition, the skins would increase the men's shares, and second trips would enlarge them further. In 1831, the brigantine *Rachel and Ellen* (98 tons), commanded by Capt. Edward Purcell and carrying 30 men, made two trips and brought in a total of 5,727 pelts. These were rendered into 82 tuns of oil, which was sold at £23 10s per tun for a total value of £1,927; half of this was shared among the men, giving each a share of £32.[99] John Nugent, a member of the House of Assembly, speaking on behalf of the sealers in 1842, stated that they risked all to make between £20 and £30.[100] A contemporary taking a more positive view of the industry, wrote in 1852 that a man could come "home to his family in less than a month with more than £30 to his credit. A labourer in England does not earn that sum in a year, and in Ireland not half of it."[101] This was an important matter; a man could earn £20 to £30 during a time in the year when there was no other employment and still leave himself free to engage fully in the summer cod fishery.

Scattered pieces of information help the reader to understand the wide range of shares received by different ships. In 1858, the crew on board the brigantine *Glide* from Carbonear, commanded by Capt. John Pumphrey, received $214.60 per man.[102] On the other hand, in 1861, the men of the schooner *Alder Davis*, commanded by Capt. James McLoughlin [McLaughlin?], made only £15 currency ($60) each. Men's shares remained fairly high in the 1850s, but the depression of the 1860s made almost irrelevant the size of any of the bigger bills[103] received. Chafe, who recorded the scattered reports available, provided the following, which indicate that at least some of the steamers paid fairly large shares during this depressed decade: in 1865 the *Wolf* paid its 120 men $92 each; in 1868, the *Lion* paid its 128 men $128 each and in 1869 paid its 120 men $120 each; and in 1870 the *Lion, Osprey* and *Wolf* paid $128, $96 and $60 each to their crews of 120, 110 and 113 men, respectively. In pounds currency, these sums ranged from £15 to £42—comparable to earlier shares but not commonly received during that decade.

Many men received little of their shares in cash because they kept ongoing accounts with their supplying merchants, and up to the 1870s, credit from the merchants was generally forthcoming. In 1876, for example, Thomas White in Bonavista began charging goods to his 'Sealer's Account' at Ryan's store, and up until 7 March, he made various purchases, including leather, tea, tobacco, cuffs, a frock, swanskin (material), hose, mitts, sole leather, grain leather, canvas, a brin bag, a box with lock and hinges, thread, flannel, a Cape Ann southwester, oil pants, sugar, a sheath knife, goggles (to protect against ice blindness) and a two-quart tea kettle. In all, he charged goods to the total of £6 16s. He went to the ice on the S.S. *Ranger* and received £12 4s 1d for his share on their first trip and £9 16s for his share of the second trip, for a total of £22 0s 1d. He was charged £2s 3d for 'Tally', 2s 3d for 'Cleaning out' and 10s for 'coaling'. In addition, he received £1 cash in St. John's, where the vessel was unloaded, and two shillings in cash when he returned to Bonavista. The remainder of his earning, £13 7s 4d, was transferred to his 'Ledger Account' at Ryan's on 18 December, and his 'Sealing Account' was closed. Michael Fennel, on the other hand, who also went out on the *Ranger* that year, charged only £3 0s 1d to his 'Sealing Account' at Ryan's, received £12 in cash and had £5 4s 1d transferred to his 'Ledger Account'.[104] As can be seen, little cash went directly to the sealers.

On some occasions, the lucky men on some steamers 'made' quite large 'bills'. The *Hector*, under Capt. Edward White, made three trips to the ice in 1871, and the crew received £38 14s 5d per man, or $155. In 1872, the sealers in the *Commodore* from Harbour Grace, under Capt. Azariah Munden, made $160 each.[105] Twenty years later, in 1892, the 224-man crew of the *Diana* (formerly the *Hector*) each made $184.30. According to Chafe, the largest share ever made by the men on a St. John's steamer was made in 1871, when the crew of the *Nimrod*, under Capt. Peter Cummins, made $208.47 per man; while the largest bill ever made in the steamer seal fishery was made in 1866, when the sealers in the *Retriever*, supplied by Ridleys and commanded by Capt. James Murphy, each made $303. These were great exceptions. Towards the end of the period, when, as has been noted, small schooners from the southwest coast began to participate in this fishery, a few individuals made a fair share. In 1910, for example, the ten men on the schooner *Mildred* each made $105.64, and the twelve men on the schooner *Cedella*, $148.15 each. However, this was a tiny operation, and it was the steamer fleet that was the mainstay of the seal fishery. The sealers on these steamers made a small and declining income after the early 1880s. For example, Joseph Akerman charged goods to his sealer's account at Ryan's in Bonavista to the sum of £5 8s 3d in 1882. His share on board the *Eagle* that year, however, amounted to only £1 9s 1d; his debt on his sealer's account of £3 19s 2d was transferred to his ledger account, and the sealer's account closed, as was the practice.[106] Soon less and less credit was advanced.

By the end of the century, income for the sealers had declined to an abysmal level. In 1895, as recorded by Chafe, the average share was just $29 per man (about £7 currency). By 1897, the average shares had declined to $11 (less than £3 currency) per man, although by 1900, they had peaked again at $43 (less than £11). During the next 14 years, that is to the end of the period under study, shares hovered around $40 per man, except in the following years: 1905, when the sum dropped to $29; 1906, when it reached $50; 1910, when it peaked at $62; and 1912, when it dropped to $26.

Not only had the sealers' shares declined drastically throughout the century, but they were burdened by charges (which, as shall be seen, led to labour problems), beginning with the cost of the berth as already discussed. As the berth charges fell into disuse (see below), sealers were overcharged for the items they purchased on credit from the steamer owner (or his supplying merchant—such as Ryans, above) before the voyage began. Usually, these supplies were essential to the sealer while ice hunting, but sometimes they consisted of necessities that the sealers would purchase to help sustain their wives and families while the hunt was in progress. The purchases that the sealers could make on credit, once they were assured of their berths (i.e., had their tickets), were known as the 'crop'.[107] In addition, the steamer owner deducted a coaling charge from the men as well. A summary published in 1891 illustrates what had happened by this time:

> Dear Sir,—A statement was published last week in the *Herald* showing all the advantages derived from the seal fishery this season; but the writer very carefully avoided showing the disadvantages which the 'bone and sinew' labor under whilst engaged at that enterprise. The amount per share may have been correct. How much of it went into the merchant's pocket? How much into the men's pockets, can be seen from the following accounts:-

	Sealers' price	regular price
1 pair sealskin boots	$7.60	$4.80
Tea, per lb.	.66	.40
Sugar, per lb.	.13	.08
Coffee, per lb.	.53	.30
Pepper, per lb.	.47	.20
Oatmeal, per lb.	.09	.05
Tobacco, per lb.	.60 to .70	.40
Shirt	.80	.50
Pan	.27	.05
Goggles	.14	.07
Stockings	.67	.35
Drawers	1.33	.60
Flannel, per yard	.46	.20
Towelling, per yard	.12	.06

> In addition to the above, there is another charge of $3.20 per man for coaling, etc., which if multiplied by the number of men on board, say 340, would give the enormous sum of $1,020; no doubt a profitable charge considering the quantity of labor you could get in this country for that amount. The amounts paid to shoemakers for sealskin boots are, $3.20 and $3.60 (not cash either). No doubt this is a most lucrative business, as far as boots are concerned, for the merchant. As regards the other goods—tea, sugar, etc., the overcharges on them can be seen at a glance and speak for themselves.[108]

Considering the fact that the men now shared in only one-third of the value of the seals killed, as opposed to one-half in the days of the sailing vessels, one can only wonder at the fact that their financial exploitation did not lead to more serious consequences than actually occurred.

Finally, it must be noted that the men were able to bring seal flippers home from the ice fields to sell and to share with families and friends. Sometimes they filled the empty salt pork barrels with both flippers and seal carcasses. The sale of flippers and carcasses took place during the few days when the vessels were unloading the pelts, and the few shillings (or dollars) received were often spent on rum for immediate consumption. Thus, seal flippers added a social as well as a financial aspect to the sealing operation.

Captains

Meanwhile, the income of the captains of the sealing vessels had increased at an even greater rate than the income of the sealers themselves had declined. In the beginning, a captain hired to take a vessel to the ice was paid for each pelt brought ashore—6d each in 1824, for example—and sometimes £5 per month besides.[109] A cargo of 2,000 pelts, not an unusual load at the time, would bring the captain at least £50—probably two or three times as much as each sealer received. By 1848, captains were being paid sometimes by quantity and at other times by weight—4d to 6d per pelt versus 1s to 1s 3d per cwt of pelts.[110] Since pelts weighed from 40 to 50 pounds (18 to 23 kg.) each, there was little change in the rate of their pay, but because larger vessels were being employed, each captain could expect to earn more—an expectation which did not apply to the men because increasing crew size brought them a diminishing share. In addition, some of the early captains were owners or part owners of their sailing vessels. Chafe reports that Azariah Munden, who owned and commanded the brig *Highlander*, brought in over 6,000 pelts in 1843, which he sold to Lawrence O'Brien for 21s per cwt.[111] Calculating the weight at 2.5 pelts per cwt, the cargo earned about £2,500 ($10,000), of which the captain's share plus one-half the remainder went to Munden, himself.

However, successful captains were to receive ever increasing shares for their performances solely as captains. In 1871, Capt. Edward White brought in a total of 31,644 pelts that were weighed in at about 682 net tons of fat with a net value of £19,799 5s 7d in three trips, and while the crew members received £38 14s 5d each, White was paid £949 6s 7d (about $3,797)[112]—over 24 times as much. This share amounted to £1 8s per net ton of fat and 4.8 per cent of the total value of the catch, or about 7d per pelt.[113] Capt. William Jackman, in the *Eagle*, brought in three loads of pelts in 1875, worth a total of $113,817.33, and if his share was equivalent to that given to White (i.e., 4.8 per cent), then Jackman received about $5,463 for his season's work. Capt. William Barbour brought in pelts worth over $100,000 in 1892, as did Capt. Samuel Blandford in 1884, which meant each of them cleared about $5,000 for their cargoes. Similarly, take the case of Capt. Azariah Munden in 1872, as seen above, whose crew of 200 men received $160 each, or $32,000 in all, indicating the total value of the cargo was about $96,000; if Capt. Munden received 4.8 per cent commission, his share amounted to about $4,600. These were generous amounts, and yet, successful captains could expect even more in annual income. In 1873, for example, Munns paid Capt. Munden £754 16s 9d (about $3,000), and yet the records show that his success at the seal fishery was very modest that year.[114] It is very likely that Munden was engaged throughout most of the remainder of the year in some other capacity by Munns, most likely in their Labrador fishery and supply trade, and that he and captains of his stature commanded and received large annual salaries, in addition to their sealing commissions.

In 1905, the steamer owners decided to regularize the system of rewarding their captains. In that year, they came to an agreement whereby the sealing captains would receive 4 per cent of the gross value of the voyage,[115] and it appears that this remained in effect until the end of the period. Capt. Abram Kean set a record when he brought in 49,000 pelts in the *Florizel* in 1910, and under this agreement, his share, although not confirmed, would be 4 per cent of the $90,800 cargo, or $3,632; one-third was divided among the men at $148.36 each. This income for a voyage that lasted about three weeks is better appreciated when compared to the top annual salaries in the colony: the governor received $10,000; Chief Justice—$5,000; and government ministers—$2,000. There were only three steamer captains whose careers included the ten-year period from 1905 (when the 4 per cent share for captains was agreed upon) to 1914 (the end of this period): George Barbour, William Bartlett and Abram Kean. Over this period, they cleared, respectively, about $19,200, $10,800, and $17,080. They did not earn as much as Edward White, Samuel Blandford, Abram Bartlett, Azariah Munden, the Jackmans and others, who were fortunate enough to be in their positions when oil prices were still comparatively high. However, in those earlier years, the crews received larger shares as well.

The ten high liners during the period from 1905 to 1914 were the following captains: Arthur Jackman in 1905; Henry Dawe (Bay Roberts) in 1906 and 1908; George Barbour in 1907 and 1911; Abram Kean in 1909, 1910 and 1913; and William C. (Billy) Winsor in 1912 and 1914. Based on a rate of 4 per cent of the value of their cargoes, their shares, on these occasions when they were high liners, were as follows: Jackman—$1,914 in 1905; Dawe—$1,999 in 1906 and $1,953 in 1908; Barbour—$2,359 in 1907 and $2,022 in 1911; Kean—$2,162 in 1909, $3,632 in 1910 and $2,782 in 1913; and Winsor—$2,401 in 1912 and $2,465 in 1914. To put these sums in context, it must be remembered that in 1910, the chief or resident physician at the General Hospital received $1,700 for a full-time position. The sealing captains were employed for about four or five weeks and were then in demand for other well-paying positions in both the public service and the business world, and it was not unknown for them to mix politics, business and sealing. For example, Capt. Arthur Jackman was the marine superintendent at Bowrings and supervised the repair and upkeep of the firm's sealing and coastal steamers;[116] Capt. Edward White, Sr., prosecuted the cod fishery; Capt. Charles Dawe owned a Labrador cod fishing firm in Bay Roberts; Capt. Samuel Blandford managed Jobs' Labrador cod fishing operation at Blanc Sablon; Capt. George Barbour prosecuted the cod fishery and was employed in the northern Labrador coastal boat service; Capt. Abram Kean prosecuted the cod fishery and was in charge of Newfoundland's coastal service for 19 years; and White, Dawe, Blandford, Kean and others were successful politicians as well.[117]

While one cannot ignore the disparity between the wages received by the captains and those received by the men, it would be wrong to conclude that the captains were somehow to blame for this development. As already demonstrated, both captains and men were employed in a capitalistic system directed by the steamer owners and even more so by the several seal oil plants; unsuccessful captains were not uncommon and did not retain their commands very long. When one examines the average shares that the captains received in any given year, the disparity between captains and sealers' shares is not as dramatic. In 1905-14, the average captains' shares were as follows: 1905—$566; 1906—$972; 1907—$828; 1908—$834; 1909—$916; 1910—$1,322; 1911—$1,098; 1912—$549; 1913—$1,040; and 1914—$996. These shares were well below the shares received by the high liners mentioned above. Finally, a brief glance at the records will demonstrate that captains who brought in the smallest cargoes received very little pay, and as mentioned, successive failures were not tolerated. The following is a list of the captains who brought in the smallest cargoes and their pay for the years 1905-14: D. Bragg in the *Southern Cross* received $57.31 in 1905; P. Blackwood in the *Leopard*, $283.15 in 1906; J. Gillam in the *Kite*, $17.94 in 1907; R. Fowlow in the *Southern Cross*, $47.51 in 1908; M. Bartlett in the *Southern Cross*, $94.63 in 1909; William Carroll in the *Kite*, $344.79 in 1910; William Carroll in the *Kite*, $280.65 in 1911; William Bartlett in the *Viking*, $40.48 in 1912; Fred Yetman in the *Kite*, $92.96 in 1913; and Wesbury Kean in the *Newfoundland*, $47.56 in 1914. While it is obvious that the smallest steamer, the *Kite*, was at a disadvantage because of its size, taken as a whole, the shares received by the least successful captains (based on the 4 per cent agreement) were often quite modest. As can be seen, sometimes well-known captains, such as William Bartlett (Capt. Bob's father) in 1912, failed to reach the seals, and Westbury Kean's unsuccessful voyage was consistent with his desperate and tragic decision to send his men far afield in search of seals. However, most unproductive captains did not become well known (and it is not surprising that all the steamers in their command were wooden walls). Needless to say, however, the standard of living that separated the average captains from the average sealers had become significant by the end of the century, while the successful captains were some of the wealthiest members of society.

Income and living standards
In concluding this discussion of income from the seal fishery, it is necessary to return briefly to the rough estimation used above of the cost of a consignment of food to a fisherman at the beginning of the century. This consignment, consisting of pork, beef, flour, bread, butter, molasses and rum, would cost about £30 6s 6d sterling (about $150) in St. John's and £37 2s 6d (about $185) in Conception Bay. Income from the seal fishery would provide about £20 (about $100) to go towards paying for these supplies. An examination of the value and tariff on the same quantities of these provisions imported into Newfoundland in 1909 shows that the total value of the whole consignment was $90.[118] Prices to the consumer are almost impossible to assess because there were so many classes of customer, each being charged a different amount depending on whether a consumer paid cash, had credit on the merchant's books or was chronically in debt. The records of the Ryans, Bonavista, suggest that the best price for a sample consignment that one could hope for as a preferred customer would be $120.[119] Buying in small quantities on credit probably resulted in customers' paying twice the value, as demonstrated above by the prices charged the sealers who wished to purchase necessities before the sealing voyage.

In addition, it is impossible to draw adequate comparisons between living standards at the beginning and the end of the nineteenth century simply on the basis of the items listed above. By the beginning of the twentieth century, the food available and in demand was much more varied and included, in addition to the above articles, such items as: varieties of plain and sweet biscuits; fresh, canned and dried fruits, including raisins, sultanas and currants;[120] cereals and grains, including rolled oats, oatmeal, bran and corn meal; varieties of sugars; nuts; cheeses; hams and bacon; vegetables; flavouring extracts; instant soups; various coffees and teas; and figs and dates. In addition, the smaller, more primitive housing with open fireplaces of earlier years had been replaced by saltbox and two-storey houses with cast iron stoves. However, when one sees the small shares of $10 to $30 that sealers received by the end of the 1890s, one can conclude that sealing was not contributing much to the ice hunters' annual incomes. Furthermore, the fewer berths available, meant that each ice hunter was required to support more family members than was the case when all fishermen and boys went ice hunting.

Living and Working conditions

General living conditions

The life style of Newfoundlanders at the beginning of the nineteenth century can only be described in broad outlines because it attracted very little attention from writers. Anspach, in Harbour Grace c.1810, described the better houses as having been built from boards and plank on brick and stone foundations with cellars underneath. He said that they were two stories high and were heated with coal fires, using fuel imported from Cape Breton Island or England. He described the houses of the average fishermen as having just a ground floor and one story at the most. (This makes it obvious that Anspach was using the English definition of a 'story' and that his idea of a one-story house would be considered in North America as a two-story dwelling, or one that could be described as one and one-half stories. Thus the two-story houses of the richer classes would, in North American terms, be portrayed as having three stories.) He went on to point out that the houses of the poorer or working class families were built with logs chinked with moss, or sometimes clapboard, on the outside and planed board inside, with a floor of planed board or small poles laid side by side. All fireplaces were situated in large kitchens and were big enough to leave room for benches inside them; in the case of the working class, wood was used for heating and cooking and, indeed, for the smoking of fish in the wide chimneys. Since Newfoundland winters were much colder and longer than southern English and Irish winters, the fireplace was, no doubt, the focal point in a Newfoundland house to a far greater extent than in England and Ireland.

Anspach, who was well-educated and had come to Newfoundland as a school master, was well-read and acquainted with current ideas and knowledge. Drawing upon his readings in the field of diets and nutrition, he was aware that fish was considered to be a second-rate food by the leading specialists of the day—midway between animal and vegetable products and "less nutritive than flesh meats, and [producing] also less red blood and strength of body." Also, Anspach was aware that bread had become central to the diet of the northern European working class. In fact, unleavened bread (usually referred to as hard bread or ship's biscuit in Newfoundland), "which could equally well be made of barley or oats, as of wheat or rye, still often remained until the period of industrialisation the only form used amongst ordinary people ... in northern parts of Europe."[121] Hard bread, originally from the West of England, and potatoes, introduced into Newfoundland in the 1750s, were well-known and easily recognized staples to an observer of Anspach's background. Therefore, it was not unusual for him to notice and record the peculiarities of the local diet. He described the combination of good health and unusual eating practices as follows:

> [In spite of the fact that fish is less nutritious than meat, it] is very remarkable that no where can a stronger and more hardy race be found than in Newfoundland, not only among the natives, but also among the strangers who have resided some years in that island; and no where is fish, either fresh or salted, in more constant or general use, even during the most laborious season of the fishery: they eat fish at breakfast with their tea, at dinner with potatoes, and again at supper with tea. Salt pork, always accompanied with cabbage or greens, is used only on particular days.[122]

He also pointed out that most people kept hens, which supplied them with eggs throughout most of the year.

Anspach described how spruce beer was made by boiling spruce boughs in an iron pot in eight or ten gallons of water and then adding molasses at a ratio of one to eighteen gallons and stirring in the grounds from the previous batch. This would be ready to drink in 24 hours and was also used as a mix with rum. He added that wine was hardly drunk at all, while a considerable quantity of rum was drunk and that "Bohea *tea, hot* from the kettle in which it is boiled, is the favourite and universal beverage, even at dinner, particularly during the winter season, as well as at breakfast and supper during the whole year."

Meanwhile, flour, hard bread and salt pork had always been the major components of the diet of the Newfoundland fishermen from the West of England and, later, Ireland. Dried peas, which could be boiled in soups or made into pudding, and dumplings made from flour and lard were also commonly consumed. It is probable that there was nothing remarkable about pork, peas, dumplings and hard bread, and consequently, he did not describe their consumption, only remarking upon the unusual features of the diet. Furthermore, he reported that pork was expensive, but all imports were more expensive at this time because of the war, which would, in turn, encourage the greater consumption of fish and potatoes. Also, according to Anspach, home-knitted clothing was in general use, although heavy cloth, leather boots and other leather products, all imported from Britain, were popular as well.

It is useful to compare the comments of Anspach with those of another writer 30 years later. J. B. Jukes travelled widely in Newfoundland in 1839 and 1840, recorded his activities and observed the seal hunt first hand.[123] Jukes kept a diary, which formed the basis of a later book in which he combined good descriptive narrative and details of daily, inconsequential events. The latter are most helpful in assisting one to acquire a sense of the people's everyday lives, particularly in terms of their food and drink. As Jukes says on several occasions, molasses tea was definitely the beverage of the colony. On many occasions he, and his guides and/or companion (Dr. Stuvitz from Norway, with whom he had gone to the ice) made tea by putting a "handful of tea and two table-spoonfuls of molasses" into a kettle of boiling water.[124] On at least two occasions, in Northern Bay and Western Bay (both in Conception Bay), he was offered spruce beer,[125] in the Codroy Valley he was given "plenty of fresh milk,"[126] and on the southwest coast, milk and wine[127]—the latter probably from St. Pierre or from the French fishing ships. Jukes ate bread and butter regularly on his trips and also "biscuit," which was obviously hard bread. On one occasion he reported: "In this hot weather I had come to loathe the salt pork and beef and lived principally on tea and biscuit."[128] On another occasion, while travelling between Topsail and Broad Cove, Conception Bay, he and Dr. Stuvitz were given "dinner and supper of salt fish, eggs and tea," and in Broad Cove itself, "we got some dinner, consisting of fresh herrings, tea, and bread and butter."[129] Shortly afterwards, in Renews, he was given a meal of "tea, eggs, fish, and fresh-baked cakes."[130] He continued into St. Mary's and then on to Placentia Bay, reporting as he went. At Admiral's Beach (on 29 May), some people gave him and his guide "an excellent supper of fresh cod-fish, which, as I had lived lately on herrings, was quite a luxury."[131] And on one occasion, while crossing Conception Bay, the vessel on which he travelled was becalmed and a Catholic priest shared his food with Jukes; they dined on "fish and vang," which was made by cutting cod and salt pork into small chunks and boiling them together.[132] The Newfoundland diet on the east coast had not changed since Anspach had reported on it.

Thus, the "Newfoundlander" became recognized as "a long and strong eater of pork, salt beef, fish and hard biscuit," whose favourite beverage was "tea, usually sweetened with molasses ... and rum only, straight from the West Indies ... his beloved invigorator."[133] It seems that rum in unusually large quantities was an accepted stimulant. Prowse, who was born in 1834, wrote as follows in the 1890s:

> A great institution on the merchant's premises, always called in the vernacular "The Room," was the periodical serving out of grog; morning, eleven o'clock, noon, and in the afternoon, all employed in the room had a glass of rum; on the Jersey and on Newman's place this continued up to my own time. Various attempts were made by reformers to alter this practice. A most worthy Scotchman—Mr. Johnson, familiarly known as Wullie Johnson (managing partner of the old firm of Baine Johnson)—tried to improve it by watering the grog; he went on diluting it until he raised a rebellion. Mr. Robert Job was the first man to make a firm stand against the practice; he was execrated at the time, but he stood steadfast, and lived to see his example followed by all his brother merchants.[134]

Prowse, drawing upon the oral sources of his period, described the change in lifestyle and habits which resulted from the spring seal fishery, and at the same time, gave his impressions of the life of an ordinary fisherman on the eve of the expansion of this new industry:

> The wonderful growth of the spring seal fishery just about this period [1820s] completely changed the social habits of the people; the work required for fitting out the vessels, building punts, repairing and strengthening the sealing schooners, kept masters and crews at work all through the

winter; what had formerly been a carnival of drinking and dancing now became a season of hard, laborious toil.[135]

Prowse continued and pointed out that the oral sources among his acquaintances had informed him that before c.1820:

> the sharemen were always clear of service by 20th September [most sources say October 10]; after that there was dancing, drinking, and card-playing every night, from house to house in the outports.... On the first of May all hands set to work again. A large number of English and Irish went home every autumn, and returned in the spring; Jersey men always left. From about 1820 the spring seal fishery changed all this.[136]

Prowse is generally correct in bringing to the reader's attention the fact that the introduction of the seal fishery resulted in an abundance of winter and early spring employment which had never before been available. He is also correct in pointing out that a considerable number of English and Irish had previously gone home every autumn. Of course they did, and before the development of the seal fishery and the unusual conditions during the period of the Napoleonic War, the people who had stayed in Newfoundland were an integral part of the West of England migratory fishery in Newfoundland waters. Prowse's phraseology understates the migratory nature of this resident branch of the West of England-Newfoundland fishery. Also, he does greatly exaggerate the amount of idle time that had been traditionally available to the fishermen who stayed in Newfoundland at the end of the fishing season. It was a rare soul—planter, shareman or servant—who could earn enough and/or save enough from the short cod fishery to keep himself in idleness from 20 September to 1 May (over seven months), and catching cod did not even begin until the schools of fish arrived in mid-June. The few planters who stayed in Newfoundland all year round engaged in subsistence farming and kept and cared for hens, sheep, pigs, sometimes cattle, often a horse, or certainly dogs. Furthermore, many of them were semi-employed in maintaining the premises of the migratory fishermen. In addition, they were extremely busy providing enough firewood for the winter, spring and following summer. The fireplaces were large and were intended to heat the house as well as the kitchen, but were so inefficient that many families had winter houses deep in the forest to which they moved during the severest winter months. Any servants or sharemen who stayed in Newfoundland after the cod fishery was over needed employment just to provide himself with food and lodging, and they sought planters who needed someone to chop and carry firewood and assist with the other chores and activities. Comparatively few found such employment.[137]

Thus, the common ice hunters at the beginning of the century were fishermen employed by planters in the summer cod fishery who now remained in Newfoundland during the winter because they could obtain employment in the spring seal fishery as well. The majority, at first, were Irish because of the large numbers of both men and women from that country attracted to the island during these early years. They lived in small dwellings around St. John's and the major harbours of Conception Bay, situated for their ease of access to wood and water and constructed of boards, poles and logs. Oral reports suggest that in the beginning, many lived in primitive huts with a hob (consisting of a half barrel filled with sand for a fireplace) and a hole in the roof to allow the smoke to escape.[138] However, a stone fireplace was the first necessity and would soon be constructed, with a thick juniper log for a lintel. This was the only source of heat and means of cooking, and it provided some of the artificial light needed. Fish, potatoes and eggs produced locally, and pork, flour, hard bread, tea and molasses imported from Europe, North America and the West Indies were the major staples. As already indicated, and as import records confirm, tobacco and rum were consumed in large quantities. The fishermen worked extremely hard during the fishing season, when it was necessary to work long hours and to work quickly in order to take advantage of the large schools of migrating cod fish near shore. Because the local climate can be both cold and wet at the same time for long periods, these fishermen were used to working on sea and on land in weather for which no suitably functional protective clothing had (nor has ever) been designed. However, the advent of the spring seal fishery added a new dimension to the meaning of the terms 'discomfort' and 'hardship'.

Work and diet on sailing vessels

The actual work involved in the seal hunt (as opposed to the net seal fishery) was uncomplicated. The vessels left the harbours and sought the harp seal herds on the Arctic ice, hoping to find them towards the end of March, when the pups (or whitecoats) were large with pelts weighing over fifty pounds (23 kg.). If the pups were found too early, they were small and provided little fat; if undiscovered, they left the ice and took to the water around the end of March, when they also began to molt and lose weight, now that they were no longer nursing; they become, as the fishermen unflatteringly called them, raggedy jackets.

The whitecoats were killed with a 'gaff', which was an iron spike with a hook fastened to a wooden staff or baton of six to seven feet (about 2 m.) long. Their skulls were delicate, and one blow usually killed them instantly. The ordinary sealer who performed this operation was usually referred to as a 'batsman', and it was this branch of sealing that made up by far the largest part of the industry. Next, the whitecoat was pelted; the skin with the thick layer of fat attached was removed with a sharp knife. The carcass was left on the ice, but it became the practice for sealers to leave one flipper attached to the pelt, and these flippers were shared among the crew back in port. The men used their tow ropes to tie two or three pelts together and haul them to the vessel or to a pan of ice accessible to the vessel. The men drove short slivers of metal, called 'chisels', 'sparables' or 'frosters', into the soles of their boots to enable them to walk without slipping, and very soon, they began to wear skin boots with thick leather soles which were made especially for the ice. After the pelts were hauled to the vessel, they were hoisted aboard, allowed to cool thoroughly, and stowed in the hold, sometimes with ice to keep them well chilled. If the men were fortunate enough to find plenty of whitecoats, they filled the vessel and returned to port. Until second trips were banned at the end of the century, they then returned to the ice to shoot the older seals among the ice floes as the herds waited for the young to be strong enough for the long journey to the high Arctic, where they would spend the summer months. Consequently, the ideal situation involved obtaining a load of young harps in March, bringing the cargo to port in good condition immediately, and then returning to the ice to seek out the old seals resting on the ice. Sometimes, a considerable kill could be made by batsmen if they managed to come upon the old seals on tight ice, where there was little chance of their escape; in fact, crews often waited in hopes of the ice tightening before leaving their vessels to kill the old ones. Naturally, the gunners played an important role in this aspect of the hunt, and gunners, followed by their 'dogs' (men who carried the powder and shot in the days of the muzzle-loading rifles and cartridges in the case of breech loaders), fanned out over the ice to attack the herd and were often primarily responsible for the success of the second trip. If the men were unable to load with whitecoats, they stayed out hunting for the young and then continued to hunt young and old until it was time to return to port. Some of the lucky crews and ships were able to make two trips to the ice in one spring, and a few captains made three, but this was very rare.[139] Some unfortunate vessels got jammed in the ice, and the men watched helplessly as the warm weather turned the fat into liquid, which was pumped overboard with the bilge.

The old harps were pelted, or 'sculped', as some called it, in the same manner as the young. The hooded seals, or hoods, were also sought in the absence of whitecoats, but these seals remained in isolated family groups and harvesting them was more time consuming. On the other hand, the adult hoods did not desert their young as the adult harps did. They remained and defended the pup, and thereby, the men were guaranteed two hugh pelts as well as that of the pup. The dog hoods were quite fierce, and a single batsman would rarely try to kill one. They were usually shot, but two batsmen working together could despatch them.

The men carried out their work with gaffs, knives and hauling ropes; in addition, each man carried a steel for sharpening his knife. In the beginning, men who brought along a gun did not have to pay for their berths, but this policy was later changed, and men were hired specifically as bow or chief gunners and went berth-free, while the men who were appointed as secondary or after gunners paid a reduced rate for their berths. However, killing and pelting, hauling and loading were often the easiest part of the sealer's voyage. But before looking at the conditions on the ice, it is necessary to know something of the sealer's life on board the vessel.

The early ice hunters were admired by contemporaries for the harsh conditions under which they worked, both on board the vessels and on the ice. Governor Waldegrave, a perceptive man who foresaw as early as 1799 that Newfoundland would become a colony, described the new seal fishery as a valuable industry and wrote that "the very mode of taking these animals is of a nature to form the hardiest race of men in the universe."[140] This note of admiration came from a man who had made a career in the Royal Navy since 1766, when he joined at the age of thirteen, and who suppressed a mutiny in St. John's harbour by threatening to sink his own ships with all hands, including the mutineers and the officers; in other words, this was a man who was not easily impressed by the rigors of the sea-faring life. Prowse described him as a "fire-eating old sailor."[141] If he concluded that the Newfoundland ice hunters constituted the "hardiest race of men in the universe," it was a rare compliment, indeed. Similarly, in 1804, Governor Gower, in his report, stated that the Newfoundland sealers exposed themselves "to the most imminent dangers ... [and that] it is certain that there is no employment so well calculated to form hardy and intrepid seamen."[142] In July, 1805, he again reported on the "extreme hardships and dangers attending it [the seal fishery]."[143] Thus, even in the context of early nineteenth century maritime and naval employment, the Newfoundland seal fishery was singled out as a particularly rigorous activity.

Anspach was also impressed by the conditions under which the seal fishery operated and described them thus:

> During the months of February, March, April, and part of May, the coast of Newfoundland is generally surrounded with ice to the distance of several leagues. The most formidable ramparts erected by military art ... require less intrepidity and experience to encounter, than those enormous floating bulwarks and the united efforts of the elements which those seas, at that time, oppose to the mariner. It is hardly possible to convey to the imagination a correct idea of the terrific grandeur which characterises this scenery. Immense fields of ice of such extent that the eye cannot reach their bounds, and sometimes impelled by a rotary movement by which their circumference attains a velocity of several miles per hour: lofty islands and mountains moving along with irregular and sometimes inconceivable rapidity, or when the comparative shallowness of the ground arrests their progress, then bedded immovable on the solid rock or earth, whilst fragments of various sizes are scattered about throughout the intervening spaces, and coming in drifts so thick and so quick as to whirl the ships about as in a whirlpool: here and there a mountain bursting with a tremendous explosion; the fields suddenly changing their directions, coming into close contact with a dreadful shock, and overlaying each other with a noise resembling that of complicated machinery, or of distant thunder. The immense pressure thus produced and the tremendous power exerted, are such as to crush to atoms or to set on fire the wood which may happen to be in their way; the strongest ship can no more stand these shocks than a sheet of paper can stop a musket-ball. Sometimes the vessel is beset and immovable, and her safety then depends only on the immediate coagulation of the surrounding ice into one uniform field; as soon as a separation again takes place, her danger recommences, and the motion and violence of the ice are so rapid and so great, the changes of direction so sudden, that her destruction appears inevitable. Some are lifted up and thrown upon the hard congealed surface by a sudden shock; others are crushed, or, at least, their hulls completely torn open; others again are buried beneath the heaped fragments of a bursting mountain. A strong easterly or north-easterly wind arises, and drives with inconceivable rapidity all this ice against the coast, where filling up the bays, harbours, and coves, it soon becomes one immense, widely extended, solid mass; until, the wind setting with equal violence to the west or north-west, this mass is broken and as rapidly driven into the main ocean: the wind changes again, the ice as quickly returns, and winter resumes its sway with increased rigour. The situation of the vessels which happen to be entangled in that ice may be easily conceived. Add to this picture a rock-bursting frost; gales whistling and howling in huge uproar, which, while on the land they shake the houses, rocking them to and fro, tear up the trees from their roots, and scatter them through the convulsed forests; at sea, they drift about with violence masses of snow and sleet, or else thick fogs, freezing as they fall, cover everything with ice, the sides, the deck, masts, and rigging of the vessels, and even the clothes of the mariners. The mere thought of such a situation, in a stormy and dark night, and on a sea covered

with islands, mountains, fields, and fragments of ice in perpetual motion, is sufficient to strike the mind with horror; and yet such a situation the Newfoundland seal-hunters court with as much ardour as vessels in other cases study to avoid it....

About St. Patrick's Day, or the 17th of the month of March, they proceed to that fishery through the most boisterous weather, struggling by all possible means to get out of their harbour and bay. It is impossible to conceive a greater degree of perseverance and intrepidity than the people of Conception-Bay in particular, display on these occasions. After having at last conquered these first difficulties and proceeded beyond Baccalao Island, their next object is to reach a seal-meadow by sailing or cutting through the intermediate fields of ice; they then run their vessel into it, the crew disperse, and whilst the gunners fire at the largest seals, the others assail the rest with clubs.... When sufficient execution has been made on a seal-meadow, or the extreme severity of the weather interrupts the operations, the dead seals are dragged on the ice to the schooner or boat; they are then *pelted*, that is, the skin with the coat of fat adhering to it is separated from the carcase, and the latter is thrown overboard, excepting a small portion as may be reserved for the mess. The voyage is continued through the ice, or through the open sea if it happens to be so, to other seal-meadows until the loading is completed, unless the state of the weather, or some material damage in the vessel makes it necessary to return sooner into port....[144]

J. B. Jukes, who went to the ice in 1840, experienced some difficulty in finding a suitable vessel for his trip. He wrote that most of them were schooners and brigs of 80 to 150 tons (much larger than those reported by Anspach), "manned by a stout crew of rough fishermen, with a skipper at their head of their own stamp."[145] He described the vessels as being "inconceivably filthy" and reported that the crew and skipper "all live and lie together in a narrow dark cabin of the smallest possible dimensions, and the fewest possible conveniences." However, Jukes discovered that the situation was just beginning to change—that some captains were beginning to set themselves apart from their men and that there were "one or two masters of ships of a superior description who reserve the after-cabin to themselves, and keep the crew in the forecastle."[146]

Jukes managed to find room on the brigantine *Topaz*, which carried a crew of 36 men, who were divided into three watches, each watch under a master watch (or master of the watch). The ship cleared St. John's, eventually found the Arctic ice and, inevitably, got jammed. Jukes wrote:

> 'Overboard with you! gaffs and pokers!' sung out the captain; and over went, accordingly, the major part of the crew to the ice. The pokers were large poles of light wood, six or eight inches in circumference, and twelve or fifteen feet long: pounding with these, or hewing the ice with axes, the men would split the pans near the bows of the vessel, and then, inserting the ends of the pokers, use them as large levers, lifting up one side of the broken piece and depressing the other, and several getting round with their gaffs, they shoved it, by main force, under the adjoining ice. Smashing, breaking, and pounding the smaller pieces in the course the vessel wished to take, room was afforded for the motion of the larger pans. Laying our great claws on the ice ahead, when the wind was light the crew warped [winched] the vessel on. If a large and strong pan was met with, the ice-saw was got out. Sometimes, a crowd of men clinging around the ship's bows,

and holding on to the bights of rope suspended there for the purpose, would jump and dance on the ice, bending and breaking it with their weight, shoving it below the vessel, and dragging her on over it with all their force. Up to their knees in water, as one piece after another sunk beneath the cutwater, they still held on, hurraing [sic] at every fresh start she made, dancing, jumping, pushing, shoving, hauling, hewing, sawing, till every soul on board was roused into excited exertion.... They continued their exertions the whole of the day, relieved occasionally by small open pools of water; and in the evening we calculated we had made about fifteen miles.[147]

This is one of the best descriptions available of the activities of the men on board of a sealing vessel which was attempting to make progress through the ice fields. These men were obviously in superb physical condition in order to continue such strenuous work from morning until night, while wet at least to the knees. In this case, the vessel covered 15 miles (about 25 km.), but in many cases, no progress was made, and often, the tides and running ice carried the vessel backwards while the men strained and struggled to the limit—all in vain. Working with frozen canvas was bad enough; hauling pelts for miles over rough and jagged ice was even worse; but for sheer agony, efforts, such as Jukes described, to force a sailing vessel through the ice fields must be without equal on water or on land.

One American writer described the conditions under which sealers worked in the early 1850s as follows:

There is perhaps no voyage requiring greater energy, courage, and hardy seamanship in its prosecution, than the seal fishery; pursued at the most inclement season—the fleet leaving port about the first of March—and subject to the dangers incident to stormy weather, superadded to sleet and snow, falling so thick and fast as to almost close the view to the surrounding dangers of ice fields in wild commotion, madly drifting before the gale, in the Northern Sea, and presenting a picture so appalling as to be almost beyond the conception of those sailing pleasantly in Southern seas.

A storm at sea is, at any time, and under any circumstance, fearful; but when the blocks become clogged with ice, and ropes will not run freely, and sails are frozen to a consistency equal to that of sheet iron, with benumbed hands, and drifting snow blinding the eyes, and shutting out all prospect ahead, then the metal of a seaman is most severely tested; and all those fearful accessories are but the ordinary concomitants of a Newfoundland sealing voyage.[148]

Therefore, the working condition of the ice hunters involved much more than the actual slaughter of the seals. The environment in which the work was carried out was such that discomfort, danger and hardship were always present, and their vessels provided, as Jukes pointed out, "the fewest possible conveniences."[149]

The information concerning life on board the early sailing vessels in the seal fishery is scarce. It was reported, retrospectively, later in the century that in the early years, "No crew would set out for the seal fishery unless everyman had his keg [of rum]."[150] When Jukes went to the front in 1840, he recorded on 11 April: "Our fresh stores had long been exhausted, as also our wine and spirits, and we were now reduced to tea and the common rum, and had not much of that left."[151] Anspach mentioned the sealers' having to pay berth money in order to cover the cost of provisions and equipment. It can be assumed that the early sealing vessels carried tea, molasses, hard bread, butter, pork and, eventually, flour to make boiled duffs. It was not uncommon for sealers to eat raw salt pork, and this meant that actual cooking could be confined to the making of molasses tea if time, firewood and/or circumstances prevented cooking on a larger scale. In 1848, nevertheless, each man was required to bring 25 sticks of firewood, which means that by then, the business of cooking was taken quite seriously.[152] However, from the very beginning, sealers also ate the flippers of the whitecoats, as Anspach was the first to report: "Some affirm that the flesh of the young cubs is very palatable; it is eaten by the seal-hunters during that fishery, and tastes something like hare's flesh."[153] Cooking facilities were usually adequate, and it had to be unusually stormy before sealers would deny themselves a hot meal of fried pork and seal meat.[154] A half puncheon filled with rocks on deck made a perfectly adequate fireplace. James Murphy writing in 1916 about "the old sealing days," described the cooking facilities on deck as follows: "Cabooses to make fire in [were] used for boiling and baking the grub. These cabooses stood on the decks of the vessels; sometimes a half puncheon was used, around which a number of bricks were placed."[155] However, the best that the men could hope for would be a breakfast of tea and bread and butter, tea and bread and butter at intervals during the day, and tea and a hot meal—possibly of boiled pork and duff or saltfish and, possibly, of seal meat—at night.

Because sailing vessels were restricted in their movements in the ice,[156] men were required to haul most of their pelts directly to the vessel. Jukes described this process and how the men used the opportunity to take short lunch breaks:

> Some of the men brought in as many as sixty each in the course of the day, and by night the decks were covered, in many places the full height of the rail. As the men came on board they occasionally snatched a hasty moment to drink a bowl of tea, or eat a piece of biscuit and butter; and as the sweat was dripping from their faces, and their hands and bodies were reeking with blood and fat, and they often spread the butter with their thumbs, and wiped their faces with the backs of their hands, they took both the liquids and the solids mingled with blood ... and after a hearty refreshment the men would snatch up their gaffs and hauling ropes, and hurry off in search of new victims: besides every pelt was worth a dollar.[157]

While on the *Topaz*, Jukes breakfasted on the fried hearts and kidneys of whitecoats and concluded, "they were very good, being just like pig's fry, but rather more tender and delicate."[158] He also commented upon the fact that all sealers ate seal meat regularly and that even Roman Catholics ate it on Fridays and other fast days on which the Roman Catholic Church forbade its members to eat meat:

> The constant employment of the men on deck, when they had nothing else to do, was boiling, frying, or roasting pieces of seal flesh and eating them. Immediately after a dinner or breakfast down below, they would come on deck and set to work at the seal by way of dessert. Their constant food both at sea or on shore being fish or salt pork, fresh meat is at all times a luxury to them, even though it be that of a seal. It is amusing enough, however, that more than one half of the men going to the ice are Irishmen, and strict Roman Catholics, who would rather undergo any privation than eat meat on their

fast-days, which in Newfoundland are Fridays and Saturdays. I had always found during the previous summer that my men, if there were no fish to be had, would confine themselves to bread and butter and tea on those days, even while undergoing the hardest labour. The good fathers of the church, however, either in pure ignorance of natural history, or by a little pious fraud, willing to indulge their flock during the cold and hardships of a sealing voyage, have come to a unanimous determination that *seals are fish*.[159]

Reminiscing in 1922, in the presence of the American writer George Allan England, Capt. Abram Kean looked back nostalgically to the days when sealers were real men and did real work under tough conditions.[160] He pointed out that men had plenty of rum in the old days but that it had to be prohibited because it caused fights. He also said he remembered the days before barrels were installed on the masts for scunners and spy masters—how previously, they had clung to the rigging. And he recalled when there were no stoves and the men cooked on the ballast (i.e., rocks) in the ballast locker, how the men slept under sails without oil skins and how the food consisted of pork and duff three times a week and hard bread and tea. In fact, despite his mixing up his own recollections with those he had heard from his uncle William and others, Kean provided glimpses of conditions at the very beginning of the industry as well as sporadically throughout the nineteenth century which were corroborated by Anspach, Jukes and later commentators. It would seem that the men fared well in terms of the quantity of food provided; in fact, it is clear from Jukes' observations that the men kept their own fireplace on deck and cooked seal meat and delicacies themselves. The food may have been basic, but it was plentiful, and with their frequent trips to the vessels with two or three pelts at a time, depending on weight, the men had many opportunities during the day to have a snack of tea and bread and butter. This is a point that is generally overlooked. Hard bread, butter and molasses tea were always available in unlimited quantities, and probably salt pork as well. Murphy also points out that "Hamburg bread and butter, and Hamburg pork, contributed to a portion of their fare."[161] In the days of the sailing vessels, when small crews hauled most of the pelts directly to the vessels, the men were, as on the *Topaz*, free to snack as often as they pleased.[162]

While the food on board the vessels was adequate and filling, if somewhat unrefined, the only redeeming feature that could be attributed to the sleeping accommodations was the fact that the men were provided with shelter at night and during storms.[163] As Jukes pointed out, the captain of the *Topaz* was one of the first to create cabin space for himself and his second hand in the after part of the vessel, leaving the forecastle to the men. However, none of these vessels had forecastles large enough to accommodate a sealing crew (which was much larger than a sailing/freighting crew). Consequently, temporary bunks, holding two or three men each, were built in the hold below the deck. The pelts were stowed in pounds directly under them, and as the cargo rose, the lower bunks were removed (to become firewood no doubt), and finally, many men would find themselves sleeping on top of the pelts. The blood, grease and smell combined with cold, wet clothing must have made sleeping and resting very unpleasant and infected cuts and bruises.[164]

One affliction or complaint that was unique to sealers was the danger of 'seal finger', or in Newfoundland terms, a 'swile finger' or 'swile hand'. This was an infection that usually occurred in the middle finger of the sealer's non-dominant hand. The finger swelled and turned red and glossy, and the swelling then extended to the hand and arm. This was accompanied by extreme pain. After a couple of weeks the pain diminished, but the swelling persisted, and gradually the finger healed in a cold rigid crooked position. It was associated with handling the pelts and sticking one's fingers through a hole in the pelt to turn it over and/or tie it on one's rope. Furthermore, it was much more common among sealers when handling older seals. (The condition was well-known in Norway as well, where it was referred to as 'speck [blubber] finger'.) In Newfoundland, there is a folk belief that the hand began to resemble the flipper of a seal, and presumably, if two or three fingers were affected, the hand would take on a curved, paw-like appearance. Heat, splints and poultices were traditional remedies, but nothing really helped until the advent of antibiotics. Fishermen very often requested amputation because the digit was not only useless but got in the way when handling cod fish lines and nets. The affliction was not understood, and, indeed, there were superstitions associated with the condition. However, modern medicine recognizes that it is caused by infection and inflammation of the joints.[165]

Ice blindness was another common hazard, because the glare of the sun from the snow-covered ice could be extremely intense at times. According to oral sources, the early method of preventing this was to smear seal's blood around the eyes or to tie a strip of seal meat across the bridge of the nose just below the eyes.[166] Later, goggles became common (and, as indicated above, cost fourteen cents a pair in 1891); however, the application of seal blood or meat remained the favourite protection against ice blindness and was used by some sealers into this century.

Sealing demanded considerable effort and was ever subject to discomfort. The living quarters were primitive, cold, unhealthy and generally uncomfortable. The food was monotonous, but usually adequate, especially when supplemented by seal organs and flippers. The task of sailing, or in many cases, hauling and chopping and manhandling the vessel through the ice, was extremely demanding. And finally, hauling pelts for miles over rough and 'hummocky' ice could be excruciatingly difficult. However, if at the end of three or four weeks, the vessel was back in port with a cargo of young harp pelts, everybody was pleased and relieved. Often, the voyage ended in failure and the men found their vessels jammed in solid ice, with not a seal in sight. The best they could hope for then was to exist on hard bread and tea until the ice dispersed and they were free to return home in time to prepare adequately for the summer cod fishery. Sometimes, the elements turned against them with a vengeance and ships and lives were lost. The living and working conditions were difficult, as observers have reported, and they were unique to this industry, but they were not unusual when compared with living and working conditions experienced by men in similar occupations. Busch, in his description of the killing of sea elephants on islands near the Antarctic continent, wrote: "Life was grim, above all for men put ashore for months on end with a cask of salt meat and some flour and molasses, to live in a shanty of rocks and canvas."[167] On the whole, although food and accommodations and working conditions left much (probably everything) to be desired by modern standards, by the standards of the community and of life in the fishing boat, they were acceptable, as long as there was the expectation of a reasonable financial reward.

Work and diet on steamers

As the previous chapter has shown, the age of steam appeared suddenly and dramatically, and affected, to some degree or other, most aspects of the sealers' lives. The early steamers were an improvement over the sailing fleet in one important respect—they could proceed in any direction when in the ice, following the leads of water as they opened, regardless of wind direction or the lack of wind altogether. Otherwise, however, they were not significantly different from the large brigantines they were quickly replacing. In 1868, for example, some of the sailing vessels—*Elizabeth* (152 tons), *Kate Cummins* (184 tons), *Gertrude* (133 tons), and *Frances* (133 tons), all from St. John's; *Thomas Ridley* (164 tons) and *William* (145 tons), from Carbonear; *Louisa* (157 tons) and *Rolling Wave* (152 tons), from Bay Roberts; and *Greyhound* (153 tons), clearing from Catalina—were not much different in size from the steamers *Hawk* (172 tons) and *Bloodhound* (153 tons) from St. John's. Nor were these steamers larger than the bigger Harbour Grace vessels which cleared from that port in 1869, including the *Isabella Ridley* (155 tons), *Mountaineer* (177 tons), *Vesta* (148 tons), *Curlew* (168 tons), *Iona* (151 tons), *William Whelan* (149 tons), *Dolphin* (173 tons), *Anastasia* (177 tons), *Topa* (148 tons) and *Eclipse* (146 tons).[168] Furthermore, these early steamers were severely underpowered by later standards, with engines of only 50 to 75 nominal horse power,[169] and consequently, were very dependent on sails. Therefore, the working conditions on board the steamers differed in one important way—they could take the men more easily through the ice fields and bring them closer to the, by now, shrinking seal herds.

The working conditions on board the steamers evolved from the time of their introduction in 1863 up to the end of the period under study, in 1914. The early steamers depended on sail to supplement their steam power, so men continued to work the canvas as before. Furthermore, each watch was required to hoist the coal from the pounds and pass it in baskets by hand to the engine room, where it was left to the coal trimmers, who kept it 'trimmed' in a pile in reach of the firemen, who, in turn, shovelled it into the firebox under the direction of the engineer in charge.

The coal dust compounded the problem of lack of cleanliness. It hung in the air between decks as the buckets were raised and lowered and the pounds were emptied; pounds were then cleaned and filled with pelts. Thus, the men attempting to sleep at night were disturbed by both the lowering of the pelts, the hoisting of the coal and the hoisting and dumping of ashes by the watch on duty. Coal was constantly moved in order to provide pounds for pelts in a planned way so that the weight of the fat was evenly distributed; in fact, if a steamer was fortunate enough to strike the seal herds early and load quickly, as many did, the coal was hoisted and dumped overboard. Thus, the dirty work associated with coal and ashes was added to the men's regular jobs of working the canvas and preparing their gaffs, ropes and knives for the hunt. However,

Figure 14 - Part of the sealing fleet with the men lined up holding their gaffs and waiting to welcome the Newfoundland governor c. 1900. After the 1898 Greenland disaster, some governors took an avid interest in the conditions on board the steamers and carried out inspections which were disguised as 'visits'. Photograph by Holloway or Holloway Studio.

they had the advantage of seeing their craft manoeuvre through the leads, taking them closer to the seals without the strenuous tasks that were always associated with handling a sailing vessel in such surroundings, as vividly described by Jukes earlier. At the same time, the advent of the steamer brought with it a greater division of labour as more specialized tasks developed and became customary.

The captain was assisted by the first mate, who was always referred to as the second hand. The second hand and captain were experienced sealers who had advanced through the ranks, except in the very early days of steam, when, as has been shown, experienced Scottish whaling officers were often hired. Both the captain and the second hand usually channelled their commands through the master watches. These latter officers filled extremely important positions because the second hand did not usually go on the ice. Each master watch was the sole authority over his watch while on the ice and was usually in charge of directing their work on board ship as well. Each master watch was assisted by a second master watch, who was sometimes called the deck router (or rowter) or stow boss.[170] The deck router was usually the one who actively directed the dumping of ashes, the shifting of coal and the stowing of pelts on board the vessel. Finally, 'ice masters' were appointed by the master watches and given

Figure 15—Men awaiting the order to leave the ship and begin the day's kill. They are carrying their gaffs, hauling ropes and nunny bags (with food and drink). A few are carrying flags, which were used to identify their ship's 'pans' of pelts from those belonging to toher ships. Although impossible to distinguish here, each man also carries a knife and steel. Photograph by Holloway or Holloway Studio.

authority over five or six men each while on the ice.[171] The 'barrel man' (or 'spy master') in each watch took his turn in the barrel on the main mast to search for seals, while the 'scunners' took turns in the barrel on the foremast to direct the steamer through the ice. As England pointed out, the scunner was in charge while in the barrel and only the captain could overrule him. As he observed, "A good scunner can do miracles in worming the ship through seemingly impossible obstructions."[172] In each watch, there was a bridge master, who passed directions from the scunner to the wheel crew.[173] The captain, second hand, master watches and bridge masters were considered officers on board. The boatswain (bosun) was in charge of the equipment and saw to it that the sealers were supplied with ropes, gaffs, twine, rifles and cartridges. He was hired directly by the ship owner and employed all year round. In most cases, the boatswain remained on the same vessel and often served the same captain for many consecutive seasons. (When he lay dying, Capt. Arthur Jackman's last words were, "Give my clothes to Mick Maddigan," his long-serving bosun and good friend.[174] Tom Carroll [1868-1961] was one of the most well-known bosuns in Newfoundland in the twentieth century. He is reported to have made 69 voyages to the ice, 66 of them on steamers belonging to Bowrings. He served the firm as bosun of the S.S. *Terra Nova* in his later years and retired finally in his late 80s while bosun of the last old wooden wall, Bowrings' *Eagle*.[175]) In addition to these positions, there were always cooks on board, a ship's carpenter and a cook and/or steward for the captain and his officers. These positions and personnel overlapped during the days of the older seal fishery; they evolved into more specialized positions as the crews got larger in the later nineteenth and early twentieth century.

However, steam brought with it, as indicated, a new class of worker as well. Engineers and firemen (or stokers) were required, and the former were recruited from Scotland until local men became qualified. In 1879, the three engineers on board the *Bear* were from Greenock.[176] Alex Mckinley, who was mentioned as chief engineer on the *Eagle* in 1910, was most likely Scottish,[177] while John Leary, who was engineer on the *Erna* when that steamer was lost in 1912, belonged to St. John's.[178] The firemen were hired from the major centres of the seal fishery, which meant that in its earlier years, they came from Harbour Grace, Carbonear, Bay Roberts and Brigus as well as St. John's. For example, in 1874 when the boilers on the *Tigress* exploded, among the 21 men scalded to death were 10 from Bay Roberts, the home of the captain, Isaac Bartlett.[179] However, as years went by, firemen became centred in St. John's. For example, the four firemen on the *Bear* in 1879 were from St. John's, as were the two trimmers. It quickly became common to fill nearly all positions—excluding captains, second hands, master watches and other positions which included in their duties the killing of seals—with men from St. John's. Returning to the loss of the *Erna*, it is worthwhile to examine the crew sent to Scotland by Baine Johnstons to bring the steamer to Newfoundland in the late winter of 1912.[180] Captain Linklater was

Figure 16—Men sculping (or pelting) seals. Each sculp (or pelt) consists of the skin with the thick layer of fat attached. One man is preparing his rope, which he will use to haul several pelts to a pile (or pan), two are sculping, and the man standing with the flag appears to be giving orders (he may be a master watch or one of the ice masters). Photograph by Holloway or Holloway Studio.

probably Scottish, but all the other officers, with the exception of Jacob Winsor, were from St. John's; similarly, nine out of the ten stokers were from St. John's, and the other, Bernard May, was from Quidi Vidi—then within a short walking distance of St. John's, and now, part of the city; of the eight seamen whose addresses are known, five came from St. John's and one each from Carbonear, Brigus and Wesleyville. This reflects the trend that developed in the steamer seal fishery as a whole: engineers, firemen, and stokers tended to come from St. John's and the immediate vicinity.

In 1910, marconi equipment was installed on the steamers *Eagle* and *Florizel*; 'marconimen' were gradually appointed to most sealing steamers, and after the tragedy of the *Newfoundland* disaster in 1914, to all. And beginning in 1901, under circumstances to be discussed later, 'doctors' with some medical background began to be assigned to the various ships.

The sealers were expected to work while there was work to be done. After a day lasting from daylight to after dark, killing and pelting seals and panning the pelts, with only hard bread and drinking water from puddles on the ice, they had to hoist the pelts aboard the steamer. With the low-powered steamers that could not manoeuvre effectively through the ice, most pelts were hauled to the ship's side; as newer, more powerful steamers were acquired, panning became common and saved the work of hauling, but resulted in the men being on the ice all day without any opportunity to have a cup of tea on board. Also, these steamers were larger, and even when the men did haul pelts to them, they were not usually permitted to go aboard, and help themselves to tea and hard bread, without orders; after all, by the late 1890s, the bigger steamers carried over 300 men (although limits were eventually imposed by legislation and mutual agreements among owners). Then the work continued through the night, as described by England:

> Night was filled with many labours, after the ice claw had been made fast [securing the steamer to the pan]. Big gangs fell to work, dragging in the pans of sculps and loading them. On deck at almost any time, even in the wee small hours, harsh toil always seemed going on, by raw flares of torchlights that half revealed frost-blackened faces, gleaming eyes, teeth glinting like gnomes.'[181]

As Sanger points out, the conditions on board the steamers appear to have deteriorated for the sealers after the transition to steam,[182] and this deterioration can be seen in the food and accommodations.

The basic food remained the same after steamers entered the industry. Hard bread and butter and molasses tea remained plentiful. However, the number of men each cook was required to feed increased with the larger steamers. In 1879, for example, the *Bear* went to the ice with 258 men on board; one cook and a steward were assigned to the captain and officers, while only three cooks were assigned to the remainder of the men.[183] When one remembers that there was a cook on small vessels, one can see that by this time, cooking had ceased to be considered a priority. Salt pork and duff or fish and brewis seem to have been provided on Tuesdays, Thursdays, Fridays and Sundays.[184] David Moore Lindsay wrote about his trip on the S.S. *Aurora* in 1884:

Figure 17 - Men hauling pelts to a pan or, possibly, to the ship (the angle of the photograph suggests the latter). Probably a photograph by Holloway or Holloway Studio.

> The owners did not overfeed the men on these trips, providing them with sea biscuits and pinnacle tea chiefly, pork and duff being served only three days a week and salt fish on Fridays.[185]

Nevertheless, as was observed by Jukes, the men continued to supplement their diet with seal meat. However, the cooking facilities could not cope with the larger crews, and cooking became a problem. On the one hand, the presence of coal meant that there was no shortage of fuel, and the small stoves in the sleeping area could be used by some, while the galley was accessible at night. However, when crews began to number up to and over 300 men, it became literally impossible for any except the most persistent to do much cooking. Therefore, after a 12-hour-plus day, the men drank their tea and ate the hard bread and butter and began their watch duty. One or two watches loaded pelts and shifted coal and ashes, while the others slept, to be awakened at midnight to do their four-hour stint. However, some men ate the raw seal meat and drank the blood. Lindsay again supports Jukes' main contention that the men lived fairly well when among the seals, except that now they were eating more raw meat:

> When a ship was amongst the whitecoats, as the young seals are called, the crew lived well, as they ate the livers, hearts and flippers of the seals. The men carried a supply of livers and hearts in their belts and ate them frozen or cooked as opportunity afforded. It is easy to see how little cooking can be done for a crew of three hundred men on a small ship. I have often seen a man tie a cord to a liver and drop it into a pot of tea sitting on the galley

stove, drawing it out when warmed up or when the owner of the pot came for his tea.[186]

Similarly, a story by H. F. Shortis, published in the *Evening Telegram*, tells of an incident that occurred c.1880 when John McCarthy, a master watch with Capt. William Adams in the Dundee steamer *Arctic*, was searching for seals. While returning to his vessel, he sat down "on a large pinnacle to eat some bread and a seal's heart."[187] Obviously, some sealers were prepared to take advantage of the food around them, and this fact compensated for the inadequacy of the food carried on the steamer.

The deterioration in food seems to have occurred in three areas. An insufficient number of cooks were hired by the companies. Unlike the smaller sailing ships, where men could have their own fires on deck and supplement the regular food, there were no reports that this ever happened on the steamers. While the galley seems to have been available, it was not possible for two or three hundred men to gain quick access to it with so little time to spare at night. A further significant problem was the fact that the men were increasingly required, as the steamers became more powerful, to pan the pelts and work away from the vessel from before dawn to well after dark. Therefore, the old practice of hauling the individual 'tows' of pelts to the vessel, as was the case in the days of sail, ceased. The practice, observed by Jukes, of snatching a cup of hot tea and some bread and butter during these trips to the vessel came to an end. That was the most unfortunate development of all. Nonetheless, only on one occasion were there reports of the poor quality of the food. It seems that the crew of the S.S. *Ranger*, commanded by Joseph Barbour and supplied by Stewarts, complained about the food they were given on the 1886 voyage. The *Evening Telegram* (later accused by Barbour of supporting another firm's hard tack) reported the incident as follows:

> A Deputation from the crew of the S.S. *Ranger* called at our office yesterday for the purpose of lodging a complaint against the owners of that ship touching the manner in which she was provisioned for the first trip. The poor fellows looked as if they were suffering for want of nutritious food, and loudly lamented the hard fate that compelled them to proceed on the second voyage. The spokesman said (and these are the exact words used by him) 'the flour we were compelled to eat was *sour*, the pork *windy*, the molasses *salty*, the butter *smitchy* and the tea *fousty*.'[188]

However, while Sanger is correct is concluding that food deteriorated in the age of the steamers, neither the quality or quantity of food was ever considered a major problem by the sealers.

Accommodations

In general, there was probably little change throughout the century in the living conditions on board sealing vessels, although they improved a little in the steamers with the use of small coal stoves and the heat from steam pipes. However, by modern standards, they were terrible, and even contemporary reports—the few that exist—call them deplorable. Lindsay made this observation while on board the S.S. *Aurora* in 1884:

> I looked into the 'tween decks and saw a horrible mess. The bunks were full of men, many playing cards, as each bunk held four. They must have stifled. For light, lamps burning seal oil were used, and the reek coming from the main hatch would almost have suggested fire.[189]

In the age of steam, a pattern for crude accommodations was established. The men slept in temporarily constructed wooden bunks between decks in an area where there was often insufficient standing room. The bunks usually accommodated four men, who brought their own mattresses; those from near St. John's usually brought brin bags of hay, while those who came from a greater distance generally brought empty bags to St. John's and had them filled with wood shavings from the cooperages in the city.[190]

After the *Greenland* disaster in 1898, the conditions under which the sealers laboured came to the attention of the local media and were widely reported during that spring. One such report reads as follows:

> You are out on the ice all day tugging at the seals and, without any proper food. Now a man can endure that, and even endure it day after day, for a succession of days, if he is strong and well, and can look forward, after the day's work is done to a good square meal and a quiet bunk. But the poor sealer has none of these. The steamer in which he is pigged together with some two or three hundred other men *has no sleeping berths whatever for the men*. There are a few rough shelves without bedclothes or bedding of any kind, about enough for one-third of the men on board. If a man wants a bed he must bring a bag of shavings on board to lie on, which he has borrowed or begged from some charitable neighbour on shore. Then he must take *his chance with the other men* for permission to lie in or on one of those rough bunks ... If a man is young and hearty he can bear the strain, and survive it. But *God help him if he takes sick at the ice and does not want to die*. An ox or a cow on board an Allen cattle ship has a thousand chances to one of recovery as compared with him, and he knows beforehand that illness on the voyage means a box of dead meat sure.[191]

In some respects, the steel sides of the later steamers made matters worse because the condensation on the inside walls dripped onto the bunks. However, at least one iron clad had improved accommodations, according to the report welcoming it to the seal fishery in 1901:

> The S.S. *Virginia Lake*, will be one of the best fitted steamers for accommodation of her crew that will leave this port for the seal fishery. She is fitted fore and aft with berths, which will never be disturbed even if she was filled with seals from stem to stern. There will be no leaving berths and sleeping in the fat, for all will be as comfortable as it is possible to make them, and 250 men will hardly be noticed on board. There are six outlets to the deck and everything has been provided for the men's comfort.[192]

However, like the situation with food, fishermen did not consider the quality of accommodations to be a sufficient problem to deter them from the fishery.

Discipline

There is very little discussion of disciplinary matters in the literature. By and large, in the earlier period, when communities provided the crews, discipline would primarily be imposed by peer pressure, and any unfair application of disciplinary measures from above would be frowned upon by the community. However, an old ballad (two versions of which exist), entitled *The Song of the John Martin* or *The John Martin*, indicates that there was sometimes violence on board these vessels.

> Come all ye jolly fishermen agoing to the ice,
> Oh, beware of the *John Martin* and don't go in her twice.
> For I was in her last spring and I'll go in her no more.
> If I cannot get a better berth, I'd rather stay ashore.
>
> It was down in the forecastle that a hell of a row arose,
> It was all about the boiling of a kettle full of brewse.
> The skipper he came forward and he swore once or twice,
> He took the kettle by the hangers and he threw it on the ice;
> I never felt so scalded since the day that I was born,
> When I saw my little piper and it floating off astern.
>
> Now when we got into the jam the swoiles were very thick,
> And the skipper he came forward with a junk of a stick,
> He said you burned all my lasses and you roasted all my pork,
> So now you mortal sons o'guns I'm going to make you work.[193]

With the coming of steam and the arrival of Scottish captains on the scene, a stricter, more 'British' type of discipline was imposed, often combined with violence. Frank Bullen, writing of his voyages as a youth on board British vessels in the 1870s, described frequent violence by the officers as a common means of maintaining order. In describing the non-British crew of one vessel he sailed in, he mentions that the captain was a genial soul and a pleasure to work for, "But I am afraid he would have had a bad time of it with a crew of Britishers. They appreciate a tight hand and are quick to take advantage of anything like easy-going on the part of their officers."[194]

Lindsay described an incident on board the Dundee whaler/sealer S.S. *Aurora* in 1884, when the men complained about the quality of the food and approached Capt. James Fairweather:

> Hearing a noise on deck, I went up. On the poop a lot of duffs were lying about like 64 lb. shot. A crowd of angry men could be seen on the main deck and facing them was the Captain. A big Newfoundlander came up the steps and, breaking a duff in two, held it up and asked the Captain to look at it. It was an awkward moment and called for immediate action. But the Captain was a man of action, so he planted a blow between the man's eyes and asked him to look at that; the man dropped back dazed and the trouble came to an end at once.[195]

However, Fairweather was probably overly violent, because in 1886, he was convicted of beating up the chief engineer of the S.S. *Esquimaux*, which was under the command of Capt. Henry Dawe and tied up to a wharf in St. John's. Captains Guy (from Dundee), Fairweather and Dawe were drinking in Dawe's cabin until late at night, when Dawe ordered the engineer to build up steam and take his visitors across the harbour to their vessels. The engineer replied that it was impossible to do so at short notice, and when Fairweather and Guy intervened in the argument, he pointed out that he was not under their command. He was thereupon assaulted viciously, according to the *Evening Telegram*, which took exception to the attack and commented on it as follows:

> On Saturday evening a reckless and unprovoked assault on a hard-working and peaceably disposed man, Mr. Page, chief engineer of the sealing steamer *Esquimaux*, was perpetrated by two individuals who, no doubt, regard themselves as the superiors of their unoffending victim, and who, with a manly regard for their relative positions, took advantage of the circumstance to inflict their abuse under the belief that they were too far above the common herd of transgressors to be held accountable for their misdeeds.[196]

Fairweather was fined the small sum of $4, and the case against Guy was dismissed.[197]

In another incident, in 1897, the S.S. *Walrus* was badly damaged while engaged in the seal fishery, and the crew, concluding that the vessel could not be saved, signed a petition asking the captain, Alpheus Barbour, to abandon ship:

> He, however, dealt summarily with the incipient meeting, promptly knocked down the ringleader with a blow he has not yet got over, and ordered the men to return to their duty and not be so faint-hearted.[198]

The ship was saved, but got no seals.

In England's book on his voyage on the S.S. *Terra Nova* in 1922, the members of the sealing crew related stories of violent captains, and spoke admiringly of the captain whom they considered the most violent—Capt. Hickson [Jackman?]. In addition, Capt. Abram Kean, on that same voyage, explained that violence as a form of disciplinary action was common in the earlier times.[199]

Violence on the part of captains, like the state of the accommodation, the quality of the food or the harsh and long hours of work, was not usually considered a major problem by the sealing crews. The 'manus' (or refusal to work), which is discussed in Chapter 6, was the common form of non-violent protest engaged in by the men, and it was usually caused by an overly long voyage and no seals. The major problems, from the point of view of the fishermen/ice hunters, concerned the question of a fair share in the voyage and the issue of safety.

Landsmen

Although the men who went "to the ice" in the spring are the principal concern of this study, one must not overlook the men (and women and children)—the 'landsmen'—who participated in the harvest of this resource from shore, whenever the opportunity arose. In certain areas along the coasts of the island and Labrador, a few fishermen caught migrating seals in nets—the winter fishery—but this little-reported-upon activity did not attract much attention or make much difference to the economy. However, the landsmen's annual hunt, on foot and in boat, was fraught with danger, while offering opportunities to make a considerable profit.

The landsmen's seal fishery was carried on in an ad hoc manner, depending on where the winds and tides drove the seal herds; it did not have the reliability of the small net fishery, but it could and did provide a bonanza on occasion in certain areas. When the ice brought seals close to shore near relatively heavily settled harbours and coves, the people in the vicinity were very adept at making the best of the situation. One report retrospectively recorded that in 1839, "The women of King's Cove hauled tows of whitecoats as well as the men; 17,000 seals were hauled by the people of King's Cove."[200] In 1843, about 40,000 pelts were hauled ashore in Bonavista Bay, "where even the women took a hand in the work, and some of them made as high as £100 for their own rope."[201] In 1862, part of the seal herd drifted into Green Bay, and during this 'Green Bay spring', "The women and dogs made ten pounds a man."[202] Twillingate also had a similar experience that year, and a local newspaper reported: "Men, women and children have been out day after day. Seal pelts are seen in every direction, piled up in thousands here and thousands there. The old people say that in 1824 it was the same."[203] This report also describes a disaster:

> I can't help telling you about the seals which made their appearance here about March 12th. The first were found about eight miles distant, northeast of Gull Island. I went off for the first time on St. Patrick's [Day] and about six miles off Gull Island, I could stand on a clump of ice and see thousands upon thousands of white coats and could hear nothing but their bawling. I took four in tow. Being so far off I found it killing work to drag them over the rafted ice. It was afternoon and I had a narrow escape from death. A man named George King with his two sons, took their tows a short distance further out than myself. I started homeward a few minutes before them and had not got far when the ice parted between us. I saw the wind hauling eastwardly and I dropped two of my pelts and I ran to get in with the crowds who were ahead of me. We all dragged our seals within a mile of Gull Island, but then every man had to drop his load and run for his life. The wind being E.S.E., the Gull Island split the ice running up the Bay and opened a space of water of about one hundred yards. It was now every man for himself, I ran over slob, I know not how and got on a pan of ice in the lake, with twenty-one men, one woman and several dogs. By paddling with our gaffs we with difficulty gained the ice connected with Long Point, two miles north of Gull Island, and so got safely on shore. The people on shore were very anxious for our safety, fearing our being carried off, which we should have been if the wind veered a little more. George King and his sons were not so fortunate as the rest of us. Two days after, they were found off Western Head, the sons quite dead and King nearly gone....
>
> In looking around I am reminded of the days I spent at the California gold digging. Great excitement prevails and this will be something to talk about for the next fifty years to come. Knives are in great demand here—none to be bought in the harbour. One man offered me a pelt worth twelve shillings for the only knife I had.

In fact, over 100,000 seals were killed in 1862 by landsmen near Twillingate and Fogo. That these could be very lucrative occasions is obvious; but the fact that it was a dangerous activity is apparent as well; and finally, it is interesting that women were ready participants, as is indicated also in the report of five years later:

> The melancholy tidings of the loss of ten women and two men off Catalina and Ragged Harbour, while in search of seals, cast a sad shade upon the cheering accounts that have reached town respecting the prospects of the

Seal Fishery. One of the recent telegrams states that the inhabitants of Bonavista and King's Cove were taking seals by the thousand; another has it that at Keels, Bird Island Cove, Catalina and Trinity very large quantities have been landed.[204]

Another bonanza struck parts of the northeast coast in 1872 which became known as 'the year of the red jackets':

> This year a most unusual event occurred. Those who remained ashore gathered the richest spoils from the seal fishery. The floes on which the young seals were whelped were driven, by the continuous eastern gales, quite out of their customary track, upon the shores and into the great northern inlets. Thus the whitecoats were brought within reach of the people along the shore, without any effort on their part; and they were not slow to avail themselves of their good fortune. The whole population along these fortunate shores turned out—men, women and children—and made an onslaught on the seals. Any weapon that came to hand was seized, poker and tongs, hatchets and tomahawks. The men killed and sculped, the women and children towed the fatty piles ashore, old-bedridden men, under the excitement of the moment, forgot their ailments and crawled out on the ice; rheumatic subjects threw away their crutches, and rushed into the prey; women 'forgot their sucking children' and left them for the time being to squeal in the cradle, while without any 'bowels of compassion' they slaughtered the moaning young of the harps. The harvest had to be gathered at once, for a change of wind would drive the booty out to sea. The ice around the shores became speedily like a reeking slaughter-house and the air was filled with 'the shivering seals' low moans. The scene was quite picturesque. Among our young 'outport' ladies, red jackets, ponderous chignons, and extensive crinolines, are the rage at present. At a distance these active damsels looked like a small army of soldiers engaged in battle—their 'thin red lines' flashing far under the rays of the sun along the glittering ice-fields, their crinolines tucked up and their stout arms bare to the shoulders and stained with blood. Truly 'those who tarried at home divided the spoil.' It is calculated that at least 100,000 seals were hauled ashore in this way. Some of the families who were pretty numerous, got as many as 150 seals; others 100 or 50 each. In three or four days, seals to the value of $300,000 were taken. These poor people, by this lucky windfall, are trans formed into millionaires..... It is not unlikely that this year will be spoken of by coming generations as 'de year of de red jackets,' the red-coated women having got the biggest prizes.[205]

Chafe reported that in 1880 that there were, "seals taken by landsmen in great numbers from Cape John to Cape Race..." and that 1881 was "another Green Bay Spring."[206]

This landsmen's seal fishery was a haphazard affair, but people in Bonavista and Notre Dame Bays, especially living near the headlands and on the offshore islands, could generally count on it. A Fogo correspondent to a St. John's newspaper wrote in 1880: "The seals have paid us a visit, but very little has yet been done with them in consequence of the very severe weather. Should the wind veer north we shall do well. If west, good-bye to the seals."[207] Apparently, the wind veered to the north and it was assumed that the Fogo fishermen were able to take advantage of the situation. A writer from Twillingate reported that same spring: "Good work has been done with seals about Change Islands and

Fogo, but not without great suffering from frost and cold."[208] A Reverend J. Embree reported on a similar event, which occurred in Bonavista in the late 1880s while he was stationed in the community, and his report illustrates that it was a highly dangerous occasion when the seals arrived in such numbers and presented such a tempting prize; again, the participation of women was noted:

> A general rush among all classes made it certain something most unusual had happened to disturb the peace of our quiet town. All who could carry a gaff, a tow rope and knife, and had strength to tow one or more seals to land, made for the ice. Then followed women and children, with baskets and kettles containing bread, cake, tea and everything in the way of eatables that could be gathered in a hurry, to meet the men and big boys when they came back with their tows,—some going after them on the ice, the more timid waiting on the shore to give the hungry, tired men food, and take a turn with them on the tow-line. A town, with 4,000 or 5,000 inhabitants astir at such a time, gives one a knowledge of how people can be excited.
>
> But all who went out that day did not return. Two young men were missing after night-fall. One disappeared through the ice. His companion being only a few steps before him, looking back to see how he was getting along, saw nothing but a great field of ice. No trace of the missing man was ever found. He slipped between the loose ice and was gone. The other exhausted, lay down and died, being alone and the last coming home. He was found next morning, in sight of his father's house, which was filled with bitter grief....
>
> Some years ago, at Twillingate, what is still known as the great haul of seals took place. For more than a week, close to the land, the people killed them and hauled them 'heaps on heaps.' One Englishman went out with the crowds, but when he saw the young seals with their mothers together, and heard the cry of the young,—which is so much like the cry of an infant as can be found in creation, so hundreds testify, and can be heard a great distance,—looking in the eyes of the little creature his heart failed him and his knife has not been stained with blood from that day to this.
>
> Braver was a young lady, a physician's daughter, who went out with the men this past spring, with tow-rope and knife, five or six miles out on the ocean from the nearest land, killed her seal and towed it home in triumph,—but not the first of womankind to do so. One woman at White Bay last spring, was reported to have thirty seals to her rope, which would be worth $60.... "Two men on the other side of the 'rent,'" [men stranded on the other side of the opening in the ice.] was the startling news, one fine evening last March.
>
> ... The strength of the current parts the ice, leaving what is called standing ice next the shore, making in a short time a river impassable, but so far from land that no help can be rendered by boat, as it is not possible to get boats so far over rough ice in a short time. Men on the inside of the rent dare not go back to help those in trouble. Indeed one cannot help the others. If the rent is narrow, and small pieces of ice can be found, they get on these and ferry over with the gaffs. Men have thrown their seals in the water, the fat floating lightly, and making a jump for life have saved themselves by one step on the seals and the other on the standing ice. Those on the outside soon find out their danger and run, leaving their tow of seals, knowing that their fate may be a hard one. What must be done? Night is just at hand, if not already come. Home so near, and yet so far away. If a sealing steamship or sailing ship be in the distance oceanward, the men must try and get on board. If there is a point of land they can reach, they must try and do so. If an island,

where men are stationed seal hunting, try that. If none of these, they make a bed of seal pelts, if they can get them, lie down on them and cover themselves up with them, waiting until morning. The wind may close the ice to land next day. When all these fail, cold and hunger soon overcome the sufferer or sufferers and many sleep their last sleep far from land and home, so far that nothing but frozen ocean could be seen. One poor fellow was picked up at Tilt Cove, a few winters ago, whose home was many miles north of that place. He had worn his feet almost bare in the struggle for life, but failed, then laid down alone, miles from human habitation, with the wild storms of winter about him, and died.[209]

Seals were plentiful close to shore all along the coast of Bonavista Bay in 1899, with hundreds being hauled ashore at King's Cove, Keels and Tickle Cove, and 3,000 taken in the area from Open Hall to King's Cove from a herd located between four and seven miles off shore.[210] By late April, the estimated number taken in that area had risen to 7,000.[211] In fact, the people of King's Cove received a bonus that year, when the S.S. *Iceland*, under Captain Darius Blandford, became jammed in the ice off that community. The cargo of fat began to run and the oil was pumped out with the bilge, where it floated, trapped in the leads in the ice floe. The residents came out and skimmed off from 12 to 20 gallons each.[212]

However, not only were the landsmen at the mercy of the wind, tide and nature in general; they were affected by their religious beliefs as well. In Musgrave Harbour, a centre of fervent Methodists, the seals arrived on a weekend. It was reported on 3 May that about 3,000 had been taken by people on the Straight Shore (the name given to the stretch of coast without suitable harbours) and that on Sunday, 14 April, thousands more had been "driven in right at their doors, but not a man stirred to desecrate the day by killing. During the night, however, the wind changed and when [Monday] morning dawned the ice was scattered, seals had taken to the water, and only a few were secured in boats."[213]

White Bay, at the base of the Great Northern Peninsula, was not visited by the whelping ice nearly as often as points and islands to the eastward, like Fogo, Twillingate and Bonavista, for example. Nevertheless, it too had its lucrative and welcome visits occasionally, and 1901 was one such occasion:

> Residents of White Bay, from Western Arm on the one side, to Jackson's Cove on the other, have reaped a rich harvest from the ice floe this spring. Some men have taken as high as two hundred seals and there were thousands on the ice. On the south side men went off daily and hauled their tows ashore, killing no more than they could take, while had they panned, thousands would have been taken instead of hundreds, as the whole bay from Hauling Point to the bottom was covered with seals. The first traders down there this spring will do well with skins and oil.[214]

The observer implies that the ice did not remain in the area for long, and it is easy to see why the fishermen killed only what they could haul ashore immediately. They were wise enough to know that time and effort spent killing and panning could all be lost if the wind changed and took the ice out of the bay before they had a chance to haul their pelts ashore. These decisions were always judgement calls, and on this occasion, the fishermen made the correct one. White

Bay was also fortunate the following year, when once again, "thousands of seals were driven in ... [and] everyone capable of hauling a seal," was out, with some getting as many as one hundred pelts.[215] White Bay, being the shape of a long slender "V", was ideal for landsmen if the ice drifted in with seals because as the ice entered the bay, it tightened, preventing the older seals from getting into the water. (It was this very feature that made entering White Bay such a risk for vessels; a vessel could follow the seals and loose ice into the bay, and if an on-shore wind continued, it could find itself jammed and maybe even crushed. The number of vessels that were caught in White Bay for weeks at a time are legend. During the spring in question here, 1902, the S.S. *Kite* was jammed there from 14 to 27 April, and even after getting clear, had to steam north along the peninsula as far as Belle Isle before turning for home.[216] One of the oldest published Newfoundland sealing songs describes the small sailing vessel from Carbonear, the *St. Patrick*, under Capt. Tom Casey, that was stuck in White Bay with its crew of 28 men "until the last of May."[217])

There is no doubt that the landsmen's seal fishery continued to contribute to the income of the cod fishermen (and later, to the income of the lumbermen) of the northeast coast. The seal provided meat for the table and food for the dog teams which were required on the many islands and in the many other places where horses had never thrived because of the scarcity of fodder; sealskins were used to make boots which met an obvious need; and finally, the seal fat could be sold for cash. In fact, the role of this fishery in that area became proportionately greater in the early twentieth century.For example, another prosperous year for landsmen, this time farther North, was the 1914 season, when landsmen living between Cape Norman and White Bay took 60,000 seals.[218]

The phenomenon of unusually large numbers of seals turning up on people's doorsteps occurred fairly frequently, if intermittently and at different locations, throughout the period under study. In addition to these bonanzas, small pockets of harps were often found by fishermen hunting from the shore, who were delighted to kill even one or two. However, by the end of the period, it is obvious that the seal herds were being depleted and the opportunities for landsmen (and net fishermen) to take a share of the seals had declined. Dr. Wilfred Grenfell wrote from St. Anthony in 1912 and complained about the depletion of this resource in an article entitled "Destruction of Seals and effects on Northern Outports":

> We have just had a delightful visit from Capt. Alpheus Barbour and his crew in the *Neptune*, and we fully realize what it means in St. John's and the one or two outport towns from which the sealing crews come, when a steamer comes home with her bunting up, and her rockets announcing she has been successful. We rejoice with those that rejoice, and such must be the result of the circulation of so much money as comes to St. John's when over a quarter of a million seal pelts are brought in. But we can't help seeing the other side—and that has a very serious aspect indeed. As Capt. Alpheus Barbour lay at the ice edge, men from all around our harbour had to actually go to his ship asking for a carcass, a carcass for dog food; yes, and for food for their families, for meat is a luxury here in the winter. To me this was a pitiful sight. There was a time when the harbour could have given them all the carcasses they might need, and when we could have sold them a load of fat to boot; There was a time when traders and merchants found it worthwhile

to supply their dealers with seal nets for the return they brought from the local fishery. The best proof that the local seal fishing is no longer remunerative, is in the fact that this is no longer done. The few men who still fish for seals get little or nothing in return for their outlay, beyond the fact that they do have some dog food and some skin boots—where others must go without because their money cannot even buy any. It has struck me very forcibly in another way also. Every year now as I leave for Labrador [in the spring] more and more of our friends on the coast beg me to put them down for a pair of skin boots, if I can. I have been these past two or three years regularly employing Labrador women to make me boots for people from Conche to Kirpon, and at the mill in Canada Bay, simply because we must otherwise send to St. John's for skins. There can be no denying this was not the case even so recently as eighteen years ago. But beyond this there are places that when I came here first were nice little settlements, where as we journeyed along our lonely rounds one found bright hearths, and heard merry voices, and got a cheerful welcome; just such a place was North West Point, in Hare Bay. Now only deserted, tumble down huts greet one there, for there are now no longer any seals to depend on, and one hurries past the spot, which to its loneliness has thus added a sense of graveyard desolation. The seal fishery along the coast and at the outport was, so it seems to me, just that annex to the cod fishery, or that compliment of it, that nature provided to supply the needs of the people when that somewhat fickle industry failed. If the sea is to be held responsible for the entire sustenance of the distant outport population, and if settlers are to be encouraged to spread and settle the outer coves of Newfoundland, then it has always seemed to me that the seals are essential as a winter balance against chronic semi-starvation. Then the meat and fat are even a "dernier ressort" when starvation threatens. If these seal herds, that are now so seriously threatened, are not permitted to recuperate what they have already lost in past years, the day is not far ahead when life will be practically impossible in the far north, unless mines and factories can be started and kept at work. There are a great number of the poorer people now living as close to the food limit as they can go; we doctors see many hungry and half-starved children, who never see milk nor half enough fats in any form to produce healthy, well developed girls and boys. Nor does there appear any likelihood at present of any other source of income being provided if the seals are exterminated. They are already almost gone from the shore and bays. It is a terrible thing to believe, and a fact one dare not state publicly, if one were not fully convinced of it. But the fact is that where the seals are all gone a large part of the population will have to follow them and go too.... It is the disappearance of the harp seals that is making it so increasingly difficult for the cod fishermen to live.[219]

Dr. Grenfell was the first to note in great detail what must have been observed by many people on the north, west, northeast and southwest coasts—the decline in the seal resource. When one recalls that fishermen and their families moved up along the northeast coast in the eighteenth century to have better access to this sea mammal, then one has to appreciate the significance to the local population of its decline. Therefore, the commercial sealers, from the southern harbours were, by the beginning of the World War I, in the process of depleting the seal resource to such an extent, that they were inadvertently destroying the local economies of the small northern outports. This development was to have tragic results by the 1920s and 30s as illness and malnutrition became rife in the isolated northern outports.

Summary

The men who went to the ice contributed enormously to the colony's economy through their exploitation of the arctic seal resources that swept along the coast annually. Their ability to survive on the only two staples available—cod and seals—enabled them to establish a relatively stable economy with an expanding population. And their prosecution of the cod fishery on the coast of Labrador maintained that continental extension as part of the new self-governing dominion they created.

The living and working conditions were difficult, as contemporaries have reported, although they were not that much different from living and working conditions in general in the colony. However, the captains' positions evolved throughout the period more than any others. In the beginning, they were the best sealers from among experienced sealing families and numbered in the hundreds, with their numbers being replenished from among the ranks of their own families, relatives and friends. The introduction of steam sharply reduced the number of captains' positions available and raised the rewards for those few who were lucky and successful. A few early captains made the transition from sail to steam, but most did not, and this class of sailer captains quickly died out. Captains' vacancies came to be filled by those who were experienced second hands, and ultimately, these enviable positions became the possession of certain families who were fortunate in having talented young men who could succeed their fathers and uncles and become part of this highly paid and highly skilled circle of individuals. Furthermore, the connections which were established between St. John's merchants and the northern outports in the 1840s and 1850s resulted in the increased dependence on captains from this area to the disadvantage of the men from Conception Bay (see Table 4.7). Throughout the age of steam, successful captains became wealthy and powerful figures. The common ice hunters also declined in numbers, but their incomes, unlike those of the captains, declined as well.

The overall conditions in the seal fishery deteriorated towards the latter part of the period under study as oil prices declined, the outport industry collapsed and the herds dwindled. While the whole economy suffered from the periodic dips in production and, especially, the overall decline in the seal fishery after the 1860s, the men and their families suffered enormously from the disasters to which the hunt/fishery exposed them. The place of these disasters in the history of this industry is such that they deserve their own study.

NOTES

1. Ryan and Small, *Haulin' Rope and Gaff*, p. 17.
2. Ryan and Small, *Haulin' Rope and Gaff*, pp. 18-9. The present writer will follow the example of the nineteenth century writers and use 'ice hunters' and 'sealers' interchangeably, bearing in mind that these terms were used for vessels as well as men.
3. See Table Intro.12. 'Bonavista' most likely refers to Bonavista Bay.

4   See Table Intro.13. It is most likely that, as above, 'Bonavista' and 'Trinity' refer to the Bays and not the small communities by these names.
5   Mosdell, *Chafe*, p. 20.
6   Mosdell, *Chafe*, p. 20. This captain's son, Captain Abram, was also a sealing captain, and the grandson, Captain Bob, was the sealing captain hired by Robert Peary as captain and navigator on the latter's Arctic exploration voyages.
7   Mosdell, *Chafe*, p. 31.
8   Mosdell, *Chafe*, p. 31. Please note that Chafe was not consistent in his use of the singular and plural in this list of names.
9   Mosdell, *Chafe*, p. 32. See also a letter from Captain Henry Thomey to the *Evening Telegram*, 30 January 1909.
10 Mosdell, *Chafe*, p. 22.
11 Mosdell, *Chafe*, p. 21.
12 Little, "Collective Action," p. 155.
13 Ryan and Small, *Haulin' Rope and Gaff*, p. 16. Originally published in James Murphy, *Songs Sung by Old Time Sealers of Many Years Ago* (St. John's, 1925). (Only short extracts have been printed here.) The *St. Patrick* cleared for the ice during the 1830s under a succession of captains. Captain T. Casey took command of it in 1838, when it was registered as having 94 tons burthen (old measurement), carried 27 men and was supplied by Rennie, Stewart and Company. It was commanded by a Captain Dwyer in 1840, then by Captain Casey again in 1841. In 1843, Lawrence O'Brien and Company began to supply the vessel and were obviously the owners. Other captains commanded it until 1845, when Casey once again became the captain; by now, its tonnage had been re-classified at 79 tons (new measurement), and it carried a crew of 26 men. Casey continued to command it until at least the end of the 1848 season (records are incomplete). See the lists of departures for the ice in the various local newspapers.
14 See Tables 3.2 and 3.3 for complete information.
15 One figure for 1836 is missing, which makes it impossible to find the number of sealers for that year.
16 Much of the following discussion has been taken from Little, "Collective Action," pp. 7-66.
17 Ryan, "Abstract," p. 118.
18 'Bawn' was a rocky surface on which fish was spread to dry. On the rocky, treeless islands off the Labrador coast it was common to dry fish on the bawn instead of trying to procure timber to build flakes. See *DNE*, p. 31.
19 Little, "Collective Action," pp. 154-5. 'Berth' was sometimes spelled 'birth' in the contemporary records. See *DNE*, pp. 40-1.
20 James Murphy, *The Old Sealing Days*, (First published in the *Evening Herald*, February—March, 1916: Reprinted as a "Project of the Newfoundland Archives," 1971, PANL), p. 16.
21 *Standard*, 1 March 1884. The correspondent writes: "This [sailing] fleet has unfortunately so dwindled away that this year it numbers no more than five vessels. Two others belonging to Messrs. John Munn & Co. ... will sail, the one from Bay Roberts, the other from Brigus."
22 Murphy, *Sealing Days*, p. 20.
23 Capt. Arthur Jackman (1843-1907) became a legend in his own time. He was big, impressive, given to heavy drinking and violence, but very successful and usually generous and popular. On one occasion, he mangled his finger in the winch, looked at it, called for an axe, laid his finger on the rail and ordered the engineer to cut it

off. When the engineer refused, Jackman seized the axe and chopped the finger off, himself. He refused to cease killing seals on Sundays, and after the law was passed making it illegal to do so, he was charged with breaking this law on his last trip in 1906 and fined. In 1922, on board the S.S. *Terra Nova*, George Allan England recorded a number of well-known stories about Jackman but inexplicably referred to him as Captain Hickson. See *The Greatest Hunt in the World*, pp. 143-8. See also Ryan, *Seals and Sealers*; James Murphy, *The Old Sealing Days*; and Mosdell, *Chafe*. My thanks to the late Dr. George Story for allowing me to see his, then unpublished, *DCB* entry on Jackman.

24 Prowse, *Newfoundland*, p. 453.
25 *Evening Telegram*, 5 January 1886. It appears that Thomas predeceased his father.
26 See Kean, *Old and Young*.
27 When Capt. James Murphy died in 1871 the *Royal Gazette*, 7 February 1871, wrote that he "had been accessory to the death of more seals than any other Newfoundlander of modern times ... 240,000 seals, each worth on average, four dollars. Notwithstanding his murderous pursuits, Capt. Murphy is described as a kind-hearted man, gentle in manner and ever ready to help a poor brother. The seals may hold a jubilee on his decease. Many a poor mother seal he left cubless. Were the lower animals immortal, it would be a fearful thing for such a man to be confronted with the shades of all the seals he had slaughtered."
28 In 1874, Capt. Issac Bartlett in the S.S. *Tigress*, rescued the crew of the Hall Arctic expedition that had been adrift on the ice since the previous autumn, following the sinking of their vessel, the S.S. *Polaris* (see *Times*, 28 March 1877).
29 Capt. Henry Thomey (1820-1911), was primarily a sailing captain, first for Ridleys and then for Munns. He averaged 4,000 pelts a year for 40 years, never had a serious accident at the ice and "never lost a man" (see *Evening Telegram*, 9 January 1911).
30 Capt. Nicholas Hanrahan was born in Carbonear and commanded Rorke's sealing ships before moving to Harbour Grace. He died in the latter town in 1898 at the age of 80 (see *Daily News*, 10 November 1898).
31 George W. Jeffers to Dr. Cater W. Andrews, 5 January 1970. Biographical information about his ancestor, Capt. Joseph Jeffers (1830-87). CWA 361 III. This information has been supplemented by newspapers and by Mosdell, *Chafe*.
32 Capt. Edward Murphy was the brother of Capt. James Murphy and succeeded his brother James as captain of the S.S. *Mastiff* (see *Daily News*, 27 February 1896).
33 Kean, *Old and Young*, p. 43. See Tables 4.6 to 4.9 for more information on the careers of captains and ships during the early age of steam.
34 *Evening Telegram*, 16 April 1894.
35 Mosdell, *Chafe*, pp. 87-96.
36 *Daily News*, 11 April 1899.
37 *Evening Telegram*, 15 February 1910.
38 *Newfoundlander*, 13 May 1847.
39 *Newfoundlander*, 8 April 1858.
40 *Newfoundlander*, 27 May 1858.
41 *Newfoundlander*, 7 March 1873.
42 *Newfoundlander*, 5 March 1875.
43 *Evening Telegram*, 7 March 1882.
44 *Evening Telegram*, 7 March 1882.

45 *Evening Herald*, 7 March 1892.
46 *Evening Telegram*, 27 February 1894.
47 *Evening Telegram*, 1 March 1893.
48 *Evening Telegram*, 27 April 1897.
49 *Newfoundlander*, 12 March 1875.
50 *Evening Telegram*, 4 March and 17 April 1893.
51 *Evening Herald*, 13 March 1900.
52 *Evening Herald*, 23 March 1901. The presence of 'doctors' on board sealing ships will be discussed later.
53 *Evening Herald*, 25 March 1901.
54 *Evening Herald*, 25 March 1901. See also *Daily News*, 25 November 1901, for a fairly extensive interview with Mrs. Levache.
55 Ryan, *Seals and Sealers*, p. 73.
56 *Evening Telegram*, 16 March 1893.
57 *Evening Telegram*, 13 March 1895.
58 *Evening Telegram*, 7 March 1899.
59 *Evening Herald*, 13 April 1901.
60 *Evening Telegram*, 9 March 1899.
61 *Evening Telegram*, 10 March 1906.
62 *Evening Telegram*, 15 March 1906.
63 *Evening Telegram*, 2 April 1906.
64 *Evening Telegram*, 7 April 1906.
65 *Evening Telegram*, 10 April 1907.
66 *Evening Telegram*, 16 April 1908.
67 *Evening Telegram*, 17 April 1911.
68 *Evening Telegram*, 1 June 1911.
69 For example, see *Evening Telegram*, 7 March 1914.
70 *Evening Telegram*, 26 April 1913.
71 Rev. Lewis Amadeus Anspach (1770-1823) was born in Switzerland and came to Newfoundland in 1799 to take up the position of superintendent of a new grammar school in St. John's. He fulfilled his duties for several years and was then appointed a missionary for the Society for the Propagation of the Gospel. He was also appointed Justice of the Peace for Conception Bay and arrived in Harbour Grace in 1803, where he resided, until his return to England in 1812. He wrote several works, but the best known and most useful to scholars of Newfoundland history was his *A History of the Island of Newfoundland* (London, 1819), from which the following discussion is taken (see pp. 464-9).
72 *Evening Telegram*, 16 February 1910.
73 Jukes, *Excursions*, Vol. 1., p. 259.
74 Tocque, *Newfoundland*, pp. 304-7.
75 Two "Sealers' Agreements" for 1871; the *Gem* and *Dart*. The cook on the former cooked for 64 men, while the cook on the latter cooked for 67 men. CWA 1047 IV i and ii. At least in the case of St. John's, an 'Ice Committee' was formed 1 March 1837 under the 'Ice Cutting Act', and it was this committee's responsibility to assemble a work crew and to oversee the cutting of a channel through the harbour ice from the docks to open water whenever ice prevented the sealing vessels from

The Ice Hunters 275

leaving port. Presumably, the crews of the sealing vessels would be called upon to follow their orders (see *Newfoundlander*, 2 March 1837).

76 Tocque, *Newfoundland*, p. 304.
77 Michael Carroll, *The Seal and Herring Fisheries of Newfoundland together with a Condensed History of the Island* (Montreal, 1873), p. 9.
78 *Evening Telegram*, 10 April 1913.
79 *Newfoundlander*, 6 March 1877.
80 *Evening Telegram*, 11 February 1881.
81 *Evening Telegram*, 7 March 1882.
82 *Evening Herald*, 5 March 1892.
83 *Daily News*, 22 February 1895.
84 *Daily News*, 2 March 1895.
85 *Evening Herald*, 18 April 1901.
86 *Evening Herald*, 28 February 1905.
87 *Evening Telegram*, 6 March 1907.
88 Peter Cashin, *My Life and Times: 1890-1919*, ed. R. E. Buehler (St. John's, 1976), p. 44.
89 *Evening Telegram*, 4 March 1910. The return of Halley's comet was a topic of discussion and speculation in 1910.
90 *Evening Telegram*, 11 April to 1 May 1911 (many references).
91 See Anspach, *History*, pp. 418-25 for a description of sealing between 1803 and 1812 in Harbour Grace. All references to Anspach's discussion of sealing have been taken from this section.
92 See Table Intro.12.
93 DNE, pp. 108-9.
94 See Introduction.
95 Prowse, *Newfoundland*, p. 406.
96 Ryan, "Abstract," pp. 118, 120, 122 and 125.
97 Ryan, "Cod Fishery," MA thesis, p. 17.
98 Public Ledger, 3, 6, 10, 13, 17 and 20 April 1827.
99 *Newfoundlander*, 10 March 1831; and *Public Ledger*, 26 April 1831.
100 During the 1830s and 1840s, Newfoundland began to switch to £ currency from £ sterling. The former was worth about 80 per cent of the latter; unfortunately, it is not always clear to which value contemporaries refer.
101 Times, 3 April 1852.
102 Except where otherwise noted, information on shares has been taken from Mosdell, *Chafe*.
103 'Bill': "The wages, or share of profit, of a fishing or sealing voyage paid to men after the deduction of expenses." *DNE*, p. 42.
104 Ryan's "Sealers' Book: 1876-1877."
105 *Evening Telegram*, 10 April 1913.
106 Ryan's "Sealers' Book: 1882-1883."
107 DNE, pp. 122-3.
108 *Evening Telegram*, 5 May 1891.
109 *Evening Telegram*, 16 February 1910.

110 CO 194/129, fols. 128-53. LeMarchant to CO, 4 May 1848.
111 Mosdell, *Chafe*, p. 40.
112 Mosdell, *Chafe*, p. 42.
113 Edward White to Jobs, 23 July 1866. In this letter, a copy of which was given to the present writer by Dr. George Story, Capt. White accepted the command of the company's new steamer, *Nimrod*, as follows: "The most important point to decide upon being the allowance per Seal, and considering the great chance with steamers of killing large numbers of old Seals, both Hoods and Harps, I therefore propose that ninepence per Seal for all good seals, old and young included or sevenpence for good young Harps and Hoods and one shilling and three pence per cwt for all old Seals...." Capt. White and Job Brothers came to an agreement and he went in command of the steamer in 1867.
114 See Munn, "Balance to December, 1873," Box 4B, Mac Lee Collection, No. 262.
115 *Evening Herald*, 8 February 1905.
116 Cashin, *Life and Times*, p. 44.
117 Ryan, *Seals and Sealers*.
118 JHA, 1910. See Imports for 1909.
119 Information provided by Mr. David Bradley, History Graduate Student, MUN.
120 See Ryan, *Fish out of Water*, Chapter 7, for a discussion of Newfoundland's imports of raisins, sultanas and currants. In essence, the Colony reduced the tariff on imports of Greek sultanas and currants, and in return, Greece reduced the tariff on imports of Newfoundland saltfish. This created a problem for Great Britain because Spain argued that under the terms of the most-favoured-nation clause in the commercial agreement between the two nations, Spanish raisins (or pasas) should enter Newfoundland on the same terms as the Greek products. The issue was still unresolved when the war began, but, in the meantime, Newfoundland's imports of these items flourished.
121 Eszter Kisbán, "Food Habits in Change: The Example of Europe," *Food in Change: Eating Habits from the Middle Ages to the Present Day*, ed. Alexander Fenton and Eszter Kisbán (Edinburgh, 1986), pp. 2-10. This volume is based on papers delivered to the Fifth International Conference on Ethnological Food Research, organized by the Institute of Ethnology of the Hungarian Academy of Sciences in October 1983. See also the article by Stephanos D. Imellos, "Hard Tack as Popular Food," pp. 74-9, which mentions, among other things, that hard tack was the typical food of seamen.
122 Anspach, *Newfoundland*, p. 464.
123 Jukes, *Excursions*.
124 Jukes, *Excursions*, Vol.I, pp. 63-4.
125 Jukes, *Excursions*, Vol. I, p. 56.
126 Jukes, *Excursions*, Vol. I, p. 160.
127 Jukes, *Excursions*, Vol. I, p. 197.
128 Jukes, *Excursions*, Vol. 1, p. 80.
129 Jukes, *Excursions*, Vol. II, pp. 5-7.
130 Jukes, *Excursions*, Vol. II, p. 17.
131 Jukes, *Excursions*, Vol. II, pp. 30-1.
132 Jukes, *Excursions*, Vol. II, p. 68.

133 R. Howley, "The Fisheries and Fishermen of Newfoundland," *The Month*, LXI (September-December 1887), 489-98.
134 Prowse, *Newfoundland*, p. 450.
135 Prowse, *Newfoundland*, pp. 450-1.
136 Prowse, *Newfoundland*, p. 451.
137 Those fishing servants who stayed in Newfoundland and worked for their room and board were referred to as 'winter men' or 'dieters'(see *DNE*, p. 140.)
138 John M. Ryan (born 1912), Riverhead, Harbour Grace, informed the present writer that he remembered, in his youth, the older people of Riverhead relating stories of their elders living in huts heated by hobs until they were able to build fireplaces and enlarge and improve on their huts.
139 Mosdell, *Chafe*, pp. 42-3.
140 CO 194/23, fols. 478-9. Report on the State of the Fishery, 22 October 1799.
141 Prowse, *History*, p. 373.
142 CO 194/45. fols. 28-9. "Gower's Report for 1804."
143 CO 194/45. fol. 116. Gower to CO, 18 July 1805.
144 Anspach, *History*, pp. 418-25.
145 Jukes, *Excursions*, Vol. I, p. 250.
146 Jukes, *Excursions*, Vol. I, pp. 251-2.
147 Jukes, *Excursion*, Vol. I, p. 263.
148 *Newfoundlander*, 22 December 1853, quoted from "Report from the Exhibition at the Crystal Palace, New York," in *New York Freeman's Journal*.
149 Jukes, *Excursions*, Vol. I, p. 250.
150 R. Howley, "The Fisheries and Fishermen of Newfoundland," *The Month* (London), LXI (1887), 489-98.
151 Jukes, *Excursions*, Vol. I, p. 320.
152 CO 194/129, fols. 128-53. LeMarchant to CO, 4 May 1848.
153 Anspach, *History*, p. 418.
154 Even small boat fishermen in Newfoundland, well into the present century, would take out firewood and make a fire in a box or tub of sand. They would cook their meals of fresh cod boiled with salt pork and then boil a kettle of water for tea. Small schooners going from the island of Newfoundland to the Labrador coast with fishing families up until at least the 1930s always carried on deck a half cask, barrel or puncheon filled with sand and rock on the deck, where food was cooked and tea prepared day and night when weather permitted.
155 Murphy, *The Old Sealing Days*, p. 14.
156 See Sanger, MA thesis, for an excellent discussion of the limitations of sailing vessels in the ice fields and the effects on the sealing activity.
157 Jukes, *Excursions*, Vol. I, p. 276.
158 Jukes, *Excursions*, Vol. I, p. 293.
159 Jukes, *Excursions*, Vol. I, p. 307. Jukes, however, was well aware that this definition of fish was not exclusive to Newfoundland. He wrote in a footnote: "I have, however, an idea that this determination is not confined to Newfoundland, but that in the old rules of the church, seals, otters, whales, porpoises, and all cetacea and amphibia are classed as fish."
160 England, *Hunt*, Chapter XII. This an excellent first-hand account of the author's trip to the ice in 1922 on board the S.S. *Terra Nova*, under Capt. Abram Kean. The

only improvements over conditions in the latter nineteenth century that one could mention here were a slight improvement in food, with the addition of 'soft' bread, lobscouse and boiled beans to the menu, and the greater concern for the men's safety while on the ice. Bowrings, which owned the S.S. *Terra Nova*, also owned the S.S. *Viking*, which blew up while at the seal fishery in 1931. On board was the American film crew under Varick Frissell, who was among the men who lost their lives. The firm never allowed observers or writers to go to the ice on board its steamers afterwards.

161 Murphy, *Old Sealing Days*, p. 14. For a while, Newfoundland imported hard bread from Hamburg, which was a market for seal and cod oil.

162 The drastic change in this regard was brought about by the introduction of steam, after which the individual watches were dropped at different locations at daybreak with orders to kill, sculp and pan until the ship returned for them, sometimes after dark—and on one terrible occasion, in 1898, not until 48 men had died. Instead of hot molasses tea and bread and butter at regular intervals, the men were forced to depend on the hard bread they carried and water sipped from melting ice or carried in a canteen. Thus began the practice of depending on their nunny bags and the little extras they could bring from home, especially small supplies of rolled oats mixed with raisins.

163 For an excellent summary of this subject, with diagrams, of this subject see Sanger, MA Thesis, Chapter VI.

164 The introduction of the big steamers with the double decks alleviated this situation to some extent, and steam heat and coal stoves reduced the heating problem. However, on board both the sailing vessels and the steamers, the stowing of pelts often went on through most of the night, as one or two of the watches continued to work. Thus, the hatches would be off as pelts were lowered by hand or by steam winch into the pounds below for stowing. Occasionally, especially in the early days of steam, the 'tween decks would be needed and the bunks would be torn down and the space given over to pelts. When the S.S. *Commodore*, under Capt. Azariah Munden, entered Harbour Grace in the spring of 1872, "She was," according to Chafe, "so deep in the water that the crew could easily wash their hands over the side." Mosdell, *Chafe*, p. 36.

165 Dr. John M. Olds, "Seal Finger or Speck Finger: A Clinical Condition Observed in Personnel Handling Hair Seals," *Canad. M.A.J.* LXXVI (March 1957), 455-7; Kaare Rodahl, "Speck-Finger or Sealer's Finger," *Arctic*, V, no. 4 (December 1952), 235-40; and *DNE*, pp. 451-52. The belief that the hand took on the appearance of a seal flipper is taken from interviews with Andrew Short (1900-89), Riverhead, Harbour Grace.

166 Andrew Short was shown this method by his father in 1925.

167 Busch, *Seals*, p. 172. See also Eric Sager, *Seafaring Labour: The Merchant Marine of Atlantic Canada, 1820-1914* (Kingston, 1989) for an excellent description of life at sea during this period.

168 See Mosdell, *Chafe*; the *Public Ledger*, 11 March 1868; and the *Times*, 3 April 1869. Complete lists of sailing ships clearing for the seal fishery are unavailable for either of these years, but the sailing ships from Harbour Grace in 1869 would have almost certainly participated in 1868; the sailing fleet was declining at this point, as has already been discussed. See Ryan, *Seals and Sealers*, and Winsor, *Stalwart Men*, for information on and pictures of the early steamers.

169 The new steamers in the early twentieth century were built of steel and were driven by engines of over 300 nominal horse power, with the newer engines delivering more of the horse power they 'nominally' promised.

170 DNE, p. 136, and England, *Hunt*, p. 51. The term 'stow boss' is not in either of these sources but was communicated to the present writer by Andrew Short. It was explained that when the watch was employed in stowing the pelts in the hold, the 'stow boss' was expected to assume command from the master watch, whose main responsibility was to the men of his watch while they were on the ice.

171 Brown, *Death on the Ice*, p. 34. See also *DNE*, p. 265, and note that 'ice master' was the term sometimes, but not often, used to describe a sealing captain.

172 England, *Hunt*, p. 46. See also *DNE*, pp. 447-8.

173 See "In the Matter of Enquiry into Disasters at Seal Fishery of 1914: Evidence taken before the Commission, November 30th, 1914—January 13th, 1915." Captain Westbury's Evidence, pp. 1-44. CWA 980 Misc.

174 The late Dr. George M. Story's biography of Captain Arthur Jackman under preparation for the *DCB*.

175 See *DNLB*, p. 49. As this entry points out, Mr. Carroll served with the crew of the S.S. *Eagle* when that steamer was chartered for work in the Antarctic in 1944-45. A mountain in the Falkland Islands was named Mount Carroll in his honour.

176 "Sealer's Agreement, S.S. *Bear*, 1879." CWA 1047 IV xi.

177 *Evening Telegram*, 7 April 1910. England reported that the chief engineer on the S.S. *Terra Nova* in 1922 was a Scotsman, McGettigan, and known as Mac. He continued: "You know the old saying that if you shout: 'Oh Mac!' into any steamship engine room in the world, you'll always get the answer: 'Here!'" See *Hunt*, p. 42.

178 *Evening Telegram*, 23 April 1912. See the following chapter for a short discussion of the loss of the S.S. *Erna*, the sealing steamer lost while en route to Newfoundland to take part in its first sealing voyage.

179 *Newfoundlander*, 10 April 1874. This tragedy is discussed in the following chapter.

180 *Evening Telegram*, 23 April 1912. The list was published when it had finally been accepted that the S.S. *Erna* had sunk and all hands lost. See the following chapter for the names of those lost on this occasion.

181 England, *Hunt*, p. 138.

182 Sanger, MA Thesis, p. 182.

183 "Sealer's Agreement, S.S. *Bear*, 1879." CWA 1047 IV ix.

184 Sanger, MA Thesis, p. 181.

185 David Moore Lindsay, *A Voyage to the Arctic in the Whaler 'Aurora'* (Boston, 1911), p. 41.

186 Lindsay, *Aurora*, p. 42. It is curious that England did not seem to be aware that the sealers ate seal hearts. He never seems to have observed their activities late at night and assumed that they simply worked and slept. One of his pictures between p. 92 and p. 93 is entitled, "Seal-hearts in the belt of a young Viking," which the editor of the 1969 edition ridiculously assumes are trophies: "The passages concerning blood are frequent and integral. Less obvious, but perhaps even more significant, are references to the hearts of the seals. How common the practice of cutting out seal hearts was, I cannot say. England has brought us one flabbergasting photograph of a young sealer—is the expression on his face prideful or sheepish?—wearing them in his belt like trophies." (p. x) Incidentally, Dr. Larry Small, Folklore Department, MUN, informed the present writer that the young man in that picture is the late Warrick Horwood, Moreton's Harbour, Newfoundland.

187 *Evening Telegram*, 16 April 1913. The point of the story is that the early Dundee steamers brought out their own sharp shooters from Scotland, but these men were unaccustomed to sealing as it was carried on in Newfoundland. In the act of sitting

188 down to eat his lunch, McCarthy apparently narrowly escaped being shot by one of these inexperienced gunners. Capt. Adams was in command of the S.S. *Arctic* during the seasons 1877 through 1883, but the story says that the incident occurred early in the history of the Dundee sealing enterprise in Newfoundland, quite possibly before 1880.

188 *Evening Telegram*, 12 April 1886.
189 Lindsay, *Aurora*, p. 47.
190 See Frank T. Bullen, *The Log of a Sea-Waif* (London, 1901). Bullen describes the conditions on board the cargo steamers and sailing vessels of the 1870s, and they were very similar to these on board the Newfoundland sealing vessels: a bag of straw for a bed; the poor quality duff, salt beef and pork; the dirty conditions; the violent discipline.
191 *Evening Telegram*, 12 April 1898.
192 *Evening Herald*, 27 February 1901.
193 Ryan and Small, *Haulin' Rope and Gaff*, pp. 20-1. One version was reprinted from Murphy, *Old Time Sealers*, and the second version from Doyle, *Old Time Songs*. The *John Martin* is listed in the departures of sealing vessels from Harbour Grace or Carbonear in the 1850s, usually commanded by a Captain Taylor. In Doyle's book it is stated that this song was composed in 1845 by Stephen Reardon of Perry's Cove and that the *John Martin* was a brig owned and commanded by John Bransfield. The lists of departures indicate that a Captain Bransfield commanded various sealing vessels during the 1840s, but not the *John Martin*. There are six stanzas in the Doyle version; three have been reproduced here.
194 Bullen, *Sea-Waif*, p. 81.
195 Lindsay, *Aurora*, p. 48.
196 *Evening Telegram*, 10 May 1886.
197 *Evening Telegram*, 12 May 1886.
198 *Evening Telegram*, 9 April 1897.
199 England, *Hunt*, Chapter XII.
200 *Evening Telegram*, 28 March 1914.
201 *Evening Telegram*, 16 February 1910.
202 Mosdell, *Chafe*, p. 42.
203 *Royal Gazette*, 29 April 1862.
204 *Times*, 6 April 1867.
205 See *Royal Gazette*, 6 July 1872, for "Correspondence of the Boston Traveller," 1 May 1872. See also Murphy, *Old Sealing Days*, p. 52.
206 Mosdell, *Chafe*, p. 43.
207 *Evening Telegram*, 6 April 1880.
208 *Evening Telegram*, 19 April 1880.
209 *Evening Mercury*, 4 January 1889.
210 *Evening Telegram*, 29 March 1899.
211 *Evening Telegram*, 29 April 1899.
212 *Evening Telegram*, 5 and 6 May 1899.
213 *Evening Herald*, 3 May 1901.
214 *Evening Herald*, 6 May 1901.
215 *Evening Herald*, 15 April 1902.

[216] *Evening Herald*, 5 May 1902.
[217] Ryan and Small, *Haulin' Rope and Gaff*, p. 16.
[218] *Evening Telegram*, 19 Mar 1914.
[219] *Evening Telegram*, 14 May 1910.

CHAPTER 5
Disasters

THE DEVELOPMENTS IN sealing which added to the problems of the industry in general and to the difficulties experienced by the men who went to the ice in particular cannot be contained within the chapters devoted to marketing, investment and working conditions alone. Sealing was affected by many limits and constraints because of the nature of the industry, and these included the forces of nature and the environment in which the industry operated. Thus, it is necessary to examine the extent to which the spring seal fishery was subjected to natural hazards above and beyond those normally associated with the sea.

There is a strong current of belief in Newfoundland literature that this fishery was not only physically difficult to prosecute, but unusually hazardous to lives and shipping. In 1905, Chafe wrote:

> There can be no doubt that one reason why the seal fishery developed and expanded very slowly was because it was realized that the enterprise could not be engaged in without great difficulty and risk.[1]

George Allan England observed and wrote about the seal fishery with dispassionate objectivity and reported as follows:

> The seal hunt is without any question the greatest in the world, not only in number of mammals slaughtered but also in point of perils from ice, blizzards, fire, explosion, drowning—a whole catalogue of hardships that only Newfoundlanders... can possibly endure.[2]

These writers are representative of the way in which twentieth century Newfoundland historiography and literature have depicted the spring seal fishery. One can also turn to the contemporary observers who reported on the origins of the new industry during its very beginnings.

It was Governor Waldegrave who first drew attention to the hardships of the fledgling spring seal fishery. As mentioned, Waldegrave had joined the Royal Navy in 1766 at the age of thirteen and was given his own command in 1775. His career in the navy was very active and his experience in the rigors of life at sea was second to none. However, Waldegrave was impressed by the dangers involved in this new industry. He wrote:

> The late introduction of a valuable seal industry is no doubt an object that requires much weighty consideration, as even independent of the wealth it offers the very mode of taking these animals is of a nature to form the hardiest race of men in the universe.[3]

Similarly, in his report of 1804, Governor Gower stated that the Newfoundland sealers exposed themselves "to the most imminent dangers... [and that] it is certain that there is no employment so well calculated to form hardy and intrepid seamen."[4] In July, 1805, he again reported on the "extreme hardships and dangers attending it [the seal fishery]."[5] Thus, even in the context of early nineteenth century maritime and naval employment, the Newfoundland spring seal fishery was singled out for its inherent hardships and risks. Anspach was equally impressed by the conditions under which the seal fishery operated, and marvelled at the dangers and hardships encountered so fearlessly by the Newfoundland seal hunters, particularly those from Conception Bay (see Chapter 4). The impressions of these three early observers—Waldegrave, Gower and Anspach—provide important evidence of the general perception that this new industry was unusually dangerous to lives and shipping. British sailing ships were not expected to operate in Newfoundland waters during the winter months. It had always been the practice for shipping to return to Britain, sail to southern markets with cargoes of saltfish, or tie up in port during these months. Navigating among ice floes was considered foolhardy, and shipping in and out of Newfoundland harbours virtually ceased when ice was present. Not only were Arctic ice floes hazardous, but, also, harbour ice which formed almost every year paralysed shipping, and freezing spray at sea made canvas practically unworkable. Newfoundland ship owners and Newfoundland fishermen began to challenge the ice in a way unmatched by their forebears, contemporaries and fellow British sailors and fishermen—with the notable exception of the British whaling fleet operating in the Greenland Sea and Davis Strait, and even these ships tried to avoid entering the ice fields.

As already discussed, the Newfoundland spring seal fishery began in the 1790s, expanded and reached its peak in the 1840s, remained on a plateau for about two decades and then began to decline during the 1860s. This decline accelerated during the 1880s, and by the beginning of the twentieth century, exports of seal oil and seal skins made up only about 5 per cent of Newfoundland's total exports, compared with 30 to 40 per cent when at its peak. However, it was at this point, in 1914, that Newfoundland's most tragic sealing disasters occurred; that year, the S.S. *Newfoundland* lost 78 men when the sealing crew spent over two days on the ice during a storm, and the S.S. *Southern Cross* was lost, with its 173 men, during a gale.

There is a place for the study of disasters in this industry[6] because they highlight working and living conditions and underline the relationship between ship owners and men. Furthermore they are of intrinsic interest to the community in which they occurred and finally, a study of this subject allows one to see how communities and societies cope with and respond to disasters.[7] Given these reasons, it is important to investigate the disasters of the Newfoundland seal fishery and to examine their place in the overall story of the seal fishery. This study will look at the various reported disasters—big and small—in chronological order, up to 1914, noting the changes in technology and other factors as the history of the industry unfolded.

Early Disasters

One of the first references to ships wrecked at the seal fishery occurs in Governor Gower's report for 1804. In this exhaustive report, Gower wrote that:

instances last spring [1804] occurred of Crews, who were taken off the wrecks of vessels that were crushed between the ice, and brought home, having procured other vessels and made a successful voyage.[8]

Twenty-five ships were lost in 1804, and in 1805, Gower gave this as the reason why the 1805 fleet had declined to 131 ships. Meanwhile, in 1805, two vessels from Conception Bay were lost but their crews were saved.[9] In 1809, a schooner belonging to Carbonear was lost, but the crew was saved; a Fogo schooner was lost also, and, in the absence of any report to the contrary, this crew was probably saved as well.[10] In 1810, the ice conditions prevented the ships from reaching the seal herds,[11] but in 1811, out of a fleet of 165 ships, 25 were lost—13 from St. John's, 11 from Conception Bay and 1 from Trinity—and 1 crew member drowned.[12] Therefore, during the period from 1804 to 1811, there were at least two seasons during which 25 ships were lost, although in comparison to the loss of shipping, the loss of life was small. However, the hardships the men experienced, and which were remarked upon by Waldegrave and others, can only be inferred at this point.

The outbreak of the Anglo-American War in 1812 diverted the governor's attention to problems of supplies, shipping, and purely naval matters, and also, there was a decline in the 'novelty' factor of the seal fishery. Consequently, subsequent governors' reports contain little commentary and few statistics concerning the seal fishery. However, occasional references indicate continued losses. A report much later in the century stated that in 1814, the catch was very low and prices high, while the fleet suffered "great damage and heavy losses."[13]

Disasters during the Age of Expansion

During the 1820s and 1830s, the sealing industry expanded, often in spite of disasters but, generally, in tandem with them. In 1820, there were "several losses; great damage";[14] in 1823, "great damage."[15] That year, the schooner *Active* from Brigus, on its way to the ice fields in a thick snowstorm, ran ashore on the island of Baccalieu, and out of a crew of 32, only 4 survived[16]—a major blow to the small community. In 1825, there was "much damage,"[17] and in 1826, during a heavy gale, the captain of the *Speedwell* and nine crew members were swept overboard to their deaths. As their ship began to sink, the remainder of the crew took to the small boats and were picked up by a nearby schooner.[18] During that same storm, the schooner *Belisarius* went down. The crew took to the boats and, after rowing for five days, managed to reach King's Cove, Bonavista Bay.[19] There were unusually heavy loses in 1829. One writer noted on 7 April that "many vessels have arrived with full cargoes, yet scarcely one has come in without the crew of some other which had been totally lost."[20] The same writer listed some of the vessels from St. John's that were lost, including: the *Fanny*, under Capt. Maurice Cummings; the *Lady Margaret*, Capt. Piccot; the *Visiter*, Capt. T. Beck; and the *Sally*, belonging to Hunters and Company. Also lost were: the *Favourite*, from Grates Cove and supplied by Bulley, Job and Cross; the *Carolina*, from Bacon Cove, under Capt. John Gushue; the *Experiment*, from Brigus, under Capt. Sheehan; and an unnamed vessel from Mosquito, under the command of Capt. Samuel Pike. On 21 April, it was reported that the *Mayflower* was wrecked, but the crew saved,[21] and a week later, a further report informed the public that five ships from St. John's—the *Brothers, Highland Laddie, Envy, Nancy* and *Industry*—had all been lost.[22]

The loss of the *Visiter*, mentioned above, was due to an explosion involving gunpowder—always a danger.[23] It appears that on the night of 28 March, one of the crew accidentally fired his gun. Although the gun contained only powder—and no shot -he had left his powder horn open and that had exploded, which caused, in turn, the supply of gunpowder in the ship's locker to explode, blowing the deck and upper structure off the ship. He was very badly injured, but the rest of the crew escaped on to the ice, carrying their careless shipmate, while the ship sank.[24]

In its "Annual Report" for 1829, the local Chamber of Commerce described that year's seal fishery:

> The season having been boisterous, numerous losses happened among the sealing vessels, consequently an unusually heavy charge has fallen upon those who are interested in the various Mutual Insurance Societies of this place [St. John's] and Conception Bay.... The Chamber, admitting the disasters among shipping to be numerous, have the satisfaction to observe, considering the great number of men employed, few lives have been lost in prosecuting the fishery.[25]

It seems that 15 ships were lost in 1829, out of a fleet that numbered about 300. Furthermore, in reference to these losses, it was reported that the Mutual Insurance Society of Conception Bay consisted of 106 vessels "in the seal fishery, nearly all first class, valued at £51,050 and insured for £49,800."[26] Each vessel was worth, on average, £482 and insured for £470.[27] It is obvious that many owners were not insuring their vessels and experienced total losses when their craft sank (although this did not compare with the suffering of widows and orphans left behind by frozen or drowned fishermen).

The year 1830, during which a new record was established for the number of seals taken in one season, witnessed the loss of several vessels—fully loaded with pelts—and their crews. A gale on the night of 27 March apparently caught these ships in open water.[28] The wreck of one of these unfortunate vessels drifted into Petty Harbour in early April and was salvaged by the local fishermen. A St. John's newspaper gave a most detailed account of this incident, obviously written by a correspondent with an intimate knowledge of shipping terminology:

> The following are all the particulars which are known relative to the wreck which has been found, within these few days past, off Petty Harbour. The strongest apprehensions are entertained that the vessel was the Schooner *Confidence*, belonging to Mr. John Piccot, which lately left this port on a sealing trip, and it is melancholy to add that Mr. P and his son, a fine youth about 18 years of age, were on board.
>
> On Thursday morning last, a large quantity of wreck was taken into Petty Harbour, which had been observed two days previously by some persons who thought it was a pilot boat at anchor a short distance inside the fishing ground at the south point of that Bay; but being again seen on the following day in the same situation, several persons were induced to go out. Upon their return the punts were laden with what is judged to have been the whole

of the standing-rigging, and nearly the whole of the running rigging and blocks, of a vessel of about 70 tons, all nearly new; together with the two lower mast-heads and iron; the bowsprit (broken at the stem); topsail-yard and fore-yard (broken); several parts of the bends or wales, to which were attached the chains, bolts, and dead-eyes of both masts; the counterpiece, with iron traveller, mainsheet blocks and mainsheet, the cheek ends of both gaffs, and part of the main-topmast, to which was affixed the gaff-topsail (apparently quite new); a square-foresail, topsail, and a large quantity of sails cut and torn up.

The whole of the above was entangled together; and part of the rigging had caught the bottom, a short distance from the shore, the broken spars appearing above water. It looked as though it had drifted across the Bay, from the Northward, but it is the opinion of several persons in Petty Harbour, that the vessel had struck upon the fishing-ground, at the *Motion*, while breaking. No other pieces of wreck have been seen in any part of that Bay.

We regret to learn that the crews of the different punts cut up the whole of the sails and rigging into small pieces, and divided it among themselves as fast as they could clear it—so that in about an hour after they had gone on shore, scarcely a vestige was to be found.—There was amongst the rest, a white linen shirt, with a cumbric frill not marked; and a cotton shirt marked "J.P." From the appearance of the whole it could not have been washing on the shore any length of time, for it was scarcely chafed.[29]

James Murphy, the local chronicler, also wrote about this wreck, but not all the details match. However, because Murphy fleshed out the incident, his account is well worth repeating:

In 1830 there was another such frost, the sealkillers were out at the ice, when the gales drove them south fifty miles or more to the outer edge of the [Grand] Banks in the strain of Cape Broyle. Picco [sic] of the Cove was out in the *True Blue*. He was a great 'swoil' killer and had 5,500 that year. On March 29 the wind ceased and the vessels made sail to work to land. There was no light on Cape Spear in those days, the ice was loose, and that night it snowed and blew dreadfully. About daybreak it was worse, and the vessels were anxious. Bill Ryan in the *Caledonia* got in safely. 'Native' Walsh from the Beach got in also. Pat Mackey was in the *Devonport*, and after running for a good while, he hove her off to sea. There was no braver man than Mackey, but he knew when to stop. Picco was coming behind him and shouted, "Aren't you going to run in, Pat?" "No," replied Mackey, "I don't think it's safe." "Tis safe enough for me," shouted Picco. "Good luck to you," returned Mackey. Picco missed the Cape, ran in, and took the land near Petty Harbour Motion. A blinding snowstorm was raging, and not a soul was saved. He had 30 of a crew, men and boys.[30]

By the 1830s, it seems that disasters and losses were being addressed by a faith in size, as bigger and stronger vessels were built for this industry. While the overall effects of the developments in the seal fishery are addressed elsewhere, it is useful to point out here that ship owners and the media thought they were moving closer to the time when they would, if not defeat the elements, at least hold nature to a stalemate. This effort was praised by a contemporary:

The description of Vessels employed in this branch of our commerce has of late years been rapidly improving; and adventurous and hazardous as these voyages have usually been, the perils incident to them will thus be considerably diminished. We may, therefore [hope] that the disasters, a number of which we have to account in every succeeding spring, will bear corresponding proportion to the means which have been adopted to avoid them and that the owners will be reimbursed for the capital which they will thus have laid out in the prosecution of so highly important a branch of the Newfoundland trade.[31]

The belief in the ability to conquer the ice in the seas around Newfoundland persisted throughout many improvements in the building of sealing, whaling and passenger ships and culminated in disillusionment with the loss of the S.S. *Titanic*, the biggest disaster to occur in the period covered in this study.

Whereas the schooner *Confidence*, above, was lost at the end of its sealing voyage, the following year, a schooner was lost on its way to the ice fields, and this loss was also graphically described in a St. John's newspaper:

> The new schooner *Azariah*, John Bonnell master, with a crew of twenty men and two boys, besides the master, sailed on the sealing voyage from Cupids, Port-de-Grave, about 7 P.M., on Wednesday the 16th instant [March]. She went down the bay with a smart wholesale breeze, at S.W. by S.; the weather was tolerably clear until midnight, so much so that the man at the helm, John Newall, could see both sides of the bay. At midnight some snow-showers came on; Newall was shortly after relieved at the helm, and went below. About half past 2 A.M., he was aroused by an alarming call for all hands; he sprang from his berth and ran upon deck, when he saw the land right over the foreyard; in a minute the vessel's bowsprit struck the cliff end on; the mainsail had been lowered, the other sails were full, the wind blowing a smart but fair breeze; the weather was thick and dark. The bowsprit presently broke off, and both fore and main masts then quickly fell with a tremendous crash, and so encumbered the deck, and hampered the pumps that it was impossible to get them out;—the wind was increasing and the vessel crashing heavily against the cliff. It appeared now necessary for the crew to leap from the vessel to the rocks, to save their lives. Four men only, out of twenty-three, succeeded in the attempt. All the rest perished. At daylight these four saw the floating pieces of the schooner, and one only of all their companions, and he in the agonies of death feebly grasping a part of the broken bow of the vessel. In a few minutes his hands resigned their hold, and he sank under the whelming waters. They remained in the cliff until about 2 P.M. on Thursday, when they succeeded in clamouring up the precipices to the top of the Island (for it was Bacalieu). In the afternoon of the same day they saw a schooner, and hailed her, but without effect, she proceeded on up the bay; the weather was foggy. All Thursday night they passed without shelter or food, except what the fir boughs afforded for the one and the ground berries for the other.
>
> On Friday morning, the 18th instant, they were providentially discovered and taken off the Island by the crew of the Schooner *Joseph*, of Cupids, James Le Drow, master, then coming in with a trip of seals from the ice. At 3 P.M. of the same day, they were safely landed at Cupids.
>
> It appears from Newall's statement, that in going down the Bay, they steered too much to the northward, thick weather came on, and there was not that vigilant look out kept, which should under such circumstances be always strictly attended to. The vessel was lost in a small cove on the S.S.W. end of Bacalieu, where there is no beach, and the cliffs are steep and craggy.

> Several of the unfortunate men were married and have left large families to deplore their untimely fate.[32]

For the small outport, Cupids, this was a disastrous incident and affected a relatively large proportion of families. However, one is struck by the casual shipboard procedures implied in this account. There may have been a good reason to leave Cupids at 7:00 p.m. instead of waiting until daybreak, but it does seem that there was extreme carelessness involved in not keeping a better lookout and in not being able to hold to a reasonably accurate course sailing out of this wide bay with a "smart wholesale breeze" from southwest by south. Human error was involved in many sealing disasters.

The loss of another vessel that year was reported in some detail, and illustrates a quite common danger to sealing vessels. The schooner *Hope*, from Carbonear and commanded by Capt. Mullowney, was returning to that port with 3,500 pelts on board, when it encountered a heavy storm on the evening of 22 April and, "while lying to in the heavy gale on the above evening, struck on a rock near one of the Wadham Islands, and almost immediately went to pieces, when the whole crew of twenty-five men (with the exception of one man who had been sent aloft to clear the topsail, to endeavour to clear the land) perished."[33]

Losses in 1833 were also considerable, if not so dramatically described as in 1830 and 1831. The season had proved a poor one because most ships missed the main 'patch' of seals and were caught in a gale late in April as they pursued older seals in the water. The schooner *Union*, from Trinity, was found waterlogged, with masts cut away and men's bodies in the forecastle. The schooner *Olive Branch*, from Greenspond, was found bottom up. The schooner *Lark*, from St. John's, was lost and the captain's box, with the ship's papers, was found floating among some wreckage. The schooner *Robert Brine*, also from St. John's, was wrecked, but the crew was saved and taken into Bay Bulls. The schooner *Selina*, from Carbonear, was abandoned, and the crew and cargo of 1,700 pelts were brought into Bay Bulls by the schooner *Ann*, from St. John's. Finally, it was reported that the schooner *Anna*, from Trinity, had managed to make Bay Bulls, despite the loss of its masts.[34]

Ships were in danger of running into reefs and cliffs, as has been seen. However, even in the open sea, loaded ships returning from the ice fields were very vulnerable if caught in a storm, because the cargo of pelts could shift in the hold and cause the ship to capsize or list so badly that the crew would be forced to abandon it. No doubt, there were many cases of this happening, because only gradually did owners learn to build stout pounds in the hold to keep the heavy pelts in place. In 1834, the *Caledonia* was returning to St. John's with a large cargo of 5,000 pelts, which had been taken in only 14 days. It encountered a severe gale, the cargo shifted, and the ship listed and began to leak. The captain and crew were forced to abandon it.[35] This type of problem was compounded when fat began to run to oil and became more unstable, as was the case with other losses that year. The Chamber of Commerce regretted "the loss of many valuable lives and much property" and continued:

> Three vessels are known to have been lost by the seals having melted in the hold, the pound boards having in consequence started by the shifting of the

cargo in the first gale of wind, and the ship so becoming unmanageable. The vessels to which this accident happened were of the very best class, and being all commanded by experienced masters, were probably secured in the usual way; it is therefore evident that farther [sic] security is necessary in fixing the pounds, which will doubtless be attended to next season.[36]

Obviously, the crews of other ships were not as lucky as that of the *Caledonia*, who at least escaped with their lives.

During the remainder of the 1830s, there are suggestions and reports that indicate other losses. In 1836, the weather during the period of the seal fishery was "tempestuous," the presence of many large ice bergs was a constant threat,[37] and an undetermined number of ships and lives were lost.[38] "Boisterous" weather in 1837 created some problems, and although there were no reports of losses, the brig *Dingwell* was seriously damaged, and a schooner from Twillingate barely reached St. John's after losing its rudder.[39] The following year, the schooner *Trial*, from Bay Bulls, was wrecked, but the crew was rescued by a Newfoundland ship returning from Lisbon;[40] and that same year, the brig *Terra Nova* was damaged in a storm and forced to abandon the voyage.[41] However, the latter 1830s were relatively trouble free.

Disasters during the Industry's Zenith

During the 1840s, the sealing industry peaked and reached a plateau. However, it was marked by other developments as well—some relating to the seal fishery specifically, of which labour unrest was the most obvious; others relating to the general economy (depression), including the potato failures; and lastly, the fire of 1846, in which most of St. John's burned.

One report states that a Carbonear brig, *Active*, commanded by Capt. McCarthy, struck a pan of ice in a storm and sank. Two men, Marshall and Walsh, climbed the rigging; fortunately, as they reached the top gallant yard, it touched a pan of ice, to which they leaped and survived, to be picked up the next day.[42] Meanwhile, the 1842 season is significant in the history of the seal fishery because of the efforts of some fishermen/sealers to gain concessions from ship owners during the winter and spring of 1842 (see Chapter 6). It was in connection with his support for the sealing crews that J.V. Nugent, member of the Amalgamated Assembly, summarized and contrasted the risks faced by investors and fishermen:

> In order, therefore, to arrive at the haunts of the seal at a time when the cubs are some three weeks old, for then are these animals easiest caught, and their fat is, at the same time, purer and in greater quantity than when they are more grown—the sealing vessels leave our southern ports about the first of March, and proceed to the northward to seek those ice-bergs and floating fields of ice, which by all other mariners are looked upon with terror and dismay, and, once coming up to the seals, they plunge into the midst of the ice.
> The intrepid seal-hunters now pour forth upon the expanse of ocean, and rush upon their prey far away from their vessel, bounding from mass to mass along the glassy surface of the frozen deep. Here you see one leap across a chasm where yawns the blue wave to engulph [sic] him. There,

another, amid the mist, mistakes a mass of slob or soft snow for an ice-pan and is buried in the ocean, whence, sometimes, he is rescued from his peril by the timely aid of his associates, if they be near, at others, he sinks to rise no more. Anon comes the thick freezing snow-drift, that shuts out all ken of neighbouring objects, and the distant ship is lost. The bewildered sealers gather together, they try one course, then another, but in vain, no vessel appears: the guns fired from the vessel are unheard, the lights unseen: night comes on and with it hunger, and the blasting wind, and the smothering snow overwhelm the stoutest, and many, very many, yielding to fatigue and mental misery, sink into despondency, and the widow's wail and the orphans' cry, are the only record of the dreary—of the dreadful death of the sealer.

We speak not of the peculiar tempestuous season in which they are engaged—the Vernal Equinox. We speak not of the vessel crushed between the icebergs, consigning all to a tremendous fate, or of the thousand other disasters to which even these *iron-bound* ships are liable, but may say, in a word, that scarce a season passes that we have not to deplore the loss of vessels, of crews, or of individuals, leaving many a bereft mother, a widowed wife and orphaned child, to heave a heart-rending sigh o'er the memory of the sealing voyage....

Never, indeed, was there an adventure in the prosecution of which are combined more of commercial enterprise on the one hand, and of nerve, of strength, of vigour, perseverance and intrepidity—manly and dauntless daring—on the other. The merchants adventurously contribute the outfit—consisting of the vessel with all her materials fully equipped and victualled. The fisherman contributes his toil, his dangers, his life—all the hopes, the fortunes, the fate of his family. Thus is the Seal Fishery a lottery, where all is risk and uncertainty, but still, the risk, we must confess, is not equally, or even proportionally distributed.

We shall take for instance one vessel of about 120 tons. In her success is involved the success of *one* merchant—he may gain £1,000 or more, if the voyage prosper. In her success is involved the success of some *thirty* fishermen—they may gain each from £20 to £30 if the voyage succeed. The merchant to run the chance of gaining £1,000 has risked a capital of perhaps £2,000. The sealer to gain from £20 to £30 has devoted an incredible amount of toil and suffering—he has risked all—his life. If the voyage fail, the merchant has still his ship, &c., he has suffered an actual loss of the provisions consumed on the occasion. If the voyage be unsuccessful the poor man returns with the loss of his labour, penniless. If the vessel founder, or be dashed to pieces in the ice, the insurance officer relieves this *one* merchant by compensating him for his actual loss. If the vessel founder, *thirty* valuable lives are lost—*thirty* widows, and perhaps *one hundred* orphans shriek their curses upon a fishery that brought upon them miseries that cannot be compensated—the grave of all their hopes—the dawn of every misfortune.[43]

Here, Nugent reminds his audience of the traditional sailor's fear of ice. He then goes much further than any other prominent commentator of this period to describe the dangers and the risks to lives and capital. Furthermore, he stresses that the risks the men face far outweigh the risks the ship owners take, yet the latter benefit most from a successful voyage.

The "Bonavista Bay Spring and another cat year" of 1843[44] was a particularly disastrous one, and one report, from St. John's, states that "about 20 vessels were wrecked... [and] many poor fellows engaged in the hazardous enterprise have unfortunately lost their lives."[45] Another report pointed out that "we have to lament the destruction of upwards of 20 sail of craft of a superior description, and a great and appalling loss of human life, together with an incalculable amount of human suffering sustained by those who escaped from shipwreck."[46] And another report, from Carbonear, is more specific and recorded the loss of the schooners *Charlotte* and *Ambrose* from that port, the *Rebecca, Dart, Relief, Mary, Trial* and *John* from Harbour Grace, and the *Despatch* from Spaniard's Bay.[47] In all these cases, the crews were saved. Another report this same year describes in detail the loss of 15 men from one of the Carbonear ships:

> It is impossible to describe the gloom which has been cast over this town [Carbonear] and neighborhood by the mournful intelligence obtained from a part of the crew of the schooner *Princess* of this port, who arrived here overland from Trinity Bay on Monday evening last. They state that on the afternoon of Saturday the 1st inst. [April] the *Princess*, Meagle [Captain], with two other sealing vessels viz., the *Mary* of St. John's and *Ocean* belonging to Bonavista, were driving in the ice along the southern shore of the above named bay—the Salvage Rocks lying to the leeward but at some considerable distance, and the wind blowing a stiff gale from the E.N.E.
>
> Shortly before midnight the ice commenced running fast, so that by half past 12 o'clock, A.M. the *Princess* was in the midst of the breakers, and there being no prospect of saving her, she was immediately abandoned, the crew making the best of their way seaward. The unfortunate men, however, had not proceeded one hundred yards when a tremendous sea broke in among the ice scattering it in all directions, by which disastrous and fatal occurrence there is every reason to fear that fifteen of our hardy and enterprising seal hunters were buried in the waves. The remainder of the crew continued upon the ice till daybreak, when they landed at Silly [Scilly] Cove with much difficulty. The vessel in the course of the night having been driven almost miraculously over the reef, was boarded by eight of the survivors and the following morning, the rest returned home with the melancholy tidings.
>
> The other two vessels above named were also abandoned; but it is expected that one of them—the *Mary*—will be preserved; the other is a wreck; crews of both, saved.[48]

This is the earliest first-hand description available of what could happen to a ship and crew when the ship was caught in 'running ice'. One is reminded of Anspach's report at the beginning of the century to the effect that the ice "could whirl about as in a whirlpool."[49] The report also vividly describes the terrible situation in which the men found themselves and the great difficulty involved in escaping from a ship in the darkness over running ice during a gale. The chances of anyone surviving were slim, and it is remarkable that the other two crews managed to escape without any losses. The hardship involved in trying to survive on that heaving ice, with the waves breaking over them, soaked to the skin and freezing, is almost impossible to imagine. In all, "losses amongst the shipping [in 1843] were unprecedentedly great."[50]

Although "heavy pans of sunken ice or small icebergs" did a "good deal of damage... to shipping"[51] during the "Spring of the Growlers" in 1844, there does not appear to have been any loss of life. According to Chafe, there were a number of serious losses in 1845—of lives and vessels. He states that the barque

Ringwood, under Capt. Henry Norman, was lost with all hands; the brig *Peerless* was lost with Capt. John Nagle and 40 men; the brig *Elizabeth Margaret* was lost with the captain and 15 men, and only one man, Frank Wiseman, survived by climbing into the rigging; and the brig *Mary*, belonging to Ridleys in Harbour Grace, lost all its crew who were out in boats, with only the 6 men on board surviving.[52] In 1846, referred to by Chafe as the "Spring of the Great fire,"[53] losses increased.

First of all, in 1846, the Conception Bay and St. John's sealing fleets had great difficulty in trying to get out of their frozen harbours, and it was after 26 March before they were able to leave for the ice fields.[54] This was a harbinger of things to come, for losses were high during the 1846 season. One report states that the *Rebecca, Mary, Louise Stuart, Elizabeth, Swan* and *William L. Black*—all from St. John's—were wrecked, while the *Tyro* from Harbour Grace, the *Amy Ann* from Greenspond and the *John and William* from Trinity were also lost.[55] A letter from Magistrate Sweetland, Bonavista, to the Honourable James Crowdy, Colonial Secretary, described the situation in the wake of some of these shipwrecks:

> Be pleased to acquaint His Excellency the Governor, that on Saturday last, the 4th inst. [April], the Masters and Crews of the *Louise Stuart*, Stanton [Captain], thirty-six men, and that of the *Elizabeth*, Nurse [Captain], twenty-nine men, both of St. John's, were cast on shore here having lost their vessels the previous night upon the shoal of Old Harry, North and East of Cape Bonavista. The *Louise Stuart*'s crew in a most destitute condition, having after the loss of their vessel been swept on the ice by the force of the current amongst the shoals of the Flowers Point, where they lost eight or nine of their number, together with their punts, clothing, etc, by the sea breaking upon them, and... they... were literally naked when they arrived at my dwelling. In the course of the same day the *Elizabeth*'s came in and pretty nearly in the same state; and on Monday were followed by the crew of the *Amy Ann*, of Greenspond, twenty-five in number, but bringing in with them their clothing. To alleviate the distress of the two first cases, the Magistrates advised providing lodgings for the invalids amongst them, and placing the remainder in such untenanted dwellings as could be obtained, and there subsisted until they could be shipped off for St. John's, and the purchasing of such articles of clothing as could be procured here and in the neighborhood for them. This has been done at a considerable expense, and the situation of all rendered as tolerable as circumstances would permit, under the presumption that we were carrying out what His Excellency the Governor would have directed to have been done could an immediate application have been made to His Excellency....[56]

Sweetland went on to explain that he had had to arrange transportation to St. John's for the crews and that, in addition, the people in his neighbourhood were looking after the shipwrecked crews of the *Rebecca, Mary* and *Tyro*. These men had, at least, managed to save their supplies. He also suggested that the Newfoundland Government build in Bonavista "an asylum for distressed shipwrecked seamen, for there is scarcely a season but some two or three crews are cast ashore at this place." Finally, he concluded by warning that the loss of sealing vessels in 1846 "will be fearfully great." In May of that year, it was reported:

> There has not been within our recollection a sealing adventure attended with so many unfortunate and disastrous consequences, as that of the present year.... [T]he average of vessels lost exhibits a material increase over that of any late years, and the loss of human life has been by no means inconsiderable.[57]

Other reports identify additional losses that spring. The crew members of the brig *Charles*, from Carbonear, had to be rescued from their sinking vessel by the *Mayflower* and were "so exhausted and frost burnt that they could not get into the boat" without the assistance of the men from the *Mayflower*. The brig *Waterlily* was lost in Bonavista Bay and its crew was brought into Bay Bulls by the *Corfe Mullen*; and the *Sir John Harvey* also sank but the crew was rescued.[58]

Losses were high, and as the writer quoted above indicates, there was a "material increase"[59] in the average size of vessels lost. For example, the *Waterlily* was 96 tons burthen and carried a crew of 38 men;[60] the *Louise Stuart*, 140 tons and 40 men; the *Elizabeth*, 120 tons and 40 men; the *William L. Black*, 147 tons and 39 men; and the *Sir John Harvey*, 140 tons and 40 men.[61] The size of the average sealing ship had been increasing since the beginning of the century; by 1846, the average burthen of sealing ships clearing from St. John's was 93 tons and the average crew was made up of 32 men. Therefore, it is apparent that some of the largest and newest ships were among those lost.

The 1846 seal fishery was a watershed year for the St. John's industry, whose participation reached its peak in that year—consisting of 44 suppliers with 141 vessels and having a total burthen of 13,165 tons and a total crew strength of 4,470 men.[62] After 1846, the number of individuals and firms sending ships to the ice declined from 44 in 1846 to 37 in 1847, and the total number of ships decreased from 141 to 96. As Table 3.2 indicates, the St. John's sealing fleet of sailing vessels never recovered from these losses. Conception Bay suffered a setback as well, and its fleet declined from 186 vessels and 5,733 men in 1846 to 156 vessels and 5,042 men in 1847. However, the fleet from Conception Bay recovered significantly, and by the early to mid 1850s, its fleet numbered over 190 somewhat larger vessels carrying nearly 8,000 men (see Table 3.3).

In 1847, there were a number of failed voyages, and one ship had to be abandoned, but the most serious disaster was the loss of the schooner *Margaret*, from Harbour Grace, which went down off Greenspond with the captain and 20 of the crew.[63] And Chafe reports that the brig "*Hibernian* []," under Capt. Hugh Nagle, was driven on the rocks near Hant's Harbour, and half the crew was lost.[64]

At least three more ships from Harbour Grace—the *Mary Francis*, *Friends* and *Success*—were lost in 1849, and one—the *Margaret Ellen*—from Cupids. The loss of the *Mary Francis* was described in detail by a local correspondent:

> One of the most miraculous escapes that we remember to have heard of for a number of years, was that of the crew of the Schooner *Mary Francis* of this port [Harbour Grace], Mr. Henry Webber, master, lost on the night of the 12th ult. [April]. The situation of the vessel at the time of the accident was about 120 miles to the eastward of the Grey Islands. The night was tempestuous, with heavy rain, and a long ranging swell running among the ice-bergs. About half past two o'clock a.m., all hands being on deck, the

schooner received a tremendous blow under water, from a large shelving piece of ice, which took her about midships. Immediately she began to sink, and in less than two minutes her top gallant mast heads had disappeared beneath the surface, the crew having had barely time to leap (many of them half naked) upon a small pan of ice, which was heaving up and down with the sea, and was scarcely sufficient to keep them above water. They now perceived that only two punts had been saved from the vessel, but while deliberating upon the course to be pursued under such circumstances, two others were observed to float up from the bottom, together with a number of oars and what is more strange, a binnacle compass. Thus provided, Mr. Webber, with his usual coolness and self possession, ordered the boats to be bailed out, and stepping into the foremost one himself, led the way through the tumbling ice-bergs for a considerable distance, the rain still falling in torrents, and the wind and sea increasing every moment. At day break they observed a schooner to leeward, just in the act of loosing her canvas, and having succeeded after much difficulty in making themselves heard, they finally got on board, and were distributed about among a number of other vessels which tendered their assistance. Had the disaster taken place a few moments later, or had the wreck floated but two yards further astern on receiving the concussion, every soul on board must have inevitably perished.[65]

Again, one can see how vulnerable the ships were, how quickly they could sink, and how difficult and narrow were the escapes experienced by some crews. In this instance, a number of crew members from the *Mary Francis* wrote a letter to a Harbour Grace newspaper, thanking the captains and crews who came to their assistance:

Mr. Abram Northcott, of the schooner *Liberator*, the Messrs. Taylor master of the *Ann*, *Princess Royal*, and *John Martin*; Mr. E. Dwyer, and Mr. Thos Thistle, masters of the *Sir Howard Douglas*, and *Echo*, of Carbonear, and the crews of these vessels respectively, for the kind, humane, and generous treatment they experienced from them after their shipwreck. They would also beg leave to express their admiration of the noble conduct of Capt. Michael Fitzgerald of the *Haidee*, and of Capt. Williams of the *Herald*, (the former of this port [Harbour Grace], and the latter of St. John's), in tendering their aid and assistance on the unfortunate occasion.[66]

The sealers—captains and crews—willingly assisted each other, despite the competition involved in the seal fishery. Thus, in 1849, when the *Sarah Jane* from Brigus was lost, the crew was saved and brought into port by the *Hare*, which had had to abandon the seal fishery early because of an outbreak of smallpox on board.[67] Furthermore, with several hundred ships engaged in this fishery and operating in a defined area, the chances were good that a shipwrecked crew would be rescued once the men had managed to get safely into small boats or onto firm ice. As the industry evolved over time, larger vessels and more experienced officers and men came into play; nonetheless, the seal fishery continued to be plagued by disasters.

'Spring of the Wadhams' and the 1850s

Although no one could have foreseen it, the 1850s proved to be the final decade of ice hunting using only muscle and wind power to propel the ships. It was a positive decade overall, with the granting of responsible government to the colony in 1855, and a fair degree of economic success in general. This decade, along with the previous two, became the "good old days" referred to later in the century, when the industry, which seemed to be holding its own with mother nature, found itself the victim of other disturbing forces. Nonetheless, the 1850s saw its share of sealing disasters.

The "Spring of the Wadhams"[68] of 1852 began on an ominous note. In February, five men were landed on Funk Island by Stephen March, a St. John's merchant, with instructions to engage in the seal fishery from shore. On 23 February, the captain of this small crew and three men went on the ice, leaving the cook ashore. About two hours later, the cook observed seas rolling towards the island and signalled the others to return. These signals:

> were for a little time unheeded by the unfortunate men, until at length one of them Thomas Beckett ran towards the island, and when within about fifteen fathoms caught the end of a rope, thrown from the shore by the cook, which he tied around his body. He was hauled ashore, but at the moment when his preservation seemed certain, a heavy breaker burst upon the two men washing the cook inshore into a pond, the receding water carrying the unfortunate Beckett out, and causing his destruction. In the meantime, the master and the two men that were with him made for the N.W. of the island, but here they were prevented from landing by the force of the sea rolling in towards the rock. The master asked the cook if he could not launch the punt, but the sea was breaking in too heavily; the soft ice around them was in a short time beaten to pieces and one huge wave bursting in upon the pan which held them in an instant dashed them out of life.[69]

The cook was picked up by the *Coquette*, one of the first sealing ships to arrive back in port that spring, and it soon became apparent that this tragedy was only the forerunner of a disastrous sealing season.

In early April, there was a major gale, and the first concrete evidence that the sealing fleet had been affected was received from the *Pursuit*, which arrived in St. John's on 7 April and reported that a crew member, James Fudge, had been swept overboard in a storm. At the same time, the *Billow* arrived and also reported the loss of a man under similar circumstances. Both of these vessels also reported that the *Henrietta* and the *Mary* had gone down but that the crews had been rescued.[70] Within a week, reports confirmed that the storm had taken a heavy toll and that the *Western Trader*, *Ajax*, *Vesta*, *Elizabeth*, *Placid* and *Christianna* had been lost.[71] Soon it was reported that the *Argyle* had lost five crew members and that the *Helen* had sunk.[72] The *Dash* and the *Caledonia* were abandoned by their crews but boarded and brought into St. John's by crew members from Lawrence O'Brien's ship, the *Kingaloch*.[73] Also the *Gleener*, from Greenspond, found the *Imauna* abandoned off Cape Freels and brought it into St. John's.[74] The weather continued stormy, and ships continued to arrive in port with reports of losses, rescued crews and abandoned ships. The *Gannet* brought in the crew of the *Christianna*; the *Gem* was brought in by crew members from the *Oresta*; the crew of the *Helen* arrived in St. John's overland from Bay Bulls; the captain and crew of the *Caledonia* were rescued by the *William*, which, was being brought into port by crew members from the *Escape*; the *Rake* then picked up some of the crew members from the *Dash*, while the remainder of the *Dash*'s crew was rescued by the *Funchel*.[75] Other arrivals reported that the *Sally, Jessie Louis, Britannia, Fortitude, Cornelia, Hope* and *Corfe Mullen*—all from Conception Bay—were also lost.[76]

Reports indicate that conditions were chaotic. Not only were there considerable losses of lives and property, but the number of vessels brought in as abandoned ships led to charges, counter charges and law suits for a long time afterwards. One report stated:

> Many vessels with valuable cargoes of seals, being driven towards the breakers, were abandoned by their crews, who made for the shore. Several of these derelicts have been taken possession of by other vessels and brought into Conception Bay, etc....[77]

The chaotic conditions affecting the seal fishery "created the deepest excitement, alarm and grief throughout the city"[78] and the colony, and forced the government to take action. The House of Assembly was convened by the speaker and it was decided to send an address to the governor, requesting that three vessels from St. John's and two from Conception Bay be despatched to the north to bring home shipwrecked crews.[79] The first of these vessels to complete its mission, the *Coquette*—which had earlier rescued the cook from Funk Island—under Capt. Houlihan, landed 100 shipwrecked men at Catalina and brought 250 more to St. John's.[80]

It appears that most of the losses resulted from a combination of two factors. Prior to the gale, the wind had been easterly for a considerable while, and the seal herd was on ice packed against the northern shores in shallow water. The vessels had worked their way into these shallow waters among the floes, and when the gale occurred, many were driven into reefs and rocks, unable to save themselves.[81] Reports do not agree on the exact number of losses, and, no doubt, no complete picture was ever assembled. However, the latest estimate, made at the end of May, when the damage had been assessed, stated that there was a "loss of between 50 and 60 sealing vessels in the northeast gales between the 5th and 12th of April." This same report pointed out that 90 men perished and "upwards of 1000 shipwrecked sealers are stated to have sought shelter at various points about Pinchard's Island and Bonavista Bay."[82] While the exact figures will never be known, there is no doubt that the 1852 season, in absolute terms, was the most disastrous up to that time. Given the number of vessels lost, the number of men involved, and the fact that only 90 died, one has to conclude that the level of ice knowledge and skill possessed by the average sealer must have been extensive, indeed.

The ensuing legal cases involving the 'abandoned' versus 'deserted' vessels continued for some time, and the evidence presented in support of each side provides additional information on the hardships endured by the men. The case involving the *Kingaloch* and the *Dash* is one such example. The men sent from the former to take possession of the latter while the storm was in progress reported to the court that they had great difficulty crossing the rolling, running, loose ice—that sometimes they fell in, sometimes spray and water washed over them, and sometimes they were forced to crawl over thin pieces of ice, and that only by perseverance were they able to succeed in boarding the *Dash*.[83] They found the ship partly filled with water, pelts scattered around, and everything in a generally chaotic state. They reported that the conditions were too bad for them to contemplate returning to their own ship, even if they had wished. They then managed to unfurl and hoist the frozen sails, narrowly avoided the rocks, ice pans and icebergs, and made their way to St. John's. This type of extremely hazardous enterprise was repeated a number of times during this season (and the courts decided in favour of those crews that salvaged abandoned ships, although they were awarded reduced compensation because they had left the area for St. John's so quickly).[84]

However, a number of factors already mentioned helped Newfoundland to absorb these losses, and investment continued and even increased a little—encouraged, no doubt, by the fact that seal oil prices on London markets had risen steadily from £25 5s per tun in 1848 to £37 3s in 1850 (see Table 1.8) and the future market conditions looked promising (and in the short term, these promises were fulfilled).

There were a number of losses and deaths at the seal fishery during the remainder of the 1850s, but nothing like the scale experienced in 1852. The *Hornet* from Harbour Grace lost a man from Bryant's Cove overboard in 1853. A William Hamond on the brig *Mary*, from St. John's, died of tetanus from an injury to his hand when his gun exploded. A punt was picked up off Catalina in 1854 with five men from Grates Cove; one had died and three were severely frost bitten. In 1855, the brig *Mary* ran down one of its own boats and six men were drowned.[85] In 1857, the *John & Maria* was lost off Cape Broyle with 24 out of the crew of 30 men[86]—certainly, the most serious loss since 1852. However, although this decade started on a gloomy note in terms of disasters, it ended much more happily.

'Green Bay Spring' and the Decline of Sail

The next decade began on an optimistic note when it was reported in 1860 that "out of about 300 vessels engaged in that voyage during the past season only two or three casualties occurred... without any loss of life."[87] And in 1861, only one man was killed when a punt fell on him.[88] However, major disasters were soon to follow—at least in terms of shipping losses.

The 'Green Bay Spring' of 1862 was another disastrous season.[89] On 12 April, it was reported that the following losses had occurred: The *Eliza* (Capt. Winsor), *Ronana* (Jackman), *Margaret* (Cummins), *Hunter* (Pike), *Rosebud* (Dawe), *Melrose* (Pike), *Emily Tobin* (White), *Christina* (Parsons), *Alma* (Norman), *Victoria* (Agerton), *Hope* (Andrews), *Jura* (Snelgrove), *Elizabeth Margaret* (Power), *Mary Anne Rossiter* (Winser [Winsor?]) and *G. M. Johnson* (Lynch). The *Emily Tobin*, *Christina*, *Jura* and *Elizabeth Margaret* belonged to Munns, Harbour Grace, and the *Melrose* belonged to Ridleys, also of that outport; the rest belonged to St. John's.[90] Other losses were recorded later that year: the *William Stairs*, *Jessie Brown*, *Caroline*, *Elizabeth Jane*, *Alma* (another Alma), *Hope* and *Prince Edward*. Then, the *Balaclava* was reported abandoned, and the *Livingstone*, leaking badly in Trinity Bay.[91]

It was a peculiar year because many ships became jammed in the ice and were abandoned swiftly and safely by their crews, who, in almost all cases, set their ships on fire. Reportedly, this was in order to eliminate them as a danger to shipping, but there was considerable suspicion that they were burned to guarantee that they could not be salvaged by others and that the owners would at least receive some insurance money.

As already noted, the loss of the *Caroline* was more widely reported and discussed than the others. It was a brig of 139 tons burthen and carried 68 men, who all escaped to shore, but not before setting fire to their vessel. They explained that they considered it a menace to shipping and, therefore, burned it and all their provisions as well. One newspaper hoped that the men could prove that they could not save any provisions without endangering their own lives. The crew acted "very thoughtlessly (to say the least) in not securing even a few biscuits for their own sustenance."[92] The same newspaper reported later in more detail:

> We have already noted the lamentable loss of shipping engaged in the seal fishery this spring. Some of them are reported to have been fired from prudential motives—the protection of life and property at sea and which might at any moment be imperilled by collision. This is all very well as long as it can be shown that prudence was the prevailing impulse. It is feared that, however, that [sic] certain parties are guilty of having, in a more wanton and unjustifiable manner, violated the strongest sections of the maritime law and it remains for the underwriters to institute the strictest investigation touching the loss of vessels this spring. It is thought that the loss to underwriters will be little short of £25,000 and that the aggregate loss to the owners of the sealing fleet will scarcely be short of £100,000....[93]

In 1852, there had been no reports of abandoned ships being set on fire and many reports of ships being successfully salvaged. In 1862, the reverse was true; it seems that only one ship, the *Elizabeth Jane*, was salvaged in that year—brought into port by part of the crew of the *Sea Flower* [*Seaflower*?][94] All other abandoned vessels were burned. That year, 1862, must be considered a disaster in economic terms, and it was a major turning point in the seal fishery and the beginning of the decline in the industry.

The St. John's sealing fleet, which is the most completely documented, had consisted of 99 vessels, with a total burthen of 12,342 tons and employing 4,542 men, in 1859.[95] In 1863, this fleet had declined to 39 vessels, with a total burthen of 4,706 tons employing 2,004 men. The fleet in Conception Bay had numbered 152 vessels, with a total burthen of 18,644 tons and carrying 7,416 men in 1861; in 1863, the numbers had dropped to 113 vessels, 14,073 tons and 5,836 men.[96] Other, smaller ports experienced a similar decline in their sealing fleets.

Therefore, the colony's sealing fleet had been considerably reduced by 1864, when disaster struck again. That spring, the fleet sailed northwestward under favourable winds through loose ice. Then, on 5 April, a storm blew up from the northeast, driving the ice into the area around Green Bay. At least 26—another report says 30—vessels were crushed and sank, and nearly all the others were jammed in the ice until well into May. One reporter wrote:

> As for the result of the voyage [i.e., catch], it is beyond comparison the most disastrous within living memory. The number of vessels fitted out for this Spring's fishery was small; and of these we are already informed of the loss of thirty. This would have been a very large proportion of casualties, had the whole outfit been equal to that of former years; but at present, it presses upon our Sealing interest with double severity....[97]

Chafe recorded later that 1,500 shipwrecked sealers—among them Captain Edward White, a leading sealing captain of this period—were landed at Greenspond.[98] The government sent the S.S. *Wolf* in late May to bring relief to the ships still jammed farther north in White Bay, but it suffered damage and had to return to St. John's.[99] The brig *Coquette* was despatched to Greenspond with provisions,[100] and gradually, the shipwrecked sealers and the remaining ships made their way home. Again, as in 1862, there are no reports of loss of life, but economically, it was a disastrous spring in terms of shipping losses, and the 'catch' was the smallest for about 50 years.

Meanwhile, in 1863, two steam ships were brought to Newfoundland to take part in the seal fishery and proved their superiority to the sailing ships. Thereafter, as sailing ships were lost at the ice, they were not replaced, and the old fleet declined rapidly. Although few lives were lost during the early years of this decade, losses to lives and ships increased towards the end. In 1868, the brig *Eclipse*, from Harbour Grace, was severely damaged, and the *Fanny Bloomer* and *Nautilus* were lost,[101] as well as four small vessels from northern ports;[102] and on the S.S. *Nimrod*, a sealer was killed when a punt fell on him.[103] That same year, 1868, the Chamber of Commerce reported that the seal fishery had been disastrous because unusually stormy weather prevailed during the season.[104]

The sailing fleet continued to suffer losses in 1869, as several ships sank. In Conception Bay, the *Elfrida* (Capt. Jeffers), from Harbour Grace, sank with seven of the crew when the vessel struck Mosquito Point in a storm.[105] Also lost that year were three vessels from St. John's—the 119-ton *Packet* (Capt. Hutton), the 89-ton *Selah Hutton* (Capt. Batterton) and the 124-ton *Renfrew* (Capt. Blandford [sic])—but no lives were lost.[106] (Also, seven men were lost near King's Cove and 20 women and "some men" in the Twillingate area).[107] Losses continued at a steady pace. In 1870, the *Jane*, from St. John's, and at least six other vessels were lost in the Greenspond area,[108] while in the waters off the southwest coast of the island, three vessels and a total of 29 men were lost.[109] It is not known whether these numbers include the schooner *Moonlight* from Rose Blanche, which belonged to Benjamin Rose and was lost with the crew of ten men—three of whom were married. It was reported from Channel, the principle port on the southwest coast, that:

> some of the wives of the lost have suffered so much that their minds are deranged.... It is sincerely hoped that our Government will do something for us in this our time of need and sore affliction. I trust it will assist the poor starving widows and orphans who are ruined, as their only means of support has been taken away from them.[110]

The following year a disaster of a different type occurred when between 30 and 40 sealers, homeward bound by sailing vessel from St. John's to Old Perlican, were lost when the vessel sank off Cape St. Francis. Besides the captain, crew and sealers, the ship carried other passengers, and there were no survivors. It has been assumed that the ship struck a rock or an iceberg.[111] The following year was uneventful, but 1872 turned out to be another disastrous year.

The sailing fleet was decimated that year and many lives were lost. The *Eneas MacIntyre, Mary Joyce, Greyhound, Seaflower, Cecilia, Dolphen* [*Dolphin*?] and *Glencoe* were all wrecked with no loss of life.[112] However, other incidents resulted in fatalities. Four crew members of the *Gertrude* were washed overboard and drowned,[113] and the *Velocipede* went down with some loss of life.[114] Other casualties were even more serious. The brig *Huntsman* from Bay Roberts "was literally smashed to pieces by the ice" near Battle Harbour and sank, taking with it the captain, Robert Dawe, his son and 40 of the crew.[115] The remaining 17 members of the crew were saved by the *Rescue*, commanded by Robert Dawe's brother, and only 3 of them escaped without broken limbs.[116] The *Dundanah* (Capt. William Jenkins) from St. John's, carrying 31 sealers, disappeared without a trace,[117] and the same thing happened to the *Village Belle* from Brigus, carrying 18 men under John Antle. When one adds the 4 men lost off the *Gertrude* and the men lost in the wreck of the *Velocipede*, at least 100 men died at the seal fishery in 1872. The magistrate in Brigus reported that 10 of the men from Brigus were married and left 30 children behind.[118] He led the appeal for contributions from the public. Also, it was reported that £836 was raised to help the families of the men lost on the *Huntsman*,[119] and £211 was distributed among the 30 families, numbering 147 individuals, who were dependents of the men lost on the *Dundanah* and *Velocipede*.[120] Apparently, the loss of the *Village Belle*, *Huntsman* and *Dundanah* left "more than sixty-one poor women... widows and upwards of 250 children fatherless, many of these... utterly unprovided for."[121]

The events of 1872 were picked up, distorted and recounted by "New York sensation mongers," who described the:

> dreadful series of catastrophes which occurred to the sealing fleet... forty vessels including four steamships... total wrecks having been dashed to pieces amid huge icebergs... and out of four thousand human souls only 175 have been accounted for.[122]

Although very much exaggerated, in terms of lives lost, this report is the first to suggest that the international community was becoming aware of, and interested in, the harsh conditions under which the seal fishery operated.

The sailing fleet continued to decline, but there are only scattered reports of actual losses, until 1878, when a number of major casualties again occurred. It was reported that the *Glengarry* from Harbour Grace, was lost[123] and the *Brighton, Ecliptic, Silver Stream, Stella Jessie* and *Cyrus* from St. John's were crushed in the ice in Green Bay and sank.[124] In all, over 14 vessels and at least "several lives" were claimed that year. The governor, in his address to the Legislature in February, 1879, remarked upon the "unusually large loss of property in sailing craft" during the 1878 seal fishery. Another report stated that 14 vessels were destroyed, most crushed in White Bay,[125] but at least one of the 14 sank in a strong gale.[126]

The sailing fleet did not disappear entirely from the seal fishery because, as has been shown, small schooner owners—especially those based on the southwest coast—began to enter this endeavour towards the end of the century, and there continued to be occasional losses of ships and men. In fact, as late as 1914, the St. John's schooner *Georgina* lost its captain and six men overboard in a storm while at the seal fishery.[127] It was not long after this that the schooners began to be fitted with auxiliary engines, and a few continued to try their luck at the seal fishery each year. However, after the losses of 1878, the sailing vessels surrendered their role to the sealing steamers. From the introduction of the first two steamers in 1863, their numbers had continued to grow, and in 1878, there were 21 steamers operating out of St. John's, with 3 from Harbour Grace.

Landsmen

The landsmen had always suffered their share of fatalities, but living in small isolated outports, their problems were only addressed in times of severe depression and want; their births and deaths were rarely noted, and it was not until the 1860s that the public begin to become aware of the number of accidental deaths among this sealing population.

In 1862, for example, it was reported that in Twillingate, George King and his two sons perished on the ice, as mentioned in the chapter above. In 1867, ten women and two men from Catalina looking for seals were carried out to sea and lost when the wind changed,[128] and in 1868, at least thirteen landsmen in small boats lost their lives in the Twillingate area.[129] The report of an inquest held that year at Change Islands on the bodies of two boys—Philip and Eli Young, aged ten and twelve, respectively, and sons of John Young—gives a good idea of the family disasters that were constantly occurring among the landsmen:

> It appeared from the evidence of Mr. Young, that on Tuesday morning the 10th March he went out in a small punt from Twillingate, to try to shoot some seals, taking his two boys with him, having rowed about six miles off, and having killed three seals and taken them into the punt. He was returning homeward but could not get round a string of ice and had to tie on his punt for the night.
>
> On the next morning he found that he had drifted down off Stone Island, some three miles North of Fogo. He then tried to row in (having no sail) to Fogo, but was unfortunately overtaken by a squall of wind and snow from the North, in which he had to run his punt as well as he could to the shore ice, about two miles from Change Islands. He then took the boys out of the punt on the ice, but the sea washed over the ice and immersed the boys several times. He then put the youngest boy into the punt and took the other boy with him some short distance to try to find some house in Change Islands—but the boy could not walk and he lay down upon the ice. Mr. Young then made the best of his way to Change Islands and got to Mr. George Morgan's at noon, being quite exhausted himself.
>
> Having made known the circumstances, a number of men went off immediately and found the two boys quite dead, one in the punt, and the other on the ice, as stated by the Father. The men put the bodies into the punt and hauled them into Change Islands, to Mr. Thos. W. Taylor's store.
>
> Verdict—died from exposure, and the severity of the weather.[130]

In 1878, a correspondent, writing from Twillingate, reported that the landsmen had done well with the seals, but sadly, six men—Gillett, Maddox, Elliot, Brenton, Phillips and Tissard—had lost their lives. "The first five were lost or drowned on or through the ice, the last was killed by an accidental gunshot wound through the leg."[131] The hazards of sealing from shore were serious and varied, whether one was on foot on the ice or in a row boat.

A complete account of losses of life in the landsmen's seal fishery will never be ascertained because of the scattered nature of settlement and the lack of communication with the media. However, 1880 was a particulary bad year for landsmen. On a Friday in late March 1880, five men left Northern Bay to go seal hunting. The ice moved out and the men were forced to keep on the move all night; on Saturday, three of them, a Fahey and two Hogans, could go no farther. The remaining two, a Hogan and a March, reached Cripple Cove Rock, where they rested before trying to reach land. March finally gave up, but Hogan managed to reach Cape St. Francis Lighthouse and safety. The S.S. *Hercules* was despatched to the scene but found only two bodies, which were taken to Northern Bay.[132] That same year, other lives were lost in the vicinity of St. John's; obviously the seals were nearby. A local newspaper reported:

> The following are the names of the unfortunate men who were driven off on the ice yesterday: Fowler, (two brothers) Cook's Town; Power, Carter's Hill; Chislett, (father and son); O'Toole, (father and son); Puddister, Hoyle's Town; and a man recently in the employ of Mr. Lindberg.
>
> O'Toole and his son and Puddister succeeded in getting ashore this morning, by what means we have not ascertained. The other poor fellows were not so fortunate, and little hope is now entertained for their recovery.
>
> During the rush for the shore, after the ice began to open, a man named Martin Murray fell through and immediately disappeared, notwithstanding the efforts of his companions to save him.[133]

Also in that year, Enoch Shearing of Canwell (near Bonavista), fell through the ice and was drowned while towing his pelts to shore.[134] In addition, "several men were disabled for a long time and in two instances death resulted from exposure to the cold" near Twillingate.[135] Finally, a man from Bird Island Cove (now Elliston), near Bonavista, died sealing, as had his father before him:

> Another fatal accident happened at Bird Island Cove. A young man of the name of Job Steeds, one of a punt's crew, when coming in with his load of seals, also went down and was no more seen. His father met a similar fate when the seals were last in there four years ago.[136]

Twillingate was the scene of tragedy in 1891 as reported in a local newspaper:

> On St. Patrick's Day the sealing boats here went a longer distance off than usual in quest of seals, when the wind suddenly changed and a snow storm came on. It was in reality a struggle for life..... Joseph Lindfield's two sons were in a punt and exposed to the full force and fury of the storm. The younger gave up and lay down to die. His brother, however, stuck to the oars and battled gallantly for life, rowing several miles, and not knowing whether his partner had succumbed to the exhaustion and cold. He at last reached shore with bleeding hands in time to save him. Life was just flickering when the boat touched land. Skinner brothers acted nobly. Newman brothers became exhausted and the Skinners took them in tow, and thus

enabled them to get in.... John Elliot, junior, had a narrow escape.... Philpott of Herring Neck died in his boat.... Six boats were missing until next day.... One boat is still missing, with Frederick Roberts and Charles Rideout, of Wild Cove. There are poor hopes of them now....[137]

And indeed, they were never found. These were isolated examples of the many hazards that landsmen faced.

The most terrible landsmen's disaster which has been recorded occurred in Trinity Bay in 1892. Prowse, who was busy on his monumental *History* at the time, described this event and explained how it, and others like it, could occur as follows:

Saturday, the 28th of February 1892 is a dark day in the annals of Trinity, a day to be ever remembered and mourned. The morning sun ushered in a lovely dawn, the sky was clear, a soft, bright, balmy air blew from the land over the treacherous sea, the light breeze scarcely ruffled its bosom. From Trinity and every harbour adjacent, boats were out by early dawn in pursuit of seals, which had been seen the previous evening. From Trinity, Ship Cove, Trouty, English Harbour, Salmon Cove and other small places the daring ice-hunters set off with high hopes and buoyant spirits to chase the wary seal; in this most exciting and dangerous pursuit the Newfoundlander recks not of danger; difficulties and perils that would affright one unaccustomed to the icefields are mere sport to the hardy native. On this eventful day seals were few and scattered; in the fierce excitement of the chase many went far out into the bay, heedless of the coming tempest; a few of the older fishermen, especially those from Trinity, more wary, and probably less vigorous, noticed the first signs of the storm, and before the ice blast came down with full force they were under the lee of the land and could row in. Two hundred and fifteen men were out on that day; the majority got safe to land after a tremendous struggle for their lives; the rest of the unfortunate fishermen, in spite of their heroic exertions, were finally overpowered; with strong arms they rowed for their lives, but the freezing icy tornado swept down upon them and paralysed their efforts; they had done all that men could do against the blizzard; they fought with the gale whilst instant death appeared on every wave. One bold crew from English Harbour, seeing all their attempts to stem the tempest were in vain, made for the ice; so chilling was the blast that before the boat had reached the floe, flying before the wind, one young fellow became paralysed with the cold; however, Newfoundlanders in a difficulty are never without resource. They climbed on a pan higher than the rest, where they made a rude shelter; their boat was broken up to make a fire; with this and some seals they managed to live through that awful night. Thirteen fishermen were found frozen to death in their little punts; eleven others were driven up the bay and perished in that dark cold night of death.... All that could be done was done to alleviate the distressed; a gallant crew from Heart's Delight saved the sixteen men on the ice; the generous Captain Fowlow put out in the schooner and succoured some and brought them home to their agonised families. Charity flowed in to the widows and orphans; kindness, open-handed liberality, tender human sympathy was called forth for the mourners, for those, alas, whose sorrow for the dead will never die, who all life long will grieve over the death and destruction of that dark day of storm.[138]

While Prowse puts the number of dead at 24, a local newspaper reported, on 29 March, that 25 men from seven communities died.[139]

Landsmen sometimes suffered dreadfully while hunting seals in punts among the loose ice floes or while on foot on tight or nearly tight ice. As has been seen, men and boys in punts risked getting blown out to sea or having their approach to shore suddenly blocked by ice. Men on foot fell through snow-covered cracks in the icy surface or discovered too late that a rent had opened between them and the shore. And they all suffered and died from the sudden drops in temperature and the rapidly increasing snow- and rainfalls, accompanied by rising winds. A landsman's situation was thus more critical, usually, than the hunter from the vessels. If the latter could not reach his ship, it would usually reach him; a landsman finding himself on the other side of the rent died if wind and tide did not create a bridge between the standing ice near shore and the running ice on which he stood. Similarly, to be in a punt when it sank was instant death, and one's chances if blown out to sea were not much better. However, the landsmen's tragedies usually involved scattered individuals or families and, except for a few, such as the 'Trinity Bay Disaster', were not widely known and thus, went unreported. Furthermore, landsmen's disasters generally affected fewer people than vessel disasters at any given time and thus attracted less attention.

Early Steamer Losses

Although steamers could prosecute the seal fishery more efficiently and safely than sailing ships, they were not invulnerable to hazardous conditions. The *Wolf*, one of the first two steamers brought to Newfoundland in 1863, was the first to become a casualty, when it sank in 1871 after being struck by an iceberg while jammed in the ice;[140] the crew escaped. On 28 April 1872, both the *Bloodhound*, from St. John's, and the *Retriever*, from Harbour Grace, sank during a storm, but the crews were saved.[141] Murphy described these losses and the rescue of the men:

> During a dark night the *Bloodhound* struck upon an island of ice, and was so much damaged that it was with the greatest difficulty that she was kept afloat until morning, when the crew had to take to the ice as she was in a sinking condition. Half an hour after the *Bloodhound* went down the men made their way over the ice to the *Retriever*, which was about two miles off, but on arriving they found that she had got her quitus from the same ice monster, and was sinking rapidly. Fortunately the men, about 308 in number, were not many miles from Battle Harbour. Two other shipwrecked crews [presumably from sailing vessels] arrived shortly after. After being there a few days they saw the steamer *Nimrod* at a distance, but failed to attract her attention by signals or by firing guns. At length they discovered an old rusty cannon lying about and, as a last resort, dragged it to the top of a hill, put in a tremendous charge, and applied a match. The gun flew into a thousand pieces, fortunately without killing anyone. The captain of the *Nimrod* [Peter Cummins] saw the flash, came near enough to send men over the ice, and on learning what happened, went for another steamer called the *Mastiff*, and between them they took off the shipwrecked crews and carried them to their homes.[142]

That same year, the *Hector* had to abandon the seal fishery after it was badly damaged.[143] Thus, the industry began to take its toll on the steamers as well as the sailing craft.

In 1874, the first losses of life on sealing steamers occurred when the *Tigress* was shaken by a massive explosion; its boilers burst, scalding to death 21 men—coal trimmers, firemen and engineers.[144] The steamer was commanded by Isaac Bartlett from Bay Roberts, which explains the fact that ten of the dead were from that port. Jobs as firemen and coal trimmers were highly sought after, and Bartlett had probably been instrumental in obtaining these jobs for his ten neighbours. That same year, the S.S. *Osprey* sank in a gale, but the crew was saved;[145] and in 1875, the S.S. *Ariel* and the *Tigress* were lost. In 1876, the *Hawk* was the first steamer to sink from the pressure of the ice while jammed; it went down "after being crushed in the ice—both sides stove in."[146] The S.S. *Micmac* sank in 1878, when it, too, was caught in an ice-jam and "a large pan of ice went right through her engine room."[147] On 6 January 1882, the S.S. *Lion*, under Capt. Patrick Fowlow, was lost while proceeding from St. John's to Trinity to prepare for the seal fishery.[148] It was a calm moonlit night when the steamer, with "all hands," disappeared, without a trace. Its loss is a mystery but was probably caused by the boilers exploding. However, no trace of wreckage was ever reported found. In 1884, the S.S. *Tiger*, under Capt. Thomas Dawe, was lost in the Gulf of St. Lawrence, "the sea completely burying her," but the crew was saved.[149] The losses continued.

Fatalities in the early steamer seal fishery were few, partly through circumstances, in that vessels that were crushed or stove in by the pressure of the ice floes were always situated so that men could escape unto the ice. Nevertheless, men were subjected to the same risks while on the ice as were men on the older sailing vessels and, to some degree, landsmen. Thus it comes as no surprise to read the sad case of a sealer who drowned in 1882, while dragging his tow to the vessel:

> Capt Wm Knee on the *Falcon* lost one man. On the 4th instant [April], while all hands were busy on the ice, two of the men met with a mishap, involving the death of one of them—a poor fellow named Joseph Parsons, belonging to Greenspond. They were on their way to the ship with a 'tow' of seals each when a spot of thin black ice over which they were passing gave way and both fell through. Some of their shipmates observed them as they went down and hastened to the rescue but their efforts were only partially successful. One of the men was hauled out of the water in a much-exhausted state and taken on board, but the other poor fellow—Parsons—had already disappeared beneath the broken slob, leaving his cap and tow of seals to mark the place where his life of honest manly labour terminated. We are not surprised to learn that 'many a tear of genuine regret' was brushed away from the soiled and sun burnt faces of the *Falcon's* hardy crew as they gathered round the treacherous spot.[150]

Even on board their vessels, the men faced other risks, some of which seem foolhardy today. In 1894, two men, Boatswain Brett and George Toms, died lingering deaths after an explosion occurred while they were "warming dynamite in the Galley."[151]

In all, 25 wooden steamers were lost between 1871 and 1898, as Table 5.1 indicates. There was comparatively little loss of life because most steamers were wrecked in tight ice, either by being crushed or by being swept along over rocks and shoals.

Greenland Disaster

In 1898, however, there occurred a terrible disaster in which many ice hunters were lost in a storm while away from their steamer, the *Greenland*. The fleet had steamed down past Cape Freels and had reached heavy ice and seals off the Funks. They began killing on 13 March; two days later, a heavy storm caught many men on the ice, but no lives were lost. On Sunday, 20 March, the ships lay along the outer edge of the heavy ice on which the seals could be seen. At daybreak on Monday, Captain Barbour put off one watch under James Gaulton and then steamed two miles further and put off the other three watches under master watches Jesse Knee, Nathaniel House and James Norris. By the time the last men were put out, the weather began to look threatening, and the *Greenland* steamed back and picked up Gaulton's watch. When the steamer turned again to retrace its track, the captain and crew discovered that the ice had swung about in a massively thick and tight sheet and now formed an impassable barrier. To make matters more desperate, in swinging about, the ice sheet had created a wide lake on the other side, which was an equally impassable barrier to the men stranded in the distance. If the weather had not deteriorated, boats could have been dragged across the ice to the lake and then used to ferry the men across; however, a furious blizzard struck. Apparently the men split up into seven or eight groups and tried to reach the ship, but with no success. On Tuesday, the storm abated, and most of the survivors were taken on board, but another storm stopped the rescue efforts until Wednesday morning, when six more survivors were found.[152] In all 48 died. Subsequently, it was reported that the *Greenland* had had a full load of pelts panned, but because these had been stolen by another crew, the vessel's trip was prolonged, and hence, the crew was caught out in the storm. Capt. Abram Kean, whose men were blamed for the theft, felt that he was being unjustly accused in this respect and reacted strongly in the local media to that effect.[153]

Besides the dead, many were injured, and three or four had limbs amputated back in St. John's, where the *Greenland* was met by family and relatives of the crew. Most of the dead—29 men—came from Bonavista Bay, while only 5 came from St. John's and the surrounding area; only 25 bodies were recovered. The tragedy for some surviving friends and relatives was accentuated by the part played by fate. Archibald Courage and three other men had fully intended to go in the S.S. *Mastiff* but had changed their minds; the four died. William Voisey of Quidi Vidi was well known as an athlete and had won many a race; he had gone as a stowaway and did not return. Two Newtrys from Carbonear also stowed away and one, Alfred, died. James Maher of Quidi Vidi had taken his father's ticket and gone out to see for himself what the seal fishery was like; he did not live to relate his experience. There were reports that dying men had their clothing taken from them before they completely expired, but these were never investigated. Indeed, the sealers were removed to their homes away from St. John's as quickly as possible after the *Greenland* was unloaded, and their evidence was never taken. Meanwhile, 18 coffins were set up in the reading room of the Seaman's Home, for the bodies of the men from the outports, while the dead from the immediate area were taken home at once. It was reported: "Many in reading the names recognize those of friends while one poor woman, Miss Norris, could be sympathized with for two brothers and four cousins were amongst the deceased, and only two bodies were recovered."[154] In all, it was a devastating experience, although it does not seem to have discouraged any significant number of men from sealing; in fact, one of the survivors of that terrible night was a young George Tuff, who was to play a fateful role in the later, and even worse, disaster which befell the crew of the S.S. *Newfoundland*.

The '*Greenland* Disaster' became a turning point as far as Newfoundland society was concerned. For the first time, the people expressed their concern publicly over the risk to lives at the seal fishery. They were no longer satisfied with the view that disasters were 'Acts of God' for which no one was to blame. Consequently, reactions to this disaster emanated from several quarters, and these are discussed in Chapter 6. For the moment, it is sufficient to point out that the vessel owners were made more aware of their accountability, and they began to pay more attention to the recruitment of ships' officers and master watches (who bore heavy responsibility for the men while they were on the ice and out of contact with their captains).

Other Steamer Losses

In the meantime, 1898 also saw the loss of the S.S. *Mastiff*, which sank in a terrible gale on 14 March. Two large sheets of ice driven by the wind and sea nipped the vessel amidships and passed through it. The crew barely escaped on to the ice at 10:00 p.m. with little more than their clothes. After the night in the freezing storm, the men were rescued by the steamers *Walrus* and *Neptune*.[155] The S.S. *Hope* was lost in 1901 when it became jammed in the ice and was forced ashore on Byron Island, where it gradually went to pieces.[156] The introduction of iron-clad steamers in 1906 forced the captains of the wooden steamers to drive their ships and men even harder in an attempt to obtain a proportionate share of the declining seal herds. This was counterproductive, and in the four-year period from 1907 to 1910, eight more wooden-wall steamers were lost, followed by two more in 1913 and 1914 (see Table 5.1).

The S.S. *Greenland* experienced a broken shaft in 1907, and while most of the men were taken off, the captain and a few crew members stayed aboard and tried to rig sails. However, the vessel drifted helplessly among the floes and began to leak. Finally, the S.S. *Newfoundland* came upon the steamer and took away the remainder of the crew. A local correspondent wrote:

> The S.S. *Greenland* ... has gone to the bottom and many mariners say this is the best possible thing that could have happened to this "voodooed" ship for she has been in trouble since first coming to the country. She has been burned to the water's edge, riots have occurred on board her several times, she has been on the rocks and her crew were frozen on another occasion.[157]

When the crews of the *Greenland* and the S.S. *Algerine* (which had had an unsuccessful voyage) went to the railway station, expecting a train to take them to their local stations at government expense because they had no money, they were disappointed. The train had been cancelled, but neither the men nor the station personnel had been notified. Eventually, the men were told to leave, and when they refused, the police were sent for; they finally ended up sleeping on the floor of the fire hall. A local newspaper, the *Daily News*, was quoted as reporting that the sealers had caused unnecessary trouble. A reply appeared in the rival newspaper, the *Evening Telegram*, in the form of a ballad, signed "Bay Man," attacking the *Daily News*, with the final verses as follows:

> On board the stout ship *Neptune*
> We came from o'er the foam,
> Prepared to leave on yesterday
> To see our friends at home;
> We were denied of passage,
> And the *Algerine*'s men also,
> Up at the railway station -
> They refused to let us go.
>
> In want of food and shelter -
> It was a sorry sight -
> They bade us leave the station,
> And then turned out the light;
> And then the police were sent for
> To turn us from the door,
> And they brought us to the fire hall,
> Where we slept upon the floor.
>
> We'll think on Donald Morison,
> For the mean and shabby way,
> He treated the shipwrecked sealers
> From bold Bonavista Bay.
> When we were cold and hungry -
> Oh, shame upon his name -
> His *Daily News* has stated
> We were the ones to blame.[158]

Meanwhile, also in 1907, the S.S. *Leopard*, under command of Capt. Robert Bartlett, was forced ashore by the ice at Blackhead, just outside St. John's, while en route to the Gulf. The men had to flee for shore over loose slob ice at one o'clock in the morning; they spent the night on the cliff and returned and retrieved their clothes the following morning before the vessel went to the bottom.[159]

As well, the S.S. *Bloodhound* had a narrow escape that year. It became jammed in the ice in Bonavista Bay and began to keel over in the storm. To many, it looked like the vessel would roll over, and they panicked and got onto the ice with their more important belongings. It was reported that there was "a peculiar assortment of goods lying there when daylight broke" including several pieces of pork, a bag of bread, some molasses and a bottle of rum. "There was no thought of life belts, put on board at the request of the Board of Trade."[160]

The following year, 1908, the S.S. *Grand Lake* was engaged in picking up pelts to add to the 19,000 already on board when the exhaust pipe became choked. During the attempt to blow it free, the pipe burst, and the "mishap was as bad as if one of the planks had been knocked off her bottom." The men saved what they could and escaped onto the ice; the steamer went to the bottom inside two hours.[161] That same spring, the steamers *Walrus* and the *Panther* were so damaged in trying to butt their way through the ice that the bows were caved in and they sank.[162]

In 1909, the S.S. *Vanguard's* main shaft broke, and the vessel drifted helplessly, being buffeted by the ice. When it began to leak badly, the men escaped to nearby steamers.[163] Similarly, the S.S. *Virginia Lake* was crushed in the ice that same year; first it lost the main shaft; and then, as the ice came together, the after part of the ship was lifted out of the water altogether, and the stern post and the rudder were carried away. As the ship settled back in the slackening ice, it began to leak badly and the men were forced to abandon it; some boarded the S.S. *Bellaventure*, while the remainder walked to the nearby mainland.[164]

The details of the loss of the S.S. *Iceland* in 1910 are scarce, although the steamer was honoured in a local ballad that was composed and published shortly after its sinking. It was pointed out in this ballad that this steamer was the last of the fleet that had sailed out of Harbour Grace. Three stanzas are as follows:

> In good old days in Harbor Grace,
> When it was in its bloom,
> John Munn & Co. rallied out supplies
> And made the fishery boom.
>
> But all the steamers now are gone,
> The *Iceland* is the last
> That links the present harder times
> With the good old days that's past.
>
> Farewell to the good ship *Iceland*,
> The wealth she brought was great;
> Although she's gone to Davy Jones,
> Thank God her crew are safe.[165]

The S.S. *Labrador*, on its way to the Gulf in 1913, sprang a leak while butting through countless miles of slob ice in heavy seas. As it began to fill with water, the captain steered for land. One fire was put out by the inflow of water, but engineers and stokers worked up to their waists in water for 30 hours to produce some steam, and the vessel struggled through the storm to reach St. Mary's Bay, where it was run ashore in Branch. It was a wreck, but all the men were saved; they then walked to Placentia and returned to St. John's by train.[166] In 1914, the little S.S. *Kite* went to the bottom as well. Although all ten steamers were lost in varying circumstances, there was no loss of life.

Erna Disaster

Another sealing disaster occurred in 1912 which is usually overlooked by Newfoundland historians because the steamer was actually en route to Newfoundland from Scotland to take part in the spring hunt when it disappeared with all hands. That winter, Baine Johnstons despatched a crew to bring back their newly acquired steamer, the *Erna*, and publicized the information that the company would have a new steamer for the seal hunt that spring. Berths were distributed through the merchants who dealt with the company and probably through others as well. In any event, the sealers with tickets came to the city, and everybody awaited the arrival of the steamer, which had departed Scotland on 1 March. When 12 March came, approximately 150 men were housed by the company in one of their warehouses in which two stoves were installed for heat; and the waiting went on. So awful was the thought of the steamer being lost with all hands that hope was not given up until 23 April, 54 days after the date of the *Erna*'s departure.

The S.S. *Erna* had been built in Greenock in 1890 and had been bought by Murray and Crawford (part of Baine Johnstons) for £14,500. On board when it went down were 33 crew and 3 passengers.[167] The passengers were a Miss Oakes, Scotland, who was coming to St. John's to be married to Mr. Eaton of the Royal Stores, Mrs. Linklater (the captain's wife) and her son. The loss of the *Erna* was made more tragic by the fact that many of the seamen, firemen and others working on the vessel had gone to Scotland primarily for the visit while work was slow at Baine Johnstons' premises. Consequently, their positions on board the vessel were not necessarily those that they normally performed. Jacob Winsor, who served as third officer, was to take charge as sealing captain when the vessel went to the ice; Caleb Winsor, who was listed as a seaman, was his brother.[168] It is not known how many dependents were left as widows and orphans, but five of the ten stokers were married, one left ten children, Jacob Winsor left a wife and four children, and Peter Jackman left a wife and five children.[169] This was the capital city's worst sealing disaster on record when one considers the number of lives lost and family members affected—and that, of course, is what is most important.

Southern Cross Disaster

In 1914, the greatest single sealing disaster of all occurred when the S.S. *Southern Cross* sank while returning from the seal fishery in the Gulf of St. Lawrence with a full cargo—the captain, George Clarke, anxious for the recognition and the small prize awarded to the first arrival. On 30 March, a wireless operator at St. Pierre reported that a sealing steamer had passed, and from the description, it could only have been the S.S. *Southern Cross*; the following day it passed the coastal steamer, *Portia*, near Cape Pine and was obviously on it's way to Cape Race, in spite of the storm which had convinced the *Portia*'s captain to take shelter in St. Mary's Bay. The *Southern Cross* was never seen again. As in the case with the *Erna*, the people of Newfoundland could not bring themselves to accept the fact that the *Southern Cross* had sunk. (This attitude was reinforced by the dreadful news on 2 April that over 50 men belonging to the *Newfoundland* had been found dead or dying on the ice.[170] The following day, this bad news was followed by even worse news, when it was confirmed that over 70 men had died in this disaster.[171] Thus the tragedy unfolding on the front actually prevented the community from being able to believe that an even worse tragedy had already occurred to a vessel homeward bound from the Gulf.) On 3 April, a local newspaper reported hopefully:

> Nothing has been heard of the *Southern Cross* since she was reported off Cape Pine on Tuesday last, and the general opinion is that she was driven far off to sea. Various reports were afloat in the city last night, one in particular that she had passed Cape Race yesterday afternoon, but upon making enquiries this and the other reports were unfortunately found to be untrue...
>
> At 5:30 yesterday the Anglo [Anglo-American Telegraph Co.] got in touch with Cape Race and learned that she had not passed the Cape, neither was she at Trepassey. A message from Capt. Connors of the *Portia* said she was not in St. Mary's Bay. A wireless message was sent by the government to the U.S. patrol steamer *Senaca*, which is in the vicinity of Cape Race, asking her to search for the *Cross*. The S.S. *Kyle* will also leave tonight to make a diligent search for her and it is hoped that something will soon be heard from the overdue ship, as anxiety for her safety is increasing hourly. If she has been driven off to sea, which is the general opinion expressed by experienced seamen, it would take her some days to make land again. The ship is heavily laden and cannot steam at a great speed.[172]

Instead of seeking shelter from the raging storm, the steamer, with 173 officers and men, gambled on reaching St. John's, and lost.[173] No wreckage or debris was ever found, and no one knows what happened during the final hours on that vessel. Oral tradition points to rotten pound boards which gave out in the heavy sea and thus allowed the cargo to shift and capsize the steamer.[174] This disaster, though greater in its extent than that which befell the *Newfoundland*'s crew, was overshadowed to some degree by the horrific details emanating from the latter tragedy.

Newfoundland Disaster

Just as the community began to express concern about the *Southern Cross*, everybody's attention became focused on the unfolding details of the disaster that had befallen the crew of the S.S. *Newfoundland*. This old wooden wall, under the command of Westbury Kean, had already been having bad luck. The ice was tight and the vessel could not penetrate the floes as could the seven steel ships in the fleet. To make matters worse, the owners, Harveys, had removed the wireless because the cost of the equipment and operator was not justified by the results in the past, so Westbury was unable to contact any of the other captains. Meanwhile, his brother Joseph was commanding the second largest steel steamer, the *Florizel*, and his father, Abram, had the largest vessel of all, the *Stephano*. Although they were employed by competing firms, Abram had informed Westbury that he would raise the after derrick on the *Stephano* as a signal for Westbury whenever the *Stephano* was among the whitecoats.[175]

After a frustrating, time-consuming and unproductive two weeks, the *Newfoundland* came up to within hailing distance of the *Florizel* and Westbury told his brother Joseph of his poor luck; Joseph relayed the news to Abram. Meanwhile, the steel fleet had been doing fairly well, but, of course, they were 'cleaning up' the scattered patches as they came upon them and leaving nothing for the slower wooden walls. Finally, on 30 March, the *Stephano* came upon the main herd and Abram ordered the after derrick to be raised.[176] The *Newfoundland* was jammed between five and seven miles southeast of this area (the correct distance was never agreed upon).

The following morning, Tuesday, 31 March, Westbury sent his entire complement of officers and batsmen under the overall command of the second hand, George Tuff, in the direction of the *Stephano*, with instructions to go aboard and take orders from Abram. It was highly unusual to send the second hand on the ice, but the seals were so far away that Westbury was reluctant to allow the four master watches—Jacob Bungay, Thomas Dawson, Sidney Jones and Arthur Mouland—to travel such a long distance and to work independently without somebody in overall charge. It is obvious, according to the later evidence of Thomas Dawson and others, that Westbury intended that his men spend the night on board the *Stephano* or, possibly, the *Florizel*, depending on which steamer was closer when darkness fell. However, it was never clear whether George Tuff had heard these instructions; it seems he was aware only that he was to take orders from Capt. Abram. They left the *Newfoundland* at about 7:00 a.m., but soon there were indications that the weather was going to deteriorate. Thirty-four men, fearing a storm, returned to the *Newfoundland*, explaining to an angry Westbury that they did not think it was a good idea to continue on towards the *Stephano* in the circumstances; they got back on board at about 1:30 p.m. Meanwhile, the main party, led by Tuff and containing all the officers, reached the *Stephano* at 11:20 a.m. and were invited to help themselves to 'dinner' of tea and hard bread. Capt. Abram testified later, that it was his understanding, at the time, that the men had spent two hours and twenty minutes walking from the *Newfoundland*—two hours less than their actual journey took.

Some managed to obtain a cup of tea, while others had to do without. Meanwhile, the *Stephano* was steaming towards a "spot" of seals, although it was never agreed in which direction it proceeded; southwest, south, southeast or some point in between. Abram then ordered Tuff and his men over the side at 11:50 a.m., with instructions to kill the 1,500 or so seals in the vicinity and return to their own vessel.

The 132 men of the *Newfoundland* found themselves on the ice with a storm coming on and unable to see their own ship, which they were expected to reach without really knowing where it was; and Tuff insisted that they obey Abram's orders and walk in the direction opposite to the vessel and kill seals before going aboard. Some of the men objected to this latter idea, but Tuff ordered them all to follow him, and they came upon a small spot of seals. By this time, it was obvious that the storm was increasing and at about 12:45 p.m., Tuff gave orders for everybody to start for the *Newfoundland* with Dawson in the lead on a course southeast by east. The intention was to cut across their morning trail and follow it back to their vessel.

Meanwhile, the 34 men who had quit and were returning to the *Newfoundland* discovered that it was only by following the steamer's whistle that they were able to find it in the thickening storm at around 1:00—1:30 p.m. The main party of men was far from the *Newfoundland,* but tragically, Capt. Westbury was convinced they were on board the *Stephano* and in no danger. And by all reports, this was the beginning of the worst storm to hit Newfoundland that winter.

Figure 18—Bodies from the *Newfoundland* disaster stacked on the deck of the S.S. *Bellaventure*.

At about 2:30 p.m., Dawson found the morning track and some of the flags the men had dropped because of their weight and clumsiness; they were now back on the heavy arctic ice, away from the loose ice where the seals were and through which the steel ships could move so freely. It seemed that they would make their ship safely after all. Almost immediately, however, they suddenly realized that they were much farther from the *Newfoundland* than they had thought—a four-hour walk at least. (Cassie Brown suggests that had they stopped there and built shelters while they had light and strength as well as pelts to burn and carcasses to eat, many lives might have been saved.[177]) Dawson led off again, but it became increasingly difficult to follow the trail as the wheeling ice pans sent the men first in one direction, then in another, and the drifting snow made matters worse. Meanwhile, on board the *Newfoundland,* the boatswain, John Tizzard, asked Westbury if he should blow the whistle. The captain saw no need for it but suggested that it be blown once or twice. Tizzard blew the whistle twice—possibly at 4:30 and 4:45 p.m.—and the lost men heard it and waited in vain to hear it again.[178]

It was at this time that the trail finally disappeared beneath the drifting snow, and the men came to a halt. Tuff ordered the watches to separate and choose large sturdy pans on which to build wind breaks from the loose blocks of ice. Mouland drove his men to pry up the blocks and build a wall that eventually extended about 30 feet, with short sides built out at right angles. Dawson's strength had almost given out after breaking trail for hours through knee-deep snow, but he managed to force his men to build a shelter about shoulder height, as did Bungay, who was joined by Tuff. Jones was not able to get his men organized, and most of them went to Dawson's pan, which became seriously overcrowded, making it difficult to jump and exercise for warmth.

In the meantime, the weather had been mild with the wind from the east, and they were soon drenched by a fierce rainstorm. Then, the wind changed to northerly and increased, bringing snow flurries; the temperature dropped quickly to well below freezing, and the shelters, which had offered some protection, now exposed them to this new storm. They managed to light fires, using the gaffs and ropes but these only burned briefly, there being no pelts available for regular fuel. By morning, the dead and dying lay all over the ice, and even Dawson, with both feet wet, finally lay down and succumbed to sleep.

Wednesday continued bitterly cold and windy, and because Westbury thought his men were on the *Stephano*, and because he, alone, had no wireless,

Figure 19 - Thomas Dawson, Bay Roberts, a master watch on the S.S. *Newfoundland* sruvived the disaster but lost both feet and suffered frost-bitten hands. A sealer for 19 or 20 years, and master watch on various ships for 15 or 16 years, he was ordered by the second hand George Tuff, to lead the *Newfoundland's* men in their exhausting and futile effort to reach their ship. Dawson questioned Tuff's decision to search for seals when the weather was worsening and was doubtful of Tuff's ability to direct them to the ship, which was out of sight.

nobody bothered to look for the missing men. The *Bellaventure* came close through the loose ice but did not see them; the *Stephano* did likewise; and, to make this horror story more terrible than it already was, when Tuff, accompanied by John Hiscock and Richard McCarthy, managed to cover about two miles in the direction of the *Newfoundland* and had only about two more to go, that steamer finally broke loose from the ice and steamed away from them.

The surviving men crawled into nooks and crannies in the rafting ice and waited through another bitterly cold and windy night for dawn. In the morning, Arthur Mouland, Tuff, Bungay and several others set out for the *Newfoundland*, which was obviously jammed again. Capt. Westbury had just climbed up to the barrel and was studying the *Stephano*, which was only two miles away, through his spy-glass, wondering why his men did not leave and return to their own ship. He swung his glass idly around and saw the several men staggering toward him. Now he knew the worst, as he slid and stumbled his way down to the deck and broke the news to the boatswain and the navigator, Capt. Charles Green. At the same time, he raised an improvised distress signal, which was seen on board the *Stephano*, and Capt. Abram despatched men to find out the trouble. When these men informed Westbury that his men had left the *Stephano* on Tuesday, the full horror of the situation became known. John Hiscock, the second master watch under Dawson, was one of the first men to board the *Newfoundland*, having seen his brother die and Dawson unable to walk. He told Westbury that had he blown the whistle more frequently, or if his father had not put them out on the ice in the first place, the tragedy would not have happened.[179]

Meanwhile, the *Bellaventure*, under Capt. Isaac R. Randell, was still in the vicinity but on the other side of the disaster area; the second hand, Abram Parsons, spotted Benjamin Piercey and Jesse Collins in the distance, making their way towards them. When the men were brought aboard and explained the tragedy, Randell ordered his men out with blankets, food and drink to go in search of the *Newfoundland*'s men because the ice was too heavy for the steamer to make much headway. By 6:00 p.m., Randell's crew had taken aboard 58 dead and 35 survivors. The *Stephano* and *Florizel* had participated in the search and had collected a number as well. When John Keels died in St. John's from the effects of his experience, it brought to 78 the number of lives lost as a result of the *Newfoundland* disaster.[180]

The photographs of the bodies of these sealers being landed in St. John's made a deep impression on the community, and an inquiry was held into both the *Southern Cross* and *Newfoundland* disasters—but given the circumstances, the inquiry concentrated on the latter. Sealers as well as officers were questioned—a total of 44 witnesses. The witnesses included John Hiscock, from Carbonear, the second master watch (deck router or stow boss) under Tom Dawson. Hiscock made an impassioned plea for the safety of the batsmen on the ice. He said, "instead of as much talk about the food aboard of a steamer and punts and dories the man with the canvas jumper should be thought a little more about."[181] In the end, a majority report concluded that Capt. Abram Kean had to bear most of the blame, but a minority report disputed this. And needless to say, the well being of the "man with the canvas jacket" was soon forgotten.

Years later, Rockwell Kent, who had lived in Brigus during the 1914 sealing season, recalled the storm, the disasters and the reactions of the families and friends of the victims. He wrote of returning from the Church of England evening service on 29 March, with his friend, Robert Percy, when he stopped to admire the new moon whose "slender crescent lay almost fair on its back." "'That's a bad moon', said Robert Percy. 'We'll have weather, for you can hang a powder horn on it.'" He remembered a poignant experience earlier in the month, when he was at the cable office in St. John's. A sealer was sending a message to his family: "I have a berth on the *Southern Cross*, sail tomorrow night." The wireless operator counted the words and informed the sealer that he was allowed one more for the same standard fee. "Then put 'Goodbye'," was the reply. He described how the older people of Brigus became restless and suspected the worst, long before the disasters were confirmed; how the wife of the captain of the *Southern Cross* had cried herself to exhaustion; how the daughter called deliriously for her father; and how the old people helped the younger ones to accept their tragedies. Like most of his contemporaries, Kent recognized the power of the elements to batter the lives of the families dependent on the sea, and felt that they could not be conquered. Still, his reminiscences reflect the combined sense of resignation and hope which helped Newfoundland communities through their many marine tragedies. When the storm had passed and the sun shone bright on 6 April, Kent wrote, that he knew, as he looked at the melting snow, that soon the grass would "resume its growing" and the town would again appear "serene and beautiful." His story provides a fitting requiem for the victims of 1914.[182]

Conditions improved to some extent after 1914. The report of the inquiry commission had made it clear that Newfoundland could no longer look upon the dangers of the seal fishery as being acceptable, and the commissioners had made recommendations concerning safety which were incorporated into legislation. One significant outcome was that wireless equipment and an operator became mandatory on all sealing vessels.

The government had become involved in legislation concerning the seal fishery as early as 1873, but most regulations had been aimed at conserving the seal stocks. In 1898, more legislation had put a limit on the number of men on each steamer, and in 1899, legislation had given sealers' wages some protection. However, legislation passed in 1916, two years after these great tragedies, was certainly aimed more specifically at the overall safety and well-being of the sealers; these measures prohibited men from being on the ice after dark and provided for rocket signals, search parties, masters' and mates' certificates, medical officers, better food, compensation and other improvements. The seal fishery had left behind the laissez-faire of the nineteenth century and finally entered the twentieth century, where men were beginning to be held accountable for the destruction of human life resulting from 'industrial accidents'.

Support for Survivors and Dependents

Throughout the period under study, there were no specific provisions made under the government's health and welfare schemes to assist the survivors and the widows and orphans left without means of support by disasters and misfortunes. However, the general public always rallied around to assist the victims in these cases.

As early as 1830, when there was a disastrous seal fishery, there is a record of the public responding to an appeal for financial contributions for the "Relief of those who lost their Friends at the Seal Fishery...." Ordinary sealers, crews and companies all contributed to raise a sum of £146 19s 11d, with Messrs. Gosse, Pack and Fryer of Carbonear making by far the largest donation of £58 10s.[183] The total sum was divided among the 64 survivors, who belonged to 21 families, with each survivor receiving £2 7s 2d (see Table 6.1). One family of eight received almost £19, and the remainder, in proportion; all payments were made to wives or mothers. Another £36 4s 5d was collected in 1831 and also divided among the same survivors.[184] Of course, from this point on, these people would be dependent on government relief.

In 1831, the leading St. John's merchants decided to establish "a fund for the relief of the widows and children of poor fellows who may unfortunately be lost, or maimed, while engaged in the Sealing fishery." However, the sealers themselves were to contribute the full amount to this plan, which would be set up by the merchants with funds deposited in the merchants' offices. The *Newfoundlander* (3 March 1831) thought that this was an excellent idea, and no one seemed surprised that the merchants would not be expected to pay into the fund. This paper wrote: "It is to the interest of merchant, planter and servant, to forward so feasible a plan, and we hope it may be warmly and successfully followed up." However, it was never implemented.

Some idea of the attitude of the government towards the victims of local tragedies can be gleaned from the complaint of the *Patriot* on 19 April 1843 that the government had authorized the expenditure of £500 to assist the crew of a foreign ship wrecked in Newfoundland waters en route to the United States, but had given only £200 to help the survivors and dependents of "upwards of twenty sail of craft" lost at the seal fishery.

However, there were often successful appeals for funding whenever the occasion required. For example, 1872 was a particularly terrible year, with the loss of the *Huntsman*, *Village Belle* and *Dundanah*, resulting in over 66 women being left widows and over 250 children left fatherless. It was reported, on 9 October, that £836 ($3,344) had been raised for the relief of the relatives of those lost on the *Huntsman*,[185] and presumably, other appeals were in progress. Similarly, when 20 men were killed in the explosion on the S.S. *Tigress* in 1874, a concert was held at the Total Abstinence and Benefit Society's new hall on Duckworth Street in aid of the dependents.[186] Also, within a week of the disaster, the fund in aid of the *Greenland* sufferers had accumulated $2,686.50.[187] The public responded to all appeals, and the media and the governors took an increasingly keen interest in them as well.

After the loss of the S.S. *Erna* in 1912, the governor chaired a meeting on 20 June to raise a fund for the relief of the survivors, and interest was very intense.

Public sympathy was even more evident after the disasters of 1914; by 27 April 1914, the disaster fund set up to provide for the survivors and dependents of the sealing disasters of that spring amounted to $88,550.42.[188] However, this type of public response was not confined to sealing disasters alone; it was common practice for society to react in this fashion to most industrial accidents and deaths during this period.

Summary

It is difficult to quantify the risk to men and ships inherent in the prosecution of the seal fishery in the absence of studies of other similar activities. However, the Newfoundland seal fishery required ships to enter the ice floes. This was not the case in any other 'fishery'. The walrus hunts of Bering Strait, the North Pacific fur seal industry and the whaling voyages to the Greenland Sea and Davis Strait all involved prosecuting 'fisheries' on rocky islands or in the open seas—although often quite near the Arctic ice. The Newfoundland seal fishery involved ships operating in the ice, following leads of water and, by the end of the century, making their own tracks through the sheer power of steel and steam. Getting jammed was a frequent occurrence and often led to the end of a ship—schooner or steamer. Other ships sank in storms, taking lives as well, and of course, storms could overtake men on the ice. For the most part, men were reasonably safe in tight ice because the very ice that could crush a ship provided a haven for the crew. Moreover, in the earlier period, the large number of ships (300-400) at the ice provided opportunities for crews to be rescued. In the later period, as in 1914, the involvement of fewer ships meant that vessels were often far apart and unaware of problems outside their immediate neighbourhood. The Newfoundland seal fishery appears, from the evidence to date, to have been a dangerous industry, deserving the reputation it had in the eyes of early nineteenth century observers like Waldegrave, Gower and Anspach, and the reputation it still retained when Chafe and England were writing at the beginning of the twentieth century.

However, this does not mean that Newfoundland's other fisheries were free of disasters. Many inshore fishermen were lost in storms, and the new bank fishery re-established in the latter decades of the nineteenth century, had its share of losses as well. In a severe August gale in 1887, for example, three banking vessels and about 45 fishermen were lost.[189] The Labrador cod fishery also experienced significant losses. Probably the greatest recorded disaster in this industry occurred during 10-12 October 1885, when an autumn gale wrecked between 70 and 80 vessels on the Labrador coast, taking the lives of about 70 men, women and children and leaving 1,500 to 2,000 people destitute.[190]

However, even among industries that were dangerous, the seal fishery stood out as unusually hazardous, and the risk to human life was accepted by Newfoundlanders until the beginning of the twentieth century, when firms, captains and officers began to be held accountable for their actions.

The batsmen were most vulnerable to disasters while on the ice and away from the vessel. The engineers and firemen were most vulnerable to scalding and even death in the engine rooms of the steamers. However, everybody's life was at risk in storms at sea, when vessels could sink with all hands on board. Societal forces, pressures, divisions, commonalities and attitudes stand out in greater relief during disasters. An examination of such tragedies can help readers appreciate the conditions under which men worked, the pressures applied by ship owners to officers, including captains, and the ineffectual response of government to what was a constant threat to lives and ships. Only after 1914 did Newfoundland society insist—somewhat ineffectively, granted—that safer procedures be followed at the ice.

NOTES

1. Levi G. Chafe, *Chafe's Sealing Book*, 2nd ed., (St. John's, 1905), p. 5.
2. England, *Hunt*, p. 17.
3. CO 194/23, fols. 478-9. Waldegrave to CO, 22 October 1799.
4. CO 194/45, fols. 17-47. "Gower's Report for 1804."
5. CO 194/45, fol. 116. Gower to CO, 18 July 1805.
6. The present writer uses the term 'disaster' in a general sense and does not try to distinguish disasters from 'accidents' and 'catastrophes'. For some discussion on this point, see John C. Burnham, "A Neglected Field: The History of Natural Disasters," *AHA Perspectives* (April 1988), 22-4. *Webster's Dictionary*, 1972, defines 'disaster' as "an adverse or unfortunate event; great and sudden misfortune; calamity." Levi G. Chafe, in his list of sealing "Disasters," published in 1923, included the burning of a ship and the loss of three lives. The Newfoundland *Weekly Herald*, 9 May 1849, described the loss of a schooner and the subsequent hardship experienced by the crew—who were all saved—as a disaster. Because the loss of a single wage earner was generally disastrous for his family, one can appreciate the difficulty in trying to define this term.
7. See an excellent paper on the study of disasters by Kenneth S. Coates and W. R. Morrison, "Towards a Methodology of Disasters: The Case of the *Princess Sophia*" (Presented to the Canadian Historical Association Meeting, Victoria, British Columbia, 1990.)
8. CO 194/45, fols. 17-47. "Gower's Report for 1804."
9. CO 194/45, fols. 199-205. "Gower's Report for 1805."
10. CO 194/49, fol. 15. Governor Holloway to CO, 30 April 1810.
11. CO 194/49, fol. 49. Duckworth to CO, 25 November 1810.
12. CO 194/51, fol. 36. Duckworth to CO, 5 November 1811.
13. *Royal Gazette*, 29 April 1849.
14. *Royal Gazette*, 29 April 1849.
15. *Royal Gazette*, 29 April 1849.
16. Mosdell, *Chafe*, p. 39. This disaster, like a number of others in *Chafe*, cannot be verified.
17. *Royal Gazette*, 29 April 1849.
18. *Mercantile Journal*, 27 April 1826.
19. *Mercantile Journal*, 27 April 1826.
20. *Royal Gazette*, 7 April 1829.
21. *Royal Gazette*, 21 April 1829.
22. *Royal Gazette*, 28 April 1829.

23 An explosion involving dynamite and blasting powder was the cause of Newfoundland's last sealing disaster, in 1931, when the S.S. *Viking* blew up, taking the lives of 24 men, including that of Varick Frissell and his American movie production team. Frissell had filmed his movie the previous year while on the S.S. *Ungava* and, briefly, on the *Viking*; he had returned in 1931 to obtain some more dramatically interesting footage, and for that reason he carried his own supply of dynamite (which was stored with the vessel's supply.) The movie was later released under the title *The Viking* and is an extraordinary audio-visual chronicle of the Newfoundland seal fishery at that time. (Frissell's earlier documentary, *The Great Arctic Seal Hunt*, is a silent film with sub-titles and also very informative.)

24 *Royal Gazette*, 7 April 1829.

25 *Newfoundlander*, 6 August 1829. (Note use of 'disasters'.)

26 *Newfoundlander*, 16 April 1829.

27 Please note that in 1853, it was estimated that the average value of a sealing vessel was estimated to be £1,000, and this included vessels of all classes.

28 *Royal Gazette*, 24 April 1829.

29 *Newfoundlander*, 8 April 1830.

30 James Murphy, *Old Sealing Days*, pp. 15-6. Contemporary newspapers do not mention Ryan or Mackey nor the *True Blue*; 'Native' Walsh was the nickname of a St. John's merchant, William Walsh, who sided with the strikers in 1842 (See Murphy, *Old Sealing Days*, p. 9); and the 'strain of Cape Broyle' means that they were on the same latitude as Cape Broyle (*DNE*, p. 538). The *Newfoundlander*, 11 March 1830, reported that the *Confidence*, a vessel of 77 tons burthen, under the command of John Picott [sic], went to the ice that year with 26 crew members. There were no lighthouses in Newfoundland (except Fort Amherst at the entrance to St. John's harbour), which made sealing particularly dangerous. The Cape Spear lighthouse opened in 1836 (see Malcolm MacLeod, "Lighthouses," *ENL*, pp. 295-303).

31 *Public Ledger*, 15 February 1831.

32 *Newfoundlander*, 31 March 1831.

33 *Public Ledger*, 6 May 1831.

34 See *Royal Gazette*, 30 April 1833, for all these 1833 losses.

35 *Royal Gazette*, 27 May 1834.

36 *Newfoundlander*, 21 August 1834. Annual Report of the Chamber of Commerce, 6 August 1834.

37 *Patriot*, 3 May 1836.

38 *Royal Gazette*, 24 April 1849.

39 *Newfoundlander*, 6 and 20 April 1837.

40 *Royal Gazette*, 3 April 1838.

41 *Royal Gazette*, 3 April 1838. Alexander A. Parsons, "*Newfoundland* Tragedy and the loss of the *Southern Cross*," *Newfoundland Quarterly* XIV, no. 1 (July 1914), 1-6. Parsons wrote that 14 vessels and over 300 men were lost at the seal fishery in 1838; this has not been confirmed.

42 Mosdell, *Chafe*, p. 37.

43 Tocque, *Newfoundland*, pp. 304-6.

44 Mosdell, *Chafe*, p. 40.

45 *Royal Gazette*, 18 April 1843.

46 *Patriot*, 19 April 1843.
47 *Sentinel*, 18 April 1843.
48 *Royal Gazette*, 25 April 1843. Quoted from the *Conception Bay Herald*, 5 April 1843.
49 Anspach, *Newfoundland*, p. 419.
50 *Newfoundlander*, 13 April 1843.
51 *Newfoundlander*, 4 April 1844. Mosdell, *Chafe*, p. 40, reports that it was known as the "Spring of the Growlers."
52 Mosdell, *Chafe*, p. 37. This report has not been confirmed.
53 Mosdell, *Chafe*, p. 40. In June, most of St. John's burned.
54 *Star*, 26 March 1846.
55 *Newfoundlander*, 16 April 1846.
56 *Newfoundlander*, 20 April 1846.
57 *Newfoundlander*, 11 May 1846.
58 *Star*, 16 April 1846.
59 *Newfoundlander*, 11 May 1846.
60 *Newfoundlander*, 20 March 1846.
61 *Newfoundlander*, 19 March 1846, contains information on the above four vessels.
62 *Newfoundlander*, 18 March 1847. (The port sent a greater number of ships in 1832, but the total tonnage was less.)
63 *Patriot*, 26 April 1847; and *Times*, 21 April 1847.
64 Mosdell, *Chafe*, p. 37.
65 *Weekly Herald*, 9 May 1849.
66 *Weekly Herald*, 9 May 1849.
67 *Newfoundlander*, 5 April 1849.
68 Mosdell, *Chafe*, p. 41, and Murphy, *Old Sealing Days*, p. 37. Murphy says that the spring of 1853 was called the "Spring of White Bay," and that of 1857, the "Frosty Spring."
69 *Newfoundlander*, 8 April 1852.
70 *Newfoundlander*, 8 April 1852.
71 *Newfoundlander*, 15 April 1852.
72 *Public Ledger*, 16 April 1852.
73 *Public Ledger*, 16 April 1852.
74 *Times*, 10 April 1852.
75 *Times*, 17 and 21 April 1852.
76 *Times*, 21 April 1852. According to Chafe, the brig *Hammer* ran ashore near Cape Broyle and 37 men died. Mosdell, *Chafe*, p. 37. Please note that no reference to this disaster has been found in the newspapers. Chafe may have intended to describe the loss of the *John & Maria* and 24 out of 30 crew members off Cape Broyle in 1857. The captain of the *John & Maria* was Carew; Chafe's captain of the *Hammer* was also Carew.
77 *Royal Gazette*, 20 April 1852.
78 *Times*, 17 April 1852.
79 *Times*, 17 April 1852.
80 *Newfoundlander*, 3 May 1852.

81 *Times*, 28 April 1852.
82 *Patriot*, 31 May 1852.
83 *Newfoundlander*, 3 October 1853. Evidence presented to the Vice-Admiralty Court in 1852.
84 *Newfoundlander*, 1 September 1853. "The decision of the Vice-Admiralty Court."
85 Mosdell, *Chafe*, p. 37.
86 *Newfoundlander*, 9 March 1857.
87 *Royal Gazette*, 14 August 1860.
88 *Royal Gazette*, 16 April 1861.
89 Mosdell, *Chafe*, p. 42.
90 *Times*, 12 April 1862; and *Public Ledger*, 4 April 1862.
91 *Royal Gazette*, 15 April 1862.
92 *Times*, 12 April 1862.
93 *Times*, 16 April 1862.
94 *Royal Gazette*, 22 April 1862.
95 *Public Ledger*, 25 March 1859.
96 See Tables 3.2 and 3.3 for more complete figures on the sealing fleets during this period.
97 *Royal Gazette*, 31 May 1864. See also Mosdell, *Chafe*, p. 42.
98 Mosdell, *Chafe*, p. 42.
99 *Royal Gazette*, 24 May 1864.
100 Mosdell, *Chafe*, p. 42.
101 *Royal Gazette*, 31 April 1868.
102 *Royal Gazette*, 28 April 1868.
103 *Royal Gazette*, 19 May 1868.
104 *Royal Gazette*, 19 May 1868.
105 *Royal Gazette*, 20 April 1869. See also letter from George Jeffers (grandson of Captain Joseph Jeffers), Farmville, Virginia, to Dr. Cater Andrews, 26 September 1962. CWA 213 VI.
106 *Evening Telegram*, 5 June 1893; and *Chronicle*, 16 April 1869.
107 *Royal Gazette*, 20 April 1869.
108 *Royal Gazette*, 26 April and 17 May 1870.
109 *Royal Gazette*, 21 June 1870.
110 *Royal Gazette*, 21 June 1870.
111 *Newfoundlander*, 26 May 1871.
112 *Royal Gazette*, 26 March and 2 April 1872; *Newfoundlander*, 7 May 1872 and 17 January 1873; and *Times*, 8 May 1872.
113 *Royal Gazette*, 28 April 1872.
114 *Newfoundlander*, 7 May 1872 and 17 January 1873.
115 *Newfoundlander*, 7 May 1872.
116 *Times*, 8 May 1872. See also, Murphy, *Old Sealing Days*, p. 42. Murphy states that the captain, his son and 41 of the crew were lost and that either 17 or 18 survived. Mosdell, *Chafe*, p. 39, says 49 lives in all were lost. It would appear, however, that 44 were lost and 18 saved.

117 *Royal Gazette*, 3 September 1872.
118 *Newfoundlander*, 3 July 1872.
119 *Royal Gazette*, 9 October 1872.
120 *Newfoundlander*, 17 January 1872.
121 *Royal Gazette*, 9 October 1872.
122 Quoted in the *Newfoundlander*, 7 June 1872.
123 *Newfoundlander*, 2 April 1878.
124 *Royal Gazette*, 30 April 1878.
125 *Patriot*, 29 April 1878.
126 *Newfoundlander*, 2 April 1878.
127 Mosdell, *Chafe*, p. 37.
128 Mosdell, *Chafe*, p. 42.
129 *Royal Gazette*, 19 May 1868.
130 *Patriot*, 1 May 1868.
131 *Patriot*, 29 April 1878.
132 *Evening Telegram*, 29 March 1880.
133 *Evening Telegram*, 5 April 1880.
134 *Evening Telegram*, 9 April 1880.
135 *Evening Telegram*, 19 April 1880.
136 *Evening Telegram*, 9 April 1880.
137 *Evening Telegram*, 19 March 1891.
138 Prowse, *Newfoundland*, pp. 520-21.
139 *Evening Telegram*, 29 February and 29 March 1892. Green Bay(two)—Solomon Penny and John Nurse. English Harbour (ten)—James and Tobias Penny (brothers); Martin Batston, William Barnes, Edward and Reuben Pottle (brothers); Isaac Batston, Artheu [sic] Batston, William Batston and James Ivany.Salmon Cove East (four)—George, Charles, Henry and William Nurse (brothers). (This is a little confusing because the report continues; "father and mother lost five sons. They have two young men left, at the herring fishery, and a boy and girl at home." Possibly this means that four sons were lost on this occasion and one other son on a different occasion—terrible tragedy in either case!] Salmon Cove West (two)—John Penny and Charles Day. Robin Hood (two)—William Stockly and Isaac Butler. Ship Cove (two)—Robert Bannister and son Charles. Trinity West (three)—John, George and Jacob Moores.
140 *Newfoundlander*, 9 May 1871; and Mosdell, *Chafe*, p. 42.
141 *Newfoundlander*, 7 May 1872 and Mosdell, *Chafe*, p. 42. The *Retriever* was one of two steamers owned by Ridleys and its loss was, no doubt, a factor in the firm's almost immediate bankruptcy.
142 Murphy, *Old Sealing Days*, p. 53.
143 *Newfoundlander*, 12 April 1872.
144 *Newfoundlander*, 10 April 1874; *Times*, 11 April 1874. Mosdell, *Chafe*, p. 43, puts the number of dead at 25.
145 *Newfoundlander*, 10 April 1874; Mosdell, *Chafe*, p. 43.
146 *Newfoundlander*, 23 May 1876; Mosdell, *Chafe*, p. 43.
147 *Royal Gazette*, 23 April 1878; Mosdell, *Chafe*, p. 43.

148 Mosdell, *Chafe*, p. 43.
149 Mosdell, *Chafe*, p. 43.
150 *Evening Telegram*, 12 April 1882.
151 *Evening Telegram*, 8 March 1894.
152 See *Evening Telegram*, 18 April 1977. Michael Harrington, "Offbeat History," for a detailed account of this tragic voyage of the S.S. *Greenland*.
153 *Evening Herald*, 9 April 1898. The men who died were as follows: George Bungy, Kenneth Parsons, Isaac Green, George Norris, Herbert Norris*, Henry Curtis, James Cheeks, Walter Norris—Newtown, B.B.; William Kelloway, James Howell, Joseph Osmond, Benjamin Bowne*, Thomas White—Pool's Island, B.B.; John Pinsent, John Thomas, Edwin Davis—Safe Harbour, B.B.; Albert Boland, Theodore Norris—Pound Cove, B.B.; Edwin Hunt, Alexander Andrews, Job Vincent**—Cape Freels, B.B.; Jacob Pond, William Blackwood—Greenspond, B.B.; Stephen Squires*—Salvage, B.B.; Michael Hennessey***—St. Brendan's, B.B.; Charles Ralph—Flat Island, B.B.; John Wicks—Wesleyville, B.B.; Frederick House—Gooseberry Island, B.B.; Thomas Ricketts—Knight's Cove, T.B.; Heber Ryan—Ship Cove, T.B.; George W. Pelly—Hant's Harbour, T.B.; Jacob Conway—Turk's Cove, T.B.; William Lauder, William Woolridge—Trinity, T.B.; Ambrose Rogers—Lower Island Cove, C.B.; Alfred Newtry, Walter Murphy—Carbonear, C.B.; William Heath, Matthew Wells, Lorenzo Wells, Archibald Courage, Noah Mortimer, George W. Pynn—Harbour Grace, C.B.; Richard Pynn, James Mallard—St. John's; William Cullen—Torbay; James Maher, William Voisey—Quidi Vidi. See *Evening Telegram*, 28 March 1898.
*In the *Evening Herald*, 28 March 1898, Herbert Norris is Heber Norris, Benjamin Bowne is Benjamin Brown, and Stephen Squires is J. Squires.
**In the *Evening Telegram*, he is referred to as John Pincent, while in the *Evening Herald*, he is Job Vincent.
***In the *Evening Telegram*, he is listed as Nicholas Hennessy, but in the *Evening Herald*, he is M. Hennebury. (Mr. Ronald Hynes—formerly of St. Brendan's—confirmed that Michael Hennessey was his name.)
154 *Evening Herald*, 9 April 1898. A man by the name of Butt from Pouch Cove narrowly escaped death. He was part of a boat's crew that picked up living and dead sealers and brought them to the vessel. The boat was overloaded and he stayed behind to wait for its return. When it did not return after a number of hours, he took off his clothes, tied them in a bundle on his back and swam across the lake which separated him from the ship.
155 *Evening Herald*, 2 and 4 April 1898.
156 *Daily News*, 6 April 1901.
157 *Evening Herald*, 28 March to 2 April 1907.
158 *Evening Telegram*, 3 April 1907. Donald Morison was a prominent member of the parliamentary opposition.
159 *Evening Herald*, 7 to 11 March 1907.
160 *Evening Herald*, 2 April 1907.
161 *Evening Herald*, 13 April 1908.
162 *Evening Herald*, 8 April 1908.
163 *Evening Herald*, 14 April 1909.
164 *Evening Herald*, 10 April 1909.
165 *Evening Telegram*, 11 April 1910.

166 *Evening Herald*, 10 March 1913; and *Evening Telegram*, 6 March 1913.
167 The names of the officers and crew are as follows: Captain—L.M. Linklater (with wife and son on board); First Officer—Capt. Chris Allen, St. John's; Second Officer—Capt. George Jackman, St. John's; Third Officer—Capt. Jacob Winsor, Welseyville; Chief Stewart—Joseph House, St. John's; Cook—H. Whitten, St. John's; Boatswain—Peter Jackman, St. John's; Chief Engineer—John Leary, St. John's; and three other engineers; (names unknown). The stokers were as follows: James Cooper, John Grouchy, John Byrne, Edward Byrne, John Collins, John Locke, John Connolly, Joseph Joyce, Joseph Jackman (from St. John's) and Bernard May (Quidi Vidi). The seamen were: Caleb Winsor, Wesleyville; Gerald Graham, St. John's; M. Palphrey, St. John's; James Murphy, St. John's; Albert Howell, St. John's; J. Finn, St. John's; Silas French, Carbonear; James Penney, Brigus; and Harold Fifield, F. Lucas, Peter Wills and Herb Balsam (addresses unknown).
168 See CWA 375 III. Alice L. Lacey to Dr. Cater W. Andrews, 19 September 1974.
169 Peter Jackman had been second hand in the S.S. *Walrus* for five years leading up to 1907, when the position of captain on this steamer became vacant. He applied for the post, but the job went to Jacob Winsor, who was also lost on the *Erna*. See *Evening Telegram*, 9 February 1907. It is very likely that Peter Jackman had accepted the position of boatswain on the *Erna* because other work at Baine Johnstons was slow and he was therefore one of the number who applied for the various positions involved in taking the steamer to Newfoundland. It seems that a number of men with the company were simply sent to Scotland, while on the company's payroll, to carry out this task.
170 *Evening Telegram*, 2 April 1914.
171 *Evening Telegram*, 3 April 1914.
172 *Evening Telegram*, 3 April 1914.
173 S.S. *Southern Cross* officers and crew:
Captain—George Clark, Brigus;
Second Hand—James Kelly, Brigus;
Chief Engineer—David Parsons, St. John's;
Second Engineer—Thomas Connell, St. John's;
Third Engineer—Thomas Hammond, St. John's;
The firemen were as follows: W. Walsh, M. Scammel, Patrick Stapleton, Gregory Brennon and John Whelan from St. John's.
Sealers: Fred Follett—Broad Cove; James Dunphy—Tors Cove; Noah Sparkes, Thomas Sparkes, Angus Winsor, James Youdan, B. Watts and John Clarke—Brigus; George Hall, John Walsh, John Conway, John Cole, M. Conway and Patrick Burke—Colliers; John Landry and Ambrose Matthews—New Chelsea; Josiah Noel, Robert Gillet, John P. Hiscock, Oscar Forward, Robert Penny, James P. Patrick, Wm J. Howe, Amos Penny, Alfred Pike, Henry Clarke, Norman Penny and George Murray—Carbonear; Eleazar Morris—Clarke's Beach; John Coombes, James Neil, John Mercer and John Robbins—Island Cove; Wm Gosse—Little Bay; Arthur Benson, Ronald Knight, William James, Lorenzo Parsons, Mark Yetman, Wm Coombes, John Bradbury, John Griffin, Wilfred Parsons, Elias James, Ernest Noseworthy, Isaac James, W. C. James, James Noseworthy, Norman Noel, James Lynch, M. Morrissey, James Bray, Herb Parr, Arthur Martin, John Callahan,* George French, Herbert Bray, Wm Webber and Thomas James—Harbor Grace; Thomas Hickey—Holyrood; Pat Dyer—Logy Bay; W. White—St. Mary's; James Quilty—Horse Cove; John Manfield, John Costello, James Walsh and Thomas Costello—Conception Harbor; Wm Norman, Kenneth Taylor and Herb Butler—Cupids; Samuel Kennedy, Wm Kearney, Joseph Morgan and Alexander Morgan—Seal

Cove; Joseph Corbett—Clarke's Beach; Thomas Bright—Breen's Cove; Martin Newell and Fred Newell—Upper Island Cove; George Smith, Art Clarke, George Vokey, Isaac Vokey, George Chetman, Nathan Chetman, Wm Clarke, Robert Gosse, Thomas Barrett, Joseph Yetman, Robert Clarke and Edward Crane—Spaniard's Bay; Wm J. Butler, Wm. Butler, Henry Leary, John Bishop, Uriah Button, Samuel Butler, Joseph Bussey, James Maley and Samuel Rideout—Kelligrews; Alfred Bussey, Thomas Bussey, Ambrose Taylor, George Patten, W. C. Butler, Henry Butler, Noah Bussey, Gordon Bussey and Joseph Batten—Fox Trap; Patrick Hearn—Goulds; John Field, Lawrence Yeo, Fred Squires, James Robertson, Allan Lindsay, Charles Quetel, James Martin, Walter Clarke, John Ebbs, John Mansfield and John Butler—St. John's; James W. Hallett—Arnold's Cove; Edward Squires, Geo. Hiscock and Alex Squires—Topsail; Walter Pierce, Charles Norman and Elias Mason—Catalina; James Foley—Grey Islands; Walter O'Roarke and Walter Carroll—Outer Cove; Albert Clarke, Wm Sharpe, Walt Lynch, Ambrose Sharpe and John W. Clarke—Paradise; James Squires—Broad Cove; James Porter, Henry Smith, John J. Stanley and Wm Stanley—Long Pond, Manuels; Alex Field, John Evans and Thomas Manning—Torbay; Abner Harris—Adeytown; Lawrence Gibbons, Sebastian Gibbons, Edward Gibbons, Thomas Gibbons, Cornelius Hemming and James Walsh—St. Vincent's; Henry Chafe—Petty Harbor; Wm Walsh—Northern Bay (Bay de Verde); Leonard Skiffington—Newman's Cove; Edward Kenny—Fermeuse; James Blundon and John Hannon—Low Point; Benjamin Robbins—Lower Island Cove; Thomas Bartlett—Turk's Gut; Noah Rowe and Jacob Rowe—Chance Cove; Edward Barrett—Tilton.
Source: H. F. Shortis, "Fugitive History," Vol. VIII, pp. 462-5. *The above named were those who received tickets to go on the *Southern Cross*; however, John Callahan of Riverhead, Harbour Grace, sold or gave his ticket to Jack Russell of the same community, who went in Callahan's place and was also lost. Also, because some men used assumed names in order to escape creditors, it is impossible to list with certainty the correct names of all sealers.

174 Edward Russell, Riverhead, Harbour Grace (1883-1985), whose brother Jack was lost in this disaster, always claimed that sealers who had previously sailed on the *Southern Cross* maintained that the pound boards were rotten and broke loose in the storm; the heavy pelts shifted to one side and the ship rolled over.

175 Most of the present discussion is taken from the information presented to the commission that was appointed to enquire into the sealing disasters of 1914. This document is entitled "In the Matter of Enquiry into Disasters at Seal Fishery of 1914: Evidence taken before the Commission, November 30th—January 13th, 1915." See CWA 980 Misc. Referred to hereafter as "Evidence: Sealing Commission." For an excellent account of the *Newfoundland* disaster, see Cassie Brown, *Death on the Ice* (Toronto, 1972).

176 Captain Abram Kean's Evidence, "Evidence: Sealing Commission," pp. 177-98.

177 Cassie Brown, *Death on the Ice*, pp. 118 and 262.

178 John Tizzard's Evidence, "Evidence: Sealing Commission," pp. 116-20.

179 John E. Hiscock's Evidence, "Evidence: Sealing Commission," pp. 139-47.

180 The following is a list of those who died as a result of the *Newfoundland* disaster as published in the *Daily News*, 6 April 1914. This list varies to a small degree from that in Cassie Brown, *Death on the Ice*, p. 267.
Raymond Bastow, John Brazil, Charles Davis, Daniel Downey*, Charles Olsen, William Pear—St. John's; Stephen Donovan—Petty Harbor; John A. Ryan—Gould's; John Butler, Valentine Butler, Bernard Jordan, Thomas Jordan—Pouch Cove; Patrick Gosse—Torbay; William Lawlor—Horse Cove, Torbay; James Porter—Manuels; John Taylor—Long Pond, C.B.; Michael Joy—Harbor Main; John

Mercer—Bay Roberts; R. Corbett*—Clarke's Beach; G. L. Whitney—Harbor Grace; James Bradbury—Shearstown, C.B.; Albert Kelloway, Joseph Hiscock—Carbonear; Ambrose Mullowney—Bay Bulls; James Ryan—Fermeuse; Joseph Williams—Ferryland; Charles Foley—Placentia; Peter Lamb—Red Island, P.B.; Benjamin Chaulk, Noah Tucker, Albert J. Crew, Reuben Crew, Alex Goodland, Charles Cole, William Oldford—Elliston, T.B.; Fred Pearcy—Winterton, T.B.; F. B. Marsh*—Deer Harbor, T.B.; Alan Warren—Hant's Harbor; George Carpenter—Catalina; William J. Tippert, Norman Tippett, Theophilus Chaulk, Jr., Abel Tippett, Edward Tippett—Little Catalina; Charles Warren*, Robert Matthews, Hezekiah Seaward, Peter Seaward—New Perlican, T.B.; Simon Cuff, Thomas Hicks, Fred Carroll—Bonavista; Mark Howell, Adolphus Howell, Adolphus Dowling*, Edgar Howell, M. Howell*—Newtown, B.B.; Robert Brown, Jonas Piccott—Fair Islands, B.B.; Percy Kean—Valleyfield, B.B.; Eli Kean—Pound Cove, B.B.; Robert Maidment, Alfred Maidment—Shambler's Cove, B.B.; Job Eastman—Greenspond; William Fleming—Spillars Cove, B.B.; Fred Collins—Newport, B.B.; David Abbott, David Cuff—Doting Cove, Fogo; Fred Hatcher—Cat Harbor, Fogo. Addresses presently unavailable: John Keels (died in hospital), Michael Murray, Pat Corbett, Samuel Martin, Benjamin March, Mike Downey, Nick Morey, Alfred Dowden, Henry Jordan, David Locke, Michael Murray, Art Mouland, "Uncle Ezra" Melendy, Henry Dowden, James Howell, Philip Holloway.

Please note that the * denotes individuals who were identified by the *Evening Telegram* and the *Daily News* as victims of the disaster but who were not identified as such by Cassie Brown. However, the last sixteen names above were not listed in the newspapers, and the last eight of these were listed by Brown as those lost whose bodies were not recovered. The fact that there are a number of names which are not listed in the different sources may be explained by the practice of sealers going to the ice using tickets issued to others.

181 John E. Hiscock's Evidence, "Evidence: Sealing Commission," pp. 139-47.
182 Rockwell Kent, *North by East* (New York, 1930), pp. 68-77.
183 *Newfoundlander*, 12 August 1830.
184 *Newfoundlander*, 21 April 1831.
185 *Royal Gazette*, 9 October 1872.
186 *Times*, 18 and 22 April 1874.
187 *Evening Telegram*, 5 April 1898.
188 *Daily News*, 27 April 1914.
189 Thanks to Mr. Fred Winsor, Ph.D. student, History Department, MUN, for use of his notes.
190 *Evening Mercury*, 24 October—9 November 1885. Forty lives were lost at the White Bear Islands, where two vessels, the *Hope* under Capt. King and the *Release* under Capt. Hayden, were lost. In this instance, the unfortunate people had boarded the vessels to leave for their homes on the island of Newfoundland. The wind rose, and the captains decided to wait until morning. The *Standard*, 2 and 3 November 1885, reported that 39 lives were lost at the White Bear Islands on that occasion, including about 30 women and children.

CHAPTER 6
A Sense of Identity

AS ONE CAN SEE, the spring seal fishery brought enormous changes to the old cod fishing station that made up the quasi-colony of Newfoundland—a colony that included the coast of much of Labrador but only part of the island, because the French Shore was an entity onto itself. (Only with the growth of Newfoundland as a society and colony in the nineteenth century did that region come to the attention of the English/Irish east coast.) At the beginning of the French Revolutionary War, Newfoundland was home to about 12,000 permanent residents, living in about 2,000 households.[1] These householders or planters provided employment to thousands of migrant fishermen, many of whom used the fishery as a stepping stone to the mainland; in fact, the passage of the Passenger Act by the British Parliament in 1803 guaranteed that Newfoundland's position was enhanced, because the cost of travel on fishing ships to the Newfoundland fishery was so much lower than the price charged by emigrant ships.

The deterioration of the situation in Ireland, combined with the enormous increase in the price received for Newfoundland saltfish in the Spanish and Portuguese markets, led to an influx of Irish to Newfoundland, seeking and acquiring employment in the saltfish industry at high rates of pay. By 1815, the population had grown to about 40,000 people, and the depression of 1815-18, exacerbated during the winters by unemployment, violence, extremely cold weather and fires, seemed destined to transform Newfoundland back to a migratory fishery, centred on a skeletal population of planter inhabitants employing migrant English and Irish fishermen. However, as already described, a successful seal fishery in the spring of 1818 changed the future of Newfoundland forever.

The direct and general impact of this industry on the economy and on demographic developments have been discussed. However, the seal fishery also had a comprehensive influence on society and culture in general and contributed to the development, by 1914, of a distinctive Newfoundland identity.

One point has been obvious from the beginning of this study and that has been the change that became apparent in the seal fishery (at least in retrospect, but also to contemporaries) around 1860. Up to this point, for the most part, the industry was noted for its buoyancy, and its impact upon society tended to be positive during this period. After 1860, with only few exceptions, the industry was faced with a problem of image and long-term prospects, which set their own more negative parameters for developments in the last half of the period up to the beginning of the Great War. It is the extent to which the influence of the seal fishery could be felt in the social and cultural areas that needs to be examined next.

Labour's Reaction

Given the fact that the growing colony was built around the expanding seal oil industry, it is not surprising that sealers were soon at the centre of protests and collective action aimed at increasing their share of the proceeds of this fishery and trade. As traditional cod fishermen, these men had never had a significant voice in determining their living and working conditions and wages. Only their proximity to the United States of America with its emigration option and some rare cases of desperate violence on the part of the fishermen guaranteed them a basic livelihood. However, as ice hunters, they were engaged in a prosperous industry, with outports, firms and planters in competition with each other. Furthermore, many lived in larger centres, such as Harbour Grace and Carbonear, and were easily able to communicate among themselves. To their numbers were added others from smaller neighbouring outports—men who came by horse and on foot from miles around to prepare the ships and equipment in February and early March. It is not surprising, then, that these men were the first to engage in collective action in Newfoundland in an effort to improve their conditions and acquire a bigger share of the profits from sealing. In addition, their actions were bound to have wider implications for Newfoundland society.

Strikes in Harbour Grace and Carbonear in 1832

The first strike reported in the seal fishery occurred in 1832 and proved to be a highly significant event. The strike began with the posting of notices on 5 January in Harbour Grace and Carbonear, which announced that a meeting would be held on 9 January to which fishermen and shoremen of both towns were invited.[2] The meeting took place on Saddle Hill, which is between the two towns, and its purpose was to put pressure on the merchants to cease paying sealers in kind or 'truck' and switch to payment of shares in cash. The meeting, attended by between 1,000 and 3,000 men (reports vary), agreed that the men were not to go to the ice unless they were promised wages in cash.

The meeting did not appear to make any impression on the merchants, so on 4 February, another meeting was called for 9 February, to which sealing masters were invited and commanded to bring copies of their sealing agreements.[3] A number of masters complied, but some merchants ignored these activities, and one of the most important merchants in Harbour Grace—Thomas Ridley—was among the latter. This brought swift reaction; before dawn on 18 February, a large body of men boarded that firms's vessel *Perseverance* with saws, axes and guns and, forcing the officers sleeping below to stay where they were, caused considerable damage.[4]

The local magistrates appealed to the governor for assistance, but although more constables were sent from St. John's, they accomplished little. The men continued to march in solidarity from one fishing premises to another, exhorting their fellow sealers to stay ashore and threatening the few who seemed to be willing to go to the ice under any agreement. It seems that the men were successful in their demands for cash payments and by 14 March, the fleets had sailed.

This strike, as Linda Little points out, was significant in that the men acted collectively and without regard to their religious differences.[5] Also, as Little reminds the reader, this strike took place in the context of a growing and prospering industry, which was a constant reminder to the men that their labour was valuable and very much in demand. Cash versus truck seems to have died out as a problem at this time—not that sealers could always expect to be paid cash, because many, especially in the later decades, were forced to allow their seal earnings to go towards their accumulated debts.

Strikes in St. John's in the 1840s

The next issue which became highly visible was that concerning berth charges. As already indicated, men had always paid a sum for their berths, with gunners paying less than regular batsmen. These charges varied from one port to the next, with the highest ones in St. John's and the lowest ones in the smallest outports. By 1842, the batsmen in the capital were paying over £3 each, with the possibility of another rise. That spring, the men carried out a protest, demanding a reduction in this charge. On 16 February, the sympathetic *Patriot* reported:

> A public meeting was to have taken place on the Barrens, yesterday, to discuss the subject of Sealing Birth [sic] Money; but the weather proved too inclement. It is, we understand, to take place to-day. This is a subject of great importance, and those engaged in the Seal Fishery should endeavour to have the question settled in such a way as that no obstacles should lie in the way of the prosecution of the voyage. The men appear determined to have some abatement, and we think *that* abatement could be made without the slightest injury to the Trade.[6]

A few days later, the men held a meeting and a parade, which were reported by the local newspaper most hostile to the cause of labour—the *Ledger*:

> This town has for the last few days been the scene of an unusual degree of excitement—a mob, variously estimated at from 500 to 1000 men, having paraded the streets with drums beating and colours flying, and visited the wharves, and the vessels preparing for the Sealing voyage, for the purpose of compelling the men to abandon the engagements they have entered into, and of coercing them into a combination to reduce the prices severally charged to them as 'berth-money.'
>
> In the prosecution of this design they have to a certain extent succeeded, inasmuch as by the use of threats, and by personal violence, they have intimidated the men into a temporary compliance with their arbitrary mandates. But several of the ring-leaders have been apprehended, and securely lodged in gaol, where they will most probably undergo two or three months' imprisonment.
>
> There is nothing illegal in a number of persons quietly and peaceably assembling together to determine for themselves the wages which they may agree to receive for their labour, nor even in their parading the streets in a body, unless under circumstances such as would naturally create apprehension in the minds of the inhabitants; but when they proceed to offer threats, and to use personal violence by way of compelling others to accede to their views and to join their number—(and it matters not whether those views be just or unjust)—it evidently becomes a conspiracy of a most dangerous

> character, which it is of the utmost importance instantly to suppress by the strong arm of the law.
>
> It is utterly impossible for us to offer an opinion as to whether the 'berth-money' charged to the sealers is in general too high or too low, and very many circumstances occur which must of necessity render the charge in a great degree arbitrary; but it is clear that if the merchant or ship owner cannot force the sealers into any given arrangement, so neither can the latter force the ship owners into a compliance with their particular views; it is wholly a matter of mutual bargain and agreement, and any attempt at unlawful coercion in the matter is destructive to the interests of all; for concessions so extorted would only lead to the renewal of fresh demands and to that description of violence and outrage not at all tolerated by the laws of the land, and in which the guilty party would inevitably involve himself and his family in much misery and distress.
>
> We trust that the examples which are about to be made of those who have been already apprehended will have the effect of deterring others from pursuing so ruinous a course.[7]

It was difficult for sealers to maintain a united front because of the short and seasonal nature of this industry. At the end of each voyage, the sealers re-entered the varied strata of the cod fishery and went to work once again at their respective jobs, from the unskilled shoreman to the relatively well-off planter. However, in 1842, the St. John's men not only prevented the merchants from raising the berth charges further, but forced them to make reductions. To return once again to the *Patriot*, reporting on 2 March:

> The local occurrences of the past week are of considerable interest. The first matter, (and one of paramount importance,) which has transpired, is the partial settlement of the disputes between the owners of Sealing vessels and the hardy and heroic body of people who man them for the fearful voyage to the Ice.
>
> The custom of the ship-owner hitherto, had been to charge every man, with the exception of a marksman called the 'Bow-gunner,' the sum of £4 or £4 10s. for his *berth*, or the *privilege* of risking his life in an enterprize which, at the furthest limit of success, could bring him not so much more than treble an ordinary seaman's pay, while under any degree of success the profits accruing to the Owner would be larger than could be obtained in any other sea-faring speculation, for which his vessel was competent. But it very frequently happened that the labouring Sealer made little more by ordinary voyages than sufficient to meet the charge of his *Berth*.
>
> This *Berth* charge had been a source of considerable dissatisfaction for several seasons, and had been tolerated, only, on the part of the Sealers from a desire to preserve harmonious feelings between themselves and the Ship-owners. The present, however, was considered a fitting opportunity to make a general stand against a charge which pressed so exceedingly heavy upon this industrious and adventuring class of men; and they forthwith made a simultaneous *strike*, and assembling peaceably on the Barrens, they drew up a scale of prices, by which they all agreed cheerfully to abide. This scale, we understand, was first suggested by Mr. Walsh, a ship-owner himself, and the reduction seemed to be a fair sum, considering the present rate of provisions and other expenses of outfit. We are happy in being able to state that Lawrence O'Brien, Esq. was the first in the Trade who complied with the requisition of his men; and this patriotic individual having set the

benevolent example, there is no doubt that it will be generally followed. The concession on the part of Mr. O'Brien, and the belief that the generality of the Trade will comply with the petition of the people, have had the effect of setting at rest the unwonted excitement to which the subject had given rise, and the business of the season is again proceeding with some activity.

There cannot be a question that the merchant should be perfectly secured for the capital he risks; but it is equally beyond doubt, that in effecting this security, the claim of justice towards the people who give their labour and risk their lives in the voyage, is of paramount importance. We should be the first to condemn the agitation of the subject, if we did not feel convinced that, under all the circumstances, the humble Sealer had no other alternative to make his grievance known, and that it was strictly a matter of right for him to combine legally with his fellow men for the settlement of the price of his *Berth* as it is for the Trade to meet at the Commercial Room to decide the price of the *fat* procured by that labour, at the hazard of the labourer's life.

But such questions, should never be allowed by the parties themselves to arrive at such a point. The interests of the ship-owners and the interests of the sealers are perfectly reciprocal, and matters of this nature should be mutually discussed and decided, without the untoward array of stubbornness on one side met by determination on the other....[8]

In 1842, Newfoundland was entering a new political era, and the Roman Catholic "Liberal" leadership had just seen the political power they had come so close to winning slip through their fingers as the British Government abolished the two-chamber system of representative government and installed a one-chamber amalgamated assembly. The Bishop and the Roman Catholic leadership were probably concerned about keeping their rank-and-file support, and because a majority of the sealers in the St. John's area would have been Catholics at this time, this probably explains the haste with which Walsh and O'Brien[9] threw their support behind them as well as John Nugent's impassioned plea on their behalf. This does not mean that the sealers did not have genuine support; the *Patriot* article demonstrates that they did—although this newspaper did have a political agenda as well. However, it does indicate that this was a particularly sensitive period, and working class support (meaning the Irish vote, in St. John's) had to be retained throughout the coming political hiatus.

In his book, published in 1878, Philip Tocque quotes extensively from a speech given by one of the leading Catholics of the 1830s and 1840s, John Nugent, who observed and commented on the 1842 strike and estimated the number of men involved at between one and two thousand. Tocque quotes Nugent as stating that the body of sealers met on the "Barrens," on the outskirts of St. John's, to protest the effort by the merchants to raise berth money to £3 10s for batsmen and £1 for the bow or chief gunners, who had always gone free. According to Tocque, the demonstration was successful in keeping the berth money at £2 for batsmen and £1 10s for after gunners, and the bow gunners went free as before. Nugent's comments on the seal fishery were made sometime after this episode (and have been quoted above), and he concluded with a plea for a fairer deal for the sealer:

> Upon the return of the sealing vessel, one half of the proceeds of the industry of the men is handed over to the merchant, in remuneration for the capital he had advanced in the first instance. The other half is divided amongst the

men, whose toil and daring procured it; but then, the merchant's half is given perfectly clear and unencumbered of all charges, of every deduction—the poor man's half is clipped and curtailed—he is, first, obliged to pay hospital dues; and, further, besides giving the merchant a full and undiminished half of the entire voyage, he is still further taxed by the merchant, to whom he is obliged to pay a sum of money, not only for the very materials used in its prosecution, but actually, a further sum for the privilege of *being allowed* to hazard his life to ensure a fortune for the merchant, and both of these latter charges combined are here both technically denominated 'BERTH MONEY.'[10]

Tocque, himself, comments that the action was successful and that although "some of the parties committed a trifling breach of the peace and were imprisoned [it] was for a short time [only]."[11] There is evidence to suggest that Captain Henry Supple was involved in leading this strike. Murphy, writing in 1916, quotes an elderly man, Mr. David King, as having said or written 14 years earlier (c.1902):

> It was organized by Henry Supple, a man of superior education, but a fisherman like myself, with whom I worked for some years. The object of the strike was to secure a reduction in the berth money. Prior to that, batsmen going to the ice paid 20 shillings for their berths, after gunners 10 shillings, and bow gunners went free. The rates were gradually raised to 60, 50 and 40 shillings [£3, £2 10s and £2] and then the men struck. They assembled at the head of King's Road, and with Bradley the fiddler, a piper and a drummer, marched through the town, visiting all the wharves, and searching the ships for those not in sympathy with them; each man had to fly for their lives [sic]. The strikers were masters of the situation, and the merchants reduced the rates to 20, 10 and 5 shillings. The town was small then and the merchants made a big showing. Besides rum was plentiful in those times, and it was not wise for the merchants to hold out too long.[12]

However, the St. John's strike of 1842 was not nearly as violent as that in Harbour Grace and Carbonear ten years earlier, and it seems to have been powerfully organized and smoothly run. Certainly, it was the first successful attack by sealers against the berth charges.

Hattenhauer, Murphy, Mosdell and Gillespie all point to 1843 as a significant year in the industry, when sealers sought major concessions from ship owners and merchants.[13] Although primary evidence is somewhat lacking, this point is borne out by the *Patriot*. On 15 February, that newspaper reported an "Important Meeting of Sealers" and went on to state:

> On Friday last a large Body of Sealers met on the Barrens, to take into consideration the oppressions and hardships to which they are subjected by the present system of the Seal Fishery, and to adopt means to obtain redress. A Committee of active men, practically well acquainted with the business of the sealing voyage were appointed to draw up a series of Resolutions, expressive of the Sealers' hardships and of their determination no longer to submit to them—but to remedy them in a legal and constitutional manner. After the formation of the Committee the meeting adjourned till Monday last, when a great concourse of persons interested in the Sealing voyage, with drum and fife and Banners, in processional order, marched through the town—and having in their route, repeatedly cheered "Our Gracious Queen," "Sir John Harvey," "The Authorities," etc., etc., proceeded to the

Barrens, and having gathered near the Military Road, the meeting was regularly called "to order," and Mr. Henry Supple, a practical fisherman, was nominated to read the Resolutions, as agreed to by the Committee appointed on Friday. This duty he readily complied with, and they all passed unanimously—The following is a copy:-

No.1.—Resolved—That in meeting to protect the rights of the SEALERS of St. John's, we have an equal desire to uphold the rights of the MERCHANT, and we strongly condemn as unjust and injurious to the character of the Sealers, any attempt to obtain their rights by any means but those which the laws of the country justify.

No.2.—Resolved—That all BATSMEN in general should have free Berths, and all others, such as Master-watches and Gunners, BOUNTY according to the station they fill.

No.3.—Resolved—That the object of this meeting is to enable the Merchant and Sealers to consider the best means of advancing and protecting the interest of both in the Seal Fishery.

No.4.—Resolved—That it is only just and reasonable that any man employed in fitting out vessels for the Seal Fishery, previous to the 25th February, shall be paid for his daily labour while so engaged and found in diet in that employ from the said date.

No.5.—Resolved—That upon returning from the Seal Fishery it is just and proper that each Crew should be allowed four days in order to dispose of the Seals to the best advantage, and that a provision to that effect be embodied in the agreement of the Sealers with their employers, but in all cases the Supplying Merchant to have the preference.

No.6.—Resolved—That as it is consistent with the laws of our country, that the labourer should be paid in CASH for his labour, we strongly urge that in all agreements between Sealers and their employers a provision be inserted to the effect that the Seals be paid for in Cash immediately after the vessel is cleared out.

No.7.—Resolved—That in order to protect as well the interest of the Merchant as the Sealers, we shall also require that in the said agreement, provisions be inserted whereby the employer shall engage to send upon the voyage three months provisions.

No.8.—Resolved—That in all cases where a Gun is lost on the voyage, by accident, one-half the price of the said Gun be subscribed by the Crew, and the other half by the Supplying Merchant, and that the same be provided in the agreement.

No.9.—Resolved—That when any man engages in fitting out a vessel for the Seal Fishery, as one of the Crew of the same, after the 25th February in each year, shall sustain in the said employ injury sufficient to disable him from proceeding on his voyage, he shall be regarded as entitled to his share of the Seals in the same manner as if he had proceeded on the said voyage.

No.10.—Resolved—That provisions be embodied in the agreement consistent with each of the foregoing resolutions, and we solemnly pledge ourselves to abide by the same.[14]

As one can conclude from the above, the St. John's men, having forced a reduction in the berth charges the previous year, were now attempting to eliminate them altogether. They also wanted men to be paid and fed while working on the vessels in preparation for the hunt, especially during the period prior to 25 February. In other words, they accepted that they should have to work on the vessels after 25 February as part of their obligation to their employers, but that the earlier period of employment should be dealt with separately. It is possible that merchants were abusing the custom of sealers' preparing the vessels for the voyage and were requiring them to do an unreasonable amount of repair, maintenance and preparation. Also, as Resolution No. 5 indicates, the merchants who supplied the vessels were engaged in finding buyers for the pelts, and it seems that the buyers were actually themselves. This meant that the price could be set artificially low to allow the merchant to increase his profits on the finished products—oil and leather—in the market place. Furthermore, the men were obviously required to go out on second trips without taking time for proper provisioning and were probably dependent on bread and tea for much of that time. There was a definite concern here about the quantity and quality of the food—one that had not arisen before—which is probably connected to the fact that larger vessels were being employed in the industry and the food situation was deteriorating (a point that has already been made). In addition, it must have been irritating to the crew to find that it was their share of the voyage from which the costs of replacement guns was taken. And finally, for the sake of justice and fairness, it was felt that men who were injured on the job while getting the vessel ready for the ice and unable to go on the trip should receive their share of the pelts procured.

As a result of this meeting the merchants and ship owners met in the "spacious Hall of Mr. Kielty, the Hon. James Tobin in the Chair," on February to discuss the "Resolutions passed at a large meeting on the Barrens of the Working Seal-hunters." The merchants were debating the issue of berth charges when the hall was invaded by a "vast number of Sealers." The crowd called on Supple to read the resolutions recently passed, and when he finished, he asked for a show of hands in support; of course, overwhelming numbers voted in support of them. The sealers then left, but there are no indications that they received any concessions at this time.[15] Regardless, what stands out in the contemporary reports and observations is the sophisticated and professional manner in which the meetings were held and the demands presented.

It seems that further protests and negotiations were carried out in 1844 as well. In 1845, when negotiating with the Brigus ship owners, Captain Supple mentioned a sealing agreement that had been reached the previous year at Kielty's Long Room in St. John's. Murphy also reports that a "Fishermen's Mutual Protective Society of Newfoundland" was organized at St. John's and that the fishermen met at Kielty's Long Room.[16] However, the contemporary reports used above suggest that 1842 was the key year as far as concessions to the sealers were concerned, and, although new evidence may demonstrate otherwise, it is the collective action itself and its success that were significant, rather than the specific year in which it occurred.

Strikes in Harbour Main and Brigus in 1845

Another major labour dispute occurred in 1845, once again, set off by the berth charges. The growing anger over these charges was probably related to the fact that the vessels had become much larger, and the sharing of expenses, from which berth charges originated, no longer could be justified to men whose working conditions were declining, partially because of the sheer size of the crews. Captain Henry Supple also played a prominent part in the labour protest on this occasion. It seems that Supple was asked by the sealers from the "South Shore" of Conception Bay—Brigus to Harbour Main area—to pay them a visit and assist them in their protest. A long detailed account of this visit was fully reported under large headlines:

> GREAT EXCITEMENT IN CONCEPTION BAY ON THE SUBJECT OF BERTH-MONEY! MONSTER MEETING ON BRIGUS POND—PEACEFUL AND SUCCESSFUL RESULT OF THE SAME!!

> A Deputation from the Fishermen of the South Shore of Conception Bay arrived here on the 26th Feb., for the purpose of inviting Captain Henry Supple of this town to visit that portion of the Bay, and direct the movement then determinedly operating, for the suppression of certain oppressive charges usually made on the sealers of that Bay, under the head of 'Outfits' and 'Berth-money.' Captain Supple, at once responded to the call of his countrymen and started for Chapel's Cove; on arriving at which place, he was met by some hundreds of people of the Shore, who welcomed him in the most hospitable manner, amid the firing of guns, and the loudest acclamations. It was evident the agitation was at its greatest height and it became the duty at once of an experienced leader to allay the storm which was evidently brewing. Capt. Supple took his steps accordingly, and immediately the multitude proceeded in processional order to Harbour Main, Capt. Supple at their head in a sleigh drawn by the people, having a neat white silk banner trimmed with green satin, ornamented with a red satin St. George's cross in the centre flying over his head. Various other flags were also waving in the breeze, and the woods resounded to the harmonious strains of the Temperance band of Harbour Main, which had arrived at Chapel's Cove for the purpose of escorting the procession down.

> ### -HARBOUR MAIN-

> Arrived at Harbour Main, a meeting (in consequence of the inclemency of the weather) was convened at the School House, which, though a tolerably roomy building, could not contain one-fifth of the people who had congregated from all parts of the Southern Shore, on the interesting and important occasion.

> On the motion of Mr. John Gorman, seconded by Mr. Peter Ezekial, Capt. Henry Supple was unanimously called to the chair, and on the motion of Mr. Ezekial, seconded by Mr. Michael Keating, Mr. John Gorman was appointed to the office of Secretary of the Meeting.

> The Chairman opened the business of the Meeting by first proposing nine times nine, and one cheer more, for Her Most Gracious Majesty Queen Victoria; next, three cheers for His Excellency Major General Sir John Harvey, Governor of Newfoundland, etc., then he proposed three cheers for the Executive Authorities of the Colony, and next nine times nine for all men who sympathize with the fishing population of the island. Each of which

propositions was vigorously sustained, and the last one rapturously encored.

The Chairman then stated the objects for which they had assembled, and regretted that they had not alighted on someone more qualified than himself to carry out the important and most serious objects contemplated by the sealing and fishing population in general. However, as they had conferred the high and responsible position upon him, humble as he was, he would endeavour, in an honourable and peaceful manner, to accomplish their wishes. One thing he could say, that if he failed, it should not be for want of determination and courage, to benefit his native countrymen and the Land of his Birth. They only required justice, and they ought to take no less. He then impressed upon the people the necessity of union and firmness, without which they could never succeed in throwing off the oppressions under which they had for so many years smarted without pity or commiseration. The Chairman concluded by calling on those who had Resolutions to propose to come forward, when, Mr. Peter Ezekial, proposed, seconded by Mr. Daniel O'Connell,—That the series of Resolutions he held in his hand be read from the Chair; which having been read were unanimously carried.

The Chairman read the following Resolutions: -

Resolved,—That the fishermen of Newfoundland will yield to no other class in Her Majesty's broad Dominions, the entertainment of more profound loyalty to the person, crown and dignity of their Sovereign Lady Queen Victoria, than they do, and that whenever occasion required it, nobody of her people have shown themselves more ready to shed their blood in the defence of her crown than have the fishermen of Conception Bay. That their attachment to the constituted authorities of the colony has never been doubted and they are determined never to lose by their conduct their loyal and peaceful character.

Resolved,—That the object of this meeting is to give an opportunity to the Merchants, ship owners and Sealers to consider and decide on the best means of settling the present difference between parties, and of advancing and protecting the interests of all engaged in the Seal Fishery.

Resolved,—That in meeting to protect the rights of the Sealers of Conception Bay we have an equal desire to uphold the rights of the merchants, and we condemn any attempt to obtain our rights by any means but those which the laws of our country justify.

Resolved,—That the charges imposed upon the sealing crews, called 'Berth-Money,' and 'Outfits' are unjust and unbearable, and ought to be discontinued—and that we consider a fair division of the Catch of Seals between the Merchant or owner of a vessel and the Crew a fair and equitable arrangement.

Resolved,—That we consider it just and reasonable that men employed in fitting out vessels previous to the 1st of March in each year, should be paid for their labour whilst so engaged, and found in Diet from said date.

Resolved,—That on their return from the Voyage it is right and proper that each Crew should, if required, be allowed four days to enable them to dispose of their Seals to the best advantage; but in all cases of sale the Supplying Merchant to have the preference.

Resolved,—That it is consistent with the laws of our country, that the Labourer should be paid in cash for his labour; we therefore strongly urge that in all agreements between the Merchant or Owner and the Sealers a provision be inserted to the effect that the Seals be paid for in cash immediately after the cargo is landed.

Resolved,—That in order as well to protect the interest of the Merchant as the sealer, that three months provisions of a wholesome 2nd quality, be put on board before proceeding on the voyage.

Resolved,—That in all cases where a gun may be lost on the voyage, accidentally, that half of the cost of the said gun be subscribed by the crew, and the other half by the Supplying Merchant.

Resolved,—That if in fitting out a vessel for the Ice, a man should, after the 1st of March, unfortunately be disabled to proceed on the voyage, he shall be regarded as one of the crew, and entitled to his full share of the Seals, as if he had proceeded.

Resolved,—That provisions be embodied in all Sealers' Agreements in accordance with the foregoing Resolutions, and we solemnly pledge ourselves to abide by the same.

These Resolutions were put by the Chair, and carried unanimously and with repeated cheers. The Chairman then informed the Meeting that it was necessary that a strong Committee should be formed to carry out the Resolutions, and to regulate their future proceedings; when the following persons were proposed and came forward to act as such: -

Henry Supple, Chairman	Jeremiah Quinlan
John Gorman, Secretary	Martin Costello
Peter Ezekial, Sen.	John Crawley
Michael Keating, Sen.	John Weach
Jeremiah Kennedy	Phelix Lewis
John Woodford	John Kennedy
Michael Hickey	John Joy
John Brick	Peter Quinlan
Thomas Wall	Philip Mahony
John Ryan	Michael Mahony
Richard Strapp	Richard Hearn

The Chairman then congratulated the meeting on its respectability and unanimity, and having again impressed upon them the necessity of obedience to the laws left the chair and Mr. Peter Ezekial being called thereto, he spoke in warm terms of the zeal of their chairman for his country, and then proposed a vote of thanks to Capt. Supple, for his judicious conduct in the chair—which was responded to in the house and outside of it, with the heartiest cheers. The meeting then adjourned, to be held next at Brigus.

-BRIGUS-

On the 27th, the Fishermen of Holyrood, Chapel's Cove, Harbour Main, Salmon Cove, Cat's Cove, Colliers, etc., etc., assembled at Brigus, and the only convenient place for so large a body of men being "Brigus Pond," a temporary hustings being erected thereon, Captain Supple was again called to the chair, and the vast assemblage formed in a wide circle around him, when he immediately explained to the men of Brigus the objects of the meeting—namely, to seek by legal and constitutional means the right of the Sealers to "free berths." On this announcement the hills of Brigus reverberated to the loud huzzas of the people—it seemed as if it were the first sound of liberty and independence that ever echoed from the Shores of the Bay—so powerfully was it responded to from shore to shore. This day was expended in awakening the people who were yet inactive to a sense of their rights. Late in the afternoon, the Chairman adjourned the meeting to the following Saturday.

On Saturday, the first of March, a very large meeting took place at Brigus, and proceeded, in peaceable procession around the town by Green's farm out to the town of Cupids, through that settlement and back again to Brigus, where the people separated, and each man returned to his home; many of them that day making a journey of between 40 and 50 miles.

It was understood by all that the "Monster Meeting" was to take place on Monday, the 3rd inst.; and accordingly on that day preparations were made by the people far and near to be in attendance. Early in the day, the people had poured into the town by hundreds—they came by sea and land, and assembling about half a mile from Brigus, formed in grand Procession. The Temperance Band of Hr. Main, consisting of a drum and fife, a clarinet, and two violins, led the way; then came the Chairman, carrying a neat and appropriate banner; next came the British Ensign, and at intervals several other beautiful flags; among them were the Temperance Flags of Hr. Main—one with the motto, "With the Divine Assistance I pledge myself." Another with the motto, "Firmness, Temperance, Union." A green silk flag with a White Cross, and, another with the figure of an Angel, and bearing the motto:

"The traveller heeds not the weary mile,
As he comes to bask in the angel's smile,
Low bending to earth his heavenly brow,
As he pledges his faith in that holy vow."

The Native Flag was amongst the most conspicuous.

The procession, six deep, then advanced into the town of Brigus and halted on the Pond. The number it is said amounted to fully 3,000 able-bodied men! The Chairman of the other meetings was enthusiastically called to the chair, and Mr. Gorman did the duty of Secretary. The cheering for the Queen, Sir John Harvey, and the Authorities being here repeated, the chairman proceeded to read the resolutions as passed above and the names of the Committee, which were appointed and again passed unanimously. At this period the meeting were agreeably surprised by the appearance amongst them of some of the most respectable Schooner-holders of Brigus, who notified the Chairman that with a view to settle matters amicably he, with the Secretary and Committee, would be kind enough to attend at the house of Robert Brown, Esq., to meet a Committee of Merchants and Schooner-holders appointed for the purpose.

Accordingly Capt. Supple, the Secretary and Committee of Sealers, agreed to accept the proposal, and immediately the Chairman addressed the people, acquainted them of the apparently happy result of their exertions, and left the meeting for the purpose of attending the house of Mr. Brown.

On arriving at the place appointed, business was at once commenced.

The Committee on the part of the Merchants and Schooner-holders consisted of the following persons:

Robert Brown, Nathaniel Munden, William Munden, Azariah Munden, John Penny, Richard Walsh, John Norman, James Wilcox, James Walsh, Patrick Brine, William Burke, Edgar Stirling, [and] Jeremiah Nowlan.

R. Brown Esq., proposed that the people abide by the St. John's rules.

Mr. Michael Mahoney objected, and thought that free berths should be allowed.

Capt. N. Munden suggested that he would reduce the price to 20 shillings for a berth.

Capt. Supple, the Sealers' Chairman, produced the Resolutions passed last year in St. John's, at Kielty's Long Room, and read them to the Committees, and suggested, on the part of the people, that the sum of 10 shillings be allowed as the price of Berths for the present.

Mr. Jeremiah Nowlan agreed with the Chairman, provided the sum, would be allowed all throughout the Crew. Outfits free.

After some discussion, the Mercantile Committee agreed to this reduction, and Robert Brown, Esq., with their consent, drew up the following agreement:-

"We agree to give our Sealing Crews their Berths at the Ice this Spring, at the rate of 10 shillings per man all throughout the crew. Persons shipped for Gunners to bring a good gun with spare lock, and use the same. Outfits free. Brigus, 3rd March, 1845.

Robert Brown, Patrick Brien, Jeremiah Nowlan, John Penny, James Willcocks, Edgar Stirling, John Willcocks, Wm. Walker, Wm. Woodford, Richard Ash, N. Munden, Wm. Cole, Wm. Burke, Joshua Curtice, Azariah Munden, Wm. Antle, Stephen Percey, Jonathan Percey, James Walsh, Wm. Munden, John Norman, Nathan Norman, Thomas Delaney, Thomas Roberts, John Antle, George Wells, Wm. Wells, James King, Wm. Andrews, John Dawe, George Andrews, Alfred Smith, Isaac Clarke, John Gushue, James Norman, James Burke, George Roberts, Thomas Spracklin, Jonathan Spracklin, Wm. Spracklin, Abraham Bartlett [and] Wm. Rabbitts.

The Chairman and Committee of Sealers, having so far brought the important business to a successful conclusion, returned to the assemblage on "the Pond" and having read to them the agreement as above, it was received with deafening applause.

The Chairman then formally dissolved the meeting, and the people peacefully and amicably departed, every man to his business or his home.[17]

The resolutions passed in Harbour Main were very similar to those passed in St. John's in 1843, but seemed to ask for just a little more. For example, it was demanded that sealers be paid for any work they were required to do on vessels prior to 1 March instead of 25 February, as was the case in St. John's. Nevertheless, this event was a considerable breakthrough for the sealers of the Brigus and (presumably) the Harbour Main area. The agreement was a definite improvement over that reached in St. John's in 1842, when 40s per batsman was set as the charge for a berth. There is an implication that a later agreement was reached in 1844 which probably reduced the sum charged in St. John's. Nevertheless, the offer of 20s, which was rejected at the above meeting, seems to have been less than that in effect in the capital. No doubt, the strike in Harbour Grace and Carbonear in 1832, when the sealers resorted to violence, created a precedent which continued to make an impression on all the ship owners in Conception Bay. In addition, the successes of the St. John's sealers in the earlier 1840s, provided encouragement to the sealers of Brigus and Harbour Main in 1845.

Certainly, this strike demonstrated, as had that of 1832, the sealers' ability to transcend religious boundaries and the effectiveness of their collective action.

Not all contemporary observers were heartened by the solidarity of the sealers. One St. John's newspaper sniffed on 15 March:

> A very large number of Sealing Vessels were seen in the ice from the Block House on Thursday and it is not at all unlikely that these are the very vessels which were protracted in proceeding on their voyage, owing to the improper interference, in the Bay, of a stormy 'peteral' [sic]—a Liberty Boy, who has lately been 'beating up' in order to coerce the 'suppliers' respecting the berth money etc.,—we trust that the seal hunters will not suffer through the arbitrary conduct of the would-be Patriot, to whom we allude and who we hear, has made a very good thing of his agitation.[18]

Thus, not all observers were happy with the evolving power of the sealers to negotiate agreements with the merchants and ship owners.

Labour after mid-century

The year 1845 had been a pivotal year in that it was the last time the sealers struck in the context of an expanding industry and a demand for labour. Furthermore, the major strikes of 1832, 1842 and 1845 had all occurred in the heart of the original spring seal fishery monopolized by St. John's and Conception Bay. There was one more disruption within this framework, and that occurred in 1853, when it was reported about:

> the 'diabolical conduct' of a number of men at Carbonear, who by way of enforcing the prosecution of the seal fishery on their own arbitrary and high handed terms, resorted to brutal violence and intimidation, so much so that the aid of the military was called in to quell the outlaws ... They must be made to feel that they cannot with impunity, place the law and the constituted authorities at defiance. It is the duty of the resident Magistrate who we presume has been an eye witness of all that has occurred to make a true and faithful report of the whole affair to the head of the government.[19]

Apparently, the issue of free berths had prompted a number of men in Carbonear to try to organize a strike, because it was reported:

> While the sealers were preparing for the Ice a crowd of people paraded the streets of Carbonear with drum and fife, demanding "free berths," and in some instances intimidating men from work on the vessels. When the first attempt was made, information having been given to the Magistrate, the constables were promptly despatched to the crowd, and in the course of the day, without riot or resistance, the obstructives were dispersed. The crews again went to work, and once or twice after, the "free berth" men repeated their efforts—which only went the length of interrupting the preparations of the sealers for very brief intervals, and were then quietly quashed as in the first instance.... It is remarkable indeed that in the whole affray about which such hubbub is made, there was no such thing as rioting; no injury worth naming as such inflicted even by the most turbulent; no stone-throwing, window-breaking or any violence of the sort.[20]

The courts attempted to deal with this lawlessness in the autumn, and this sparked some riotous conduct in Harbour Grace, both near and within the court house. Windows were broken, but some observers thought that the media over-reacted by making it appear that a serious riot had occurred. It was thought by some that the resulting bad publicity would only discourage investment.[21] This illustrates clearly the difficulties in dealing with labour problems during a turning point in the history of an industry. In the 1850s, the seal fishery had reached its zenith, and there was competition among the mercantile establishments of St. John's, Harbour Grace, Carbonear and Brigus as each fought to retain its share of the industry. Negative publicity regarding the security of investments would not help the cause.

The last half of the century was a period of labour peace—a sad commentary on the depressed state of the industry rather than a happy reflection of improved conditions. With the exception of the occasional 'manus' (refusal to work), the sealers accepted the reality of their situation and the fact that berths were difficult to acquire in the best of times. In addition, the politically aware and volatile working class of St. John's and Conception Bay was decimated by the depressed 1860s and by the depression that began in the early 1880s and lasted until 1900. The collective experiences of the early strikers of the 1830s and 1840s had been acquired in a period of growth and relative prosperity. The seal fishery, politics and religion[22]—sometimes in conjunction and sometimes at cross purposes—had taught them the effectiveness of collective action. Furthermore, the very nature of the seal and Labrador cod fisheries allowed many of these fishermen to live in concentrated, comparatively urbanized areas; they were largely urban and working-class in their way of life and attitudes. However, by the latter part of the century, political, religious and business leaders had, intentionally or inadvertently, succeeded in creating such religious antagonisms between the different denominations, but especially between Catholics and Anglicans, that common economic interests became blurred and obscure. At the same time, as the century advanced, greater and greater numbers of independent boat keepers—inshore fishermen—from the northern bays travelled to St. John's annually to participate in the spring seal fishery. These were confirmed individualists from small communities—often populated by only their own families. They were often Methodists or Low Church Anglicans, who believed in each man standing on his own two feet. They did not have a tradition of collective action, but looked to their families and churches for support. All these factors resulted in a decline in labour protests among sealers.

St. John's strike of 1902

In 1902, however, there was one final major strike. On the morning of Saturday, 8 March, the men of the S.S. *Ranger* left their vessel, led by Albert Mercer of Bay Roberts, because they had heard the rumour that the price paid for fat was to be reduced to $2.40 per cwt from the 1901 price of $3.25.[23] The men marched along Water Street and were joined by the crews of the other steamers until about 3,000 men were on the move towards Government House. The sealers—under their leaders Mercer, Robert Hall from Halls Town near Clarke's Beach, Conception Bay, and Simeon Calloway from Pool's Island, Bonavista Bay—demanded to see the governor, Sir Cavendish Boyle, to whom they spoke. He suggested that they appoint a committee to meet with representatives of the steamer owners, which was agreed, and Edgar Kean, Jacob Bishop and N. Waterman were named to join the three leaders on this committee.

The sealers asked A. B. Morine, a prominent lawyer and politician, originally from Nova Scotia, to assist them, and the meetings with representatives of the steamer owners began. The sealers' demands were more extensive than at first articulated. They wanted to share one-half the pelts, as in the days of the sailing vessels, instead of just one-third, which had become the practice; they demanded that the price of fat be increased to $5 per cwt, given that it had been worth over $6 in the early 1880s; they objected to the $3-per-man coaling fee

Figure 20—A meeting of sealers during the 1902 strike (photograph shows the intersection of Water Street and McBride's Hill).

deducted from each sealer's share; they also objected to the 33 per cent surcharge on the necessities that sealers bought on credit from the steamer owners before leaving for the ice (these items would probably include a knife, steel and pan and made up the sealer's 'crop'); and they asked permission to live on board the steamers for the several days they were in St. John's waiting to be signed on and that they be fed after they were signed on, which was usually a day or two before sailing. The owners had no problem with sealers living and eating on board the steamers before the voyage, and in some cases, it was allowed by an informal arrangement; they argued, however, that the coaling fee was only fair because "... in the days of the sailing vessels the crews outfitted the vessels, trimmed the punts, made and bent the sails ... etc. Now the crews of steamers have nothing to do, the ships are coaled, cleaned and ready for them";[24] they contended that the charge on the crop was necessary to offset the number of times the men and vessels returned empty and the crops had to be written off; in fact, the "men were not obliged to take crops, and owners would be better pleased if they did not"; the problem of outport men incurring expenses by travelling long distances "could be obviated by dispensing with them and getting men nearer St. John's"; they stated that they could not give the men one-half the pelts and continue to operate; and finally, they likewise argued that they could not afford to increase the price of fat to $5 per cwt because "... the price of oil was never known to be so low as today, and skins were practically valueless."

The men responded that the $3 charge was not imposed in the days of sail and that there had only been a charge for berths "... until the famous strike started by one Peter [sic] Supple against these charges, led to their removal." Furthermore, the men pointed out the following: that sealers had received one-half the voyage in the early days of steam, but that share had been reduced to one-third; then the second trip was prohibited by the government, and this reduced the men's earnings further; the men would coal the steamers if given a chance; there were not enough experienced sealers near St. John's to take the place of those from the northern bays; and finally, the price of fat paid in St. John's was set by the oil plant owners (who engaged in the sealing as well) and not by world markets.

These negotiations continued all day Saturday and nothing was resolved. The situation remained tense but calm Sunday, but after midnight, the men began to leave their vessels with their bags and boxes. The captains had orders to leave the wharves at 4:00 a.m. but found that their vessels were severely undermanned because, in most cases, the firemen had left with the sealers. Some steamers were unable to raise their anchors, while others depended on the engineers to fuel the fires. The *Aurora*, *Terra Nova*, *Iceland* and *Kite* managed to steam out into the harbour and anchor. Still, there were no further concessions from the owners, not even when the men agreed to accept $4 per cwt for the fat. As a result, many sealers walked to the railway station to find transportation home, but railway employees prevented all but the handful with tickets from boarding. It would appear that during Monday night, some working-class home owners kindly took in a few of the sealers for the night.[25]

On Tuesday, the owners guaranteed the price of $3.25 per cwt of fat, and some sealers began to waver, which made the others more adamant than ever that all would stand together. A party of strikers boarded the *Virginia Lake*, *Neptune* and *Diana* and expelled the few men on board. Meanwhile, Capt. Arthur Jackman boarded the *Terra Nova* without incident, although his officers were turned back; Jackman prepared to leave the harbour (from his anchorage out in the stream) and announced his intention to sign on a crew in Cape Broyle and elsewhere. At the same time, Capt. Blandford, on the *Neptune*, began to contemplate leaving as well, probably because he and Jackman were old rivals (Jackman was considered the 'commodore', having commanded steamers since 1871, while Blandford, who had commanded his first steamer in 1874, was leading in total number of pelts taken). In any event, when it became known that the *Neptune* was leaving the wharf, about 300 strikers boarded it and:

> took one of the heavy hawsers and ran it up the wharf to the cove, cut away several of the boats, damaged the rigging and did other things not anticipated. The men hauled on the hawser and tried to hold the ship but Capt. Sam put on full steam; the strain was fearful and the strength of the men such that the rope snapped."[26]

The *Neptune* then left for Bonavista Bay to sign on a new crew.

Meanwhile, the government was becoming concerned with the fact that about 3,000 angry men were roaming St. John's, with only a handful of police officers to restrain them. Therefore, the owners were advised to compromise, and by the middle of the afternoon of Tuesday, 11 March, the men were offered $3.50 per cwt of fat and the elimination of the coaling fee; the crop (for those who needed one) would become a standard $9 advance, for which $12 would be charged.

According to a newspaper account, by Tuesday afternoon the strikers had begun to lose the sympathy of the public; "... merchants feared for their property ... and shopkeepers scented a riot and talked ominously of the night that was to come."[27] This newspaper reported that by 3:00 p.m. on Tuesday, only the extremists were holding out: "The sensible ones realized that they had done very well, and town folks remonstrated with them against being too grasping ... [and told them] very plainly that they should be satisfied with what they had won." With dissension and division among the sealers and a lack of public support, the police moved in and began to clear the entrances to the wharves and to escort the men who wanted to go aboard. The "extremists" managed to clear the sealers from the *Diana*. "Then word was passed that the *Vanguard* was crewing, and away rushed the 'extremists,' on the double, to Grieve's, to clear her out also. But here they were checkmated, for the police held the gate." A short time later, under police escort, the men of the *Vanguard* began to board; the vessel soon had "half her crew and intelligent onlookers saw that the back of the strike was broken."

> The extremists saw this too, for with a cheer one squad went east to get boats to attack the men on the *Southern Cross* while another gang went to get gaffs to attack the police. The former found the *Southern Cross*'s crew too big a contract to tackle. The latter were met at Bowring's by more police. Then they rushed for Bennett's to get weapons from the *Ranger*, but all these had been wisely removed from every ship, and when they came back again they found the whole strike over.[28]

By 6:00 p.m., the majority of the men had accepted the compromise offer, and soon, all were in the process of boarding their vessels. Both the *Terra Nova* and *Neptune* returned and took on their crews, and the strike was over.

In contrast to earlier actions, this strike was an act of desperation under trying circumstances, rather than the act of a group of men whose labour was in great demand. Certainly, the returns to sealers from the seal fishery had dropped off substantially. The records show only the average shares paid to sealers, beginning in 1895, when the amount averaged at about $29 per man. During the next seven years ending in 1902, the average shares were as follows: $20, $11, $29, $33, $43, $33, and $34 in 1902; with the exception of 1902, each sealer had the $3 coaling charge deducted from his share. In addition, before leaving for the ice, many men bought supplies on account at highly inflated prices from the companies' stores and agreed to have the debt deducted from their shares. The *Evening Telegram* drew its readers' attention to this situation in 1891 (see Chapter 4) when it published two lists of prices—one, indicating regular prices, and the other, the price charged sealers on credit before the voyage. For example, sealskin boots were priced at $4.80 and $7.60, respectively; drawers, at $0.60 and $1.33; and a shirt, at $0.50 and $0.80.[29] In 1898, the prices were: boots, $2 and $3.50; pants, $1.65 and $3; and drawers, $1.10 and $1.80.[30] This left very little money for the sealers to bring home to their families after four to six weeks of hard and dangerous work.

The 1902 strike was unique in several ways. It was a strike by largely migrant workers, who came to St. John's every spring to participate in the seal fishery. The situation differed considerably from the Harbour Grace and Carbonear strikes in 1832. There, the men were in their own home towns, with food, shelter and moral support at hand. The same situation prevailed in St. John's in 1842 and 1843, when the sealers in the capital went on strike, and likewise in Harbour Main and Brigus in 1845. However, the 1902 strikers were isolated from the city in which they struck, and although it was reported that the working-class population helped some of them with lodgings on Monday night, for the most part, they are known to have rested and slept on their bags and boxes on the streets. It seems apparent that by Tuesday, the St. John's people wanted the sealers on their ships and away to the ice. The livelihood of seal skinners, coopers, dock workers and oil plant workers depended on this industry; the incomes of shop clerks and tavern owners, likewise. Therefore, the sealers could not retain the public support they needed, and, no doubt, this added to their growing antipathy towards St. John's, which could be clearly detected by now (see below). As pointed out, this was a strike dominated by rural family-fishermen—the independent inshore boat owners. Earlier in the century, Conception Bay and St. John's, with their large Labrador cod fishery operations, were concentrated urban settings where bigger proportions of hired fishermen lived—or at least raised their families. That environment was conducive to the development of unions and to successful industrial action.

However, there can be no doubt that the strikes of the 1840s influenced the strike of 1902. It was mentioned during the negotiations, even if the men got Henry Supple's first name wrong (few would have heard it, anyway, because he was always known as "Captain" Supple or "Mr." Supple). Also, it was probably no accident that the man reported to have started the strike was Albert Mercer from Bay Roberts—a formerly prosperous port located between Harbour Grace and Brigus, where memories of previous strikes would have been retained. The same can be said of another of the leaders, Robert Hall, whose home town was also near Brigus. The oral traditions about the successful strikes of the 1830s and 1840s—only 50 to 70 years earlier—would have been alive on board the steamers in 1902.[31] And just as the strike of 1902 was influenced by earlier actions, so it, in turn, became a conduit which transferred the experience of collective action in general, acquired in Conception Bay and St. John's since the 1830s and 1840s, to the northern bays, where it created a receptive and knowledgeable environment for unionism and was adapted some years later by William Coaker in building his Fishermen's Protective Union.

The strike of 1902 was a turning point in another sense. It demonstrated to the sealers their relative lack of power and effectiveness in St. John's, where they were without food, living accommodations and friends. Their antipathy towards the St. John's merchants, and towards St. John's in general, must have added to their feelings of alienation and isolation from the capital. Thus, when the northern outport sealers were heard from again in a serious way, they were part of a populist uprising culminating in the formation of the Fishermen's Protective Union, which neither requested nor received any favours from the capital.

Labour during post-1902

The idea of a sealers' strike came up on several more occasions but did not result in any significant action. In 1906, a rumour spread to the effect that the price paid for fat would be reduced from its 1904/05 level of $4 to $3.75; the crews of the *Bloodhound* and *Vanguard*, then unloading in Harbour Grace, telegraphed the crew of the *Diana* in St. John's that they were holding out for $4 and asked for their support.[32] However, the men were assured that the price would not be decreased, and the matter was dropped. The following year, 1907, the crew of the *Grand Lake* refused to unload their cargo of fat until the price was settled. Three master watches went to Harveys' office and asked Mr. Harvey to settle the price of fat. He pointed out that it was impossible to do so at that point, but that the price would not be less than that paid in 1906, and if there was any increase, the men of the *Grand Lake* would receive it as well. That effectively put an end to the issue.[33] In January 1911, there were reports that the sealers would strike for half of the pelts taken. However, the steamer owners were reportedly ready to leave their steamers tied up, and nothing more was heard in this connection.[34] That same year, the crew of the *Beothic* threatened that they would not sell their one-third of the pelts for less than $5 per cwt (they were being offered $4.50). However, the men did not press the issue.[35] Again, in 1913, the crew of the *Stephano* refused to unload the cargo of pelts until they received $4.75 for their share instead of the $4.50 offered. Work was halted for several hours, but when Bowrings refused the request, work resumed.[36] The idea of a strike on an individual steamer continued well into the twentieth century, but this kind of strike was reminiscent of the manus rather than the great strikes of the 1830s, the 1840s and 1902.

However, the strikes and threats of strike made a lasting impression on the St. John's steamer owners. The price for fat was increased, and they were much less intransigent in their responses to public criticism and in their dealings with sealers and, later, the Fishermen's Protective Union. In 1905, the principal steamer owners in St. John's negotiated an agreement (mentioned above) which brought into harmony the various arrangements already in operation. They concurred that all agreements between companies and sealers would be as identical as possible. They also decided that they would reduce the number of sealers they hired to about 75 per cent of the number allowed by the legislation of 1898 to enable those hired to make a larger share. The crop was to remain at $9 (for sealers who wanted to avail of the service), but the men would have to repay $12 in order to compensate for those who could not repay anything due to a failing voyage. It was also recognized that the men were sometimes at a loss for a place to stay and eat before the steamers left for the ice. Therefore, the owners agreed to announce in advance the date on which each steamer would sign its crew, and the men would be expected to arrive in St. John's in time for this event, not before. Given the fact that most were, by now, travelling by train and since it was taken for granted that all had 'tickets' (i.e., berths) to the ice before they left home, this seemed a reasonable arrangement. For their part, the owners were satisfied to allow the men on board the ships immediately after signing and supply them with food. There would be no coaling charge, berths would be free, one man's share would be divided among the firemen and one among the officers as a bonus, and, with these exceptions, the crew would continue to share one-third of the voyage. Captains would receive 4 per cent of the gross value of the voyage. Crews could be brought to St. John's or picked up by steamers in the outports. The owners were ambivalent on this point: on the one hand, they feared what three or four thousand disgruntled men could do under effective leadership; but on the other hand, the merchants, bar owners,

Figure 21—Seal skinners at work removing the fat from the skins. These men were experienced at jobs requiring strong arms and skills with knives and blades; many were butchers or coopers during the other eleven months of the year. A good skinner could 'skin out' 400 to 500 young harps in a ten-hour day. Each skinner was assisted by an apprentice (see above), who served a long apprenticeship before being permitted to join the seal skinners' union and seek employment as a seal skinner.

tradesmen and others wanted the business sealers brought with them. Where feasible, owners agreed to supply medical men for the larger vessels, and they concluded by agreeing that no steamer would sail before 8:00 a.m. on 13 March in order to give time for the whitecoats to reach their prime weight of 55 to 60 lbs. (avoir., or 25 to 27 kg.) This agreement was signed by or on behalf of E. R. Bowring, A. Harvey and Company, Jobs, Baine Johnstons, James Baird and H.

D. Reid. The only representative absent, John Browning, had already given his assent.[37] There is no doubt that the strike of 1902 had given the owners reason to draw up a mutually satisfactory agreement so that they could not be split and dealt with individually by organized sealers. However, the public interest shown by the newspapers, the governors and the general community was also a factor in the preparation of a formal agreement among the vessel owners.

Seal Skinners

One body of men whose work and activities were a direct part of the seal fishery were the seal skinners, and a discussion on labour and strikes would not be complete without mentioning the seal skinners' union. This organization, founded in 1854 or 1855 because of dissatisfaction with rates of pay for seal skinners, was the second union to be founded in Newfoundland.[38] These men, who cut the fat from the pelts, worked in the industry for only a few weeks each spring; during the remainder of the year, they were engaged in other employment, often as butchers or in cooper shops. Originally, the union may have represented seal skinners in Harbour Grace as well as in St. John's but by the end of the nineteenth century it was well-known only in St. John's, where it operated a strict apprenticeship system, requiring every trainee to spend at least five years as an apprentice.

Seal skinning attracted new technology in a way that the coopers' and shipwrights' trades did not, and in 1900, Bowrings introduced a skinning machine. This provoked a strike, and when the machine proved unable to handle the work effectively, the company gave in to the demands of the union and dismantled it. A local newspaper commented:

> This morning Bowring Brothers tried the new skinning machine at their Southside premises, Mr. Turner superintending the work. The machine proved successful as far as taking off fat, and did so with surprising rapidity, but it is said many of the skins were badly cut by the knives, proving that only by great care could the work be accomplished. The seal-skinners at work on the S.S. *Terra Nova* at once went on strike and all work was suspended, they refused to work unless the machine was entirely removed from the building. The men were very much put out at the using of the machine and especially as they contend it was done at a time contrary to law. Bowring Bros., however, do not desire to cause any friction with these men who are earning their living from them, so the machine was dismantled during the morning, removed from the premises and the work of skinning proceeded with at 2 p.m. as usual.[39]

Jobs decided to experiment with a skinning machine in 1901 and set one up to skin pelts being brought in from the outports, leaving the steamer cargoes to be skinned by the local skinners. The machine became disabled, and the skinners were offered the work of finishing the remaining pelts. They refused to do so unless the company would agree to dismantle the skinning machine and not set it up again. Jobs refused, however, and the machine was repaired.[40] In fact, the seal skinners held an increasingly precarious position vis-a-vis encroaching technology, and they were finally replaced altogether by machinery after the Great War.

Manuses

In addition to the strike, which was used only rarely, the sealers sometimes resorted to the 'manus', which was a refusal to work.[41] There were a number of instances when crews refused to continue sealing—a refusal usually caused by the men's desire to quit an unprofitable and long voyage in order to return home and prepare for the summer cod fishery. Without giving it a name, a local newspaper decried this practice in 1848, the earliest reference that has been found:

> In the first place, we would deprecate not only those habits—indolence and inactivity, but the determined stand which, it is notorious, crews are in the habit of making whenever there has been, according to their own peculiarly accommodating views, no chance of falling in with the seals and that in opposition with their skippers to whose better and more experienced judgement they should certainly give way. Is it not a notorious fact too that many a schooner master have been obliged thro' fear and intimidation, to bear up and abandon the voyage at a time when the brightest prospects for making at least a saving trip, presented themselves? Ay and so long as the stand is tolerated we may expect repetitions of similar conduct.
>
> We are astonished that some honorable member has not made an effort toward the adoption of an enactment to render it imperative upon the crews of sealing vessels to prosecute the fishery to a certain and positive period—a period that could not possibly affect their prospects and engagement for the cod fishery. The Legislature should have seen the necessity of such a step to meet the exigencies of the case and which is so desirable for the faithful prosecution of that most valuable branch of our Newfoundland trade. If something of the kind be not done, it is folly to expect masters to have that control which it is so desirable they should possess over a large and self-willed body of men.
>
> Convinced as we are that much good would result from the adoption of the measure we have named, it is not unlikely that we shall renew the subject on the next opening of the Legislative Assembly. The 'hands' themselves as well as the trade, would ultimately be the gainers, and consequently we are of opinion that the matter will merit the consideration of the executive and might form a prominent feature in the opening speech. It is not for us to adopt a dictatorial tone, but we would respectfully and earnestly draw the attention of the enlightened head of the government to our humble remarks on what we conceive to be a very important subject.[42]

Occasional reports of manusing continued throughout the period. Shortis describes an unusual but singularly imaginative and effective manus:

> About sixty years ago [c. 1860], a certain brigantine belonging to McBride & Kerr, sailed from St. John's for the ice-fields with a crew of fifty men. The Captain was not very popular, being somewhat of a martinet. However, they managed to get North as far as White Bay, but, unfortunately, the seals were very scarce, and the captain became very domineering and abusive towards the crew—compelling them to perform unnecessary labors and making everything quite uncomfortable. At last, patience ceased to be a virtue, and the crew decided to hold a council to decide upon the best means to remedy the existing state of affairs. At first they thought of adopting the well-known plan, so often resorted to by mariners, of forwarding a "round robin" to the captain announcing what they intended to do. But after consideration they

thought this would not do, so a happy idea entered their minds and they quickly put it into execution. Although the Captain was of a most tyrannical and disagreeable disposition, there was, at least, one thing upon earth he was most fond of, and that was a middle-sized black dog, which was always to be seen at his heels afloat or ashore. The crew decided to use this member of the canine species to convey their decision to the Captain in the cabin. They induced the dog to enter the forecastle, and rigged him up with a canvas jacket, pants and on a little cap—printed in large black letters was the word "Manus." The moment the dog was released, he bounded towards the cabin and commenced dancing around his master, who immediately saw the ominous word "Manus" printed on the little cap. He saw that it was all up with him, and knowing that the crew were determined to take extreme measures, he ordered the ship to be steered South, and arrived in St. John's in due course. Whatever the reason was he did not bring the crew to court. Probably he thought that such a course would have given him an unenviable notoriety and subjected him to the chaff and contempt of the other skippers....[43]

In 1862, Captain Lewis of the Schooner *Mary* brought eleven of his crew before court for taking this action; three were sentenced to 28 days imprisonment, and the remainder, to 14.[44] The year 1862 seems to have had more than its share of manuses, probably because of the long, unproductive voyages. Again, the *Times*, no supporter of labour, indignantly editorialized:

'Manus' is a term which is well and distinctly understood by the 'lads' who indulge in the iniquitous imposition, and of which the commanders of vessels from this port, this spring, have had to complain. We are astonished the legislature does not put a stop to it by a wholesome enactment.[45]

However, on 3 May, the *Times* corrected itself and pointed out that there was, indeed, legislation against it, carrying a maximum penalty of one month's imprisonment for those convicted. It should be noted that many times, captains precipitated the manus by staying out too long, and consequently, very few cases were brought to court.

Manusing seems to have been unique to Newfoundland. It was the nineteenth century equivalent of the modern sit-down strike and differed from a mutiny in that there was usually no attempt to take over the vessel. The men often piled their gaffs and tow ropes on the deck and refused to go overboard after seals. In some cases, they would allow others to continue working and would agree to work the canvas, themselves, to take the vessel back to port; in other instances, sealers would manus and then threaten their mates with serious injury to stop them from going overboard.

The last significant manuses in this period occurred in 1914, when some steamers attempted to continue the hunt after news of the *Newfoundland* disaster had circulated—indeed, after some of the crews had been involved in the recovery of the dead and dying. Cassie Brown points out that the men on board the *Diana, Eagle, Stephano* and *Bloodhound* were shocked at this insensitivity and manused. In fact, the crew of the latter refused to even move coal to the firemen and allowed the steamer to drift south with the ice until they were assured they would return to port. Understandably, despite threats from Capt.

Abram Kean in command of the *Stephano*, and from others, those who manused were not prosecuted on this occasion.

Societies and unions

To place sealers' strikes and manuses in perspective, it is helpful to look briefly at the history of unionism in Newfoundland, in general, during the nineteenth century. The first 'society' of tradesmen, the Mechanics Society, was founded in 1827 under the leadership of a cooper. This society represented a variety of trades and charged an entrance fee of 18s and monthly dues of 1s. The resulting fund provided sickness insurance and death benefits to its members.[46] By the 1840s, the various trades were beginning to assert their independence, and in the 1850s, the shipwrights' and seal skinners' societies began to take on the role of unions whose objectives went beyond those of benefit societies. In the 1890s, the number of unions increased dramatically, and by 1902 (the year of the sealers' strike), there were at least 20 in St. John's.[47] Unions had also become more active, and between 1890 and 1899, there were at least 49 strikes. These were, however, craft unions; it was not until 1903 that the longshoremen of St. John's banded together and founded the Steamboat Labourers' Union (which became the Longshoremen's Protective Union in 1904).[48] This union was the first of the industrial unions with a large membership of unskilled labourers. However, well before they became formerly organized the longshoremen had been known to take frequent industrial action and to strike. In the year 1902, for example, the longshoremen went on strike on seven occasions, compared with the sealers' one. It is curious, but understandable, that neither the craft unions nor the longshoremen offered support to the sealers in 1902. Probably, these roughly-clad, weather-burned men from the northern outports were now viewed by the St. John's working class as lower-caste workers, engaged in a bloody, messy, unsightly task—essential to the economy but not a trade that most urban folk cared to dwell upon or associate with. And yet the craft unions and the longshoremen had, no doubt, learned much from the successes of the more urbanized sealing strikers of the 1830s and 1840s.

Summary

The seal fishery created an environment, temporary though it may have been, that was conducive to the development of a sense of identity and solidarity among its workers and fostered the concept of collective action. An industry employing hundreds and thousands of men in the same occupation and receiving the same share, regardless of effort (for the most part), was a different type of operation than the old migratory cod fishery, carried out by seasonally employed servants thousands of miles from their homes in England, with an eye to obtaining passage to New England. The seal fishery was also a different industry from the family-based inshore cod fishery on the island, and also different from the planter Labrador cod fishery which was prosecuted from thousands of isolated fishing stations along the coast of Labrador. It was poetic justice that an industry which was fuelled and driven by the demands of the Industrial Revolution should also become the breeding ground for modern industrial action among the fishermen of Newfoundland.

There is no doubt that the fishermen who engaged in the annual spring seal hunt were the first Newfoundlanders to become aware of their strength as a group, and their early successes quickly established their power in Harbour Grace, Carbonear, Brigus and St. John's—the main centres of the early sealing industry. At the same time, it must be remembered that large numbers of these men, both Catholics and Protestants, were taught to act and vote collectively by their leaders in the campaign for representative government—a campaign which had few supporters in the northern bays. Then later, in the 1830s, the Catholics of Conception Bay and St. John's—most of whom were sealers and Labrador fishermen—were organized by their political and religious leaders in a successful attempt to acquire political power. Again, the effectiveness of intelligent and successful collective action would not be lost on these men. Meanwhile, as the industry grew, so, too, did the demand for labour and the power of collective action. However, the decline in the importance of the seal industry—beginning in the 1860s and accelerating through the 1880s—combined with the end of the sailing ship era, practically destroyed the big outports and forced fishermen who did not emigrate abroad to travel to St. John's and compete with thousands of other economically deprived men for the shrinking number of sealing berths available. This was not a situation that encouraged collective action.

The strike of 1902 demonstrated briefly the power of the sealers, but by now, they were poorly led, migrant workers in an often unfriendly town, acting from a position of desperation rather than strength. Their brief success did little to improve their situation; although their incomes improved following the strike, the failure of the cod fishery in Norway, which had created an increased demand for seal oil, was the most important reason for this. Nevertheless, the experiences of the northern sealers in St. John's before during and after 1902 made it easier for William Coaker to organize them into the Fishermen's Protective Union.

It is difficult to draw definitive conclusions regarding the effect of the sealers' strikes on Newfoundland society in general and on the labour movement in particular. However, one can assume that the strikes of the earlier years of the century educated both business and labour in matters involving collective actions, negotiated agreements and respective points of view. Furthermore, it is reasonable to conclude that those effectively organized strikes influenced the craft societies, and it is no surprise that shipwrights and seal skinners were the first to organize their own unions in the 1850s. However, it is obvious that the situation had changed drastically by the time the sealers struck in St. John's in 1902. These were migrant men from the more northerly outports without any real local support and largely alienated from the urban city dwellers. Whereas the earlier strikes had, no doubt, encouraged unionization in St. John's, this last strike emphasized the differences that now existed between the 'townie' and the 'bayman'. Therefore, the sealers who returned to the north after the 1902 sealing season were to rechannel their actions. They soon found their voice in a populist uprising—the Fishermen's Protective Union—led by William Coaker.

Community Reactions

The depression of the 1860s that affected both the cod and seal fisheries made people aware of how vulnerable the Newfoundland economy was. The first reaction of the politicians was to advocate confederation with Canada, but this plan failed to win the support of the electorate in the 1869 election, partly because the economy began to improve that year. The first reaction of the fisherman-ice hunter was to migrate to Boston on a temporary basis, thus beginning a trend that was to continue until the 1920s and which led to the emigration of large numbers of Newfoundlanders. The second response of government was its attempt to diversify the economy. In the meantime, observers began to react to the conditions under which the ice hunters worked in this shrinking and somewhat archaic industry.

Government

Although the cod and seal fisheries (and the new lobster fishery) created prosperity in the 1870s, the 1880s brought a return to depression. It was obvious, for example, that the seal herds were in decline and that seal oil was rapidly losing its value. There was no longer any doubt that the economy would have to be diversified and that the government would have to borrow to assist in this diversification. The centre piece of the new economic policy was the building of a railway to open the interior and provide access to the land-based wealth that most were convinced was there. Not all agreed with the policies of borrowing and railway building, and the greatest opposition came from the St. John's merchants, who feared that it would change their relationship with the fishermen and raise the wages of their workers as well. Consequently, William Whiteway, who became Prime Minister in 1878 and who was the driving force on the Newfoundland political front for the next 20 years, came under attack for his railway policies and, in return, attacked his opponents.

It was a confusing situation: Whiteway accused the St. John's traditional fish-trade merchants of wanting to keep the fishermen tied to their truck system—dependent and poor; the merchants accused Whiteway of being anti-fishery and financially irresponsible. The newspapers gave space to the views they supported and used the interests of different groups in order to editorialize. The plight of the fishermen became the focus of much of the debate which ensued during the 1880s and 1890s. Commentators became increasingly concerned about finding a solution to the problem presented by the declining seal fishery, and their concerns were translated by the government into legislation.

The government began to play an active part in the operation of the seal fishery in 1873, when "An Act to Regulate the Prosecution of the Seal Fishery" was passed that came into effect at the beginning of 1874.[49] This Act, as discussed earlier, was intended to protect the seal herds from over-exploitation and prohibited steamers from sailing before 10 March, banned sailing ships from sailing before 5 March and prohibited the killing of seals before 12 March. In 1876, further legislation provided for the recovery by sealers of their wages. This was followed in 1879 by legislation changing the departure date of sailing vessels to 1 March and prohibiting the killing of cats (i.e., seals whose pelts weighed less than 28 pounds). In 1887, limits were put on the length of the season; no seals were to be killed after 20 April, and no vessels were to leave port on a second trip after 1 April. This was further amended in 1892 to prohibit second trips altogether, and in 1893, an Act was passed prohibiting the killing of seals on Sundays. In 1895, under pressure from merchants and fishermen, the prohibition of second trips was repealed and the length of the season was extended to 5 May. Assistance to outfit vessels for the seal fishery was granted by legislation in 1898, and other legislation that same year banned second trips for all time and prohibited steamers from leaving port before 10 March. The maximum number of men that each vessel was allowed to carry was set at three men for every seven tons of gross registered tonnage up to a maximum of 270 men. But only in 1914, following the election of William Coaker and the Unionist (see below), did the government become concerned about the living and working conditions of the sealers and try to ameliorate them.

Evening Telegram in the 1880s and 1890s

The government and the public in general became more aware of the seal fishery in the latter nineteenth century—its shortcomings, its problems, the lack of regulation. Cause and effect became intermingled; complaints were articulated and the government responded; and these responses drew more comments. Thus, as organized labour lost its power to influence events, it was replaced by public opinion, particularly as expressed by the *Evening Telegram*.

One embittered commentator, in a letter to the editor of the *Evening Telegram*, lamented the decline in the seal fishery, exploitation by the Dundee vessel owners and the lack of government action. He wrote:

> Dear Sir,—It is well to see that some degree of public interest and a considerable amount of public indignation has been excited here lately on account of the wholesale destruction of our valuable Seal Fishery. Once a prime source of our annual revenue this branch of the commonwealth has been gradually reduced to such a condition that it is now of little or no value to the people of this colony, and it has become a matter of crying necessity that our Legislature should seriously interfere on its behalf before the voyage has been blotted out of existence altogether. Looking to the mean and niggardly way our own people are treated by those who now prosecute this voyage it is not too much to say that in no other country in the world would these parties be "allowed" to carry matters with such a high hand, and were it not that they have influential "friends at court," I venture to say the abuses complained of would have been put down long ago.
>
> Look for instance at the way Newfoundlanders are treated by the Dundee people who prosecute the voyage from this port. When first these parties fitted out from here for the seal fishery they were glad to enjoy equal rights with our own sealing merchants, and to take their men here, paying them a fair remuneration, as well as their sealing masters. By degrees, however, they have managed to reduce to a minimum any benefits they "leave" in Newfoundland; and having learned the way from our people to take seals, and where to look for them, they have gradually supplanted our native sealing-masters by foreigners, and are now playing the same contemptible game with the men.
>
> When these Dundee steamship owners began to fit out from here the men they brought with them could not so much as tie a "tow" of seals, but having learnt the trade, they now ungratefully fling our people aside altogether,

> and would not leave a penny in Newfoundland of all the wealth they take out of it if they could manage otherwise....
> ... Must we stand by and see every means of subsistence taken away from our native population by hungry cormorants....[50]

As pointed out by that writer, in the beginning, the participation of the Dundee steamers had been welcomed, but with the decline in the industry, their owners began to be viewed as villains in some quarters.

In similar fashion, when the captain of the S.S. *Hector*, Edward White, Jr., complained that his crew had panned 5,000 pelts about 12 miles off Twillingate, only to see them taken by about 2,000 landsmen, a local commentator supported the landsmen's actions:

> Of course they "took them all," and they were perfectly justified in doing so. What legal or divine right, we should like to know, have sealing steamers to "kill, pan and claim," indiscriminately, right under the eyes of shoremen, without giving the latter a chance at all? We submit that crews of steamers possess no such right, and that it is a piece of gross injustice on their part to run in close to land and take the bread out of the mouths of poor people who live along the shore and depend upon the harvests of the sea for the support of themselves and families. Let crews of steamers kill their tows and haul them on board. This privilege they are entitled to. But they must not suppose that they hold a "patent right" to kill and pan every seal on the icefields, to the deprivation of their fellow-countrymen who may not have the advantage of prosecuting the voyage in powerful steamers.[51]

While panning was frowned upon by most non-sealers as being wasteful, this is the first reported case of panned pelts being "forcibly" taken by landsmen, and it is indicative of the desperate actions that these men were compelled to take in order to survive. The sympathy with which their action was viewed by a St. John's newspaper illustrates the crosscurrents in public opinion that could produce support for the men on the steamers one day, but then conclude that the landsmen were more in need of assistance on another day.

Similarly, that same year, this newspaper came out in support of the petitions from Conception Bay asking for legislation to prohibit the use of steamers at the seal fishery. Apparently, the issue was raised publicly by the Attorney General, who presented these petitions in the House of Assembly. The *Evening Telegram* asked:

> Have not these steamers had the monopoly of our seal fishery long enough? Rights of property! Such trash! Have no one else rights but the merchants? Who remembered the owners of the sailing vessels, who were brought to begging by the introduction of steamers, for their loss? Who remunerates craft now lying idle on account of steamers?... I may answer no one! Why? Because they are not merchants, therefore their interests are to be neglected. Knowing how they were represented, the Hr. Grace petitioners did not expect much support to their petition from their own members, one of them being absent, master of a "sealing steamer," one other holding his seat by the order of the firm owning the same steamer and solely under that firm's control, and the other member, I am told acting as solicitor for the Dundee steamers.[52]

If there is one thread running through these commentaries on the seal fishery during 1886, it is the fact that the blame for the problems in the seal fishery was laid at the feet of the government and the sealing firms.

After Whiteway regained the government in 1889, the *Evening Telegram* turned its attention to the steamer owners again. In 1891, this newspaper complained about the fact that the St. John's sealing steamers were destroying the seal herds in the Gulf of St. Lawrence and the local schooner owners were being deprived of their living.[53] Also in that year, the House of Assembly passed a bill that would have extended the opening of the seal fishery for steamers from 12 March to 15 March in order to give the sailing craft a better chance to make a paying voyage. However, the bill did not get pass the Legislative Council. The *Evening Telegram*, which had supported the bill, was furious and condemned the Legislative Council for defeating it under headlines:

ANOTHER BLOW AT OUR FISHERMEN, NO PITY FOR THE POOR "TOILER OF THE SEA":

Intense indignation prevails here today owing to the action of the Legislative council last night in "kicking out"—as one of the councillors called it—the Sealing Bill which passed the popular branch of the Legislature on Friday last. Yea they "kicked it out," and the same men would kick every fisherman out of the country, if they could do it, and if they could live without the fishermen. Already the harsh and oppressive attitude of the class to which these Councillors belong has driven the very cream of our once magnificent fishing population into other lands, and now the four votes they command in the Upper Chamber would complete the work of depopulation by eliminating every gleam of hope from the poor fisherman's heart and home and future.

The operation of such a Bill as that just rejected by the Council would infuse new life and energy into our "toilers of the sea." In a few years the hum of industry would be heard throughout the land, in the building and equipment of a fleet of sailing vessels for the seal-fishery. Shipwrights, blacksmiths. sailmakers and all other tradesmen would find plenty of profitable employment. And with increased times and improved earning power of the people, our revenue would be sufficient for railway and all other progressive purposes. But, as it is, all these bright possibilities are rendered more remote from realization than ever, and the fisherman must suffer in silence the severe blow thus inflicted upon his prospects by those who ought—if they possessed any bowels of compassion—to do all in their power to conserve his interests. It is right that the public—the fishermen, labourers and tradesmen of the country—should know the names of the four men who voted against the Sealing Bill. Here they are: Messrs. Pitts, Mackay, Rendell and Angel. Fishermen of Newfoundland, remember these names— PITTS, MACKAY, RENDELL and ANGEL—and understand for yourselves the motives by which they were actuated last evening in opposing the measure. They are all personally interested. Some of them, as shareholders in steamers, expect to make over a hundred per cent on their money this spring, while the remainder find it pleasant and profitable to do as they are told by the principals in business.

We hear a great deal from those Legislative Councillors about "vested interests." They are extremely careful to protect the interests of all those who have capital invested in sealing steamers. Oh, yes. That is quite natural! The steamer-owner must be treated as gingerly as possible. But no word is heard in defence of the vested rights of the poor fishermen of Fortune and Placentia Bays. Treat the latter as ruthlessly and as heartlessly as you like. Confiscate their property, throw themselves into prison and allow their children to starve. But don't dare to touch the sacred interests of the inflated monopolists who now presume to set aside the will of the people's representatives, and to sanction only such measures as are conducive to ring interests and ring rule. Things are come to a pretty pass here when four irresponsible individuals in the Upper House can treat with supercilious contempt the publicly expressed wishes of the whole country.[54]

This was very strong language and illustrates the bitterness felt by the newspaper and by others towards the steamer owners and the monopoly over the seal fishery that they had managed to establish by this time. It also demonstrates a sympathy for the fishermen which has a genuine ring to it.

The *Evening Telegram* kept up the pressure and published articles and letters to the editor accusing the steamer owners of unfair exploitation. "The Country Aroused; Indignation in all Districts," reads one headline,[55] and "How Fishermen are deprived of their Earnings; Sealing Monopolies and their Overcharges," reads another. This latter headline introduced a letter to the editor which demonstrated the excessive prices that the merchants charged the sealers for their necessities and recommended that the legislative council, which largely represented mercantile interests, be abolished.[56] A few days later, the newspaper published an account of how several sealers successfully sued Bowrings for overcharging them.[57] Complaints were also expressed about steamer crews killing thousands of old hoods in the Gulf early in the season and leaving the young, which were too small for commercial purposes, to perish. The "Western people" complained about this and, said the paper, "They had good reason for doing so, because the evil was wrought on the same ground where the Western schooners usually calculate on getting their seals."[58] On another occasion, news was received that Twillingate was surrounded by seals but that the landsmen were forced to stop killing them because the price had dropped so low. The newspaper reported:

> It is very hard on the poor seal hunter, after he has risked his life on the icefields, to be deprived of the fruit of his toil by the cupidity of the trader and merchant. There has been no depreciation in the price of seals to warrant local buyers in offering only half the value for them.
>
> No sooner does the "toiler of the sea" here get a chance to gather a good harvest than the grab-all propensity of those with whom he has to deal manifests itself, and he is heartlessly victimized. Thus it is that the Newfoundland fisherman finds it so difficult to improve his circumstances.
>
> At present he does not get for the whole "pelt" anything like the value of the skin itself. Ordinarily your seal skins in the London market are worth from $1.20 to $1.40 each. This being the case, what a piece of injustice it is to offer the sealer for both skin and fat not even the value of the skin alone! Truly, "man's inhumanity to man makes thousands mourn."[59]

The ordinary sealer had found a strong and stirring champion in the *Evening Telegram*, who drew the public's attention to the practices and conditions of the seal fishery in a way that had never been done before. While it is obvious that this newspaper had its own political agenda, its attention to the seal fishery was invaluable because it exposed the industry to public scrutiny. The attacks on the government ceased after the 1889 election, when the Whiteway party was returned to power and the Thorburn merchant-supported party defeated. However, attacks were continued on the merchants and on the mercantile-dominated legislative council. And, this newspaper exhibited a strong bias in favour of the outports—reflecting, no doubt the views of its editor, Alexander A. Parsons, from Harbour Grace.[60]

As already pointed out, the 1890s were disastrous years for the Newfoundland economy, and although the Whiteway party won the election of 1893, it was a Pyrrhic victory, with many of the members being forced to resign during 1894 because of accusations of election irregularities during 1893. Then, in December 1894, Newfoundland's two banks were forced to close;[61] Whiteway managed to retrieve his electoral victory, but the economic situation remained hopelessly depressed. In 1897, the Tory party, under James Winter, won a decisive victory; one plank in the party's platform was the promise of a bounty on vessels built for the seal fishery. The intensity and radicalism of the *Evening Telegram*'s campaign was to increase with the defeat of Whiteway's Liberal party and the victory of the Tory party.

Evening Telegram and the Greenland Disaster

Disasters had always underscored living and working conditions in the industry and provided interested persons with the opportunity to subject these conditions to minute public examination. The tragedy that struck the crew of the S.S. *Greenland* in 1898 presented the *Evening Telegram* with a cause which was quickly taken up and relentlessly pursued. On 28 March, just after information was received that members of the crew of the *Greenland* had lost their lives, the newspaper stated its position on health and safety in the seal fishery:

> In placing before our readers the particulars of this shocking disaster, we do not attempt to hold any individual or party responsible for it. Rather we are inclined to blame the system or method by which the industry is prosecuted. We admit that there must always be a large element of danger connected with the sealing voyage, even under the best circumstances; but the question arises: Is it not the duty of all concerned to minimize that danger as much as possible. Unfortunately, up to the present time, no effort in this direction has been made. On the contrary, the voyage is made with reckless daring that seems to defy the very forces of nature. From two to three hundred men are taken on board one of these small steamers, almost regardless of their condition, physically or otherwise. No medical examination takes place before they embark, no doctor accompanies them, and no precaution is taken to guard against accident or outbreak of sickness. Everything is left to chance or a merciful Providence, without any of these human safeguards which should always be provided in such cases. All things considered then, the wonder is that there are so few disasters and so little loss of life in connection with this industry. Then, again, look at the risk of sending two or three hundred men out on the icefields away from their ship, without any protection in the event of a sudden snowstorm or other occurrence rendering it impossible for the men to get back to their ship, or the ship to get to the men! There they are, with neither food nor protection from the frost and snow. The present disaster calls loudly for drastic legislation in this respect. Let a sealing law be enacted providing for medical examination of crews before entering upon the voyage, making it compulsory to have a doctor on every

ship, and above all, rendering it absolutely necessary that the men be provided with food and tents, or special covering of some kind, whenever they go beyond hailing distance of their vessel. The lives of our hardy sealers and fishermen ought to be protected as much as possible, regardless of the cost to the trade or State; and the importance and urgency of such protection becomes greater and more apparent every year.[62]

During the spring of 1898, the *Evening Telegram* attacked the system under which the seal fishery operated from every possible angle and in a variety of ways. On 30 March, it published an editorial that explained the safeguards that existed to protect fishermen in England, ensuring that no similar disaster could occur there. Such a voyage emanating from an English port would operate under strict legislation requiring the following: a rigid inspection on the eve of the voyage; restrictions on the number of men carried, keeping in mind the tonnage, capacity and accommodations; adequate sleeping berths; a fully furnished and equipped lazaretto or floating hospital with qualified personnel; men whose responsibility it would be to look after the crew on the ice and to make sure that rough shelter and protection was always available; and insurance policies to cover all men and all interests.[63] The newspaper concluded that simple precautions such as these would have prevented the *Greenland* disaster or, if not, would at least have saved the men's lives.

The following day, the newspaper called for an enquiry and claimed that "public opinion suggests the advisability of beginning such investigation without any unnecessary delay."[64] The same issue carried a letter to the editor asking if Capt. "Abraham" [sic] Kean had played any role in the disaster or in its aftermath. Kean was not only the Tory MHA for Bay de Verde but was also a minister in Winter's government, and it was widely believed that Kean's crew had stolen the pelts belonging to the *Greenland*, which had forced the men of that ship to remain at the hunt when they could have been returning to St. John's. Thus, the thieves were indirectly to blame for the 48 deaths. The charges against Kean were intimated over the next several issues of the newspaper, and on 9 April, he was criticized for displaying bunting and flags when entering port in the wake of the *Greenland*. He was accused of having no respect for the dead, no sympathy for his fellow-man, and of being unfit to be a member of the House of Assembly or cabinet.

Editorials and letters to the editor continued to hammer away at the system and the government. Questions were asked about why the sealers from the *Greenland* were rushed off to their homes so quickly after their arrival in St. John's and why the master watches were interviewed and not the batsmen. Reports of inhumane treatment by steamer owners were described, and complaints about the lack of legislation to protect the sealers were made. Other letters reported on overcrowding and the lack of proper food, bedding and medical aid. Although cited earlier, the newspaper's summary of conditions bears repeating:

> You are out on the ice all day tugging at the seals and, without any proper food. Now a man can endure that, and even endure it day after day, for a succession of days, if he is strong and well, and can look forward, after the day's work is done to a good square meal and a quiet bunk. But the poor sealer has none of these. The steamer in which he is pigged together with

> some two or three hundred other men *has no sleeping berths whatever for the men*. There are a few rough shelves without bedclothes or bedding of any kind, about enough for one-third of the men on board. If a man wants a bed he must bring a bag of shavings on board to lie on, which he has borrowed or begged from some charitable neighbour on shore. Then he must take *his chance with the other men* for permission to lie in or on one of those rough bunks ... If a man is young and hearty he can bear the strain, and survive it. But *God help him if he takes sick at the ice and does not want to die*. An ox or a cow on board an Allen cattle ship has a thousand chances to one of recovery as compared with him, and he knows beforehand that illness on the voyage means a box of dead meat sure.[65]

On 13 April, the *Evening Telegram* took issue with the two 'government' newspapers, the *Evening Herald* and the *Daily News*. Both newspapers, according to the *Telegram*, were interested only in supporting the government and the system; they were both upset with the *Telegram* for taking the lead in raising subscriptions for the survivors and families of the dead and for questioning Capt. Kean's interpretation of events. But the *Telegram* noted:

> Captain Kean's conduct in connection therewith has been called in question. Captain Kean is a member of the Tory Executive. The *Daily News* is the organ of that executive. Any blame that may be attached to Captain Kean would reflect upon the members of that body, individually and collectively. Hence the wild and frantic efforts of the *News* and its little editorial brother of Prescott Street [the *Evening Herald*] to let the whole matter of that disaster rest where it is, without any adequate enquiry whatsoever.[66]

Then on 19 April, the *Telegram* reported that the sealer Joseph Hussey, who took Baine Johnstons to court for grossly overcharging him, lost his case because Judge Conway did not want to meddle in the practices of the seal fishery.

The *Evening Telegram* was incensed by the operations of the spring seal fishery and by the cavalier way in which the government ignored the conditions under which the sealers laboured. While it is correct to agree that the *Telegram* was anti-government and consequently opposed to the editorial policies of the *Daily News* and the *Evening Herald*, one must realize also that the *Telegram* was fast becoming the champion of the sealers—in fact, it was the only publication that spoke in support of them.

Nevertheless, there was a growing antipathy in certain circles towards traditional fish and seal merchants, and it was being encouraged by the *Evening Telegram*—and rightly so, in that the hazardous and deteriorating practices of the seal fishery had been allowed to continue far too long. The rich were getting considerably richer, while the poor were getting poorer—and dying in the process. Moreover, this bias against merchants provided the foundation for the outport bias against St. John's. The sealers were no longer hailed from both St. John's and the outports, as they had in the 1830s and 1840s; now, most came from the outports. For example, out of the 48 men who died in the *Greenland* disaster, 2 were from St. John's and 3 from nearby Quidi Vidi and Torbay; 43 came from the northern outports. Yet St. John's monopolized the other jobs in the seal fishery—cooks, firemen, coal trimmers, seal skinners and such. It was only natural that the interests of the merchants and St. John's become synonymous in the eyes of the sealers. Many of the issues and problems peculiar to the 1902 strike that was to take place four years later were underscored during the spring of 1898, when the deaths of 48 men attracted the full attention of the public. Furthermore, anti-mercantile feelings extended beyond the industry, itself; this antagonism was generally felt by those who were convinced that the future of Newfoundland lay in industrialization and that the colony's dependence on the fisheries, especially that of the cod and seal, had to be reduced.

In 1899, as the spring seal fishery drew near, the *Evening Telegram* attacked the government for its failure to pass legislation to protect the health and lives of the sealers in a hard-hitting editorial:

> We are quite aware, and we wish to make admission of this awareness at the outset, that the sealing adventure is, in many respects, an enterprise apart by itself; that it has many features that are peculiar to it as a unique industry, and that those who engage in it as operatives do so in full knowledge of the exceptional risks and peculiarities by which it is attended.
>
> It is not our purpose, therefore, to propose that this industry—especially in its declining days, as the present no doubt are—should be vexatiously hampered with restrictions that are either unnecessary, unwelcome or of doubtful utility. No one can possibly have any interest in such a proposal as that....
>
> First, there ought to be a compulsory official inspection of all steam vessels proceeding to the ice and carrying large bodies of men as to (1) the condition of the hull, boilers and machinery, and (2) boat-accommodation and facilities for saving life in the event of disaster to the vessel. Such inspection ought to be made by two independent authorities, and under heavy penalties for misrepresentation.
>
> Second, there ought to be such an inspection as to the quarters reserved for the accommodation of the crew (1) with regard to their sufficiency as sleeping accommodation, (2) as to their sanitation, (3) as to their comfort, in view of the fact that, for possibly two months on a stretch, they are to be the only home and refuge for two hundred or three hundred human beings.
>
> Third, there ought to be such an inspection as to the food or dietary provided for these workpeople, (1) as to the quantity, (2) as to quality, (3) as to variety and accessibility, as that the men may know what they have to depend on, upon their part, in a joint enterprise, wherein, otherwise, they are denied, all control over such necessary and reasonable knowledge.
>
> Fourth, there ought to be certain precautions adopted for the humane and speedy relief of pain and rescue of life on board the vessel in case of sudden accidents such as gunshot wounds, fractures of the limbs, hemmorage, etc., and also for coping at their early stages with any outbreaks of epidemic disease among the crew. These should include (1) the setting aside of a space as a lazaretto or temporary hospital, (2) the provision of ample surgical and medical appliances and remedies, and (3) the attendance of some qualified person—who need not be a diploma'd M.R.C.P. and S.—to afford some more or less some scientific relief to suffering sealers when far away from the amenities of home....
>
> A happy go-lucky dependence on a merciful Providence, which has stood us in such good stead in many a past sealing-voyage, is no excuse for continued neglect of such moderate and reasonable precautions as decency, humanity, charity, and Christianity alike require.[67]

These suggestions, if followed, would have improved conditions in the seal fishery considerably, and they no doubt served as a later guide to both the government and the Fishermen's Protective Union.

Governors take an interest

Strangely enough, in spite of the public outcry over conditions in the seal fishery, the government did not take any steps to remedy the situation, with the exception of a small piece of legislation passed in 1899 to limit the amount of money that could be recovered from sealers' wages for supplies. The governors, on the other hand, began to take a personal interest by the turn of the century.

Governor Henry E. McCallum (1898-1901) was the first governor to express an interest in the sealers themselves; this was probably a result of the publicity surrounding the *Greenland* disaster, which coincided with his arrival. In any event, Governor McCallum invited the sealers and captains to meet with him in the local ice rink before the 1899 season, where he pointed out that the over-crowding that had been a problem would be relieved somewhat by the legislation of 1898, which limited the number of men a steamer could carry.[68] He visited every vessel and examined the men's sleeping quarters and their food. He showed a keen interest in their welfare, expressed the wish that he could accompany them and must have created an impression on the steamer owners as well as the men.

The following year, 1900, McCallum continued to show interest in the conditions of the sealers. He met and spoke with them, examined their living quarters and food supplies and then turned his attention to providing them with medical care.[69] He instructed his secretary to contact Dr. Roddick[70] in Montreal, who agreed to find medical practitioners who would go to the ice on the steamers for the sealing season. McCallum then turned his attention to the steamer owners and proceeded to convince them to hire these medical practitioners to go to the ice. Bowrings agreed to send two—one in the *Aurora* and the other in the *Terra Nova*; Jobs decided to send one in the *Neptune*; James Baird, one in the *Labrador*; and W. B. Grieve, one for the *Iceland*. The companies agreed to pay each a fee of $100 and travelling expenses to the extent of $150. Dr. Roddick was to choose the doctors, and the program was due to begin in the spring of 1901. The plan was initiated on schedule, and four medical men—Dr. Burns and Messrs. Kelly, Baird and Macnamara—arrived in early March to go to the ice.[71] Unfortunately for the sealers, in 1901, the Colonial Office transferred Governor McCallum to another post.[72]

It does not seem that McCallum's successor, Cavendish Boyle, shared McCallum's interest; indeed, he seems to have been somewhat put out by having to deal with the sealing strike of 1902 soon after his arrival, which may explain his antipathy towards the sealers. Governor William MacGregor, however, who arrived in 1904 and who was to stay until 1909, continued the practices established by McCallum. In 1906, for example, he made a formal visit to the steamers, accompanied by the Hon. Eli Dawe, Minister of Marine and Fisheries. He went on board the *Labrador, Leopard, Bloodhound, Vanguard* and *Eagle*, and the visit on each steamer was hosted by one of the senior partners of each firm, including the Hon. James Baird, the Hon. Edgar Bowring, Mr. Henry Bowring and Mr. J. S. Munn. He "examined everything minutely, more particularly the arrangements for cooking men's food, and expressed himself as being thoroughly satisfied...." When he had finished for the day, he "spoke highly of the arrangements for the prosecution of the fishery and was surprised at the excellence of the food." It was reported that he would complete his tour of the steamers the following day.[73] MacGregor also intervened in 1907, when it was reported that the crews of the *Algerine* (which lost its rudder and took no seals) and the *Greenland* (which had sunk) experienced some difficulty in getting their transportation home. When he heard this, he notified Mayor Gibbs that he would contribute £10 to any fund that was being raised to help these distressed sealers.[74] However, it was no longer necessary, because the men had been provided with their train fare; in fact, the following year, the steamer owners agreed, at the request of Prime Minister Bond, to provide transportation home for any sealers who did not make any money on board their steamers during any sealing season.[75]

The governors continued to take an interest in the seal fishery through the remaining years under study. It was reported in 1908 that Governor MacGregor visited the *Virginia Lake* and inspected the steamer and the men's quarters. He stayed for dinner, which was also attended by R. G. Reid and Capt. Samuel Blandford, and they all drank a toast to the success of the *Virginia Lake* and their host, Capt. Jacob Kean, who was in command.[76] In 1910, MacGregor's successor, Sir Ralph Williams:

> paid an unexpected visit to the S.S. *Aurora*, laying at Bowring's wharf. He was received by Capt. Greene, Messrs. J. A. Munn and F. W. Hayward who if taken by surprise tendered the hospitality of the staunch Newfoundland sealer. His Excellency visited all parts of the ship including the men's quarters, saw for himself what conditions were and at once made himself popular with the crew. From the bridge he made an encouraging speech to the men lined up on the deck, and when leaving the ship His Excellency and Miss Dean were cheered time and again.[77]

It seems that the governor was genuinely interested in seeing for himself the conditions on board the steamer and had, therefore, visited it without warning. Following this surprise visit, he and his "principal staff" visited the *Florizel*, *Adventure*, *Bellaventure*, *Bonaventure* and *Beothic* and inspected "all parts of the vessels and expressed satisfaction at the arrangements made for the comfort of the men."[78] It should be noted, however, that only the *Aurora* was one of the old wooden walls; the others were the newer steel steamers.

Thus, it is apparent that most governors took an interest in the lives and work of the sealers after the disaster of 1898. Although they could do little of a concrete nature to improve the lot of the sealers, they did place the steamer owners in the public spotlight because where the governor went, the press followed. In addition, the steamer owners did not wish to be seen by the governors as crass money grubbers and were, hence, forced to make their steamers as presentable as possible. Out of this comes an impression that the governors did not trust the merchants—this is one theme that can be detected throughout most of Newfoundland's history. Consequently, it was no accident that the sealers marched to government house and demanded to see the governor in 1902 at the beginning of their strike, for they doubtless hoped to find there a sympathetic ear. It was unfortunate for them that Boyle had just arrived and was somewhat of a nervous disposition; one feels that either McCallum or MacGregor would have been much more receptive to their complaints.

Fishermen's Protective Union

In the fall of 1908, William Coaker, a farmer living on an island in Dildo Run, close to Herring Neck in Notre Dame Bay, called a public meeting in Herring Neck and launched the Fishermen's Protective Union.[79] Coaker had been born in St. John's and had been employed at Twillingate as the agent of a St. John's firm when the business went bankrupt in 1894-95. Instead of returning to St. John's, Coaker became a farmer and began to read and to think about the problems of the outport fishermen and how their situation might be improved. He decided that the three main problems the fishermen faced involved the truck system and its implications, illiteracy and a general lack of education, and a lack of political representation. Therefore, his basic reform plan was threefold and called for the development of co-operatives, education and communication through a newspaper, and the sponsorship of local political representatives. He began to implement his program by speaking to the fishermen in the small outports and soon organized local and district councils. The movement expanded rapidly, and, as Noel has written, it "became more than a union; it became a crusade."[80]

The Fishermen's Protective Union was somewhat slow in taking up the cause of the sealers.[81] It was not until the third annual convention at Greenspond in 1911 that a committee was appointed to report on what steps should be taken to "improve the conditions of the men who engage in the seal-fishery."[82] The committee reported that iron steamers should be sheathed with wood and that hatches should be partitioned off from the crew's bunks on all steamers. Furthermore, it recommended that soft bread should be provided to the men on Mondays, Wednesdays and Fridays; hot dinners, consisting of pudding, pork, beef and potatoes, should be provided on Tuesdays, Thursdays and Sundays; pea soup with onions, potatoes and turnips should be supplied on Saturdays; and breakfast should consist of beans and pork. It was also recommended that cooks be required to cook only, that a room and medical aid be provided for the sick, that the use of guns and panning to be prohibited and that the men be returned home by the steamers at the end of each voyage. However, these proposals made up only one article—No. 15, "Amended sealing laws as outlined by the FPU agreement"—of the 23 planks in the organization's 1911 platform.[83]

Coaker wasted no time in negotiating with the steamer owners and, in 1912, reported that most of them had lived up to the commitments they had made to him.[84] According to the *Evening Telegram*, these were as follows:

Provisions:-

That if it be found practical, soft bread to the amount of not less than a one-pound loaf three times each week shall be served out to each member of the crew, and if this be found impractical, soft bread shall be supplied as nearly this extent as can be arranged.

That beef, pork, potatoes and pudding shall be supplied for dinner three times each week.

That stewed beans shall be supplied for breakfast daily.

That onions, potatoes and turnips shall be some of the ingredients in the pea soup supplied on Saturdays.

That the allowance of butter shall be continued as hitherto, and that the fat taken from the [cooking] boilers by the cooks shall be used in the puddings.

General:-

The owners agree to issue instructions prohibiting cooks from performing any work other than that comprised in their duty of cooking, and from carrying ashore any fat or any other provisions belonging to the ship.

The owners agree to provide in the steel ships a room for the accommodation of any men who may become sick or disabled, and to arrange that where practicable, men that may become sick or disabled in the wooden ships shall be transferred to the steel ships belonging to the same firm, so that they may have use of those rooms and the benefit of medical care and attendance.[85]

In addition, the sides of the steel ships, located in the vicinity of the men's accommodations would be sheathed with wood to reduce condensation, partitions would be constructed to reduce drafts, and the bunk area would be heated with steam pipes. Also, while wooden ships would be allowed to use rifles in the hunt, the steel ships would not; owners would try to insure their pelts so that the sealers would receive payment in the event that a steamer was lost; the killing of adult females would be prohibited; the price of fat would be fixed before the season began; and wooden ships would leave port after 8:00 a.m. on 13 March, and steel ships after 8:00 a.m. on 14 March. As one can see, these clauses were taken from the internal report of the FPU that was drawn up in 1911. However, although Coaker was satisfied with these promises, they were not binding on the steamer owners. Nonetheless, at the union's annual meeting held in December 1913, this time in St. John's, Coaker said he was pleased with the fact that so many steamer owners were living up to the agreement "in reference to food and accommodations in ships."[86] Moreover, he expressed the hope that the wooden steamers would introduce baking outfits in order to be able to "provide food as acceptable as that found on steel ships."

Coaker made other efforts on the sealers' behalf. He began the practice of holding a public meeting and addressing the sealers in St. John's each spring before they went to the ice. In fact, A. B. Morine, continuing his interest in the sealers, addressed, them, himself at the meeting in March 1913. In February 1914, Coaker, now a Member of the House of Assembly, introduced a Sealing Bill which passed unanimously but was later amended in the Legislative Council. The main provisions in the Act, as it finally passed in March 1914, were as follows: wooden sheathing to reduce condensation in steel ships; portable iron-frame berths; some protection from draughts in the sleeping quarters; the heating of sleeping quarters with steam pipes; a room for the ill and disabled and, where practicable, a doctor; and a prohibition on the use of rifles.

There were also regulations concerning food, which included the following stipulations: that one pound of soft bread be served to each man three times a week; that beef, pork, potatoes and pudding be served three times a week; that fish and brewis and stewed beans be served alternatively for breakfast; that onions, potatoes and turnips be among the ingredients in the soup to be served on Saturdays; that fresh beef be served once a week and, if not available, that canned beef be served instead; and that cooks were to do no work other than cooking. However, it is obvious, from evidence presented to the commission that investigated the *Newfoundland* disaster, that some sealing companies ignored the FPU agreement and the new legislation. For example, Joshua Holloway, a seven-year veteran of the seal hunt, testified that the *Newfoundland*'s crew were given tea, hard bread and butter every morning for breakfast and every evening for 'tea' or supper; dinner on Monday, Wednesday, Friday and Saturday consisted of the same meal. Only on Tuesdays, Thursdays and Sundays did Holloway received a cooked dinner of salt pork and boiled duff.[87] Jesse Collins, another witness, agreed with Holloway but added that he remembered being served salt meat and fresh meat on a couple of occasions and canned meat at one meal.[88] Cecil Taylor, a three-year veteran, remembered receiving soft bread twice a week, pork and duff on Tuesdays and Thursdays, thought that they had pea soup "sometimes," reported being served fresh meat on the first Sunday at sea, and ate hard bread and tea at all other meals.[89] Robert Winter, the chief cook on the *Newfoundland* and with six years experience as a cook, tried to deflect the blame from himself and the company for the state of the food and cooking facilities. He pointed out that his four cooks provided the men with pork and duff on Tuesdays and Thursdays, with pea soup made of pork, peas, turnips and onions on Saturdays and "fresh meat on Sundays as long as we had it." He added that the men were given raw, unwatered saltfish on Mondays, Wednesdays and Fridays and salt pork and a pint of beans each on Mondays, Wednesdays and Thursdays. The men were expected to cook these provisions in a galley that could accommodate only 25 men at one time and consequently, it often took them four hours—4:00 a.m. to 8:00 a.m.—to cook breakfast. He placed the blame for this situation on the men: "This is how they wanted it. The first meal or two going out we cooked bruise [brewis] and everything. We fulfilled the law the first meal or two going away and then they came to me and said they would rather take it whacked out [allocated]."[90] It seems likely that Winter whacked out supplies to the men in addition to the three or more cooked meals but in actual fact they would, like Holloway, be forced to depend on hard bread and tea for most of their nourishment. Beans and saltfish need extensive soaking and, in their cramped quarters with limited supplies of water, soaking beans and saltfish would be virtually impossible. It is obvious that the *Newfoundland*'s chief cook was primarily interested in fulfilling the letter of the FPU agreement and the letter of the 1914 legislation which had not specified that food was to be cooked before being served. Thus the FPU had little impact on the quality of food provided on the sealing vessels.

Another regulation in the 1914 legislation required that the men who became sick or disabled on wooden steamers be transferred to a steel steamer with a medical officer. Finally, the regulations governing the date of departure for steal steamers and wooden steamers and the date on which killing could begin were all spelled out in detail. When reporting to the annual meeting in Catalina in the fall of 1914, in the wake of that year's disasters, Coaker made it clear that he wanted to prohibit the "right of property in panned seals" and that steamer owners should be fined $1,000 for each man who died on the ice from exposure.[91]

Strangely, the FPU did not attempt to increase the men's share of the voyage to one-half, which had long been desired by the sealers, and neither did the union attempt to fight for a higher price for the pelts, which remained at $4.50 from 1908 until it rose to $4.75 in 1914. Just as surprising, the FPU did not challenge the combine of three oil plants which set the price of fat. Instead, the union concentrated on food, comfort, safety and the prohibition of the panning of pelts, whereas income had always been more important to the men. In fact, the men would, no doubt, have objected to the prohibition of panning if their opinion had been sought, because it would mean that sealers would have to haul all the pelts to the steamers and would result in smaller cargoes and harder work.

While it is obvious that the FPU was concerned about the lot of the sealers, this issue was not a priority, indicating the extent to which sealing had declined in importance by 1912-14. Coaker, understandably, concentrated his efforts on the industry whose basic resource was considered inexhaustible—the cod fishery. Therefore, the saltfish trade and all that accompanied it occupied most of his attention. In fact, both the publicity provided the sealers by the media, especially the *Evening Telegram*, and the concern expressed by the governors, especially McCallum, MacGregor and Williams, made it easier for the union to negotiate the few basic, non-monetary concessions obtained from the employers. However, the readiness with which the fishermen from Trinity, Bonavista and Notre Dame Bays accepted the views express by William Coaker and his FPU would indicate that the experience of the 1902 strike had helped to demonstrate to them the power of collective action. Thus the influence of the 1902 strike must be viewed as a significant factor in the establishment of the new union and its political wing, the Unionist Party.

Summary

The seal hunt and the deterioration in the men's working conditions and pay became the focus of public attention during the latter part of the 1800s and the early years of the twentieth century. Surprisingly, the *Evening Telegram* (and other newspapers to a lesser extent) and the governors (after 1898) became the principal champions of the men, while governments largely ignored the situation as they sought to industrialize and diversify the economy. Coinciding with the growth in public awareness of the annual spring ice hunt was the growing interest in it cultural significance.

Culture and Reputation

The seal fishery, as one would expect, had a direct impact on the folklore and culture of Newfoundland.[92] Unlike the cod fishery, which was an English/British migratory fishery that became transplanted to Newfoundland, the seal fishery was a local, home-grown industry. Therefore, the stories, songs, tools, activities and language of sealing are 'Newfoundland' in a way that their counterparts in the saltfish industry could never be.

Narratives and vignettes

Unfortunately, the oral stories of the nineteenth century seal fishery have nearly all been lost. However, a few were written, and thereby saved, by a number of local chroniclers.

Henry William LeMessurier (1847-1931)—born in St. John's, MHA and civil servant, for two years editor of the *Evening Herald*, and author of the song "The Ryans and the Pittmans"—wrote and published many local stories on all aspects of Newfoundland life. The following is not a sealing story, but a tall tale concerning frosty weather on board a sailing vessel:

> I remember when I was a youngster hearing my grandfather and father tell about the olden time taverns, which later were termed grog shops and then dignified by the name of saloons, where gathered many of the leading people of the mercantile class, and retailed stories of the day. At one particular tavern several captains on one evening were boasting of strange occurrences that had happened to them. One said 'Come Captain Holmes, I am sure in your experience you have seen and heard some queer things'. 'That I have', said Captain Holmes, 'and the most remarkable event that I ever met with was during last fall when we were a-coming to the westward. You'se all remember that I had a very long passage from Poole, and got driven off twice over the Banks, in the month of November and didn't get here with the fall's stock until the 2nd of December. My word for it, but it was the coldest voyage ever I made. We was on short commons, and our water was getting so low that I had ordered the last cask locked up, and served it with my own hands. Two nights before we reached port it froze guns, and after four o'clock in the morning, being my watch below, I turned in cold enough, just taking the precaution to put a small tot of spirits in me to liven up the temperature. I couldn't have been very long asleep when I woke up with a start and heard pistol shots near me, followed by the noise of the bullets striking the roof of the cabin. I was afraid to stir, thinking I might be hit, but as the noise ceased after a time, I had the courage to strike a light and look about, but I could see nothing, and at length concluded that it must have been some noise on deck, so I turned in and went to sleep again. In the morning, when I was called, and commenced to dress, I found every sparrowbill drawn out of my boots and looking up saw them stuck in the top of the cabin roof. Gentleman, the frost had been so great that it had actually drawn them all out, and it was the noise of them coming out of the boots that sounded like pistol shots, and their striking the top of the cabin was like the noise of pistol balls—and that was a hard frost, I tell you'.[93]

Henry Francis Shortis (1855-1935)—born in Harbour Grace, editor of the *Terra Nova Advocate* and, later (1912-1935), "historiographer" of the Newfoundland Museum—was an assiduous recorder of Newfoundland stories and folk life. Several examples relating to the seal fishery follow:[94]

The Brigantine *Fanny Bloomer*

When we were youngsters it was really wonderful the knowledge we possessed of ships. It was our whole study, and not alone, to this day, we can remember their names, but also the names of the captains and many of the crews.

Our chief source of amusement was to get on board one of the sealing vessels, run up the ratlines, on to the truck and place our cap thereon, and come down by the back-stays or jib-stays; and the boy who would not accomplish this feat was looked upon as a lubber and would be frowned down on by his associates as being no good.

We all had our favorite vessel, but really I do not know why we chose her. It may be that sentiment played a prominent part in our selection, by her having such a "catchy" name as the *Nancy Lee, Annie Laurie, Nora Creina,* or it may be attributed to the popularity and success of the master. Even to this day in St. John's, you will hear the old-timers refer with pride to Capt. John Barron's *Dash,* John Silvey's *Elizabeth,* Terence Halleran's *Arthur O'Leary* or *Zambezi* [*Zambesi*], Glindon's *Sonora,* Graham's *Glance,* White's *Evanthes,* Duff's *Guitar,* etc.; but few, if any, have such a grip on the minds of the veterans now remaining than has the famous little brigantine—*Fanny Bloomer*—the subject of this brief sketch....

Now what our ancestors did for our own Country is of more interest to me than all the feats of prowess accomplished by the Greeks, Roman and Carthaginians in the days of their prosperities and victories. The end of it all was that in all probability the dust of the mighty conqueror was used to stop a bung-hole.

But it is not so with our heroes who performed their feats during our own life-time, and were equally, or more so, as worthy of commendation as were the heroes recorded in the Iliad, Plutarch or elsewhere. But the history of our own country is not to be taught by the regius or emeritus professors in the universities, for they know nothing at all about it. There is not one of them, who knows the rig of a vessel much less to be "on a pan" sculping seals, and if they were given a tow of seals they would not know how to lace them up, much less to sculp and haul them, the probability being that they would haul them "against the grain."

No, the best Alma Mater I have ever found for obtaining the history of our family and its resources was in a sail-loft, the cook-room or around the kitchen fire during the winter months, when the celebrated veterans who prosecuted the seal-fishery as well as the cod fishery, related their experiences for a period of half a century, as well as the experiences of their ancestors handed down from father to son, generation after generation. Not alone would you get a general outline of the mode of conducting the fisheries, the dangers, escapes, etc.; but you would be enlightened in every detail, and the chief source of regret to me is that we had not the *Evening Telegram* or some other popular and enterprising daily newspaper to take down and record the facts as related by those who participated in the events, and were the means of building up the country to what it is today.

There are hundreds of the readers of the *Telegram,* who, on viewing the head-line of this article—*The Fanny Bloomer*—will put on their thinking caps and exclaim, "I was in her one spring with Capt. Arthur Jackman," or "I heard my father tell all about her, she was indeed one of the favorites of the fleet," or similar expressions.

The *Fanny Bloomer* was a Colonial built vessel, put out of hand for the famous seal-killer, Capt. John Silvey, in 1853—the year after the Spring of the Wadhams.

She was a pretty model, strongly built and a fast sailer. At the sealfishery, she carried a main-topsail, as was usual with all our brigantines, for working through the ice. Many of them carried a main top-gallant sail. These sails were more quickly handled than the fore-and-aft main-sail in working through the ice. Some of the orders roared out in trumpet tones by the famous skippers are very interesting, and now almost obsolete and forgotten, such as: "Brace forward the main-yard," "shove up the lee bow," when the men would be on the "rams" (two 25 feet large sticks lashed together in the shape of a triangle) well supported by grab-ropes for the men's safety.

The first Spring the *Fanny Bloomer* went to the seal fishery she returned with 4,000 prime whitecoats. Capt. Silvey was in command for six springs being very successful. After the failure of the great firm of Barnes & Co. (St. John's), Capt. Silvey bought the brig *Elizabeth*, the largest vessel going to the ice, and he gave the *Fanny Bloomer* in charge of his brother-in-law, Capt. John Flynn, who also commanded her on foreign voyages during the summer months.

In the year 1856, the *Fanny Bloomer*, Capt. Flynn, and brigantine, *Mary Belle*, Capt. James Day, one time member for the St. John's West and Chairman of the Board of Works, were in Liverpool, England, together, and as both were being towed down the river Mersey, side by side, by the one tug boat, Capt. Day hailed Capt. Flynn, exclaiming: "I'll bet you £25, I'll be in St. John's before you"!

"I can't bet you that much," said Capt. Flynn, "because I have no money to pay you, only out of my wages; but I'll bet you £10."

"Done," said Capt. Day.

They came down St. George's Channel together, kept in sight for a couple of days, and did not see each other again until mid-passage, when they signalled.

They again lost sight of each other, until they made the land, when both vessels were about three-quarters of a mile apart when day-light broke. This was between Bay Bulls and Petty Harbor. As soon as they discovered each other, on went studding sails and every inch of canvas—hoisted respective house-flags (Bowrings and Tessiers) to foremast head, a signal to the cape (the two flags went up together on the one staff), and they ran for Cape Spear. The wind was blowing a strong breeze from the West-South-West, and the two vessels rounded Cape Spear, bound for the Narrows of St. John's—not a hundred yards distance between them.

The race was up, and it was declared a draw by mutual consent. The pilot boarded Capt. Day's vessel first, as she was to windward of the *Fanny Bloomer*, and the two vessels beat in the Narrows together. As they left Liverpool, side by side, so they arrived in St. John's after a passage of 33 days.

Capt. Flynn was two Springs to the sealfishery in the *Fanny Bloomer*, after Capt. Silvey bought the brig *Elizabeth*, which was lost near the Gut of Canso in 1871. After Capt. Flynn left her, Capt. Silvey took charge of her again, in 1861, and died on the 14th. March on board of her at the sealfishery. After Capt. Silvey's death Bowring Bros. bought the *Fanny Bloomer* from Mrs. Silvey (Capt. Flynn's sister), and she sailed from Catalina to the ice, but was not very successful.

The next year, 1863, old Capt. Thos. Jackman (father of the famous Captains William and Arthur) was given charge, but unfortunately received an injury at the ice-fields by being struck by the "tiller," and the heroic Capt. William took charge, being his first time in command. He did fairly well, and then went in the brigantine *Sarah Ann* out of Bowring Bros.' employ. The year Capt. Snelgrove was in the S.S. *Hawk* the first trip, (1867), Capt.

Wm. Jackman took charge of her the second trip, and then commenced his phenomenal success in the *Eagle*, to the end of his life—his death occurring on the 25th. Feby., 1877. In the meantime Capt. Arthur Jackman took charge of the *Fanny Bloomer*, and opened up a career at the sealfishery second to none in the country. In the year 1870, Capt. Arthur took a load of seal oil and skins to Liverpool, and the *Fanny Bloomer* was sold there, after which she was employed in the coal trade, plying between Cardiff, Wales, and Waterford, Ireland. But I have never been able to ascertain what was the subsequent end and final resting place of this most historic ship—famed in song and story.[95]

Hard Winters of the Past when the Mercury almost Froze

As the severe weather for the past week or so [23 February] has occasioned much talk and discussion on the matter, I desire to forward a few records of the following weather of long ago.

The winter of February, 1843 familiarly known as that of "The Three Suns,"—a curiosity that aroused the fears and excited the minds of many of the inhabitants at the time—the glass consecutively registered for some fifteen days, from 9 to 10 below zero [F.].

In February, 1857, fourteen years later, much similar weather set in about the middle of the month, and continued uninterruptedly till the 27th. During this spell an incident worth recalling occurred in the burning of a considerable portion of the brig *Margaret*, owned by Capt. Peter Feehan, at the wharf of Messrs. McBride and Kerr. The vessel had been partially outfitted for the sealing voyage when the unfortunate happening occurred, on the night of February 21st. However, as the time was getting short, the sailing being then on the 1st. of March, an unusual rush was made to effect repairs. Shipwrights, joiners, riggers, etc. were hurriedly engaged to perform this work, during which intense frost prevailed, making it quite difficult in expediting the job. Finally the good old firm of McBride & Kerr, who were Agents for the *Margaret*, and anxious that the ship should be ready in time, thoughtfully, and with the desired results also, placed a hogshead of Jamaica rum on the wharf for distribution among the workmen. The "stings" was given in charge of a trusted employee, who systematically dispensed four "welts" daily to each man, together with a "grog-bit" of No. 1 Hamburg bread. Consequently everything went swimmingly while the frost lasted, and till the vessel was ready to sail.

February, 1876 was an unusually frosty month, the thermometer registering for a considerable time on and about zero. On the 5th., however, the mercury dropped to 16 below zero in the city. The first week in February, 1874, a severe snap also struck the city and suburbs, the glass on one occasion, February 7th., recording 10 below zero. On February 16th., 1875, during a frost spurt, the glass at the Block House, Signal Hill, registered 23 below; while a few days later, Feby. 20th., it recorded 20. On February 25th., 1889, an intense cold wave swept over the city, and the temperature on that day is recorded for 12 below.

On March 1st., 1897, during a cold spell, the thermometer showed 19 below in St. John's and vicinity. However the lowest temperature on record, as generally asserted, was that of the 10th. February, 1883, the night of the Wool Factory Fire—forty years ago today.[96]

Loss of the Brigantine *Eric*: 1878

There are many records of the tragedies of the sea more affecting and impressive in detail, more mournful in circumstance, and more fatal in the recounted result than that of the brigt. *Eric* at Bird Island Cove, and the noble rescue of her suffering crew; but this last instance met with providential aid, seconded by the noble and courageous efforts of men, rises to a lord with the highest and most impressive examples of heroism on record. We shall best do justice to the case by a simple recital of the facts with all their native effectiveness.

The brigantine *Eric*, Capt. Penny, owned by Messrs. Rorke & Sons of Carbonear, with a crew of sixty-four men left port on a sealing voyage on the 5th. day of March, 1878. All went well till the 9th. of the same month when one of the crew had his leg broken whilst engaged before the bows of the vessel. On the 22nd. of March all the crew were engaged taking seals, when a snow storm suddenly came on. The crew with difficulty returned to the ship, but not the whole company. Two poor fellows were lost, and sank in their ocean grave or perished miserably on the ice. Of those who returned, fifteen were frostbitten, and all were more or less suffering from exposure and cold to which they had been subjected for a period of over thirty hours. Between this time and the 25th., about 300 seals were taken, on board, and the crew being at this time almost utterly disabled, the vessel was put about and steered for home. During the next two days some 200 seals were taken on board by the less disorganized portion of the crew. On Tuesday, 2nd. of April the *Eric* spoke the S.S. *Eagle*, and ascertained true position, and got correct course for the land. On the following Friday observed breakers ahead and almost immediately made the Southern Bird Island, and as soon as practicable let go both anchors. At this time the gale was increasing and blowing directly from the South East. The following Saturday, it being apparent that the vessel could not ride out the storm, it was decided to cut away the masts, and afford every facility to the hull to live out the furious gale and heavy sea that were mercilessly belabouring the doomed vessel. All that time there was no human possibility of landing the crew with any chance of safety. The spectators from the shore looked on in helpless blank agony. The crew on board the wrecked and storm-lashed vessel crowded tumultuously on the deck, or moved to and fro with feverish anxiety—but we will not venture an analysis of their tremulous, hope-lit despair-darkened passions. To the spectators from the shore the vessel with her unhappy crew seemed to all human probability destined to a dismal end. She lay rocking and surging in the trough of the sea, her crew were moving restlessly about her decks, that now alone offered a frail protection from the mad seething waters beneath. No help could reach them from the shore, for no boat could live amid those moon-stirred waters. During the long dreary hours of Saturday night and down to the dawn of Sunday morning, fires gleamed ashore, symbols of hope and also evidence of watchfulness on the part of those who were safe on land, and of determination to let no opportunity slip that might be available for the salvation of the agonized sufferers on the frail planks of which had been once a vessel.

Preparations of a nature to cover every possible contingency were meanwhile being made on shore. Coils of cordage were got ready, ladders, boat hooks, life-buoys, etc., were all at hand, awaiting the moment when as was supposed, the entire crew would make a last despairing effort to reach the shore, and consign themselves to the rude mercy of the waves. On shore was a scene and spectacle of intense and sorrowful anticipation. On board the fever called living must have been burning with fierce consuming fire. Moments were measured by hours. Hours under the sickening influence of mingled hope and despair stretched away into days; but a benign Providence was controlling the destiny of that tempest-tossed ship. Slowly, like a far off loom, approaching came a pale dim streak of ice into sight. Driven before the gale it rapidly neared the land. Soon it came down on the surging vessel, but was skilfully warded off from the chains and the bows by the crew, who with poker and gaff, worked as men only can work under the frenzied pressure of mortal terror or mortal agony. The ice came in, surrounded the vessel, stretched away to the shore and also, even as oil, smoothened the troubled waters. A signal was made on board indicating that the crew of the *Eric* would make a supreme effort to reach the shore. Preparations were redoubled on the land. It was seen that about twenty of the men had consigned themselves to the ice and were swiftly nearing the breakers of the landwash. Nothing but death stared the unfortunate men in the face; but the quarry of death was still afar off. A number of volunteers, under the guidance of High Constable McGowan, of Catalina, descended the precipitous rocks, and having thrown off all impediments of clothing, succeeded in reaching a small island rock to which they attached their ropes. When the apparently doomed men drifted within about twenty yards of the rock, the ropes were thrown to them and the lives of all were providentially saved. As one poor fellow was being drawn up the beetling crag, a huge wave overtook him and pitilessly smashed his leg. Almost immediately another storm wave rose over the island rock and nearly swept away the whole volunteer party. One of them named Oldford had his leg broken and was otherwise seriously mutilated. At this juncture High Constable McCowan was taxed to the uttermost to keep his men steadily together as each moment seemed to be the herald of death to all. Ten of the volunteers escaped injury of a more or less serious character. But the motto of all seemed, practically to be, "Death before surrender"! They would not desert their self-appointed posts of danger whilst a unit of human life remained to be saved.

There were now about sixteen men floating in towards the shore on a small string of ice, but from their vantage ground they could observe on shore that another large patch of ice was approaching the vessel, which would afford a more safe conveyance to the shore. These men were therefore ordered back to their vessel, and before two hours had elapsed, the ice to the windward forming a breaking water and the leeward ice a bridge, the whole remaining, portion of the *Eric*'s crew were enabled to walk into Middle Brook. Despair and intense anxiety now gave place to exultation and thankfulness to Providence for the happy rescue from a fearful death of sixty human beings. Amongst the volunteers to whom all praise and recognition for their noble services are due, the names of Mr. Arthur Tilley and Constable Bailey stand out in bold relief. These were the valorous lieutenants of High Constable McGowan, no doubt ably seconded by the rank and file of the volunteers, whose exceptionally noble conduct on this occasion are worthy of substantive appreciation by the authorities. Poor Oldford, maimed and mutilated should not be forgotten by those who reverence the nobilities of human nature. He is a poor man and had only his life and limb to give in the cause of humanity, and of these he made a willing oblation. That the paramount part played in this heroic drama by High Constable McGowan will meet with the guerdon so richly due him, we have no doubt, and that it may come to him without delay is our sincerest wish. The crew of the ill-fated *Eric* are now safe at home in the midst of their families. The vessel is floating in debris and broken plank and spar along the seething foam-fringed shore of Bird Island Cove.[97]

Perfect Disciple on the Ice-Floes
A True Story of Newfoundland

My chief delight when I sit down to write for one of the journals is to give readers something connected with the good old days of our sealing fleet. I do this because it was my good luck to be born just at the time that fleet was "under full sail" and when old skippers of the previous generation were about setting their courses for their happy destination o'er that unknown sea, whence no traveller returneth. I always had my ear cocked for the nightly stories during the winter months, of those hardy mariners of a previous generation—whose iron constitutions withstood the hand of time over the allotted space of life, and whose faculties and descriptive powers were not impaired, even up in the nineties, and in some cases, over the century, as witness skipper Tim Curtis, of Carbonear, who was 106 at the time of his death. The last time I saw him he was shingling the back house—had not a grey hair in his head, and as he would say himself, could spy a mosquito on Bell Island. One had only to sit on the settle over the dogirons, light the pipe and take notes. Nor would it be necessary for even the Disciple of Higher Education, of the present day, to rectify the language in which the stories were couched. As for practiced information on general affairs, foreign or local, the "scholar" of today, like the man on the house, would not be in it.

We have heard a great deal of the bravery, fortitude and subordination of the respective European and American armies during the respective campaigns which they have fought in the different parts of the world; but these soldiers have from youth been drilled into this subordination. They have drill-masters whose duties it is to qualify them for their future avocation and with inducements, under various ways, to face the inevitable. But with the Newfoundland sealer of the old days, they had no person to teach them to brave the hardships of the ice-floes—it came to them naturally. It is only when caught in a corner that their heroism and hardihood was displayed. In the first coming of the Scotch Steamers, over thirty years ago, the ships would always bring out thirty of forty sharp-shooters; but while those men could draw a bead on an old or young hood or harp, they were not at home on the ice, and invariably found wet jackets in stepping from one pan to another. In such instances, the hardy Newfoundlanders would "copy" on the heads of the Scotchmen, secure good footing on the other side, and then haul them out, and help them on board ship. Witness the experience of foxy John McCarthy, a man Capt. Adams said was worth a dozen ordinary hands. One time John wandered away in search of seals—he was master watch—and was foremost on the pan. After a travel of many miles he returned towards the "Arctic", but feeling hungry he sat down on a large pinnacle to eat some bread and a seal's heart—when bang! and a bullet went only a few inches from his head. He darted behind the pinnacle; but if he made the least movement to peer around the ice, bang! and the thud of a bullet was near his ear. His feelings can be better imagined than described. At last he took off his canvas jumper, guernsey frock, and placing them on his gaff, moved them about the pinnacle. After a short time a voice was heard hailing, to come out that it was alright. The owner of the voice was a sharpshooter, and took John McCarthy and his fur cap for an old seal. During my forty-five years at the fishery, said John, that was the narrowest escape I ever had. But the idea of taking me for an old seal. But to come to the discipline displayed by Newfoundlanders. Many years ago, the famous Capt. Azariah Munden, was to the sealfishery in the *Atlanta*. Seals were fairly plentiful but scattered. After having placed the different watches (about 30 men under a master watch) in their respective positions, Capt. Munden said to Curtis and his men, "Now, boys, work around here; but on the peril of your life do not change your positions, no matter how the weather may come. I have to go North East, to see if there are any improvements in the seals; but I will return for you before nightfall."

Curtis and his men went to work panning seals, and were doing very well. Suddenly the wind shifted, a blinding snow-storm with keen frost came on and the men had to huddle together, kept walking, and beat their hands to keep themselves warm. Night was coming on, and their prospects were becoming more gloomy and terrible each moment. At last one or two of the crew began to murmur and say, that they would not remain there to be frozen to death, that they would go in search of the vessels.

"For God's sake, boys," said Curtis, "remain where you are. Did not the Captain tell us he would come for us by nightfall?"

Still dissatisfaction increased, and the crew were in open manus. They were hungry, cold, wet and saw no hope of avoiding death. Still, Curtis held out, entrusted, commanded the dissatisfied ones to trust to the captain—that he would come for them. Every hope was given up by all except Curtis and all hands were preparing for the next world. Prayers first heard at the mother's knee were wafted aloft by weak voices, when suddenly the creaking of blocks were heard, commands given out in quick succession to starboard, clew up the royal, etc., etc., and through a rift in the blinding snowstorm appeared the bows of the *Atlanta*, and the first man to appear was Capt. Azariah Munden, who turning to his crew on board ship, exclaimed: "All right, boys, here they are, safe and sound. Let the cook get everything ready to make all hands comfortable, and we shall tie on to this iceburg till morning."

Ah! was there not something almost supernatural in the faith of that man, Curtis? By his unshaken confidence in the skill of his commander, he saved the lives of thirty men. What a lesson it teaches! Think over it readers! Could any, but a Newfoundlander be the hero of such an act?

This is but one of the many instances of the hardihood, daring and full confidence in their captain displayed by the sealers of the old school.[98]

Shortis wrote many of his stories of Newfoundland history for publication in the local newspapers. In 1913, he wrote a story about Capt. Munden, Munns' most prominent sealing captain. The essence of the story (which can be found verbatim below in this chapter) is that in 1872, Munden had ordered his men to walk towards the shore in White Bay, which was blocked with ice, because he was convinced that there were seals between his vessel and the land. To emphasize his seriousness of purpose, he allegedly told his men to bring back green boughs from the trees on shore if they could not find seals. In other words, he would not be satisfied until he was certain that his men had searched the ice all the way to the shore. The story ends with the men returning with pelts. This has become part of the folk history of Munden's career. However, in this case, Shortis' story elicited a response from a sealer who had been on board the

Commodore with Munden that spring which differed from the Shortis version in one important respect and is an illuminating story in its own right:

The Sealer's Response

Mr. Shortis is in error on the green bough incident. It is true that the Old Viking sent men on a long tramp toward Cape John and repeated his order the following day, with the additional instructions to bring the green boughs before he would give up his firm belief that the main patch of seals [sic] were between the ship and the shore, but the men returned without either boughs or seals, as we were too far from the shore for any man to reach it in one day.

... [W]ithout the least sign of seals apparent to any other person on board but himself he forced the *Commodore* in the same direction as the men were sent, and where we all thought we were going to be jammed for the remainder of the spring, which proved to be the case. We well remember some of the strong language used by some of those who were then veteran sealers, such as 'the d-n old mad-man' and the 'cursed old fool,' but before twelve hours of our getting clear their opinion of the old mad-man was very much altered, for about an hour before sun set that evening, we were in the whitecoats as thick as ever they have been since, and by dark we had the deck piled to the rail with whoppers averaging 52 lbs. The next day we panned about 9,000, but it was almost impossible to work for it was a day which will be remembered by many as the SHEELAH BLIZZARD, and the most awful blizzard for many years. Several people in this country lost their lives. We sailed from Hr. Grace that year on Feb. 27th, and steamed north nearly to the Groais Islands, where we got jammed as after events proved, not far from our goal, but with no sign of anything encouraging.

From St. Patrick's Day, when we first started the slaughter, we were 27 days in succession on the ice without a break, not even Sunday, for rest. This was caused by the ship not moving during that time, so we had to haul all our seals from the pans. We finished loading about three miles off King's Cove, Bonavista Bay, and in the afternoon previous to the day we got clear, three women came alongside. The old captain gave them a tow of seals each and sent them on their way rejoicing.

They appeared to be quite capable of taking care of themselves and were able to get along over the rough ice fairly well, as that was before the invention of the hobble skirt.

We got clear and bore up for home on Apr. 13th, with 1,500 seals in tow, but when getting near home we got these on board also, which put the *Commodore* when upright, first deck on a level with the surface of the sea. The weather being very smooth we arrived at Harbour Grace at 8 a.m. on April 15th. We then commenced discharging our cargo and made a bill of $160 or £40 it was at that time.

The getting ready for the second trip was rushed with all speed, and we left on the 21st, a very foggy morning, and ran ashore at Bear's Cove. The ship being so much damaged that it was impossible for us to proceed on the voyage, which caused us great rejoicing as we did not want to go out so late in the season, but the man who refused to go the second trip need not apply for a berth the next spring.

We got off the rocks about sunset and lay in the stream at Harbor Grace that night, but even then our troubles were not all over, for a very serious difficulty arose which was to devise some means of disposing of the surplus stock of sagwa [rum], for the jars were full to the stopper, but even this was not too much for the inventive minds on board the *Commodore*, for one bunk's crew got over the difficulty by boiling pea soup using rum instead of water and when all was ready invited their friends to the banquet, so during that night and the next morning one had to be very careful in getting around to avoid falling over dead men.

Next day we got to Capt. Stevenson's wharf at Ship's Head, and discharged our sealing gear. So ended one of the most eventful sealing voyages on record, and may we all who take interest in the sealfishery never forget the old Vikings who, although in many cases being illiterate men, through their pluck, determination, endurance, good judgement and common sense, did so much to help build up the many commercial enterprises which were started in their day.

We never heard what bill the King's Cove women made, but as seals were seven dollars and fifty cents per cwt. that spring, they must have done nearly as well as the people of Green Bay did the spring when it was said the seals were so plentiful and so near the shore that 'women and dogs made fourteen pounds a man.'[99]

One final entertaining and informative selection from Shortis:

...I know of only three men who made three trips in the one year in sailing vessels, viz. Wm. Whelan and Wm. Rabbits of Brigus and John Murphy of Harbor Grace—the latter in the *William* in the fifties. Capt. Murphy was a remarkable man in his way, and it was said no man in the Island had a more thorough knowledge of our coast than he had, yet he possessed such an imperfect knowledge of the English language that he could not give orders by naming the ropes, but had them marked in different coloured flannel, and when required would order his crew to "pull the red rag," "blue rag," or whatever particular brace or sheet he wished attended to. He is also said to have his punts so arranged that the old punts were on the port side, and the new ones on the starboard, so that when he wished the helm put a port he would shout "old punt," for starboard "new punt," and for steady, "caboose." He was a county of Cork man. As I stated before many of them had no learning whatever, and kept their accounts by marking down hieroglyphics that were unknown to any person but themselves, and in one instance, a certain well-known skipper got himself in a serious dilemma after marking down a large circle or O opposite the name of a dealer who purchased a cheese, and after some time, in reading off the account from his quaint notes, was not certain whether it was a cheese, a bake-pot or a grind-stone the man had from him, until the matter was finally settled by arbitration. There are some good stories told about those famous Vikings, for instance, the first time that Tom Nowlan took charge of the *Greyhound*, Skipper Tom was rather out of place on the quarter-deck of such a large square-rigger, and when she was all ready to proceed on her voyage, Tom looked forward amongst the crew and exclaimed, "Well, boys, you know what to do as well as I can tell you." And another instance somewhat similar comes to my mind, when a well-known skipper was asked by the helmsman, "What course shall we steer, Skipper?" The answer was thundered forth, "The same course as we steered last year, darn ye."

There is one individual whose history goes back until it is lost in antiquity. No one can tell for certain how it originated or whence he came. Nevertheless he has always existed, and is as strong and vigorous at the present day as he was in the days of the "jowlers" of historic fame. I refer to that ubiquitous individual—the jinker. No Newfoundland sealer is without a thorough and often melancholy knowledge of the jinker. For to him is often attributed the misfortune of losing a trip of fat, when the seals seemed within the very reach of the crew, one of whom he was unfortunately a component part. From my childhood I have been familiar with stories of the ill-omened individual. This much must be borne in mind in connection of our sealers with regard to this unwelcome customer on ship-board, that wherever he went, no matter what prospect of a voyage was in sight, the fact of the presence of the jinker entirely disheartened them, and they were never disappointed when luck forsook them. In fact, it would be a matter of the greatest notoriety, if a vessel having him on board, had been successful. I have never heard an instance cited when a jinker's ill luck had left him.[100]

These tales and vignettes provide a sample of the impact that the seal fishery had on the oral literature of the colony. They must have supplied an enormous amount of entertainment and information to the audience in the setting of the Newfoundland home, especially during periods of relaxation and particularly during the Christmas holidays. In addition, these story tellers were oral historians who educated the non-sealers about the seal fishery. Although it is unfortunate that so few examples were saved in print, the above suggest the richness of this material. The story by the sealer who had been with Captain Munden in 1872, for example, is illuminating as well as entertaining. It describes the unlimited quantities of rum carried during that period; it shows that the price of fat was double that paid in 1902; it suggests the unpopularity of the second trip but the fear of the consequences if one refused to go; and it comments on the prevailing style of women's dresses. It is information like this that shines a light on the daily lives of the ice hunters.

Honour the heroes

Another aspect of the dramatic growth of interest in the history of the seal fishery that can be observed in the media, beginning in the latter part of the nineteenth century, was the phenomenon of the 'heroic' obituaries that were written about the sealing captains, casting them in the role of heroes of a vanishing industry, age, culture and way of life. Some examples follow:

Death of the Hon. Capt. White

It is our painful duty today to record the death of another much esteemed and universally respected member of the community. We refer to the "removal hence" of the Hon. Edward White, whose translation from time to "the great beyond" took place this morning at the family residence, Southside. His illness had been of short duration, but while it lasted he suffered intensely up to within a day or two of "the final struggle," when tired and worn out nature became exhausted, and the vital spark gradually and almost imperceptibly expired. Of him it may be truly said, he lived a blameless, worthy life, and his death was that of a consistent Christian man; the last scene imparting fresh grace and beauty to the words of the poet -

> "The chamber where the good man meets his fate
> Is privileged beyond the common mask of virtuous life:
> Quite on the verge of heaven."

Of Captain White's public life little need be said now. Few names are more familiar to the people of Nfld than the name here mentioned. Whether we regard him as a friend, a citizen, one of the most successful seal-hunters the colony has ever produced, or as a member of the Legislative Council exercising his wisdom and experience for the general good of the Commonwealth,—in fact it matters not in what position we find him, there he is, or was rather the same sincere, guileless and patriotic Nflder—erring if ever, only on the side of charity. Capt. White lived to attain the ripe old age of 75 years, during which time he surrounded himself with all the unostentatious comforts and social relations necessary to a happy and enjoyable existence. He also lived to see the different members of his large family well provided for and comfortably settled in life, a blessing which none but the thoughtful and affectionate parent knows how to fully appreciate. So that it may be truly said of the deceased gentleman, "He was gathered to his fathers in the midst of a grand old age, taking with him the blessings of his day and generation." To the bereaved family we tender our heartfelt sympathy.[101]

The Last of a Long Line of Well Known and Famous Seal Hunters

Last Week, by the death of Mr. Jas. Whelan, there passed away from among us,—the people of Brigus,—the last male representative of the family of the late William Whelan, the most successful seal killer that perhaps Newfoundland ever produced. His famous trips in the little *Hound* are spoken of to this day; he often made 3 trips in a single spring, his men sharing from 50 to 60 pounds being quite usual. Old "Skipper Bill Whelan" as he was called, for in his day seal killers ignored the modern title "Captain", had 5 sons, 2 of whom, John and William, pre-deceased himself; the 3rd son James, we followed to his last resting place on Monday, the fourth son Stephen was drowned on the Labrador a few years ago and the youngest son George died in the U.S. some time before. All these sons, excepting George who was a doctor, followed the occupation of their father and were in their day successful seal-hunters and fishermen. Mr. James was in his 79th year.[102]

Captain Arthur Jackman Died
His Name a Household Word—A Famous Seal-Killer

At 12:30 this morning Captain Arthur Jackman passed away, fortified by the rites of the Catholic Church. He had been ill about two weeks and up till a few days ago his friends did not believe his sickness would have a fatal termination. Last Monday the doctors attending him advised that a priest be called in. Very Rev. Dean Ryan, Father McDermott and Rev. W. Jackman, his nephew have in turn been attending his bedside every day since. Yesterday when told that death was at hand he was quite reconciled and prepared to meet his maker. Deceased was known widely all over Nfld as a great navigator and seal-killer. He was born in Renews in 1843, and would be 64 years of age. He leaves a wife, one daughter (Mrs. Hewster, of Quebec), two sisters and numerous other relatives to mourn their loss.

At the early age of 22, he commanded a vessel at the seal fishery. In 1871 he made his last voyage in the *Fanny Bloomer*, and took command of the S.S. *Hawk*. Every spring following he had command of a steamer up to 1906. His ships were the *Hawk, Falcon, Narwhal, Resolute, Eagle, Aurora, Terra Nova* and *Eagle*. He also prosecuted the Greenland whale fishery in 1894.

He commanded the coastal steamer *Curlew* on the western route for several years, and made hosts of friends. His coastal work was almost entirely free of accidents to his ships. Few men have added more wealth to the country than Capt. Jackman. The result of his work at the seal fishery for 37 years is 552,510 seals, in value about $1,000,000. This would be an average of 15,347 seals for each spring. Though brisk and rough in manner at times, he had a kind heart and was of a charitable disposition. He was noted for his kindness for the sick men of his crew at the seal fishery and would nearly always tend on them himself, going to the forecastle or hold to minister to their wants. Capt. Jackman will be greatly missed, especially now on the eve of the seal fishery.[103]

The Passing of Capt. C. Dawe

In the demise of Capt. Charles Dawe Newfoundland loses one of the most respected and eminent citizens. For a considerable time past citizens have watched with anxiety the gallant struggle, made by a gallant man, against dread diseases; and when the sad tidings came that the struggle was hopeless and the end must come soon, deep sympathy was evoked for one who was a manly man and who in health and sickness fought a good fight in whatever he turned his hand to. The end came last midnight and closed a career of pre-eminent usefulness to his native land.

Capt. C. Dawe was born on Feb. 28th, 1845, at Port-de-Grave, Conception Bay, and throughout his life has been associated with the staple industries of the colony. He was a man of splendid physique, strong constitution, and of great energy and enterprise. In his early years he went to the ice and obtained command of the *Huntsman* and the *Rolling Wave*. Thirty-four years ago, in his 30th year, he began a very successful career as master of sealing steamers which lasted nearly a quarter of a century. As a sealing captain he was accordingly popular, a popularity partly due to his success at the ice, but more particularly to his big hearted consideration of the men he commanded....

Successful as Capt. Charles was as a seal killer, his chief and greatest success was in building up with his brother a great Labrador trade. When others feared to expand this branch of the fishing industry the firm of C. and A. Dawe, Bay Roberts, entered into this business with the greatest pluck and enterprise; and where others failed, the firm of which Capt. Charles was principal made of it the greatest Labrador business in the Colony. This courage and enterprise will be a lasting feature in the life work of the gallant captain.

But Capt. C. Dawe did not limit his energies to his own private business, he was a man of great public spirit, and spent nearly thirty years in the public life of the colony. In 1878 he was elected as a member for Hr. Grace and continued to be its representative until 1889. In 1893 he was elected for Port-de-Grave, and in the Goodridge administration of 1884 he was placed in the Executive Council. In 1897 he was re-elected for Port-de-Grave and was a member of the Winter Executive. In 1900 and 1904 he stood for Hr. Grace, and headed his team, but did not succeed in getting elected. After the death of Mr. McKay he was elected for Port-de-Grave. Capt. Charles Dawe was always in the forefront of the fight for his party, and obtained their confidence and support, insomuch so that after his re-election for Port-de-Grave he became leader of the Opposition in the present House of Assembly....[104]

Death Like a Passing Breeze Ends a Useful and Honorable Career.
Hon. Captain Blandford Passed Away This Morning

At 8 o'clock this morning after a useful and honored career Hon. Capt. Samuel Blandford, M.L.C., one of Nfld's most noteworthy men passed into that sleep which is the awakening of the soul. The summons came suddenly as a passing breeze.

On Friday last he was seen on the street. Yesterday he dined at his residence with a number of his old friends among the sealing Captains before their departure for the icefields. He was in his usual health and spirits, and recounted with them the adventures and perils of voyages past, little thinking how soon the call would come to him.

The Hon. Capt. Samuel Blandford was born at Greenspond, August 10, 1840, where he was educated and is now nearly 69 years of age. He worked with his father as a blacksmith, and at the age of 16 assumed full charge of the business, which not only included the profession of vulcan but also a large fishery and supply business. He early took to the sea, and in 1864 assumed command of the brig *Hebe*, engaged in the sealfishery. He subsequently commanded the brig *Renfrew* and the brig *Isabella Ridley* in the same enterprise. In 1873 he was in the S.S. *Tigress*, Capt. I. Bartlett, which picked up part of the crew of the *Polaris* which had been adrift on the ice for 5 months. In 1874 he took charge of the S.S. *Ospray* which had previously been engaged in the mail service between Halifax and St. John's. He subsequently commanded the steamer *Iceland*, from 1876 to 1878; *Eagle*, 1879; *Esquimaux*, from 1850 to 1883; and since then the *Neptune*. In the latter ship he brought in the 2 largest loads of seals, both as to the number and weight, ever landed. He was manager for Job Bros. & Co., at Blanc Sablon, Straits of Belle Isle, one of the earliest fishing establishments connected with the trade of Nfld. In August, 1893, he went to Cape Chidley in the S.S. *Nimrod*, the first steamer persecuting the cod fishery on the Labrador, and going 100 miles farther north than any other vessels fishing for cod. Capt. Blandford also commanded the S.S. *Hercules* from 1873 to 1875, which vessel conveyed the circuit courts during that period. He also commanded the S.S. *Plover* in the maritime coastal mail service from 1875 to 1883, and was one of the most successful and popular coastal captains the colony ever knew. In 1889 he was elected to represent his native district, Bonavista, in the Assembly, declined to contest the district in 1893, was appointed to the Legislative Council. Hon. Capt. Blandford was twice married—Miss S. A. Edgar, of Greenspond, and a few years ago Mrs. Wilkie, of Halifax, N.S. He retired from the fisheries some years ago and up to the time of his death occupied the position of ship's husband with the Reid Nfld Company.

Hon. Capt. Blandford was a man possessed of genuine benevolence and sterling worth. He was a devout member of the Church of England and foremost in all movements connected with that body. He took a particular interest in the boys of the Church of England Orphanage and many springs took them for a trip on the harbor in his steamer before leaving for the sealfishery. Hon. Capt. Blandford is survived by a widow, 2 sons, Hon. S. D. Blandford, Minister of Agriculture and Mines, and Edgar, now in Montreal, and one daughter, Mrs. H. J. Duder, for whom general sympathy will be felt in their bereavement. The funeral takes place at 3 o'clock on Wednesday from his late residence Hopedale, Circular Road.[105]

Old Sealing Capt. Crossed the Bar

Capt. Peter Gosse of Torbay, and an old time sealing captain, died at his home, Gower Street, at noon yesterday after an illness of long duration. Deceased who was a native of Torbay was in his 79th year, and never fully recovered from the effects of a fall received on the Labrador about 5 years ago. He was one of the good old Newfoundland Vikings, and in days gone by was master and owner of the schooner *Maggie* in which he was most successful at the seal fishery. He also bought the brigt. *Gipsy* and sailed her to the icefields bringing in many heavy trips of fat in that vessel. He leaves a widow and several relatives including Mr. W. Gosse, late Road Inspector, who is a brother.[106]

The Late Captain Henry Thomey

There passed away on Saturday at Harbor Grace in the person of Henry Thomey a very remarkable Newfoundlander, one of the old Vikings, a veteran sealing master, and the son of another skipper, Arthur Thomey, as well known and as successful as his celebrated son. Henry Thomey had attained the great age of 91 years but on his visit recently to St. John's he appeared to enjoy wonderful health and the spirits of a boy. The Ridleys, in whose employ, the Thomeys served for two generations thought a great deal of their skippers. One of the great events of the spring was Thomey's arrival to Ridley's, or the famous Az. Munden to Munn's. Thomey was great as a teller of yarns about the ice.

They were splendid men these old sealing skippers, and nearly all bay men. Thomey's success at the ice in the *Isabella Ridley* was phenomenal, and he made good trips in both the S.S. *Commodore* and the S.S. *Greenland*. Whilst Thomey was remarkably successful, averaging 4,000 for 40 years, he never lost a man....[107]

Captain Azariah Munden

He was fifty years a sailing master as his tombstone will testify. He often had a word about the brig. *Highlander*, no doubt the remembrance of his early manhood, as this was the first ship in which he took charge, but he is better remembered with the *Four Brothers, Three Sisters, Atlas, Alert* and *Atlanta*.

His ideas were progressive, and if he had his way Harbour Grace would have had the first Newfoundland steamer ever going to the seal fishery, but as the old firm of Punton & Munn had been so successful with sailing ships, and had so much money invested in them, it was several years before they invested in steam, but when they did it was a compliment to Captain Munden, that they called her the *Commodore*. In the *Commodore* his success was greater than ever, and in 1872 he brought in the greatest record voyage to that date numbering 31,314 seals, the weight of 655 tons, which for tonnage of a vessel has never been surpassed and never likely to be. The registered tonnage of the *Commodore* was 290 tons, and the value of that one trip was $94,927.

The following year when Captain Munden went to Aberdeen to supervise the construction of a new steamer for Punton & Munn, that was to beat everything afloat, Mr. John Munn told him to christen her the *Admiral*. The compliment was again being forced on him, but he wouldn't have it, lest people think it was pride on his part, and his name for the new steamer was the *Vanguard*.

The height of his glory was in the old steamer *Commodore* and then age and infirmities came on him, and his old sailors often said, 'It was a pity he ever grew old'....

Perhaps no story related about him is more typically of his knowledge of the seals and that fishery, than the following:

It was the year 1871 [sic, 1872], he had forced the *Commodore* up into White Bay, apart from all the fleet. When he found her jammed and could get no further, he sent his Masters of Watch out to look for seals. They returned that night with information that there was not a seal to be found in White Bay, and no use wasting time there. He told them this was not what he thought; or he would not have wasted so much time in forcing the *Commodore* where she was. In the morning he told the Masters of Watch to go out again and not to return until they brought him green boughs from the shore. They were back again before evening not with spruce boughs, but each man had his tow of seals, and a load for the steamer was assured....

The seal fishery is brought down to a science now, when steamers like the *Stephano* can bring in 36,000 seals in seventeen days, but the romance of the sealers will always be bound up with the old brigs, that for fifty years held sway in this country....[108]

As can be seen these men were esteemed for their capacities as sealers and for their contribution to the wider community. Only in the case of Capt. Edward White's obituary was the sealing record given little emphasis, but Capt. White had been given considerable space in the media when he retired from sealing a short time earlier, in 1883.[109] The public had begun to honour the men who had made sealing great and who, along with the industry, were passing from the scene.

General recognition that the "good old days" were passing and that the captains and owners of the wooden sealing ships were fading from the scene was evident by the end of the century. Chafe, Murphy and Shortis were just more vocal than others in their statements to that effect. In 1901, one other well-known writer, M.A. Devine, complained about the end of the outport middle class in a powerful and moving ballad:

The Outport Planter

The times baint what they used to be,
'bout fifty ye'rs or so ago,
And he hooked a coal from the bar-room stove,
and set his T.D. pipe aglow.
The b'ys be changed, the men be changed,
their place supplied by fraud and ranter,
But deadest of all the burr'ed past
is the dead and gon' outharbor planter.

He's gon' wid gansy and coatin' pants;
with Hamburg boots and ne'e'r a collar;
He's gon' wid cook-room, pork and duff;
gon' wid the good, old pillar dollar.
Gon' wid his chare at Christmas time;
gon' wid his rum in the red decanter;

> He's chareful v'ice and breezy song
> are burr'ed low wid the outport planter.
>
> But when 'counts be squar'd at the final day,
> and into the ledger the Lord is sarchin',
> He'll say, 'I find you cussed a sight,
> and once in a while you stuck the marchan',
> But you clode the naked, the hungry fed;
> so go up first with the harps and chanters;
> The place resarved for all good men,
> and honest, squar outharbor planters.[110]

Songs

Songs, too, were part of the cultural influence of the seal fishery. Songs were composed to meet certain specific purposes or to highlight significant events. Capt. Tom Casey was ridiculed in song for jamming the *St. Patrick* in White Bay. On another occasion, sealers were warned away from the *John Martin*, which seems to have been commanded by a particularly harsh captain. The unfortunate Capt. Terry Halleran was unfairly lampooned when he failed to live up to the expectations in his ability to command the S.S. *Esquimaux*. Similarly, the unnamed politician who guaranteed berths to sealers if he was elected was mocked when he failed to fulfil his campaign promises. The changes in the outports with the disappearance of the outport planters who owned or commanded the small sailing vessels at the ice fields and the Labrador cod fishery were the subject of another well-known ballad. And there were numerous ballads celebrating voyages and men. For example, almost immediately after his death, Capt. Arthur Jackman was honoured in a local newspaper by three ballads.[111]

The topics dealt with in poetry and song during the period under study are varied, and the surviving published sources indicate the importance placed on certain subjects.[112] Tragedies and disasters lead in numbers, and songs such as the "Trinity Bay Tragedy," "Died on the Ice Fields," "Loss of the *Maggie*," "The Wreck of the *Maggie*," "*Southern Cross*," two on the *Newfoundland* disaster, and seven on the *Greenland* disaster illustrate this. Individuals provided the subject matter of other songs, as with Capt. Arthur Jackman, Capts. Abram Kean and Bob Bartlett. The vessels themselves were also popular subjects, and individuals on board them were often identified, as in the case of "The *Nimrod*'s Song," where no less than 11 men are singled out for mention. Similarly, the firms, their vessels and captains are honoured in several laudatory pieces. In addition, three songs celebrate the sealers' strike of 1902, giving considerable praise to A. B. Morine—which indicates that in these cases, as in many others, the songs were written by observers and not by participants. In fact, Johnny Burke, an independent and individualistic local songwriter and author of one of these songs, did not condemn the sealing conditions but, instead, praised the fact that a fair settlement had been reached. A few were written for their entertainment value as comic ballads, and quite a number show influences of other, older traditional songs: "The Fisherman's Son to the Ice is Gone" and "The Girl I left Behind Me" are two examples of the latter. At the same time, an examination of published sealing songs indicates that the surviving songs are primarily from the period beginning in the late 1800s. It is impossible to assess whether this was because earlier compositions have been largely lost or because using the seal fishery as the subject for songs and ballads became common only when commentators became aware that the old seal fishery was basically over. Certainly, the latter point must provide a significant part of the explanation. However, the songs were written and composed to be performed; as England observed in 1922, the sealers on the S.S. *Terra Nova* sang sealing songs, among others, for their own entertainment, and what was true on board the vessels was certainly true in the kitchens and at the 'times' on shore during the winter months. Finally, whatever the motivation of the writers, composers and singers, the songs served to educate non-sealers about sealing—its terminology, practices and conditions.

Other cultural aspects

Although the annual seal hunt was a condensed period of intensive labour, there were periods of inactivity on board the vessels. In the early days, these occurred primarily when the vessels were jammed in solid ice; however, the men were divided into watches and each watch had its turn below. In addition, the men usually lowered sail and secured the vessels each night. In their spare time, the men played cards, sang, told stories and scrimshawed their powder horns. (According to Jukes, already quoted, it seems that the men in the earlier period also spent an inordinate amount of time cooking and eating seal meat.) With the introduction of steam, the work associated with canvas was at first reduced and later eliminated. Men had more spare time, and (as already seen) less opportunity to cook. Very likely, their pastimes became a more significant part of life at the ice. Reports from the early twentieth century indicate that they played checkers and cards, some sang songs or told stories and jokes, others read books and magazines, the occasional sealer would have a cat[113] to skin, and a number of men engaged in wood carving, particularly the carving of wooden buckets. A stick of wood, about 2 inches thick and between 12 and 24 inches long, would be carved in such a way as to produce a short decorative stick with a hanging bucket on each end. These were brought home to hang as ornaments in the sealers' kitchens or parlours. In some cases, they were put on stands and made even more elaborate. It was an attractive leisure activity, for it did not require much wood nor special tools. A piece of broken gaff provided the raw material, the sealer's knife was the tool, and the carver's talent and patience did the rest. These leisure-time activities were an important part of the sealer's life.

Also, the sealer had certain personal possessions which were constant reminders of the seal fishery to his family and the community at large. While his sealing chest would generally be put out of sight on his return home, his tools could be seen on his store loft by those who visited him. Also, the wooden buckets hanging in his kitchen were constantly on public display. Moreover, affinity between sealers and non-sealers was further reinforced by the songs and stories which served to bring the men's experiences at the seal fishery to the community. The seal flippers and meat the men bought home became a most popular local delicacy, and recipes for this dish added to the oral culture.

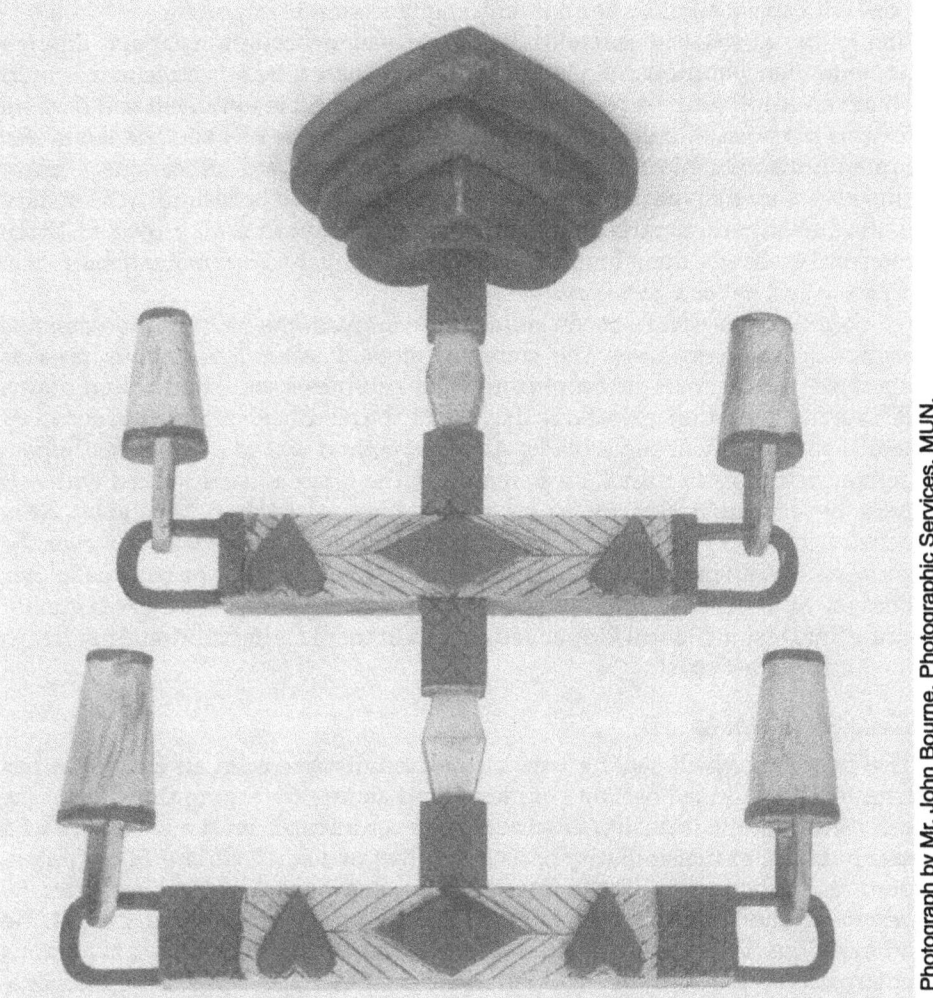

Figure 22—Two sets of wooden buckets on a decorative stand—each carved from a single piece of wood—similar to those made by sealers during periods of inactivity. The above curio was purchased in an antique shop by the author—carver unknown.

Flippers

Although the sale of flippers at the wharves when the vessels returned from the ice was an economic enterprise, it was also a social and cultural activity. In fact, one of the most significant connections between the sealers and the community at large centred around the cooking and eating of seal flippers. From the beginning of the industry, sealers made a habit of leaving one flipper attached to each pelt. The flippers (sometimes pronounced 'fippers') were sold locally when the ships returned to port, and the money was shared among the men. This dockside sale of flippers took on the air of a festival—the sealers making their sales and buying rum from the proceeds; the docks lined with buyers as well as with men, women, children and dogs just looking on. This was a small perk with social significance but only slight economic importance.[114] In addition to the sales to the general public, most sealers brought meat and flippers home to their families and neighbors. In many cases, two or three men would obtain an empty salt beef barrel from the cook and fill it with meat and flippers for this purpose. All along the northeast coast, the vessels and the landsmen brought this valuable source of fresh protein ashore (and indications, already suggested, are that the depletion of the herds resulted in a decline in the quality of the diet on certain parts of the northern and western coasts). However, to the men on the vessels, their families and neighbours, it always remained a seasonal delicacy and never a necessity.

Seal flippers always retain some of the fat from the pelt and this must be removed before cooking. The common method, after the obvious fat was removed, was to soak the flippers in water and bread soda; this brought out a distinctive coloration in the fat so that it could be readily identified and cut away with a sharp knife or razor blade. Another method was to parboil the flippers before cooking. After the fat was removed, the flippers were cooked with salt pork by a combination of frying, simmering and baking. Vegetables were usually cooked in the same pot, and in most cases, a pastry was put over the entire dish. With proper care and attention, the final product was quite delicious. Because of its scarcity, the care needed in preparing it, and the superb quality of the final result, the seal flipper had an important place in traditional foodways on the northeast coast.

Cruelty to seals

The extent to which sealing was a cruel activity has been an issue that has haunted the cultural heritage of Newfoundlanders for a long time. Since the early years of the industry, observers have commented on the pitiful cry of a seal pup and of the similarity of its cry to that of a small child. Philip Tocque, born in Carbonear in 1814, was the first to describe his feelings after he witnessed the killing of seals while at the ice on his father's ship in 1834. He admired the fishermen who were able to earn a living from such a hazardous enterprise, and he was very sympathetic to their living and working conditions, but he found the actual killing quite distasteful:

> The seal fishery is not only a dangerous and hazardous enterprise—it not only causes the sighing of the widow and the orphan, and is thus surrounded by physical calamities—but it is, moreover, a nursery for moral and spiritual evils. It has a tendency to harden the heart, and render it insensible to the finer feelings of human nature; it is a constant scene of bloodshed and slaughter. Here, you behold a heap of seals which have only received a slight dart from the gaff, writhing, and crimsoning the ice with their blood, and rolling from side to side in silent agonies; there you see another lot, while the last spark of life is not yet extinguished, being stripped of their skin and fat, while their startings and heavings make the unpractised hand shrink with horror to touch them. The seal fishery, in too many instances, may be looked upon as a sink of iniquity, where almost every principle of morality is laid prostrate, and the heart shrivelled up to the narrow dimensions of pounds, shillings and pence. The love of gain is engrafted on the heart of the seal-hunter, and this feeling predominates over every other, regardless

of the unhallowed means by which it is gratified. The sanctity of the Sabbath is disregarded and violated to an awful extent....[115]

What Tocque was observing, however, was a scene that could be observed in any European slaughter house in the mid-nineteenth century—an operation that his sensibilities would prevent him from observing. The point that many commentators have failed to appreciate is that the spring seal fishery was a hunt only while the men sought the seal herd. If they were successful in finding the herd at the most advantageous time, the hunt became a slaughter, and the ice became the abattoir.

J. B. Jukes also wrote of the revulsion he felt when observing the "kill" at first hand in 1839:

> As this morning I was left alone to take care of the punt [small open boat] while the men were on the ice, the mass of dying carcasses piled in the boat around me each writhing, gasping and spouting blood into the air nearly made me sick. Seeking relief in action, I drove the sharp point of the gaff into the brain of every one in which I could see a sign of life. The vision of one poor wretch writhing its snow-white woolly body with its head bathed in blood, through which it was vainly endeavouring to see and breathe, really haunted my dreams.[116]

This was an unusual scene because seldom would men load their punts, of very limited capacity, with dead seals that took up more room than pelts alone, and these seals would have to be pelted at some point. In addition, Jukes pointed out that "notwithstanding all this the excitement of being out in the punt ... clearing the ice of seals as we went along ... the hunting spirit which makes almost every man an animal of prey, kept me in the punt till a late hour in the afternoon." He also described shooting and wounding a number with his double-barrelled gun and how, although he "wounded several severely, most of them got away." He was surprised that even the young ones were difficult to kill with his (obviously under-sized) gun and would "carry off a good deal of shot."

> [For example] I knocked over one young one that was shuffling off a pan, but notwithstanding this he popped into the water, and swimming about ten yards, crawled onto another pan, when I gave him the other barrel: he again got into the water, and on his crawling out one of the men fired a sealing-gun at him, but the contents only striking him about the tail he again got away, and it was only by shooting him in the head as he raised it from the water, that I succeeded in killing him at last.[117]

Predictably, Jukes did not view his sport of shooting at the seals indiscriminately as cruel, even as he watched most of them swim away bleeding. He was a product of his class and culture, which viewed shooting wild life for sport as a pleasant and harmless pastime; similarly, the Newfoundland sealers were a product of their way of life and viewed their hunt and slaughter of seals as an economic necessity. It is very likely that the sealers viewed Jukes' actions as wasteful and destructive; it is probable that he was also perceived as immature and childish; and it is possible that the sealers actually thought him cruel.

Another account, in 1873, vividly described the scene of bloodshed and slaughter to be experienced at the ice fields. The unknown writer (who has been cited earlier) makes it clear that while the slaughter was terrible to observe and participate in, it was an economic necessity to the fishermen involved:

> To one not hardened to it, the whole process is said to be a most painful one to witness. The moans of the young seals—the agonies of the mothers at seeing them slaughtered—the fierce battles sometimes waged by the old dog hoods, who often make such sturdy resistance that they require two men striking them alternately with their gaffs to kill them—the whole ice strewn with the skinned carcases, still preserving their original shape, and almost quivering with life,—present a scene which nothing but the hope of large profits and quick returns could harden men to.... As an industry, the most painful feature is the cruelty which is inherent to it; but the remedy for this lies with the consumer, and not with those who supply the demand. As long as people burn seal-oil, and wear "kid" boots made of sealskin, will the slaughter of the innocents continue.[118]

There was a quantum leap in the negative publicity about sealing after Governor Henry Blake took up his brief tour of duty (two years) in Newfoundland in 1887. In early 1889, it was reported that Lady Blake had organized a branch of the Society for the Prevention of Cruelty to Animals in St. John's, "which may include the seal." Apparently, she hosted a bazaar at Government House which raised $800 for the society.[119] Lady Blake did not stop there, however, and when she and her husband left the colony, she published a damning description of the Newfoundland seal fishery under the title "Seals and Savages."[120] In this article, Lady Blake attacked the cruelties perpetuated against the seals and drew heavily upon the reports of Jukes and Tocque, already cited, to support her criticism. At the same time, she also condemned the vessels as "floating hells" and the part played by the clergy for blessing and offering up prayers for the sealers before their voyage and during their absence. She suggested that nothing could be "more brutalising and degrading to those engaged in it than such a fishery." She caustically wrote:

> On the west coast of Ireland the peasants believe that the souls of old maids go into the bodies of seals. It would be well if a similar superstition extended to Newfoundland, as at present the impression there seems general the 'seals are like fish and have no feeling.' Even a sealer would probably have some hesitation about 'scalping' (as they term skinning) alive a spinster aunt or maiden sister.[121]

Lady Blake wrote disparagingly about much of what she had observed while in Newfoundland and concluded:

> Now-a-days trading interests are supposed to override all other considerations, and to the Moloch of Commerce the health, morality, and happiness of millions of human beings are too often ruthlessly sacrificed; therefore efforts to mitigate the cruelties inflicted year after year on numbers of helpless and harmless animals will to many people appear quixotic and useless.... May it not be ... too much to hope that those who possess superior enlightenment and education will sooner or later awake to the crying sin of cruelty which, if the will were present, would easily be addressed? Surely, setting aside the sufferings of what we are pleased to call "lower animals," the wholesale brutalisation of large numbers of ordinary unthinking human beings is no light matter, and some blame has justly attached to a community where the labouring classes were allowed to retrograde from the humanis-

ing benefits supposed to be reaped from civilisation. But better days are dawning in Newfoundland. A Society for the Prevention of Cruelty to Animals has been established in the colony; the leading men there are awakening to the evils of which we have spoken, and it is to be hoped that their efforts to put down cruelties and unnecessary barbarities may be crowned with success.

Lady Blake was an unusual person; given her position in the colony, to host a founding meeting of a branch of the Society for the Prevention of Cruelty to Animals in Newfoundland, with every indication that she wished to see seals included in its mandate, was a very brave act, indeed, and certainly beyond the traditional role of the wife of a nineteenth century British governor. Then, her attack on the Newfoundland industry and the colony, itself, while her husband was serving as governor of Jamaica indicates that she was an extremely independent person. Newfoundlanders did not view these characteristics as strengths; nor did they admire her for her honesty and forthrightness. In response to her views, they, in turn, attacked her.

It was pointed out by the local press, and correctly so, that Lady Blake's information was based on information drawn from Tocque and Jukes, which was over 40 years old. Therefore, the argument continued, even if what the two writers in question had written was true in the past, the information was out of date and did not apply to Newfoundland sealing in the 1880s. In other words, it was not fair "to charge the present generation with the transgressions of their ancestors."[122] The commentator contended:

> In ruder times, great cruelties were practised by the hunters, trappers and fishermen who first came into contact with the Red Indians of this Island. Would it not be unjust to charge their inhumanities on us, and to hold us responsible for their deeds of blood, or represent us as animated by a similar spirit? We have all had very rough ancestors—even the most delicate and refined lady who now shudders at the sufferings of a seal or dog. Our fathers of the Stone Age were unkempt savages, lived on raw meat and thought little of knocking a man on the head, if he came in their way. A Society for the prevention of cruelty to animals would not have obtained many subscribers in those days when life was a "free fight," and our poor relations were splashing and floundering amid rough surroundings, hardly able to keep their heads above water....
>
> The point we wish to make is this:—that educational and religious influences have been at work among us during the last 50 years, and that, in consequence, a more humane spirit has been evolved, so that such cruelties as Lady Blake refers to—skinning seals alive—are utterly unknown or extremely rare....

In addition, the newspaper took the opportunity to refute Lady Blake's comments on the clergy and on the working and living conditions of the sealers. However, throughout the condemnation of Lady Blake's article, it is obvious that what rankled most was the title, which referred to the Newfoundland sealers as "savages" and to the community as a whole as "savage" for allowing and participating in this activity. After all, as most realized, the seal fishery was the foundation on which the leading business, political and religious leaders had built the colony.

Sir Henry Blake, who obviously realized that his wife's article could have serious repercussions on his career and jeopardize any hope of his remaining on good terms with the colony of Newfoundland, wrote to the St. John's newspapers immediately upon being informed of the publication of the offending article. His only argument was that Lady Blake had submitted her article under the title "On Seals" and that the editors had changed it to "most effectively arrest attention."[123] To prove that this was so, he included the proof sheet of the article which contained Lady Blake's original title. This assuaged public opinion considerably, although the article continued to be condemned.

Newfoundland was in the depths of a depression in the late 1880s and did not begin to recover until the early years of the twentieth century; this helps to explain the lack of concern about such matters as cruelty to seals. Only the debate over whether there should be a ban on the shooting of old seals kept the issue alive during these latter years. Many thought the practice wasteful, and most thought that it was contributing to the decline in the overall resource. Some actually pitied the wounded seals that escaped to die slowly and painfully. Capt. George Whiteley opposed the shooting of old seals and summarized the views of many in a letter to the editor of a local newspaper which emphasized the cruelty of the practice:

> During three springs in the *Diana* with A. Barbour, master, we shot the greater part of our trip of seals each year. We had twenty-five of the best gunners in Bonavista Bay, and the best average we would make was one seal out of every four stopped, what I mean by stopped is with the bullet in some part of his body, in other words or hit; 'if the seal is not hit in the head he will go off,' and this was considered good shooting.
>
> It is not easy to shoot old seals with the rifle, the running and jumping, especially if the ice is loose, makes it difficult to shoot accurately.
>
> In the *Neptune* for several springs the average was even lower than one for several shots, for various reasons, one of which was Capt. Blandford was too impatient for old fat and did not give his gunners much time, consequently we had to run and fire, and that made it very hard to shoot straight. Capt. W. Winsor and myself shot away four hundred and fifty cartridges in one day for 90 seals, in this way, running and firing.
>
> I have seen thousands of old seals go off before a company of gunners armed with modern rifles with blood streaming out of the bullet holes in their bodies, some with as many as three spurts of blood from parts of the body, and still with enough life left to crawl to the edge of the pan and so get away, only to die of their wounds and be eaten by the sharks.
>
> The awful cruelty of indiscriminate shooting of old seals at a time when every animal is supposed to need protection most, is to my mind a very serious proposition and one that needs to be looked after if we would save the sealfishery for the next generation.[124]

One can not help wondering what Capt. Whiteley, an experienced hunter, would have thought had he observed Jukes enjoying the sport of indiscriminately shooting at seals earlier in the century.

There is no doubt that the slaughter of seals, once the patch was located, was not a pretty sight, but, as the common saying goes, "that is why slaughter houses don't have windows." The pop of gaffs on skulls and the moans and cries of the seals were not pleasant sounds; when one adds the contrast of blood on snow, the overall scene must have been rather frightful. However, in the context of the practices carried out in the slaughter of the elephant seals and the North Pacific fur seals, the Newfoundland practices were usually no less humane than other examples of mass slaughter.[125] Finally, one must give the last word to Dr. Wilfred Grenfell, the English physician and reformer who made the improvement of the health care of the people of northern Newfoundland and the coast of Labrador his life's work, and who was not reluctant to criticize any aspect of Newfoundland life:

> Now the killing of young seals has been frequently described as brutal and brutalising, and the seal hunters depicted as inconceivable savages, and this not only by shrieking faddists or afternoon tea drinkers. But, to my mind, the work is not nearly so brutalising as the ordinary killing of sheep, pigs, or oxen, driven terror-stricken to the shambles, which are already reeking with their fellows' blood.
> Here [at the ice] the animal is too young to feel fear, and evinces no sign of it; no animal [white coat] is wounded and left to die in vain. The 'crying' of the young seals is not from fear, but hunger. One often hears them crying all night for their dams, while they never give tongue at man's approach.[126]

Obviously, cruelty committed by ice hunters against seals is a subject about which there will never be any consensus.

Arctic explorations

In contrast to their reputation for cruelty, at least in some circles, Newfoundlanders found themselves increasingly admired and in demand for their ability to navigate and work in arctic ice. The experience of the captains and men in navigation through ice floes was second to none in eastern Canada and the eastern United States. The Canadian government chartered Newfoundland sealing ships for summer surveys; for example, in 1903, the S.S. *Neptune*, commanded by Capt. Samuel Bartlett, was hired by Ottawa to undertake a voyage of sovereignty in the Arctic.[127] In addition, the Hudson's Bay Company used them to supply its eastern Arctic posts, and, most importantly, United States interests anxious to reach the north pole turned to Newfoundland for captains, crews and steamers. As international competition to reach the north pole increased, it was the demand for Newfoundland's assistance that brought world-wide attention to the colony.

Newfoundland's participation in several major expeditions was responsible for its reputation as a community of ice navigators. In the spring of 1873, Capt. Isaac Bartlett (c.1821-1906) from Brigus, in command of the sealing steamer *Tigress*, was involved in the rescue of the crew of the U.S. steamer *Polaris*, who had been forced to abandon their vessel while engaged in the Hall expedition.[128] (Charles Hall's expeditions to the Arctic in 1860-62 and 1864-69 were the first serious American attempts to get involved in Arctic exploration. Hall died in the fall of 1871 on the *Polaris* in the first year of his third expedition.) Lieut. A. W. Greely was sent north by the U.S. government in 1881 to live in the Arctic temporarily and explore. The Newfoundland sealer S.S. *Proteus*, under the command of Capt. Richard Pike, Harbour Grace, was chartered by the United States government to transport the Greely expedition to Lady Franklin Bay in the Arctic, with the agreement that Capt. Pike would return in two years.

However, on the return voyage in 1883, the *Proteus* was crushed by the ice and sank; Pike and the crew walked 600 miles to safety.[129] The following year, the Newfoundland sealing steamers *Bear* and *Thetis* were purchased by the United States government and sent to the Arctic in an attempt to bring back the Greely expedition; Capt. Levi Diamond (1833-1920) from Catalina served as ice navigator on the *Bear*, and Capt. William Norman from Brigus was ice navigator on the *Thetis*.[130] Greely and six others were alive when the party was rescued in June 1884.[131] Certainly by the 1880s, Newfoundland was recognized by the Americans as essential to their plans for further arctic exploration.

Robert E. Peary followed in the footsteps of the others. He went north in 1886 in the S.S. *Eagle*, under Capt. Arthur Jackman[132] and explored Greenland. Additional trips were made in 1891-02, 1893-05, 1896, 1897, 1898-1902, 1905-06 and 1908-09. Peary, in his well-known account of his explorations, does not identify the men and ships he employed in great detail; or when he does, it is incidental, except in the case of his most famous captain, Newfoundland's Bob Bartlett. Peary travelled north on the Newfoundland sealer *Kite* in 1891,[133] and later on the *Falcon*. He employed Capt. Harry Bartlett (Bob's uncle) during the early 1890s; in fact, the *Falcon* was lost with Capt. Harry and all hands while returning from Philadelphia in 1894, after landing Peary, his wife and child safely home.[134] Then Bob's other uncle, Sam, went as captain of Peary's ships and was succeeded by his brother John in this capacity.[135] In 1898, John, in command of Peary's ship the *Windward*, invited Bob to accompany the expedition on the four-year voyage planned for that year. This Bob did, and in 1905, was invited by Peary to take command of his steamer the *Roosevelt*. In writing about this voyage, Peary described the Bartlett family's skill in Arctic exploration and their contribution to this field in admiring terms. He hailed them as great ice navigators and continued:

> A great uncle [Isaac] was master of the *Tigress* when that ship picked up the drifting floe party of the *Polaris* expedition; two uncles, Samuel and John, were respectively master and mate of the *Panther* in which Hayes and Bradford visited Melville Bay; recently Captain Sam was master of the Canadian government steamer, *Neptune* [sic], which wintered in Hudson Bay; and both these, as well as Harry, a younger uncle, had been masters of my ships during one or the other of my several voyages north.[136]

On that same trip during 1905-06, Moses Bartlett, Bob's cousin, was mate, and the boatswain, John Murphy, and the steward, Charles Percy, were both Newfoundlanders, as were all the crew and firemen, with the exception of one of the latter. This was the final reconnaissance voyage, and in 1908, Bob commanded the *Roosevelt* for the concluding voyage, which ended with Peary's announcement that he had reached the Pole. Peary, when boasting about the fact that this was an 'American' triumph, inadvertently gave Newfoundlanders more credit than he intended:

> It is a great satisfaction to me that this whole expedition, together with the ship, was American from start to finish. We did not purchase a Newfoundland or Norwegian sealer and fix it over for our purposes, as in the case of other expeditions. The *Roosevelt* was built of American timber in an American shipyard, engined by an American firm with American metal, and

constructed on American designs. Even the most trivial items of supplies were of American manufacture. As regards personnel almost the same can be said. Though Captain Bartlett and the crew were Newfoundlanders, the Newfoundlanders are our next-door neighbours and essentially our first cousins. This expedition went north in an American-built ship, by an American route, in command of an American, to secure if possible an American trophy.[137]

Meanwhile, as pointed out, other members of the Bartlett family and of the sealing-captain fraternity commanded ships for Peary's rivals, and many of the ships used were unemployed Newfoundland sealing steamers. In addition, Newfoundland steamers, captains and crews were used to deliver supplies to various Arctic exploratory depots, and records of these voyages are widely scattered. There is no doubt that Newfoundland sealers—both men and ships—contributed greatly to Arctic exploration.

Summary

The Newfoundland seal fishery created a distinctive culture and reputation for Newfoundland and Newfoundlanders. It was an adult male activity and had no place in children's games, which meant that a boy's first trip to the ice was an initiation into manhood. Also, in both the expressive and material culture, its influence spread throughout the whole population, even to the postal service, which issued stamps illustrating seals or the seal fishery.[138] By 1914, sealing had gained a reputation for being harsh, dangerous and cruel; still it contained an element of romance and adventure, and the men who succeeded at it, whether as captains, officers, gunners or batsmen, were viewed with respect by the local community and by those interested in Arctic travel and exploration. As the first lines of a number of sealing songs illustrate, these men were described variously as "rugged" Newfoundlanders, "hardy fishermen," "jolly ice-hunters," "jolly fishermen," "jolly seal men," "sons of Newfoundland," "hearty Newfoundlanders," and "men of the Viking breed."[139] By the beginning of the twentieth century, sealing was having its greatest impact on the culture of the colony, as writers reminded their readers of the great days of the old industry and of the sterling reputations of the leading captains. In fact, its decline as an economic enterprise was matched by its growth as a significant cultural feature of Newfoundland.

General Summary

Therefore, besides its economic importance, the seal fishery defined Newfoundland and Newfoundlanders in the nineteenth century. It created urban centres and led to the growth of cooperation in an environment that always valued individualism centred on the traditional, independent cod fish plantation; this cooperation was expressed chiefly through collective action on the part of fishermen who were unhappy with their share of the proceeds of the industry. Although the work of the sealer was unpleasant and repugnant to casual observers, it was necessary if the fishing station was to become a colony, and there is no evidence to suggest that it was any more unpleasant or cruel than similar enterprises elsewhere. While the independent outports that grew up around the seal fishery, and its partner the Labrador cod fishery, declined in the latter half of the period, the folk traditions did not disappear. At the same time, Newfoundland ice hunters—both captains and men—acquired a reputation for Arctic expertise that was second to none. All combined to form the basis for a rich cultural output in prose and poetry by the end of the nineteenth century. By 1914, the 'Newfoundland' identity was, in large part, the result of the seal fishery.

NOTES

1. In the fall of 1792 it was reported that Newfoundland's European population was as follows: 1,996 masters, 1,602 mistresses, 5,306 children, 6,726 men servants, 833 women servants and 697 dieters (young men who lived and worked with families during the winter in return for room and board) for a total of 17,160 inhabitants. See Ryan, "Abstract."

2. Most of the following information has been summarized from Linda Little's thorough and excellent account of this strike in her MA Thesis. Strangely, neither Prowse, Chafe nor Murphy refer to this significant event. See also Bill Gillespie, *A Class Act: An Illustrated History of the Labour Movement in Newfoundland and Labrador* (St. John's, 1986).

3. The following masters were singled out and asked to appear: Robert Hunt, George Joyce, Nicholas Moran, Richard Manshale, James Forward, John Bransfield, Richard Bransfield, Felix McCarthy, William Pye, Richard Cole, Richard Hopkins, Francis Ash, Henry Taylor, Edward Dwyer, John Penny, William H. Taylor, John Lynch, Patrick Gordon, Daniel Leacy, James Henly, William Moores, William G. Taylor, Nathaniel Taylor, George Penny, John Peady, Henry Ash, John Power, John Taylor, Edward Taylor, Edward Barrett, Francis Taylor, William Davis, Charles Penny, Joseph Taylor, William Howell, Francis Howell, Thomas Finn, James Pierce, Edward Bemister, John Howell, William Udell, Charles McCarthy, James Gillett, John Long, William Penny, Richard Ash, John Snook, Terence Kennedy, Patt Scanlon, Samuel Taylor, Joseph Taylor, John Soaper, Richard Clarke, Matthew Murphy, Edward Guiney, William Brown, Dan Scanlon, William Oates, Thomas Oates, David Oates, John Nichole, Nicholas Nichole, James Roach, Thomas Pike, Jeremiah Broderick, William Thistle, William Penny, James Howell, John Luther, Francis Pike, Edward Pike, Charles McCarthy, Joseph Taylor, Willaim Yetmen, Thomas Butt and Rich Taylor. At least 22 of these 77 men were listed as the owners of the vessels they commanded. In three more cases, the masters were listed as co-owners. See GN2/2, Harbour Grace magistrates to Crowdy, 18 February 1832 (enclosure). See also Little, MA Thesis, p. 158.

4. Gillespie, *A Class Act*, pp. 13-4. Gillespie writes, "The perpetrators of the act were members of an organization called 'The Fishermen of Carbonear and Harbour Grace.' The organization was not a union. Rather it was a fleeting coalition of workers who united to settle a single grievance." The source for the title of this organization is not indicated, although Gillespie credits Little as the source of much of this discussion. Little refers to "the fishermen of Carbonear and Harbour Grace" without quotation marks (p. 154), and the notices that were posted by these fishermen during the incidents discussed and which Little includes in her Appendices were signed variously: "a Labourer" (p. 241); "the Carbonear Men" (p. 245); and "the Carbonear and Harbour Grace Men" (p. 246).

5. The religious antagonisms which were to plague this area were still in the future, and such cooperation along class lines and across religious lines was unusual.

6. *Patriot*, 16 February 1842.

7   *Public Ledger*, 25 February 1842.
8   *Patriot*, 2 March 1842.
9   James Murphy refers to them as Native Walsh and Larry O'Brien. Murphy, *Old Sealing Days* (St. John's, 1916), p.7.
10 Tocque, *Newfoundland*, pp. 304-7.
11 Tocque, *Newfoundland*, p. 307.
12 Murphy, *Old Sealing Days*, p. 9. The fact that Murphy quotes King as stating that this was the "first sealers' strike" suggests that it was the first one in St. John's—1842. If this is true, it would help to explain why Supple was called upon in 1845 by the people of Conception Bay (see below).
13 R. Hattenhauer, "A Brief Labour History of Newfoundland" (Mss., CNS, MUN, 1970), p. 99; Murphy, *Old Sealing Days*, p. 9; H. M. Mosdell, *When was That?* (St. John's, 1923 and 1974), p. 116; and Gillespie, *A Class Act*, p. 15. Mosdell gives a list of strikes as follows: 1838 [St. John's?], 1842 [St. John's?], 1843 (St. John's), 1845 (Brigus) and 1902 (St. John's). He leaves out 1832 and says of 1838 that the charges of berth money of 60s for batsmen, 50s for after gunners and 40s for bow gunners were reduced to 20s, 10s and 5s, respectively, because of this strike. It is possible that he confused 1838 with 1842 or 1843. Also he mistakenly reports that the agreement of 1902 provided $3.20 per cwt for fat, which would have been less than the men had received in 1901; the agreement provided $3.50 per cwt. Hattenhauer lists the sealing strikes as follows: 1832, 1842, 1843, 1849 and 1902; he says that Capt. Supple led the 1842 strike (pp. 98-100).
14 *Patriot*, 15 February 1843.
15 *Patriot*, 22 February 1843.
16 Murphy, *Old Sealing Days*, pp. 10-1. This study does not intend to examine any of the societies that could be considered mutual aid societies or charities, although they were founded, at least to some extent, in response to the implications of the growth and decline and the very nature of the seal fishery. Among these that are excluded are: the Association of Newfoundland Fishermen and Sharemen (c. 1829); the Fishermen's Mutual Protective Society of Newfoundland (early 1840s); the Committee of Charity (c. 1840); the Indigent Sick Society (c. 1840); the Society of United Fishermen (1862); and other similar committees and societies. And the discussion of the Fishermen's Protective Union, which has been thoroughly examined by others, will be brief.
17 *Patriot*, 19 March 1845.
18 *Times*, 15 March 1845.
19 *Times*, 12 March 1853.
20 *Newfoundlander*, 17 March 1853.
21 *Express*, 12 and 15 March 1853.
22 For further information on the political and religious struggles in St. John's and Conception Bay from the beginning of the 1830s, see: Prowse, *Newfoundland*; Gertrude Gunn, *The Political History of Newfoundland: 1832-64* (Toronto, 1966); and the various Master of Arts theses on these topics completed at MUN, including those by John P. Greene, Leslie Harris, E. C. Moulton, W. D. MacWhirter and E. A. Wells.
23 The monopoly imposed on the manufacture of seal oil by only three firms became an issue in 1901, with the *Evening Herald* arguing that any system where the price of the fat was set by the several buyers was basically unfair. It came to light that K. R. Prowse offered $3.45 per cwt of fat while the combine of oil plant owners

paid only $3.35. It was felt that the men lost about four or five dollars each because of this. However, the three major firms purchased the Prowse firm and closed it. The only solution, according to the *Evening Herald*, was for the sealers to acquire their own oil plant. It is interesting that this was not considered a priority by the Fishermen's Protective Union. See *Evening Herald*, 17 and 18 June 1901. Also, see Briton Cooper Busch, "The Newfoundland Sealers' Strike of 1902," *Labour/Le Travail*, XIV (Fall 1984), 73-101, for an excellent discussion of this strike. See also the detailed account of the unfolding of events in the *Evening Herald*, 10, 11 and 12 March 1902.

24 See *Evening Herald*, 10 March 1902 for the following information.
25 *Evening Herald*, 11 March 1902.
26 *Evening Herald*, 11 March 1902.
27 *Evening Herald*, 12 March 1902.
28 *Evening Herald*, 12 March 1902.
29 *Evening Telegram*, 5 May 1891.
30 *Evening Telegram*, 19 April 1898.
31 During the 1980s, the present writer was informed by two elderly sealers, Andrew Short and Edward Russell, both of Riverhead, Harbour Grace, that the strikers of 1902 were determined to pull the *Neptune* out of the harbour and up on Water Street. This seems improbable, but it illustrates the persistence of oral tradition, particularly when one realizes that only a handful of traditional sealers were around to pass on the oral history in the 1980s.
32 *Evening Telegram*, 31 March 1906.
33 *Evening Telegram*, 30 March 1907.
34 *Evening Telegram*, 27 January 1911.
35 *Evening Telegram*, 10 and 11 April 1911.
36 *Evening Telegram*, 31 March 1913.
37 *Evening Herald*, 8 February 1905.
38 Hattenhauer, "Labour History," p. 93, says 1854 or 1855. James J. Fogarty, son of one of the founders of the union, says 1855; see James J. Fogarty, "The Seal-Skinners' Union," *The Book of Newfoundland*, Vol. 2, p. 100. For information on other unions in trades associated with the seal fishery, such as the Shipwright's Union and the Coopers' Union, see Hattenhauer, "Labour History," and Bill Gillespie, *A Class Act*.
39 *Evening Herald*, 16 April 1900.
40 *Evening Herald*, 21 May 1901.
41 See H. F. Shortis, "Sealing in the Old Days," *Newfoundland Quarterly*, I (March 1902), 9-10. "The most reliable authorities in the old days place its origins to a row between some Irish youngsters, led by one Mickey McManus and the captain of a pink-stern schooner of about one hundred years ago, and hence the word has come down to posterity, as the word boycott is now, and will be in use for years to come." This origin has not been confirmed, but not even the *DNE* has been able to suggest anything better. In discussions, it has been suggested by Professor Emeritus William Kirwin and the late Professor George Story (two editors of the *DNE*) that the term is derived from menace—i.e., to threaten.
42 *Times*, 22 March 1848.
43 Shortis, "From the well-stored Mines of the Tradition of Newfoundland" (CNS, copy from Vol. III, 86, "Fugitive History," PANL.)

44 *Royal Gazette*, 29 April 1862.
45 *Times*, 30 April 1862.
46 Gillespie, *A Class Act*, pp. 16-9.
47 Gillespie, *A Class Act*, pp. 21-2; and Hattenhauer, "Labour History," pp. 113-50.
48 Gillespie, *A Class Act*, pp. 32-3.
49 See the following Acts of the Newfoundland Government: Cap. 9, 1873; Cap. 7, 1876; Cap. 1, 1879; Cap. 1, 1883; Cap. 23, 1887; Cap. 1, 1889; Cap. 2, 1892; Cap. 22, 1893; Cap. 10, 1895; Cap. 2, 1898; Cap. 4, 1898; Cap. 17, 1899; Cap. 2, 1903; and Cap. 19, 1914.
50 *Evening Telegram*, 4 March 1886.
51 *Evening Telegram*, 1 April 1886.
52 *Evening Telegram*, 4 June 1886. The three members for Harbour Grace in 1886 were: Charles Dawe, Port de Grave, captain of the S.S. *Vanguard* (1885-88), which belonged to John Munn and Co., Harbour Grace; Joseph Godden, Harbour Grace; and James S. Winsor, St. John's, lawyer.
53 *Evening Telegram*, 28 April 1891.
54 *Evening Telegram*, 30 April 1891.
55 *Evening Telegram*, 1 May 1891.
56 *Evening Telegram*, 5 May 1891.
57 *Evening Telegram*, 11 May 1891.
58 *Evening Telegram*, 15 May 1891.
59 *Evening Telegram*, 26 March 1894.
60 Alexander A. Parsons (1847-1932), born in Harbour Grace, was editor of the *Evening Telegram* from 1879 to 1904 (see *DNLB*). In 1865 or 1866, he went to the ice in the vessel *Union*, commanded by his uncle. During these years, two vessels by that name sailed from Harbour Grace, commanded by Capts. Pike and Parsons. See Alexander A. Parsons, "Our Great Sealing Industry," *Newfoundland Quarterly*, XIV (April 1915), 6-9. Also see *Courier*, 11 March 1865 and 17 March 1866.
61 *Evening Telegram*, 10 April 1895, reported one direct effect of the bank crash on the sealers. "Sealers and Specie—The banks are very busy lately giving out gold and silver to the sealers, whose confidence in paper money has been so recently and so rudely shocked. There is quite a boom in purses, men frequently carrying two or even three stocked with shining silver, which seems even more popular than gold. There is something solid ... about the silver, whilst the bit of gold for $10.00 is only a circumstance."
62 *Evening Telegram*, 28 March 1898.
63 *Evening Telegram*, 30 March 1898.
64 *Evening Telegram*, 31 March 1898.
65 *Evening Telegram*, 12 April 1898.
66 *Evening Telegram*, 13 April 1898.
67 *Evening Telegram*, 7 March 1899.
68 *Evening Telegram*, 9 March 1899.
69 *Evening Herald*, 11 January 1901.
70 According to Murphy, *Old Sealing Days*, p. 59, Professor Roddick taught at the "famous grammar school at Harbour Grace" and married Emma Jane Martin of Bareneed, Conception Bay. They were the parents of the "well-known physician of Canada, Sir Thomas Roddick."

71 *Evening Herald*, 9 March 1901.
72 *Evening Herald*, 11 January 1901.
73 *Evening Telegram*, 7 March 1906.
74 *Evening Telegram*, 18 April 1907.
75 *Evening Telegram*, 22 February 1908.
76 *Evening Telegram*, 9 March 1908.
77 *Evening Telegram*, 10 March 1910.
78 *Evening Telegram*, 12 March 1910.
79 The three best sources for information about Coaker and the Fishermen's Protective Union are: J. R. Smallwood, *Coaker of Newfoundland* (London, 1927); John Feltham, "The Development of the F.P.U. in Newfoundland" (MA Thesis, History Department, MUN, 1959); and Ian D. H. McDonald, *To Each His Own: William Coaker and the Fishermen's Protective Union in Newfoundland Politics, 1908-1925*, ed. James K. Hiller (St. John's, 1987).
80 S. J. R. Noel, *Politics in Newfoundland* (Toronto, 1971), p. 88.
81 Most of the following information has been taken from W. F. Coaker, comp., *Twenty Years of the Fishermen's Protective Union of Newfoundland: from 1909—1929* (St. John's, 1930).
82 Coaker, *Twenty Years*, p. 31.
83 Coaker, *Twenty Years*, pp. 29-43.
84 Coaker, *Twenty Years*, pp. 44-63. This, the fourth annual convention of the FPU, was held in Bonavista in December 1912.
85 *Evening Telegram*, 20 January 1912.
86 Coaker, *Twenty Years*, p. 68.
87 Joshua Holloway's Evidence, "Evidence: Sealing Commission," pp. 80-7.
88 Jesse Collins' Evidence, "Evidence: Sealing Commission," pp. 88-9.
89 Cecil Taylor's Evidence, "Evidence: Sealing Commission," pp. 90-2.
90 Robert Winter's Evidence, "Evidence: Sealing Commission." pp. 111-5.
91 Coaker, *Twenty Years*, p. 81.
92 The following sections will mention a few highlights in the folklore and cultural history of the seal fishery. For an excellent and extensive treatment of this subject, see John Roper Scott, "The Function of Folklore in the Interrelationship of the Newfoundland Seal Fishery and the Home Communities of the Sealers" (MA Thesis, Folklore Department, MUN, 1975).
93 H. W. LeMessurier, *Old Time Newfoundland*, ed., C. R. Fay (Typescript, St. John's, 1955), pp. 13-4. There is a note by the editor, Fay, to the effect that a "sparrowbill" is a "small headless wedge-shaped iron nail used in the soles and heels of boots and shoes." Sealers used "sparables" to stud the soles of their boots so they would not slip on the ice (see *DNE*). For information on LeMessurier, see Cuff, *DNLB*.
94 See "H. F. Shortis," CNS, MUN. There are 57 items listed under his name, and these are typed copies of items in the collection of Shortis papers and volumes in the PANL. See H. F. Shortis, *Fugitive History of Newfoundland*, Vols. 1-5, 7-8, (PANL). See also Cuff, *DNLB*.
95 Shortis, "The Brigantine *Fanny Bloomer*" (CNS, copy from Vol.4, 243, PANL, n.d.). A small section of this story, dealing with historiography in general, has been omitted. The punctuation in these stories is often inconsistent; where this is so, the

present writer has made slight changes in order to prevent it from being distracting.
96 Shortis, "Hard Winters of the Past when the Mercury almost Froze" (CNS, copy from Vol. 5, 380, "Fugitive History," PANL.)
97 Shortis, "Loss of the Brigantine *Eric*: 1878" (CNS, copy taken from "Fugitive History," Vol. 5, 71, in PANL.)
98 *Evening Telegram*, 16 April 1913.
99 *Evening Telegram*, 10 April 1913.
100 Shortis, "Fugitive History," Vol. 1, (23), PANL.
101 *Evening Telegram*, 1 June 1886. In a few cases, the seven longer obituaries that follow have been edited to save space.
102 *Evening Telegram*, 14 April 1894.
103 *Evening Telegram*, 13 January 1907.
104 *Evening Telegram*, 30 March 1908.
105 *Evening Telegram*, 8 March 1909.
106 *Evening Telegram*, 9 February 1910.
107 *Evening Telegram*, 9 January 1911.
108 *Evening Telegram*, 5 April 1913. Capt. Munden was born on 3 April 1813, which explains why Shortis wrote about him in April 1913.
109 See the *Standard*, 27 January 1883, for example. This tribute to Capt. White emphasises his success over a career of "forty-five voyages to the ice-fields," his industry, perservance and principles and "the fact that he never allowed his men to take a seal on Sundays."
110 The *Christmas Review* (St. John's, 1901), p. 9. This ballad has eight stanzas; only the first two and the final one are included here. The author is given as one Absalom Hobbs, but Gerald S. Doyle, comp., *The Old Time Songs and Poetry of Newfoundland* (St. John's, 1927), credits M. A. Devine with the authorship of this ballad under the title "The Outharbour Planter". It is likely that Devine was using a pen name. Note: The "pillar" dollar was a Spanish one-dollar silver coin with an image of the pillars of Hercules (i.e., the Straits of Gibraltar) stamped on one side. Spanish dollars worth five shillings (local currency) circulated in Newfoundland. As late as the 1950s the Newfoundland silver twenty-cent piece in wide circulation was referred to locally as a shilling. "Stuck the marchan'" in this context describes the process whereby the outport planter occasionally cheated his supplying merchant—who was considered fair game for anyone clever enough to succeed in cheating him. Also, it should be noted that the merchant often 'stuck the planter'.
111 *Evening Telegram*, 4 and 5 February and 8 March 1907.
112 Unless otherwise stated the following examples will be found in Ryan and Small, *Haulin' Rope and Gaff*.
113 A still-born seal pup was referred to as a 'cat'. A cat did not lose its hair and could be skinned carefully and cured and stuffed. They were on display in many homes and offices. Please note that the term 'cat' also applied to very young live whitecoats.
114 With the establishment of a Chinese community in St. John's in the late nineteenth and early twentieth centuries, a market appeared for the penises of adult seals. This demand resulted in a small increase in the income of those sealers who were lucky enough to kill some of the old harps and hoods. The Newfoundland sealers were ignorant of the fact that this organ has long been prized in the orient as an aphrodisiac; they observed that most Chinese men operated laundries and con-

cluded, with impeccable logic, that the penises were used to make starch. Interview with Andrew Short.

115 Philip Tocque, "The Seal Fishery of Newfoundland" in *Littell's Living Age* (Boston), XXVII (1850), 186-7. It seems to the present writer that Tocque made his trip to the ice in 1843, and not 1834, as this account states. See the discussion of 1843 and the references cited in the previous chapter. See also the discussion of Lady Blake's article below. For present purposes, it is not an important point.

116 Jukes, *Excursions*, Vol. I., pp. 290-01.

117 Jukes, *Excursions*, Vol. I., pp. 289-90.

118 "Newfoundland," *Blackwood's Magazine* (Edinburgh), CXIV (1873), 63-9.

119 *Evening Mercury*, 4 January 1889.

120 Edith Blake, "On Seals and Savages," *Nineteenth Century* (London), XXV (1889), 513-26.

121 The term 'sculp' was generally used; less frequently used was 'pelt'; and rarely, but occasionally, 'scalp.' See *DNE*.

122 *Evening Mercury*, 3 May 1889.

123 *Evening Mercury*, 22 May 1889. Governor Blakes's letter is dated 20 April.

124 *Evening Telegram*, 7 March 1914.

125 See Busch, *The War against The Seals*, for methods of slaughtering seals and sea elephants which involved clubbing, lancing and shooting; see Murray, *The Vagabond Fleet*, for similar descriptions of the slaughter of the North Pacific fur seal and walrus.

126 Dr. Wilfred Grenfell, "The Seal Hunters of Newfoundland," *Leisure Hours* (1897-98). Photocopy in CNS, MUN.

127 Thanks to Ms. Sarah Barron, MUN, Ph.D. student, for the use of her notes.

128 *Times*, 28 March 1877. The U.S. government gave Capt. Bartlett $300 in 1877 as payment for his role in this rescue. (Mosdell, *Chafe*, p. 42, incorrectly states that this rescue occurred in 1870.) According to an item in the *Times*, 3 March 1877, quoted from the *Glasgow Herald*, Capt. William Adams, Sr., in command of the Scottish sealer/whaler *Arctic*, was primarily responsible for this rescue; correspondence from the U.S. Secretary of the Navy, George M. Robeson, U.S. government to Capt. Adams, "in recognition of his services to the officers and crew of the *Polaris*, whom he picked up and brought home from the Arctic expedition in 1873, they have transmitted to British bankers the sum of $300 for the purchase of a gold pocket chronometer. The government has also authorized Capt. Adams to inscribe on the chronometer the fact that it is a token of the gratitude of the United States for his kindness to the officers and men of the *Polaris*." On the other hand, Robeson's letter to Isaac Bartlett reads: "Sir: As an expression of the gratitude to you from the Government of the United States for your kindness in 1873 to the officers and men of the *Polaris*, I am directed to inform you that your draft on Selizman Brothers of London, Bankers, for three hundred dollars will be paid by said house on presentation thereof. Should you wish to invest the money in a gold pocket chronometer with an inscription to the effect that it is a token of gratitude as above expressed you may do so without hesitation." When he described this incident that involved his great-uncle Isaac, Capt. Bob Bartlett did not mention Capt. Adams' role in this rescue; probably he was unaware of it. See Robert A. Bartlett, *The Log of Bob Bartlett* (New York, 1928), p. 59.

129 Ryan, *Seals and Sealers*, p. 21.

130 Ryan, *Seals and Sealers*, pp. 23-4. The *Bear* was then transferred to the U.S. Coast Guard Alaska patrol and, in 1929, was purchased by the city of Oakland, Califor-

nia, and turned into a maritime museum. It was at this time that the *Bear* was used in the filming of Jack London's *The Sea Wolf*. Admiral Richard E. Bird used this steamer on expeditions to the Antarctic, and in 1947, it was purchased by a Canadian company and abandoned; later, while being towed to Philadelphia in 1963, the *Bear* sank in a storm. The *Thetis* returned to the Newfoundland seal fishery in 1917 and prosecuted this industry regularly until it was damaged so badly during the 1936 season that it had to be beached near St. John's. Information on William Norman provided by Ms. Sarah Barron.

131 There is some dispute about whether Pike was supposed to collect Greely in 1882 or 1883. The *Encyclopedia Americana* says 1882, but Newfoundland sources indicate 1883. See Ryan, *Seals and Sealers*, p. 21.
132 Ryan, *Seals and Sealers*, p. 74.
133 Robert E. Peary, *The North Pole* (London, 1910), p. 12.
134 Bartlett, *Log*, pp. 59-60.
135 Bartlett, *Log*, p. 57.
136 Robert E. Peary, *Nearest the Pole* (London, 1907), p. 4. An account of his 1905-06 expedition.
137 Peary, *The North Pole*, pp. 31-2.
138 Much of the above information has been taken from Shannon Ryan, "The Seal in Newfoundland Culture" (Paper presented to the American Folklore Society, Philadelphia, November 1976).
139 Ryan and Small, *Haulin' Rope and Gaff*, pp. 169-71.

Conclusion

WITH THE EXCEPTION of the period between 1689 and 1713, when the French and their allies invaded Newfoundland, wars generally resulted in a growth in the number of resident planter fishermen and a decline in the number of migratory fishing ships engaged in the Newfoundland cod fishery. The wars of 1793-1815 produced this effect, particularly during the latter stages. The extraordinarily high demand for saltfish in Spain and Portugal after 1808 and the suspension of the fisheries of other nations resulted in high prices and caused the Newfoundland cod fishery to expand, as resident planters enlarged their operations by hiring large numbers of Irish men and women at generous wages. In order to take full advantage of the increased demand for saltfish, the planters of old established fishing centres such as St. John's, Harbour Grace, Carbonear and Brigus built small vessels and sent crews to fish, during the summer months, on the eastern side of the vacant French Shore (i.e., the Great Northern Peninsula), which the English called the 'North Shore'. The demand for fish oils in Britain grew and, simultaneously, planters discovered the extent of the seal resources present each spring off the coast of Newfoundland. Thus, vessels and men from St. John's and the harbours of Conception Bay were able to prosecute both the seal fishery and the North Shore (later Labrador) summer cod fishery. As a result, significant numbers of employee-fishermen (or 'fishing servants', as they were more usually called) could live and work year-round in Newfoundland—a situation true previously for only the wealthier employer-fishermen (or planters) dependent on migrant labour. With the acquisition of two new industries and a permanent work force, the planter resident fishery now changed from an important branch of the British migratory fishery to the centre of a growing local economy and society, and the number of people living in Newfoundland increased from 17,000 in 1792 to over 40,000 by 1815; more importantly, the 6,000 to 8,000 migratory fishing servants of the former year (and their successors) were part of the permanent population in 1815. By then, as all observers were aware, Newfoundland had become a colony, although another ten years would pass before the British government recognized this fact.

The link between population growth and the development of a viable sealing industry in Newfoundland cannot be overstated. During the period from 1800 to 1811, Newfoundland's population increased slowly from about 20,000 to about 25,000. Numbers then increased dramatically during the latter stages of the Napoleonic and Anglo-American Wars and had risen to over 40,000 people by 1815. However, this growth, stimulated by wartime conditions, came virtually to a halt over the next five years, and by 1820 the population had barely increased to number 42,535 souls. Then a strange demographic pattern emerges. The number of inhabitants rose swiftly to 55,504 in 1825—despite the almost complete collapse of the Newfoundland-Spanish saltfish trade.[1] By 1830, the population had risen to 60,088 people, to 74,993 by 1836 and to 96,296 by 1845. In 1857, the first census of responsible government, and the first to include the few people on the coast of Labrador, indicated a total population, excluding the French Shore and native groups, of 124,228. Thus, from slow beginnings at the turn of the century, population growth accelerated, particularly after 1820, to reach a six-fold increase by 1857—actually a conservative estimate because population figures for earlier years always included thousands of single migrant fishermen who had no plans to remain in Newfoundland permanently. In fact, a more useful indication of population growth can be obtained by examining the increase in the number of households between 1792 (before the wars became a factor) and 1857. In the former year, there were 1,996 masters (i.e., planters), 1,602 mistresses and 5,306 children living on the south, east and northeast coasts of Newfoundland—about 2,000 households. Towards the close of the eighteenth century, after 300 years of constant European contact, the only demographic result was a small population providing the infrastructure for a migratory cod fishery. By 1857, however, Newfoundland was home to over 124,000 people—about 19,000 families living in about 17,800 houses—a tremendous increase in the number of inhabitants and a nine-fold increase in the number of households. This growth in population was the result of Newfoundland's discovery and commercial utilization of a second major resource—the seal herds.

Furthermore, this increase in population took two directions, both related to the seal fishery. The concentration of people in St. John's and Conception Bay continued, as more and more men could be employed in the seal fishery (and the Labrador cod fishery) without owning local beach property (fishing rooms)—an absolute necessity for traditional inshore cod fishermen. Thus, not only did the population increase five- or six-fold between 1800 and 1857, but the populations of St. John's and Conception Bay grew disproportionately and were home to 52 per cent of the entire population of the colony, with about 10,700 families living in 9,700 houses.[2] Meanwhile, growth continued on another front. North of Conception Bay, there were many uninhabited islands and harbours, and inshore fishermen who could not find suitable fishing rooms in their places of birth moved to these areas, keeping in mind that the outer islands and capes brought them closer to the Arctic ice and the seal herds.[3] Some residents of Conception Bay joined this migration as well, and soon Newfoundland fishermen had taken up residence on the Northern Peninsula and in the Straits of Belle Isle. Once again, demographic patterns reflect the growth of the sealing industry.

As already indicated, the spring seal fishery provided the necessary ingredient for year-round settlement: winter employment. This winter employment forced new businessmen, who had prospered during the Napoleonic War, to reside in Newfoundland over the winter in order to manage their sealing operations. The old system, whereby partners and principal agents of West Country merchants could, at the close of the cod fishery, retire to Britain in the fall of the year and return the following spring, had to be abandoned, with the exception of some of the mercantile houses on the northeast and south coasts. Most of the latter went bankrupt, during the depression in the cod fishery over the following decades, because they did not (and often could not) become sufficiently involved in the spring seal fishery and were thus at the mercy of the depressed foreign saltfish markets.[4]

Increased year-round settlement created an atmosphere conducive to local government. Dr. William Carson had become a permanent resident and was probably the most widely read advocate of local government. Even men who spent much of their time in Britain and Ireland, such as Patrick Morris (another prominent supporter of a local assembly), came to view Newfoundland as their place of permanent residence. Prowse argued that those who fought for representative government were essentially 'Newfoundlanders' who supported their cause against West Country merchants—the traditional enemies of settlement and government. Matthews disagreed and contended that supporters of representative government in Newfoundland were people with traditional Whig and British/Irish middle class interests who would have supported, and indeed did support on many occasions, political reform anywhere they lived in the British Isles or the British Empire.[5] Neither view takes into account the trade in seal oil and seal skins which was becoming so important during the 1820s and thrived up to and throughout the 1850s. These products were in growing demand in Britain, and the merchants, planters and fishermen of St. John's and Conception Bay dominated their production and processing.

Merchants in this area had guaranteed markets for their seal oil and skins; on the other hand, there were no guaranteed markets for saltfish any more—certainly, none that were both reliable and profitable. The traditional West Country merchants, who lived in Great Britain and ran their saltfish operations through agents, were thoroughly conscious of their dependence on Britain for favourable treatment in foreign saltfish markets. They therefore felt that increasing political independence in Newfoundland would mean a lessening of the ties binding Newfoundland's saltfish trade to the mother country. They fiercely opposed the development of local government, but as pointed out above, their business operations soon collapsed, and they were no longer part of the equation. By contrast, sealing merchants concentrated their greatest efforts on supplying oil and skins to the British market and could see no reason why local self-government could hurt this trade. Thus, in the 1820s, opposition to local government came from traditional West Country firms in the saltfish trade. The merchants of St. John's and Conception Bay, on the other hand, became increasingly confident that the demand in British markets for oil would lead to the growth of local wealth and population in Newfoundland, regardless of what happened in foreign saltfish markets, and these were the firms that began to dominate both the cod and oil trades.

For example, in 1826, well before the Roman Catholic reform bill and the unruly campaign of the 1829-30 period, merchants and planters in St. John's and Conception Bay—heavily dependent on the seal fishery—were involved in building a local governmental apparatus.[6] That year, a meeting was held in St. John's which elected a committee of 30 men to lead the crusade for major changes in local government. While the committee soon split on what these changes should be, there is no mistaking the fact they were thinking in terms of permanent settlement of relatively large proportions. These were not the few merchants who had settled in the 1780s to run their West Indies trade; nor were they the type of principal inhabitant or planter to whom Newfoundland had been playing host ever since the early 1600s. In petitions and counter-petitions during 1826, these men identified themselves, and often the firms with which they were involved, but unfortunately it is impossible to ascertain the participation of either the men or the firms in the 1826 seal fishery. However, by examining the first published list of suppliers in this industry—that of 1836—one can see that many of the firms supplying the sealing fleet in the latter year were those who had taken part in the 1826 petitions. While there were, no doubt, changes in personnel over that ten-year period, it is obvious that many firms and men were still involved, or had become involved—often heavily—in the seal fishery in 1836.[7]

Therefore, permanent settlement and the self-government which accompanied it were direct results of the establishment of the spring seal fishery. This point was not fully appreciated by Matthews or Prowse, although the latter did describe how "the spring seal fishery ... changed the social habits of the people ... [altered] social customs [and] largely increased the importance of the outports."[8] However, he did not take his observations to their logical conclusion—that these changes were part of a larger evolutionary development which encouraged, and indeed necessitated, local government. Regardless of the state of the saltfish markets, Newfoundland would not have acquired representative government if the cod fishery had remained the sole industry. In fact, given the unfavourable conditions in the saltfish markets during this period, the number of resident planters would have declined to a couple of thousand who would soon have come under the administration of Quebec or Nova Scotia. However, the spring seal fishery changed the situation entirely, and in 1855, when Newfoundland received responsible government, it was as politically advanced as any other British colony.

Even the geographical scope of the colony was profoundly affected by the growth of the sealing industry. The Labrador coast had been placed under the authority of the governor of Newfoundland in 1763 and, in 1774, turned over to the jurisdiction of the government at Quebec. However, the Newfoundland governor continued to exercise authority over the cod fishery and the Moravian missionaries. In 1809, as the result of complaints that Quebec was not exercising any control over the area, the whole of the Atlantic coast of Labrador, including part of the coast in the straits of Belle Isle, was returned to Newfoundland's control. Nevertheless, by and large, the high hopes held by Governor Palliser in the 1760s and by other governors in later years, of developing a strong English migratory fishery on that coast, never materialized. It was the extraordinarily high prices paid for saltfish in the latter years of the Napoleonic War which encouraged wealthier planters in St. John's and Conception Bay to send small vessels to the island's northern peninsula and Labrador coast to fish for the summer months. In the 1820s, when the French prevented Newfoundlanders from fishing on the northern peninsula, they began to fish exclusively on the Labrador coast.

However, it was not a remunerative industry, as the Chamber of Commerce reported in 1834:

> That the fishery at Labrador being exceedingly precarious derives its value chiefly from the employment it has given of late years to the Sealing Vessels

and men after the termination of the Seal Fishery; and was, as is doubtless known to His Majesty's Government, but little pursued from this Island until [precipitated] by the late treaties which ceded to France and America a right to occupy a part of our shore to which the Sealing Vessels had formerly resorted for the Cod Fishery.[9]

Without the seal fishery, Newfoundland would not have been able to develop a cod fishery on the Labrador coast, and control of that area would have passed to other hands.

For example, in the 1860s, when the Newfoundland government began to exercise effective customs regulations and law and order on that coast, other fishing interests from the United States of America and from Nova Scotia protested that Newfoundland did not have the authority to do so. Britain supported Newfoundland on this issue, but there is no doubt that the government in St. John's would not have been able to exercise jurisdiction, and indeed would have had no reason to attempt it, without the presence of the Newfoundland fishing fleet on that coast. And while the Newfoundland-Labrador cod fishery was a large operation, its successful prosecution was entirely dependent on the employment of men and vessels engaged in the annual spring seal fishery. Thus, the eventual shape and extent of the colony, and later province, of Newfoundland were the direct result of the development of the local spring seal fishery in St. John's and Conception Bay.

Newfoundland's growth as a social, economic and political entity was inextricably linked to the fortunes of the seal fishery. Not surprisingly, then, the industry's decline in the latter half of the period under study created severe repercussions for the colony. In the 1860s, depressions in both the cod and seal fisheries forced people to re-examine the viability of political independence, and in 1869 confederation with Canada was narrowly defeated (partly as a result of last-minute improvements in the economy). After a brief recovery in the 1870s, Newfoundland's traditional economy declined. In the 1880s, successive Newfoundland governments began borrowing to build a railway across the island, with branch lines to the larger centres, and to construct a dry dock in St. John's. In the 1890s, iron ore mines opened on Bell Island, and early in the twentieth century, a pulp and paper mill was built in Grand Falls. Despite some economic diversification, the depression of the 1880s and 1890s, which included the bank crash of 1894, once again almost cost Newfoundland its independence (but Canada demanded too many concessions). The growth of the seal industry had created Newfoundland as a political entity, and the decline of this industry almost destroyed its political independence. Indeed, when the depression in international trade occurred in the 1930s, there was no seal fishery left, and seal oil was no longer needed to lubricate and illuminate British industry. There is no doubt this was a contributing factor when, in 1934, the status of Newfoundland as a Dominion was terminated, and the British government appointed a Commission of Government to administer the island of Newfoundland and the coast of Labrador which had made up the Dominion. The seal fishery, which had created the colony of Newfoundland in the 1820s and 30s, no longer thrived in the 1920s and 30s, and nothing equivalent had been found to replace it.

This theory is further substantiated by population trends after the seal fishery had reached its peak. Although overall population of the colony continued to increase during the latter period of this study, growth was much slower. Between 1857 and 1911, the population grew from 124,228 to only 242,619, or just less than two-fold, over a 54-year period. The casual emigration of Newfoundlanders to the United States that had always gone on increased significantly in the 1860s. Like their ancestors who migrated to Newfoundland from the British Isles prior to c.1815, most of them did so temporarily, returning to their families in the winter. Gradually, however, more and more stayed, and Newfoundland began to resemble the 'old world' in the sense that it became a source of immigrants for the 'new'. Thus, Newfoundland found its very survival in jeopardy, despite considerable government spending on a trans-island railway, branch-line railways, roads, shipbuilding, fishery projects and welfare; regardless of major successful mining and pulp and paper ventures; notwithstanding the surrender by the French of their rights on the French Shore; and in spite of massive (but, unfortunately, undocumented) sums of money remitted to their families by Newfoundlanders working in the United States.[10] Despite these added expenditures and economic opportunities the population of the heart of the old outport sealing industry in Conception Bay actually decreased. The number of people in Harbour Grace declined from about 15,000 in 1884 to less than 12,000 in 1911; Carbonear's population declined from 6,200 to 5,100 during the same period; and Port de Grave's population dropped from 8,700 to 7,000. Although St. John's was able to compensate for the overall decline in the seal fishery, by monopolizing the sealing industry and the fish trades in general, and by the development of light industry, Conception Bay was never able to discover or develop a significant substitute for the seal fishery and lost many of its inhabitants to emigration.

Thus, when examining Newfoundland's development as a colony and dominion, and its political collapse in the 1930s, it is essential to consider the vicissitudes of the sealing industry. In seeking reasons for the growth of a permanent resident population in the early nineteenth century, the international scope of the French Revolutionary and Napoleonic Wars (and not forgetting the Anglo-American War of 1812-14), which kept most of the world at war for a generation, attracts the attention of the historian and, indeed, attracted the attention of the contemporary observer. However, it was the rapidly increasing demand for oil, caused primarily by technological breakthroughs in lighting, which guaranteed Newfoundland's transformation from a migratory fishing station to a colony. Lighting and other urban, industrial and domestic requirements in Britain created a demand for seal oil—a demand that could only be satisfied by fishermen turned ice hunters, operating from the coasts of Newfoundland. Whereas the high price of saltfish had allowed and even encouraged fishing servants to remain in Newfoundland all year-round during the prosperous war years, it was the expanding seal fishery that allowed them to continue to do so after 1815. The seal oil industry survived the introduction of gas lighting because of the limitations of the latter and because other markets for seal oil were growing; but the introduction of petroleum into lighting and lubrication, and the development of electricity were major blows to seal oil and all fish and vegetable oil industries. In addition, the adoption of free trade by Britain in the 1840s and the importation into that market of animal and vegetable oils from other sources—as Tables 1.5 and 1.6 show—along with the general decrease in the value of commodities, forced down the price of Newfoundland seal oil by about 50 per cent during the last half of the nineteenth century. The market place that had bestowed blessings upon Newfoundland and provided for the establishment of a prosperous society did not furnish any guarantees, and by the end of the nineteenth century, seal oil was only a minor part of the international oil trade.

However, Newfoundland was only one of many societies to be affected by trends in the world market for oils. The demand for both oil and leather, but particularly the former, wrought changes throughout the world—some beneficial, some less favourable in the long term. Seal and whale oil production helped support the economies of the new colonies in New Zealand and Australia and, indeed, in the Falkland Islands and South Georgia; Arctic whaling added to the knowledge of that region as well as to the growth of inter-racial relations and an increasingly mixed race population; American whaling in the Pacific resulted in America's acquisition and colonization of the Hawaiian Islands. However, American walrus and seal hunting in the north Pacific and around Alaska destroyed local food supplies, leading to starvation and the decline in the local native population; and the demand for palm oil distorted completely local societies and cultures of West Africa. Newfoundland was also strongly affected by the rise and fall in world demand for seal oil and skins.

At the same time, it must be recognized that the decline in the importance of the seal fishery was not caused only by the decreasing demand in the market place. As Tables 2.5 and 2.7 show, Newfoundlanders overexploited the seal herds, and there was a definite decline in the numbers taken annually. During the period from 1811 to 1835, there was a steady increase in annual production, from an average of about 125,000 seals in the earliest five-year period to about 500,000 in the final five-year period. There was a slight decline to an average of about 457,000 seals in the five-year period ending in 1840, but an increase to over 525,000 on average annually during the following five years. In the 1860s, on the other hand, annual average production amounted to just over 300,000 seals and dropped to less than 270,000 annually in the late 1880s, despite the introduction of steamers into the industry. After a slight recovery, production declined to an average of about 233,000 seals annually by the end of the century, and even the introduction of powerful steel steamers beginning in 1906 did not succeed in reversing the decline. Production hovered around 260,000 seals annually during 1906-10 and just under an annual average of 250,000 seals during 1910-14. But once again, as has been seen, the harp seal population was only one of many species of sea mammals harvested so that the industrial world could continue to be lit and lubricated.

The seal fishery, in partnership with the cod fishery, created a society and an economy that were reasonably viable and stable up to about 1860. However, from that point on, this economy began to deteriorate as Newfoundlanders were forced to become increasingly dependent on the cod fishery alone. As had always been the case, this inshore seasonal industry could not support a population of employee-fishermen. Most planters disappeared—forced to become small family fishing operations, with women and children compelled to play a larger part in the short summer cod fishery as people attempted to live off this single industry. Newfoundlanders were not successful in finding a substitute for the seal oil industry, despite intensive efforts to do so, and the government was forced into debt, businesses failed, and people emigrated.

The expanding seal fishery in the immediate post-1815 period imposed a special regional and demographic profile on Newfoundland as certain centres dominated the new industry. From the very beginning, wealthier planters in St. John's and Conception Bay, with their cod fishery infrastructures already in place, dominated the industry, in conjunction with the prosecution of the Labrador cod fishery, using small schooners and, later, larger brigantines. Hundreds of vessels from Harbour Grace, Carbonear and Brigus, and reduced but significant numbers from other harbours, created balanced outport economies based on seal oil and Labrador saltfish. The introduction of steamers in the 1860s spelled the end of the sailing fleet, and only two firms outside St. John's, both in Harbour Grace, were able to afford the increased investment that the use of steam required. Thus, as the seal fishery declined, its operations became centred in St. John's, divided among a few large businesses who monopolized it by the beginning of the twentieth century.

The sealing industry had employed an increasing number of men and captains from its inception until the late 1850s. There, men found relatively highly paid employment and could, and did, advance into the ranks of officers and captains. As ships became larger such upward mobility became rarer and ended after the introduction of steam. Then, the numbers of employed sealers shrank and their incomes declined, while a lucky few highly skilled captains experienced an astronomical rise in their incomes. Even more importantly, as indicated, the seal and Labrador cod fisheries had produced a class of planters. Each owned a vessel, or had a share in one, and employed men in the spring ice hunt and the summer cod fishery. This planter class was destroyed by the introduction of sealing steamers, leaving a social and economic void in the outports.[11]

The sealing industry was a dangerous one, and many lives and vessels were lost. This was particularly evident in the latter decades up to 1914, as competition among the captains, and indeed among the crews, forced all concerned to take unreasonable risks. This culminated in the deaths of about 250 men in 1914 and provided one more ominous signal that the industry, as a serious component of the economy, was finished.

The men reacted to the general prosperity of the seal fishery by demanding a greater share in profits and, up to the middle of the nineteenth century, successfully challenged the authorities and ship owners and gained important concessions. In the latter years of the century, the scene was a dispirited one, and pure desperation drove the sealers to launch one final strike in St. John's in 1902. This was only minimally successful because by then, sealers were migrants from the outports and largely alienated from the St. John's citizenry.

This alienation demonstrates the extent to which the relationship between St. John's and the outports had deteriorated by the beginning of the twentieth century. The St. John's labour movement had been built on the experiences and successes of the sealing strikes of the 1830s and 40s; by 1902 St. John's working-class interests differed from those of the fishermen/ice hunters. The origins of the Fishermen's Protective Union, founded in 1908, can also be traced indirectly—via the 1902 strike—back to those earlier strikes. Thus, by the beginning of the Great War, the most densely populated area of Newfoundland was home to two working-class interests—one rural, the other urban. Their interests had begun to diverge and, despite their common roots, each pursued its aims independently of the other.

Meanwhile, in the latter 1800s, as the sealing industry declined and as the outport economy and society nearly collapsed, interest in the seal fishery grew. Governments tried to bolster the industry through various pieces of legislation while, at the same time, trying to reduce reliance on it by diversifying the economy, but with only minor success; inevitably, there was out migration of population as many fishermen moved to the United States of America. The media and the governors took notice of the situation, and campaigned for measures to alleviate the conditions of the sealers and return the industry to its former place in society. However, the governments of the day were not persuaded by these efforts and very little changed.

Only in the fields of literature and culture does one see a positive side to the seal fishery emerging at this time. Writers began to extol the greatness of the sealing industry and to honour in story, song and obituary the men who had built and prosecuted this industry. Heroism, tragedy, humour and pathos were all recorded, as were many isolated adventures, activities, incidents, statistical records and personal achievements—everything that could be gleaned from written and oral sources. By 1914, the impact of the seal fishery on the cultural life of Newfoundland was greater than that of any other activity.

However, the seal herds had been decimated, with St. John's continuing the carnage—not only of the resource but of the outport economies as well. The old sealing centres had been completely destroyed in the 1860s and 70s, with Harbour Grace managing to hang on—at a reduced level—until 1894. The assault on the herds continued, and it is obvious that by the end of the century, even landsmen and schooner fishermen were finding it increasingly difficult to supplement their income from this resource. Thus, even subsistence sealing, as practised by landsmen on the Northern Peninsula and on the coast of Labrador, collapsed, as Dr. Wilfred Grenfell pointed out. He was the first to note in great detail what must have been observed by many people on the north, west, northeast and southwest coasts—that the seal resource was in decline. When one recalls that fishermen moved up the northeast coast of the island in the eighteenth and early nineteenth centuries, specifically to have greater access to the seal herds, one can better appreciate the significant effect that the decline in these herds had on the local population. By the beginning of the Great War, St. John's firms were not just competing with each other; they were in the process of destroying the local economies along the coast that had always depended on subsistence sealing, leaving commercial sealing to others. The seal provided meat for the table, skins for boots, food for the dog teams (which were required on the many islands and in the many places where horses could not thrive because of the scarcity of fodder) and finally, seal fat could be sold for cash. A way of life was disappearing.

In this development, one can find an explanation for the poverty and malnutrition that plagued many outports during the Great Depression (1930s). It is likely that the reduction in the number of seals reduced the amount of fresh meat available to the people of the scattered, isolated outports that had depended on it, and thus reduced their ability to continue the lifestyle their ancestors had developed. Local economies of the west, northwest, northeast and Labrador coasts were marginal at best, and the people had no other resources to substitute for the declining seal population. Like their counterparts in Alaska and Siberia, who watched helplessly while their walrus resources were wiped out, these Newfoundlanders were the victims of the invasion of the commercial ice hunters. By 1914, the stage was set for the distress that was to occur in many isolated outports in the 1930s.

The seal fishery had created the political and geographic entity that came to be known as Newfoundland and which stretched from Cape Race to Cape Ray to Cape Chidley. It had provided the raison d'être for political 'advancement' paralleling the rest of British North America. Without this industry, the island of Newfoundland and the coast of Labrador would have found themselves under the administration of neighbouring colonies (probably Nova Scotia and Quebec, respectively) because the English and Irish fishermen would have abandoned the Newfoundland fishery during the post-1815 period, when the international saltfish trade practically collapsed. The political evolution of Newfoundland returns once more to the question implied in the conclusion of *Fish out of Water*:

> Although the Great War [1914-18] and the Great Depression [1930s] contributed to the Colony's economic collapse, developments during the nineteenth century demonstrated that Newfoundland's political independence was incompatible with the commercial reality of the international saltfish trade.[12]

To reiterate, the development of the seal fishery in the early 1800s placed local power in the hands of those merchants involved in the industry. Merchants not involved in sealing were, in almost all cases, opposed to representative government in the 1820s, while the sealing interests supported it. It is obvious, given the fact that Britain was the principal market for seal oil, that sealing merchants did not feel the same concern about marketing and did not fear becoming more politically independent from Britain. Merchants dependent exclusively on the sale of saltfish in foreign markets were totally dependent on Britain's commercial and consular services and connections, and viewed political independence for Newfoundland as a dangerous option. However, the sealing interests were in control in the 1820s, and by the time it became obvious that sealing was not to retain its place in the economy, the move toward greater political independence had acquired a momentum of its own and there could be no retreat.

In spite of the dismal end of the seal fishery—and by 1914, the end was at hand—one must give credit to this industry for the emergence and growth of the colony of Newfoundland, which included both the island and the coast of Labrador. Clearly, in order to satisfy its appetite, the industrial and technical revolution devoured Newfoundland's resources in the same way it consumed the oil resources of the far-flung corners of the world; but the discovery of almost inexhaustible petroleum supplies changed the demands of the market place. At the same time, the breakthrough in electricity and further advances in the chemical industry brought other changes to the international oil markets. It was not just the collapse in international trade that led to the collapse of the Newfoundland economy in the 1930s. Rather, it was the rapid expansion of oil-based industrialization in the world's leading nations which left Newfoundland, and much of the underdeveloped world, behind. As a result, in 1934, unable to produce much the world really needed, and unable to negotiate the sale of what they did produce, Newfoundlanders surrendered their independence in an attempt to survive—a sad ending, but perhaps inevitable.

The fact remains, the seal fishery created and maintained a society, a culture and a colony and left a legacy that has enriched Canada's tenth province. While the work of the men and women in the Newfoundland cod fishery must never be underestimated, cod alone did not lead to the growth of permanent settlement on a viable scale. It was only after they became ice hunters that the migrant fishermen settled and literally sculped a colony from the Arctic ice floes.

NOTES

1. See Ryan, *Fish out of Water*, for a discussion of the failing saltfish trade during this period. All statistics have been taken from Ryan, "Abstract," Ryan, "MA Thesis," and *Fish out of Water*, unless otherwise noted.
2. *Newfoundland Census* (1857).
3. See Sanger, "The Evolution of Sealing" and Patricia A. Thornton, "The Demographic and Mercantile Bases of Initial Permanent Settlement in the Strait of Belle Isle," in Mannion, *Peopling*, pp. 136-51 and 152-83.
4. See Ryan, *Fish out of Water*.
5. Keith Matthews, "The Class of '32: The Newfoundland Reformers on the Eve of Representative Government," (Mss. in CNS, MUN, 1974).
6. CO 194/72, fols. 158-87. Governor Cochrane to CO, 27 May 1826.
7. The men and firms who were active in this 1826 St. John's movement and whose participation in the 1836 seal fishery is clearly documented by the published accounts of the number of vessels they supplied in that latter year include: W. H. Thomas, whose firm supplied three; Henry Hawson, the agent of Newman and Co., which firm supplied twelve; George Kerr and James McBride of McBride and Kerr, which supplied ten; Patrick Morris, who supplied four; John Brine, who supplied two; James Stewart, whose company supplied four; a representative from C. F. Bennett and Company, which supplied three; a representative of Codner and Jennings (possibly William Codner), which firm supplied four; a representative of John Harvey, whose firm supplied two; a Mr. Bulley from the firm of Bulley, Job and Company, which supplied four; Mssrs. Mudge, who supplied two; William Warren, who supplied two; John Sinclair of Hunters and Company, which supplied eight; and a representative of Baine Johnstons, which firm supplied fifteen. From the similarities in surnames, it is obvious that there were other men and firms with similar connections covering this ten-year period, but for the purposes of the argument being made here, these examples should suffice.

   The earliest and most reliable information on firms involved in the St. John's seal fishery can be found in the *Public Ledger*, 15 March 1836.
8. Prowse, *Newfoundland*, pp. 450-1. See also Leslie Harris, "The First Nine Years of Representative Government," (MA Thesis, History Department, MUN, 1958).
9. CO 194/87, fol. 273. St. John's Chamber of Commerce to CO, 30 May 1834.
10. Edward Russell (1883-1985) from Riverhead, Harbour Grace, discussed this issue extensively. He described how his father worked in the United States, returning home for the winter. He explained how countless Newfoundlanders migrated to

East Boston, Delaware and Gloucester to fish, work on the docks and dredges, and, in the case of women, in offices, restaurants and factories. He said that "East Boston was the home of Newfoundlanders. Some of them were there a long while and owned their own houses; some of them boarded and used to go back and forth to Newfoundland." See *Evening Telegram*, 18 June 1988, "Edward Russell: A Newfoundlander in the Boston States, 1903-12." Edward Russell worked in the United States during the period from 1903 to 1912, returning home most winters.

11 Some planters survived into the twentieth century but they were less wealthy and usually owned only fishing boats and premises on the Labrador coast. They hired fishermen for the summer fishing voyage, but then they, themselves, sought other employment during the other seasons; sometimes in the mines and lumber camps in the winter and on the sealing steamers in the spring.

12 Ryan, *Fish out of Water*, p. 257.

APPENDIX

Table Intro.1
Newfoundland: 1696-97

Place	Men	Fishing Establishments	Boats	Fish
St. John's	300	59	125	6,270
Bonavista	300	40	20	5,000
Carbonear	220	22	56	2,500
Old Perlican	130	19	27	13,900
Bay Bulls	120	13	20	10,000
Renews	120	7	8	4,000
Port de Grave	116	14	20	10,000
Ferryland	108	12	16	8,000
Harbour Grace	100	14	15	7,500
Bay de Verds	85	14	10	11,000
Brigus	70	11	12	6,000
Petty Harbour	60	14	16	8,000
New Perlican	60	9	11	6,000
Holyrood	43	6	7	4,500
Quidi Vidi	40	9	9	4,900
Fermeuse	40	7	8	4,000
Scilly Cove	40	4	7	4,300
Musquito	35	3	5	2,500
Brine's Cove	30	4	6	3,000
Hants Harbour	30	4	5	3,000
Crocker's Cove	30	4	5	2,500
Colmare	30	3	5	2,500
Aquaforte	25	4	5	2,500
Portugal Cove	25	3	3	2,100
Trinity	24	2	4	2,000
Kelun Cove	22	3	9	2,000
Hearts Content	20	4	4	2,400
Freshwater	20	2	4	2,000
Torbay	18	3	4	2,400
Faylim Cove	18	3	3	1,500
Brigus by South	15	3	3	1,500
Witless Bay	15	2	3	1,500
Capelin Bay	12	2	2	1,000
Harbour Main	12	1	2	1,000
Hearts Ease	12	1	2	1,000

Cupids	11	2	2	1,000
Bay Roberts	10	3	3	1,000
Cape Broyle	5	1	1	500
Total	1,581	195	246	100,000

Source: Prowse, *Newfoundland*, p.698, quoted from "Census of Population, etc. of Newfoundland from Pere Baudoin's Journal, 1697." To the above table the following note was appended: "To this must be added settlements not named—Fogo, Twillingate, and other scattered population in the northern part of the Island: about 150 men; 30 fishing establishments; 15 boats; and 4,000 quintals of fish." The above column 'fish' refers to cwts (or quintals) of fish. Prowse thought that Colmare is probably Toad's Cove. Faylim Cove and Kelun Cove are unknown to the present writer. It is assumed that the large amount of fish in Newfoundland during the winter of 1696-97 was the result of the disruption of shipping because of the war.

Table Intro.2

Newfoundland Saltfish: Quantity and Prices

1793-1803

| Year | Quintals | Prices | |
		Minimum	Maximum
1793	437,460	9s.	14s.
1794	510,310	10s	15s.
1795	486,435	13s	17s. 6d.
1796	445,471*	-	-
1797	374,940	12s.	13s. 6d.
1798	485,764*	10s.	11s.*
1799	466,332*	-	14s.*
1800	517,348*	-	15s.*
1801	-	14s. 6d.	22s. 6d.
1802	461,144	-	27s.*
1803	536,188	12s.	15s.

Source: Ryan, "Abstract." *Information missing from Ryan, "Abstract," can be found in BT 6/92, fol. 164. "Exports."

Table Intro.3
Prices in Newfoundland: 1802

bread	18s. to 28s.	per cwt.
pork	£5.15 to £6	per barrel
flour	£2 to £3	per barrel
butter	9d. to 1s. 1d.	per pound
rum	3s. 3d. to 4s.	per gallon
molasses	2s. to 2s. 6d.	per gallon
salt (Portugal)	15s. to 21s.	per hogshead
salt (England)	9s. to 15s.	per hogshead
codfish (merchantable)	22s. 5d. to 27s.	per quintal
codfish (Madeira)	14s. to 19s.	per quintal
salmon	75s.	per tierce
seal oil	£24	per tun
train oil	£24	per tun

Source: CO 194/43, fol. 122. Governor Gambier to CO, 22 January 1803.

Table Intro.4
Prices in Newfoundland: 1804

District	Bread cwt	Flour barrel	Beef barrel	Pork barrel
	s s	s s	£ s d	£ s d
St. John's	20-28	44-50	4 0 0	4 10 0
Ferryland	25-36	52-60	- 0 0	5 15 0
Placentia	26-28	50-55	4 10 0	6 10 0
Conception Bay	25	46-60	4 15 0	6 10 0
Trinity	25	50	5 0 0	6 0 0

District	Butter lb	Molasses gallon	Rum gallon	Sugar lb
St. John's	1s	6s 6d	6s 7d	10d-1s
Ferryland	1 6	7	6 6	1
Placentia	1 4	7 6	7 8	1
Conception Bay	1 6	7 9	7	1s-1s4d
Trinity	1s3d-1s6d	6s-7s6d	6	1

District	Pitch barrel	Tar barrel	Salt (British) hogshead	Salt (Foreign) hogshead	Tobacco lb
St. John's	60s	42s-50s	12s-15s	18s-21s	1s
Ferryland	6d*	3**	17s-17s	21s-22s	1s6d
Placentia	7d-8d*	40s-60s	19s	22s	1s3d-1s6d
Conception	6d*	2s**	-	-	1s6d
Trinity	5d*	2s**	15s-18s	22s	1s6d

*Price given is per lb.
**Price given is per gallon.

Source: Co 194/45, fols. 20-38. "Gower's Report of 1804."

Table Intro.5
Wages in Newfoundland: 1804

Fishermen

District	Boats Master	Mid-Shipman	Fore-Shipman	Captain	Greenhand
	£	£	£	£	£
St. John's	40-50	30-40	30-40	30-35	14-18
Conception Bay	50	40	36	32	20
Ferryland	40-45	34-38	28-32	18-24	12-14
Trinity	30	28	26	23	20
Trepassey	50	34	26	21	16
Placentia	30-40	25-30	20-26	16-20	10-14

Shoremen

District	Master of Voyage	Splitter	Salter	Header	Youngster
	£	£	£	£	£
St. John's	40-50	30	26	24	14-18
Conception Bay	36	28	26	-	24
Ferryland	40	30	26	24	12-14
Trinity	30	28	26	22	20
Trepassey	40	30	26	22	16

Placentia	30-35	30-32	25-28	25-28	10-12

Source: CO 194/45, fols. 17-47. "Gower's Report of 1804."

Table Intro.6
Wages in Newfoundland: 1810*

Occupation	Wages
Shoremen:	£
- Master of Voyage	45 to 50
- Splitter	45 to 50
- Salter	40 to 45
- Header	30 to 35
- Youngster	15 to 18
Fishermen:	
- Fishermen	50 to 55
- Old Hands	26 to 30
- Young Hands	18 to 22

*"Sealers have from 1/2 to 1/3 their catch of seals. The master of Sealing vessel gets £5 per month and 6d. for each seal."

Source: CO 194/48, fols. 51-129. Governor Duckworth to CO, 25 November 1810.

Table Intro.7
Prices in Newfoundland: 1810*

Product	Winter	Summer	Fall
bread (cwt)	50s	40s-50s	50s-52s6d
flour (brl)	70s	55s-65s	75s
potatoes (brl)	12s-13s	6s	14s
butter (lb)	1s 6d	1s3d-1s4d	1s4d
pork (brl)	147s	140s	140s

molasses (gal)	3s-6s	2s10d-3s6d	2s10d-3s6d
salt (hhd)	-	18s-25s	12s
coffee (lb)	1s 9d	1s9d	1s9d
soap (lb)	1s 6d	1s6d	1s6d
longlines (doz)	36s	36s	130s
cordage, "twice laid" (cwt)	80s	70s	70s
tea (lb)	4s	4s	4s
olive oil (gal)	12s	14s	14s
rum (gal)	3s6d-5s	3s6d-5s	3s10d-5s
sugar (cwt)	56s	56s	56s
fish (cwt)	14s6d	13s-14s6d	13s-14s6d
salmon (tierce)	-	75s-80s	70s-75s
oil (tun)	£30	£25	£26-28

*"This is barter price, when cash is paid some articles are lower."

Source: CO 194/48, Fols. 51-129. Governor Duckworth to CO, 25 November 1810.

Table Intro.8
Prices in Newfoundland: 1813

	In quantities	Single
Bread (cwt)	65s	75s
Flour (brl)	111s	120s
Pork (brl)	189s	210s
Beef (brl)	110s	120s
Butter (lb)	14d	17d
Oatmeal (cwt)	45s	50s

Source: CO 194/54, fols. 73-4. Society of Merchants, St. John's, to Governor Keats, 16 July 1813.

Table Intro.9
Value of Seal Oil produced by Inhabitants: 1723-1802

Year	£	Year	£	Year	£
1723	6,025	1758	3,560	1793	8,410
1724	1,890	1759	3,350	1794	2,416
1725	-	1760	1,330	1795	14,000
1726	6,305	1761	-	1796	-
1727	-	1762	-	1797	8,130
1728	-	1763	230	1798	-
1729	-	1764	3,304	1799	-
1730	3,562	1765	5,109	1800	-
1731	3,699	1766	3,587	1801	13,180
1732	1,382	1767	8,832	1802	-
1733	1,382	1768	12,664		
1734	1,310	1769	5,374		
1735	3,379	1770	12,358		
1736	-	1771	5,509		
1737	-	1772	13,406		
1738	2,221	1773	26,388		
1739	3,052	1774	17,605		
1740	2,496	1775	6,947		
1741	9,750	1776	485		
1742	3,694	1777	-		
1743	5,060	1778	4,140		
1744	-	1779	4,550		
1745	3,940	1780	-		
1746	2,760	1781	400		
1747	-	1782	-		
1748	1,580	1783	-		
1749	1,016	1784	3,382		
1750	1,882	1785	4,292		
1751	3,139	1786	6,071		
1752	2,330	1787	5,435		
1753	1,450	1788	7,126		
1754	3,054	1789	11,688		
1755	400	1790	3,190		
1756	-	1791	3,190		
1757	4,800	1792	11,920		

Source: Ryan, "Abstract."

Table Intro.10
Newfoundland's Population by District: 1789

District	Population
Conception [Bay]	5,636
St. John's	4,420
Bay Bulls	836
Ferryland	1,177
Placentia [Bay]	819
St. Mary's [Bay]	304
Trepassey	386
Burin	486
St. Lawrence to Harbour Breton	594
Trinity Bay	2,023
Bonavista Bay	1,435
Fogo and Twillingate	1,010
Total	19,126

Source: CO 194/21, fol. 264. "Report on Newfoundland for the Year ending October 10, 1789."

Table Intro.11
Newfoundland's Fishery and Trade by District: 1789

District	No. of Vessels	No. of Men in the Fishery	No. of Cwts of Fish Made	Tuns of Train Oil Made
1	19	1,049	58,200	116
2	9	789	44,000	150
3	14	1,050	44,300	96
4	8	466	21,110	91
5	13	690	32,650	132
6	63	2,349	98,705	307
7	27	1,136	48,450	194
8	270	4,741	321,678	574
9	55	2,438	59,768	341
10	42	1,536	30,330	194
11	10	1,139	49,600	126
12	12	891	14,000	51
Total	542	18,274	782,791	2,372

District 1 - Harbour Breton, Great St. Lawrence, Little St. Lawrence
District 2 - Burin, Mortier, Auderin etc.
District 3 - Great Placentia, Little Placentia
District 4 - St. Mary's
District 5 - Trepassey
District 6 - Ferryland, Renews, Ferneuse, Caplin Bay
Distric 7 - Bay Bulls, Etc.
District 8 - Petty Harbour, St. John's, Torbay
District 9 - Harbour Grace, Carbonear
District 10 - Trinity, Bonaventure, Bay Verds, Old Perlican
District 11 - Bonavista, Greenspond, Kings Cove, Keels
District 12 - Fogo, Twillingate and Other Islands

Source: CO 194/21, fol. 264. "Report on Newfoundland for the Year ending October 10, 1789."

Table Intro.12
Seal Fishery: 1803

Winter Seal Fishery (Nets)

District	Men Employed	Seals Taken	Value £	Tuns of Oil Made
Fogo	200	2,000	1,000	20
Bonavista	235	1,350	675	25
Trinity	58	1,111	555	13
Total	493	4,461	2,230	58

Spring Seal Fishery (Ships)

District	Vessels	Tons	Men	Seals	Value £	Oil Tuns	Period Fishing
1	11	431	306	6,675	3,300	80	Mar 20 to May 15
2	12	572	135	4,466	2,013	43	Mar 30 to May 20
3 ships	21	969	241	19,012	7,717		
boats	14	430	128	7,632	3,402	333	Mar 30 to May 24
4	19	739	188	11,222	6,720	160	Mar 22 to May 16
Total	77	2,141	998	49,007	23,152	616	

Districts: 1 = Bonavista; 2 = Trinity Bay; 3 = Conception Bay; and 4 = St. John's

Source: CO 194/23, fol. 517. Governor Gambier's "Account of the Seal Fishery for 1803."

Table Intro.13
Spring Seal Fishery: 1804

District	Vessels	Shipping Tons	Men	Seals	Tuns of Oil Made
St. John's	35	1,589	430	17,600	220
Fogo	11	400	108	1,330	16
Greens Pond	6	180	70	2,000	25
Bonavista	12	407	120	3,583	77
Trinity	14	628	152	6,093	85
Conception Bay					
Schooners	43				
Shallops	25	2,470	729	40,704	510
Ferryland	3	109	30	1,464	16
Total	149	5,783	1,639	72,774	949

Source: BT 6/91, fol. 15. Gower's "Account of the Spring Seal Fishery, 1804."

Table 1.1
Train Oil Imports into England and Wales: 1700-71

Year	Tuns	Year	Tuns	Year	Tuns
1700	1,411	1724	2,089	1748	3,237
1701	1,321	1725	1,712	1749	3,290
1702	2,449	1726	1,469	1750	3,756
1703	753	1727	2,720	1751	3,932
1704	1,078	1728	2,671	1752	4,202
1705	-	1729	2,473	1753	2,868
1706	702	1730	2,422	1754	4,018
1707	579	1731	2,246	1755	3,960
1708	1,032	1732	3,502	1756	3,394
1709	1,123	1733	2,739	1757	2,882
1710	1,323	1734	3,343	1758	2,764
1711	974	1735	2,721	1759	3,284
1712	-	1736	2,645	1760	2,620
1713	1,072	1737	3,493	1761	3,258
1714	1,039	1738	2,811	1762	2,549
1715	1,486	1739	3,016	1763	5,059
1716	1,139	1740	2,462	1764	5,292
1717	855	1741	3,130	1765	5,592
1718	1,283	1742	2,062	1766	6,184
1719	1,098	1743	2,572	1767	5,724
1720	1,602	1744	1,944	1768	7,076
1721	2,050	1745	2,154	1769	6,844
1722	1,819	1746	2,789	1770	6,787

Source: Schumpeter, *Trade Statistics*, pp. 52-5. Figures taken from Table XVI, "Quantities of Selected Imports into England and Wales for the Years 1700 to 1771, 1775 and 1780."

Table 1.2
Train Oil Imports into England and Wales: 1772-1808

Year	Tuns	Year	Tuns	Year	Tuns
1772	8,147	1784	7,280	1796	11,437
1773	8,844	1785	10,138	1797	13,600
1774	9,837	1786	11,736	1798	12,132
1775	8,392	1787	14,801	1799	13,013
1776	5,678	1788	14,791	1800	13,236
1777	5,872	1789	12,971	1801	15,681
1778	5,193	1790	11,726	1802	18,037
1779	3,426	1791	10,168	1803	17,722
1780	5,155	1792	10,739	1804	21,939
1781	3,480	1793	11,662	1805	23,681
1782	4,184	1794	10,899	1806	22,338
1783	5,435	1795	10,025	1807	22,346
				1808	22,682

Source: Schumpeter, *Trade Statistics*, pp. 56-9. Figures taken from Table XVII, "Quantities and Values of Selected British Imports from 1772 to 1808." (Figures for England and Wales, 1772-91; figures for Great Britain, 1792-1808.)

Table 1.3
British Imports from Newfoundland: 1772-1808

Year	Train Oil (Tuns)	Seal Skins	Year	Train Oil	Seal Skins (Tuns)
1772	3,667	19,100	1790	3,249	41,646
1773	4,051	12,769	1791	2,185	32,128
1774	2,390	4,265	1792	2,623	33,434
1775	2,473	5,879	1793	1,510	31,702
1776	2,895	18,295	1794	2,944	27,765
1777	2,793	38,232	1795	2,739	40,676
1778	2,614	30,025	1796	1,860	36,772
1779	1,826	12,046	1797	1,943	39,647
1780	2,542	21,138	1798	2,369	55,673
1781	1,187	21,051	1799	3,234	92,210
1782	1,414	8,392	1800	3,174	62,819
1783	2,106	38,614	1801	3,975	110,466
1784	3,009	23,055	1802	2,837	47,742
1785	3,500	15,745	1803	2,899	64,004
1786	2,620	27,145	1804	4,382	113,355
1787	2,800	42,103	1805	5,034	72,994
1788	3,228	46,907	1806	5,672	128,852
1789	3,486	34,406	1807	6,960	170,509
			1808	6,288	157,041

Source: CUST 17, Vols. 1-30, PRO. These are listed as imports into England. Wales is included in the figures, but Scotland is listed separately until 1798, when the figures begin to include imports into the latter, and the term 'Great Britain' is used from that point on. The absence of the Scottish returns from the earlier listings does not affect the table because Scotland was not a major market—importing, for example, only 83 tuns of Newfoundland train oil in 1784 and 271 tuns in 1789. Although most of these returns are described as imports from Newfoundland, towards the end of this period, the terms 'British Colonies in America' or 'Colonies in America' replaced the term 'Newfoundland'. However, the present writer has concluded that very few skins and very little oil came form any other source. (All figures for train oil have been rounded off to the nearest whole tun; and in each case, the year ended on January 5.)

Table 1.4
British Imports of Seal Skins and Train Oil from Newfoundland: 1809-53

Year	Seal Skins	Train Oil (Tuns)	Year	Seal Skins	Train Oil (Tuns)
1809	93,495	4,647	1831	504,087	10,375
1810	134,147	6,167	1832	214,622	8,538
1811	-	-	1833	310,055	10,068
1812	133,594	5,805	1834	327,693	7,725
1813	-	-	1835	322,054	9,270
1814	129,600	5,755	1836	113,736	7,097
1815	132,508	5,916	1837	262,634	8,508
1816	136,724	5,938	1838	336,634	8,396
1817	41,997	4,745	1839	268,569	7,930
1818	156,752	4,883	1840	522,248	11,446
1819	227,113	5,873	1841	278,426	10,194
1820	184,593	6,392	1842	303,896	8,560
1821	242,458	7,815	1843	644,624	12,097
1822	242,015	7,497	1844	458,172	9,166
1823	172,573	6,039	1845	427,851	9,701
1824	227,357	8,499	1846	255,046	6,520
1825	216,994	5,350	1847	435,025	9,989
1826	273,570	6,866	1848	527,795	10,979
1827	370,813	7,673	1849	298,794	8,694
1828	217,474	6,524	1850	455,019	10,442
1829	225,830	6,627	1851	492,755	10,453
1830	376,162	10,153	1852	525,721	9,561
			1853	498,674	10,597

Source: CUST 4, Vols. 5-58, PRO. This table does not include a category of imports listed as coming from the 'British Fisheries' in Newfoundland. There were probably two reasons why this heading was used in the original Customs Records: there was a small and dwindling British cod fishery in Newfoundland waters which produced a certain amount of train oil; and some Customs officials were still confused about Newfoundland's standing as an evolving colony and mistook products emanating from the resident fishery for commodities produced by the traditional British migratory fishery. It must be realized that up until it received representative government status in 1832, Newfoundland's future as a colony versus a fishing station was in doubt, at least in the eyes of many officials. In line with this latter point, the last entry for imports from the 'British Fisheries' was recorded in 1833 (see CUST 4, Vol. 28). The first separate account of imports from the 'Coast of Labrador' were recorded in 1817, but there was no breakdown by item, simply the total value, which was just over £3,528 that year—a figure which is largely meaningless, given the practice of assigning outdated values to imports (see CUST 4, Vol. 12). In 1827, the Customs officials began to include imports from Labrador with those from the island of Newfoundland under the heading 'Newfoundland and the Coast of Labrador' (see CUST 4, Vol. 22).

Table 1.5
Comparative British Train Oil Imports: 1831-1914

Part 1: 1831-58 (Tuns)

Year	Train Oil Spermaceti	Train Oil Other	Total*	(From Newfoundland)
1831	6,816	17,853	24,669	10,971
1832	7,540	23,234	30,774	9,199
1833	5,484	27,391	32,875	10,596
1834	6,276	19,057	25,333	8,211
1835	7,645	16,552	24,197	9,920
1836	7,028	12,460	19,488	7,594
1837	6,312	15,491	21,803	9,085
1838	6,483	21,798	28,281	9,301
1839	5,815	16,384	22,199	8,330
1840	5,289	20,292	25,581	11,939
1841	5,227	18,053	23,280	10,650
1842	4,173	13,300	17,473	8,792
1843	6,333	17,624	23,957	12,470
1844	5,006	15,838	20,844	9,375
1845	6,996	17,945	24,941	10,301
1846	4,554	12,329	16,883	6,961
1847	6,286	16,663	22,949	10,161
1848	4,712	17,254	21,966	11,206
1849	5,662	14,350	20,012	8,892
1850	5,792	15,568	21,360	10,544
1851	-	16,816	22,219	10,453
1852	-	14,500	19,906	9,561
1853	5,180	15,757	20,937	10,645
1854	4,648	14,184	18,832	9,403
1855	3,645	13,712	17,357	8,131
1856	5,136	13,158	18,294	7,989
1857	5,410	15,765	21,175	12,337
1858	4,613	14,832	19,445	10,096

Part 2: 1859-1907 (Tuns - Including Spermaceti)

Year	From Nfld.	Total	Year	From Nfld.	Total
1859	8,513	20,608	1883	5,800	17,156
1860	6,610	17,029	1884	6,317	17,489
1861	7,218	19,176	1885	5,308	18,380
1862	6,377	18.383	1886	5,395	15,834
1863	6,178	14,265	1887	5,695	17,698
1864	3,865	14,701	1888	5,737	16,871
1865	4,699	13,147	1889	7,042	20,956
1866	5,543	13,811	1890	5,522	20,307
1867	7,110	15,945	1891	6,581	21,969
1868	6,538	13,991	1892	7,169	21,121
1869	5,330	15,264	1893	5,114	19,939
1870	8,680	19,706	1894	6,079	24,213
1871	11,065	24,679	1895	7,603	24,597
1872	7,407	18,719	1896	5,156	21,632
1873	8,329	17,886	1897	2,564	18,129
1874	5,965	17,051	1898	4,805	20,673
1875	6,802	19,359	1899	5,449	20,358
1876	6,308	16,684	1900	5,882	21,323
1877	8,633	19,365	1901	5,005	24,384
1878	8,138	20,656	1902	7,783	23,470
1879	9,416	20,196	1903	6,765	22,141
1880	5,515	15,231	1904	9,283	24,459
1881	7,772	18,030	1905	9,566	25,508
1882	6,565	15,945	1906	9,208	27,808
			1907	6,472	26,929

Part 3: 1908-12 (Tuns - Train, Blubber and Sperm)

	1908	1909	1910	1911	1912	1913	1914
Norway	3,334	2,903	3,487	1,978	8,836	5,664	6,717
Denmark	608	600	1,168	536	286	377	410
Iceland	4,167	5,147	3,488	2,767	1,408	1,258	49
French W/A	-	-	-	-	1,732	3,951	-
Azores	95	89	132	80	39	92	90
Portuguese W/A	3	-	2,004	5,165	5,442	2,262	
Japan	5,467	3,510	3,650	2,453	2,096	5,055	6,594
USA	1,040	3,885	2,369	3,053	4,111	1,552	1,676
Chile	1,836	2,280	3,672	2,262	1,860	2,101	1,960
Brazil	810	1,107	644	854	48	12	250
Argentina	2,946	2,051	2,524	1,813	3,707	1,908	999
Whaling/N	1,776	1,550	1,293	853	766	798	1,167
Whaling/S	-	-	-	2,883	2,607	8,509	10,692
Other Foreign	496	964	371	434	1,651	4,081	2,062
Total Foreign	22,578	24,086	22,848	21,970	34,312	40,800	34,928
Natal	394	512	5,901	6,618	8,829	5,594	2,987
Canada	2,960	4,466	3,181	2,096	6,900	3,713	747
Falklands	-	622	6,861	21,402	10,064	8,500	5,733
Newfoundland	**5,513**	**5,882**	**6,833**	**5,142**	**4,406**	**5,285**	**3,452**
Other British	692	653	477	1,734	3,519	7,034	4,690
Total British	9,559	12,135	23,253	36,992	33,718	30,086	17,609
Grand Total	32,137	36,221	46,101	58,962	68,030	70,886	52,537

Source: *Trade of the United Kingdom* (London, 1831-1914). A few gaps in this source have been filled by using statistics from the CUST 4 series, PRO. Please note that W/A stands for West Africa, Iceland includes Greenland, N stands for Northern and S stands for Southern.

*Total in Part 1 includes spermaceti oil and "other" train oil; it does not include train oil from Newfoundland.

Table 1.6
Britain's Principal Oil Imports: 1854-1907 (Value in Sterling)

Year	Coconut	Olive	Palm	Rape Seed	Train
1854	511,626	745,828	1,731,021	622,182	1,076,692
1855	555,610	1,411,950	1,762,607	416,753	1,117,286
1856	274,449	1,124,755	1,691,407	407,210	1,165,410
1857	490,289	1,044,054	1,868,484	587,156	1,112,923
1858	375,870	1,201,561	1,513,109	371,725	921,259
1859	380,875	1,001,041	1,545,089	392,660	1,108,052
1860	458,143	1,247,902	1,796,465	546,174	878,868
1861	645,778	984,985	1,579,953	360,520	1,014,585
1862	445,621	1,211,306	1,724,310	529,953	1,014,981
1863	753,321	1,138,336	1,419,536	559,342	826,343
1864	716,175	958,397	1,121,370	502,548	853,991
1865	411,260	1,684,852	1,450,409	595,669	746,058
1866	256,989	1,353,518	1,606,797	555,079	825,176
1867	308,745	1,244,296	1,568,194	686,730	904,253
1868	488,463	1,186,828	1,891,573	818,294	658,238
1869	572,804	1,493,285	1,583,701	766,241	861,788
1870	392,657	1,185,950	1,583,830	594,933	954,710
1871	357,260	1,858,779	1,820,698	404,386	1,087,734
1872	822,257	1,193,064	1,805,153	793,941	855,590
1873	460,681	1,559,183	1,713,829	673,215	766,927
1874	241,561	1,017,461	1,792,299	629,654	751,359
1875	411,565	1,559,068	1,508,299	637,131	917,701
1876	377,480	1,089,176	1,529,360	811,421	735,621
1877	376,952	1,483,786	1,598,166	569,967	754,812
1878	201,973	1,028,757	1,167,161	467,375	810,891
1879	398,065	1,179,021	1,344,788	508,975	589,304
1880	540,233	907,308	1,519,701	517,812	477,357
1881	369,779	1,220,650	1,202,571	535,878	578,663
1882	210,054	947,154	1,240,866	476,807	531,663
1883	365,716	1,198,935	1,315,559	366,489	604,746
1884	396,288	715,964	1,408,753	388,148	530,805
1885	276,224	981,348	1,217,816	368,130	520,412
1886	214,346	791,245	1,050,459	396,126	361,947
1887	248,218	758,348	943,126	374,120	373,275
1888	245,867	672,614	945,896	412,438	323,680
1889	277,602	817,495	1,091,922	461,691	440,360
1890	261,683	785,779	1,000,535	618,490	419,401
1891	264,228	733,435	1,186,705	609,890	454,704
1892	191,380	762,516	1,169,490	555,832	415,181
1893	202,885	652,041	1,458,642	718,837	389,864

1894	360,737	894,151	1,237,072	654,065	413,974
1895	321,550	522,811	1,320,690	721,807	406,448
1896	249,633	612,876	1,204,679	694,995	365,190
1897	265,668	515,531	1,001,368	698,694	280,877
1898	344,108	608,122	975,427	689,934	349,348
1899	545,642	553,286	1,037,265	879,171	346,996
1900	667,204	461,084	1,086,555	1,038,564	389,712
1901	594,154	581,893	1,370,645	1,193,577	453,809
1902	719,357	657,956	1,679,610	913,642	473,218
1903	994,676	495,711	1,423,120	-	428,548
1904	800,525	508,458	1,503,698	-	425,811
1905	866,712	441,486	1,358,776	-	417,933
1906	803,417	543,885	1,531,812	-	475,601
1907	857,879	538,888	1,926,155	-	529,947

Table 1.6 (continued)
Petroleum: 1859-1907 (Value in Sterling)

Year	Petroleum	Year	Petroleum
1859	18,601	1881	1,952,444
1860	62	1882	1,721,019
1861	23,558	1883	2,170,298
1862	349,987	1884	1,711,313
1863	690,698	1885	2,289,525
1864	455,913	1886	2,091,276
1865	429,999	1887	2,103,599
1866	696,639	1888	2,565,598
1867	367,722	1889	2,588,947
1868	298,856	1890	2,397,187
1869	442,858	1891	2,685,368
1870	535,272	1892	2,446,906
1871	614,017	1893	2,546,760
1872	433,472	1894	2,484,976
1873	991,841	1895	3,368,904

1874	993,418		1896	3,732,056
1875	775,241		1897	3,335,271
1876	1,431,286		1898	3,733,397
1877	1,776,506		1899	4,654,989
1878	1,212,326		1900	5,559,259
1879	1,382,534		1901	5,070,702
1880	1,309,279		1902	5,193,582

Petroleum: 1903-1907
(Value in Sterling)

	1903	1904	1905	1906	1907
a)	1,364,666	1,396,432	1,440,469	1,631,921	1,643,142
b)	2,967,588	3,309,008	2,658,156	2,665,504	2,551,297
c)	683,153	749,311	787,624	723,641	805,165
d)	222,743	296,590	443,457	690,098	979,395
e)	38,151	70,861	87,939	123,820	88,286
f)	19,050	16,831	5,479	10,075	-
t)	5,295,351	5,839,033	5,423,124	5,845,059	6,067,285

Please note: a) = Lubricating Oils; b) = Lamp Oils
c) = Gas Oil; d) = Spirit; e) = Fuel Oil; f) = Crude; and
t) = Total.

Source: *Trade of the United Kingdom* (London, 1854-1907). These statistics provide valuable evidence of the principal trends in the British oil trade after 1853. The values shown here are those which refer to 'Total Imports' and include the value of oil re-exported. Please note the following: The train oil records included imports of sperm oil and blubber; 'Rape seed' changed to 'All Seed' in 1874 and was broken down into several sub-categories after 1902, making it inconvenient for the present writer to express the information clearly and concisely in a single table; sperm oil was recorded separately for a brief period—1893-99 (inclusive)—and the present writer combined these records with train oil for the purpose of this table; in 1905, 1906 and 1907, coconut, olive and palm oils were divided into 'unrefined' and 'refined', and again, the present writer has combined them; beginning in 1899, it became the practice to sub-divide petroleum imports into 'illuminating' and 'lubricating'—the present writer has combined the values of both for the years 1899 to 1902; in 1903, it became the practice to sub-divide petroleum even further; finally, after 1907, the customs and excise records began to create a number of sub-categories in all these commodities, making it virtually impossible to express the information in a clear and accessible manner—hence, the decision to terminate this table in 1907.

Table 1.7
Newfoundland Train Oil Prices: 1854-1914

Year	Price £ s d	Year	Price £ s d	Year	Price £ s d
1854	40 12 0	1874	34 16 5	1894	18 2 2
1855	46 6 10	1875	34 17 0	1895	17 3 5
1856	47 0 0	1876	34 13 0	1896	18 19 0
1857	42 6 7	1877	33 14 7	1897	17 19 10
1858	34 17 2	1878	30 5 5	1898	19 0 0
1859	33 8 10	1879	26 5 2	1899	17 7 0
1860	35 4 10	1880	28 16 7	1900	19 16 2
1861	37 2 2	1881	29 1 2	1901	24 8 4
1862	45 0 10	1882	31 11 10	1902	20 17 0
1863	50 13 5	1883	32 4 0	1903	21 0 2
1864	51 16 2	1884	30 6 5	1904	17 3 0
1865	47 9 5	1885	27 5 7	1905	15 8 10
1866	45 12 7	1886	22 18 0	1906	17 8 2
1867	39 7 0	1887	20 12 7	1907	20 18 10
1868	37 10 0	1888	19 2 2	1908	19 7 2
1869	41 17 5	1889	20 3 10	1909	17 16 7
1870	33 17 7	1890	21 16 0	1910	21 17 5
1871	32 15 7	1891	22 0 7	1911	24 12 7
1872	36 0 10	1892	19 6 10	1912	20 19 0
1873	34 2 7	1893	21 3 0	1913	21 19 2
				1914	20 19 5

Source: *Trade of the United Kingdom* (London, 1854-1914). The prices of train oil per tun have been calculated from the declared value of train oil imports into Great Britain from Newfoundland and the Coast of Labrador.

Table 1.8
Oil Prices on the London Markets: 1832-80 (Pounds and Shillings per Tun)

Year	Seal £ s	Sperm £ s	Cod £ s	Whale £ s
1832	35	69 14	-	32 14
1833	-	-	-	-
1834	26	65	-	22 15
1835	-	-	-	28
1836	49 10	81 3	50	39 4
1837	44	81	42 15	35 12
1838	37 9	81 13	37 2	28 15
1839	33 7	90	31 3	28 5
1840	28 12	99	28 3	25 10
1841	33 4	-	27 10	31
1842	39 12	90 10	29 15	35 5
1843	36 15	69	35 5	36 10
1844	34 5	90	34 5	32
1845	33	86 13	31 10	30 10
1846	28 14	81 4	27 14	26 6
1847	28 5	89	27 15	26 15
1848	25 5	77	24 10	25 10
1849	33	77	26	31 10
1850	37 3	84	33 7	34 4
1851	31 10	85 10	37	32 5
1852	33 15	87 10	37	32 5
1853	34 5	88 5	34	37 10
1854	42 5	106 5	40 10	42 10
1855	50 13	129 13	44 10	47
1856	50	107	46	48 10
1857	44 15	91 10	47 10	45
1858	37 5	87	34 5	35 10
1859	34 10	92 10	32 10	33
1860	35 8	100	36 15	34 3
1861	38	95 15	36	36 10
1862	43 5	88	43 10	38 18
1863	46 15	81	57 15	45 10
1864	48 5	69 10	52 10	49
1865	44	102 10	50 10	46
1866	44 18	123 12	48 8	46 10
1867	40 10	106	38 15	45
1868	36	100	38	37
1869	38	96	42 10	38
1870	41 10	94	43	39

Year					
1871	36		82	35	34
1872	37	10	84	37 10	36
1873	35		94 5	38	35
1874	-		-	-	-
1875	35		94 5	35 10	34
1876	35		94 5	41 15	34
1877	35		88	39 10	-
1878	32	10	66 10	31	-
1879	28		55 10	27 5	-
1880	30		60	-	-

Source: *Mark Lane Express*, 1832-80. This Table consists of the average prices of the top quality fish oils sold from the London warehouses as published in the *Mark Lane Express* beginning in 1832. The number of price quotations varies from year to year. All prices are given in pounds and shillings per tun, rounded off to the nearest shilling. The *Mark Lane Express* contains only infrequent reports from Liverpool, Bristol and Hull.

Table 2.1
Adult Population of Newfoundland and Occupations: 1911

Number	Occupation
1	Archbishop
3	Bishops
245	Clergymen
1,395	Teachers
46	Lawyers
119	Doctors
1,326	Merchants and Traders
4,641	No. Engaged in Office or Shop Work
1,468	No. Engaged Solely in Government Services
2,915	No. of Farmers
40,880	No. of Fishermen and Others who Cultivate Land
5,376	No. of Mechanics
43,795	No. of Males engaged in Catching and Curing Fish
23,245	No. of Females engaged in Curing Fish
2,821	No. Engaged in Lumbering
2,260	No. Engaged in Mining
1,204	No. Engaged in Factories and Workshops
14,811	No. Otherwise employed
242,619	Total including children and the unemployed

Source: *Census of Newfoundland and Labrador*, (St. John's, 1911). Statistics from Labrador are included in this table.

Table 2.2
Adult Population of Newfoundland and Occupations: 1857

77	Clergymen
71	Doctors, Lawyers
1,697	Farmers
1,973	Mechanics
694	Merchants, Traders
39,805	Engaged in Fisheries
20,887	Seamen and Fishermen
334	Engaged in Lumbering
65,538	Total Adults Employed
122,638	Total Population

Source: *Census of Newfoundland*, (St. John's, 1857).

Table 2.3
Newfoundland's Principal Imports: 1856-60

Products	1856	1857	1859	1860
	£	£	£	£
Oxen and Cows	22,669	17,010	16,540	19,000
Beef (salted)	11,152	6,050	4,064	5,662
Bread (hard)	101,772	81,876	62,043	53,839
Butter	69,195	77,711	63,164	58,615
Candles	5,025	6,894	7,579	5,612
Canvas	?	20,335	13,952	13,224
Coals	20,739	23,461	18,817	20,792
Coffee	4,881	5,600	5,905	4,314
Cordage	22,394	33,786	21,866	17,802
Cotton Manufacts	65,203	?	?	?
Flour	267,932	183,387	207,735	196,830
Fishing Tackle	26,720	36,353	38,616	31,620
Hardware	37,869	54,554	44,036	54,927
Leatherware	53,165	88,465	74,213	72,998
Meat and Poultry	5,377	4,203	9,113	9,074
Molasses	63,524	80,043	60,597	59,525
Pork (salted)	95,949	100,596	100,541	92,193
Potatoes	11,482	4,056	4,773	7,161
Vegetables	?	1,965	2,152	4,432
Salt	33,166	38,505	23,852	23,812
Soap	6,538	7,384	8,671	8,332

Rum	25,678	28,802	25,398	27,551
Sugar (unrefined)	29,181	34,961	27,001	18,896
Tea	20,125	27,429	51,779	48,171
Tobacco	10,972	17,475	16,050	15,613
Woollens	98,179	257,475	235,367	214,196
Canvas	?	?	13,952	13,224
Staves	6,158	18,459	18,166	4,416
Board and Plank	9,880	15,360	21,005	16,839
Grand Total	1,271,604	1,413,432	1,323,288	1,254,128

Source: *JHA*, 1857-61. Some headings change from year to year, which makes it difficult to trace the importation of products over a period of time. Please note the following: "Cotton Manufacts" are listed only in 1856; "Canvas" is listed in 1857, 59 and 60 only; in 1856, one finds "Woollens and Slops"; in 1857, "Woollens, Silks, Linens, Cottons and Slops"; in 1859, "Woollens"; and in 1860, "Woollens, Cottons etc." "Ironmongery" appears only in 1856, and for purposes of this table, it has been included in "Hardware" for that year. In 1856, "Potatoes and Vegetables" were listed together, but separately in other years. In 1856, there is a listing for "Fresh Meats," but for the other three years, the heading is "Meat and Poultry." Also, due to the systems of headings and sub-headings used, the present writer did not bother with rearranging the lists in alphabetical order. The originals, for example, placed "Oxen and Cows" under the heading "Animals," "Rum" under "Spirits," etc. Excluded from this table because of space were many imports, usually of lesser importance: bacon and ham, other vegetables and grains, glassware, paper, pitch and tar, oakum, other spirits, wines and beers, other animals such as horses, sheep and swine, and miscellaneous items. Values in pounds sterling.

Table 2.4
Newfoundland Vessels: 1803-33

Year	North Shore	Coasting	Sealing
1803	47	25	-
1804	99	39	-
1805	156	41	-
1806	-	-	-
1807	131	48	-
1808	95	51	-
1809	90	63	-
1810	-	-	-
1811	107	40	-
1812	111	39	-

1813	116	73	-
1814	129	59	-
1815	106	76	-
1816	104	83	-
1817	110	85	-
1818	153	75	-
1819	214	89	-
1820	263	72	-
1821	261	69	-
1822	174	71	-
1823	237	85	-
1824	251	44	-
1825	204	80	-
1826	239	76	-
1827	302	-	290
1828	300	-	296
1829	243	-	278
1830	286	-	293
1831	-	-	349
1832	274	-	407
1833	223	108	369

Source: Ryan, "Abstract." Please note the following: "Vessels on the North Shore" was the heading used by the governors until 1827, when it was reported that there were 302 vessels on the "North Shore and Labrador." Beginning in 1828, the North Shore was no longer mentioned, and "Labrador" became the heading. Therefore, the figures in the above table under "North Shore," from 1828 to 1833, inclusive, actually refer to the vessels on the Labrador coast. As can be seen, in 1827, the governors began to report the number of vessels engaged in sealing. As has already been pointed out, sealing ships went fishing for cod on the North Shore and, later, to the Labrador coast, after the seal fishery was finished. Coasting vessels were employed in the coastal trade; usually transporting goods from St. John's to the outports and returning with fish and oil. As the North Shore, Labrador and seal fisheries developed, some of the ships in these fisheries also found employment as coasters during the fall and winter.

Table 2.5
Newfoundland's Production of Seal Skins: 1803-60

Year	Number	Year	Number
1803	53,468	1832	469.073
1804	106,739	1833	425,084
1805	-	1834	-
1806	-	1835	-
1807	153,175	1836	384,321
1808	141,237	1837	-
1809	92,231	1838	375,361
1810	-	1839	437,501
1811	143,004	1840	631,385
1812	116,290	1841	417,115
1813	133,847	1842	-
1814	110,275	1843	651,370
1815	121,282	1844	685,530
1816	147,009	1845	353,202
1817	-	1846	265,169
1818	165,622	1847	436,831
1819	279,670	1848	521,604
1820	194,260	1849	306,072
1821	265,192	1850	440,828
1822	368,336	1851	511,630
1823	218,853	1852	-
1824	166,424	1853	521,783
1825	284,944	1854	-
1826	282,793	1855	293,038
1827	371,163	1856	361,317
1828	217,448	1857	496,113
1829	199,179	1858	507,624
1830	536,757	1859	329,185
1831	601,742	1860	344,202

Source: Figures for the 1803-33 period taken from Ryan, "Abstract"; 1836-50 from the *Newfoundland Blue Books*; and 1851-60 from the *JHA*. The "Abstract," which is a summary of the reports in the CO/194, has two sets of figures towards the final years leading up to 1833, when these reports come to an end—"seals taken" and "seal skins exported"—and from 1821 onwards the reports end on 30 June each year instead of 10 October, as was the case from the beginning. There are discrepancies between the figures for "taken" and "exported" during the 1820s and up to 1833, probably because of the change in the statistical year end from 10 October to 30 June. After all, the seals would be taken well before 30 June, but very few skins would be exported until after that date. Thus, it would seem likely that the number of seal skins exported would refer, in most cases, to the previous years's catch. However, because they are more complete, the present writer uses "seals taken," and no problem arises, except in the case of the year ending on 30 June 1829. In that year, "seals taken" numbered 199,179 (a low number) while "seal skins exported" numbered 536,757. It is difficult to reconcile the two sets of figures because it is not known when the skins of any particular catch were actually exported. By the latter part of this period, the seals under discussion are "seal skins exported."

Table 2.6
Newfoundland Oil Exports (Amount and Value): 1803-60

Year	Seal Oil (Tuns)	Seal Oil Price (Value £)	Seal Oil (Tun £)	Cod Oil (Tuns)	Cod Oil Price (Value £)	Cod Oil (Tun £)
1803	-	20,942	-	2,029	-	23-27
1804	-	38,914		2,529	-	21-23
1805	1,172	28,269	24	3,703	-	21-22
1806	-	-	-	2,040	-	-
1807	3,293	60,364	18	3,205	-	16-18
1808	1,118	24,596	22	2,044	36,792	18
1809	1,209	26,576	22	2,600	-	-
1810	493	14,391	29	1,307	32,675	25
1811	1,581	44,925	28	3,375	74,250	22
1812	1,456	38,380	26	3,453	-	21-26
1813	1,584	50,161	32	4,054	-	26-32
1814	1,263	36,296	29	4,127	-	20-26
1815	1,397	53,609	38	4,299	-	22-30
1816	1,760	43,780	25	3,556	-	16-18
1817	431	14,430	33	3,372	-	17-35
1818	1,809	63,392	35	2,658	-	22-35
1819	3,253	95,831	23	3,807	-	24-30
1820	2,219	53,480	24	4,488	-	18-24
1821	3,005	64,743	22	4,276	-	18-24
1822	4,590	87,328	19	3,632	-	14-18
1823	2,975	55,949	19	4,012	-	14-18
1824	2,053	39,862	19	3,902	-	14-18
1825	3,611	71,645	20	-	-	14-22
1826	3,734	80,950	22	4,365	-	16-25
1827	4,726	108,872	23	3,804	-	16-20

Year							
1828	3,297	66,386	20	3,591	-	16-20	
1829	3,131	-	20-24	-	-	18-20	
1830	7,110	-	21-24	-	-	18-20	
1831	8,761	-	22-25	-	-	18-24	
1832	6,072	-	19-25	-	-	18-29	
1833	8,639	-	20-27	-	-	18-25	
1834	9,128 total (oil)	-				-	
1835	11,786 "	-				-	
1836	9,154 "	£241,502 total	average price			26	
1837	8,252 "	-				-	
1838	8,626 "	249,428 "	"			29	
1839	8,906 total	245,269 "	"			28	
1840	12,725 "	305,197 "	"			24	
1841	10,609 "	266,832 "	"			25	
1842	-	-				-	
1843	14,989 "	335,975 "		"		22	
1844	10,280 "	309,263 "		"		30	
1845	6,887	128,050	19	3,528	108,990	31	
1846	3,065	86,855	28	3,558	91,697	26	
1847	8,493 total	226,272 total		average		27	
1848	6,508	160,909	25	3,737	87,622	23	
1849	4,190	109,890	26	4,301	96,442	22	
1850	6,235	193,289	31	3,108	93,240	30	
1851	6,965	208,524	30	3,415	103,264	30	
1852	-	-	-	-	-	-	
1853	8,138	273,751	34	3,506	114,435	33	
1854	5,567	218,390	39	3,200	118,288	37	
1855	-	-	-	-	-	-	
1856	5,009	216,006	43	3,990	162,313	41	
1857	7,166	265,131	37	4,822	159,130	33	
1858	6,478	200,808	31	4,968	134,139	27	
1859	5,566	166,970	30	4,302	111,839	26	
1860	4,865	145,959	30	4,575	118,949	26	

Source: Information for the period up to 1833 has been taken from Ryan, "Abstract"; for 1834-54 see the *Newfoundland Blue Books*; and for the later period, see the *JHA*. Please note the following: All pounds are in sterling; in some cases only the total number of tuns of oil (seal and cod) is known; during the period from 1803 to 1833, the prices of seal and cod oil are given as a range from minimum to maximum; in the case of seal oil, both the total value and the total quantity were recorded for the period from 1803 through 1828 so that, irrespective of the range, one can calculate an average price, and that has been done here; in the case of cod oil, the total value was not recorded in the early period, only the price range and the total quantity, and therefore, it was not possible to calculate the total value with any degree of accuracy; there are two sets of figures for 1823 and 1825 - the C.O. 194 set is used in the table, while the figures from the *Blue Books* are listed below in this note; the quantity of seal oil produced in 1805 was recorded as 1,172 and 1,493 tuns in the C.O. 194 reports - the former figure is used here; in 1810 the figures for seal oil and cod oil do not include that produced in Bonavista, Conception Bay, Fortune Bay, Burin and Bay of Bulls - this explains why the figures are so low, given the importance of Conception Bay; in 1824, the figures for Fogo and Trepassey were not recorded, although this affected the total very little; in 1826, figures for Trepassey and St. Mary's were not recorded - but again, these omissions would not have affected the results to any great extent.

| 1823 | 6,400 total | £120,334 total | average price | | £19 |
| 1825 | 9,343 total | 214,889 total | " | 23 | |

Table 2.7
Newfoundland Seal Exports: 1861-1914

	Oil		Skins		
Year	Tuns	Value £	Number	Value £	Total Value £
1861	4,865	145,959	375,282	56,292	230,045
1862	3,407	136,263	268,624	40,294	176,557
1863	4,146	186,568	287,151	43,073	229,641
1864	1,605	76,247	125,950	18,893	95,140
		$		$	$
1865	3,268	751,574	242,471	181,852	933,426
1866	4,425	708,000	269,029	201,771	909,771
1867	4,923	787,627	399,041	319,233	1,106,860
1868	4,829	772,640	335,858	268,686	1,041,326
1869	5,294	899,937	334,958	234,958	1,234,895
1870	4,983	847,067	265,189	265,189	1,112,256
1871	6,943	972,020	486,262	486,262	1,458,282
1872	4,179	585,060	231,244	231,244	816,304
1873	6,167	801,710	449,727	472,213	1,273,923
1874	6,374	612,290	392,228	509,887	1,122,177
1875	4,837	634,160	370,679	481,882	1,116,042
1876	4,684	636,990	341,292	443,679	1,080,669
1877	5,954	762,112	431,373	323,529	1,085,641
1878	5,905	708,600	419,220	293,454	1,002,054
1879	6,233	598,368	457,855	320,498	918,866
1880	3,967	614,885	261,508	209,206	824,091
1881	6,325	759,030	408,479	367,631	1,126,661
1882	3,301	409,324	178,812	178,812	588,136

1883	5,341	662,253	322,603	322,603	984,856
1884	3,969	460,404	266,290	319,548	779,952
1885	3,441	344,100	238,596	214,736	558,836
1886	3,571	257,112	272,656	272,656	529,768
1887	3,360	228,497	230,355	230,355	458,852
1888	3,594	287,520	286,464	286,464	573,984
1889	4,444	373,317	335,627	302,064	675,381
1890	3,719	334,710	220,863	220,863	556,968
1891	4,465	414,584	364,854	364,854	780,807
1892	5,301	397,575	390,174	468,209	865,784
1893	2,914	203,980	175,217	116,456	221,944
1894	4,403	274,924	284,060	227,248	503,852
1895	4,349	304,786	302,958	378,698	685,216
1896	3,282	228,734	297,651	372,063	602,529
1897	3,270	245,250	195,040	117,024	363,467
1898	?	218,279	190,262	129,840	348,119
1899	?	252,036	276,879	136,563	388,599
1900	?	433,605	203,858	162,330	595,935
1901	4,652	424,632	327,163	282,895	707,527
1902	3,945	379,445	528,150	420,869	800,314
1903	4,375	453,684	341,395	325,137	778,821
1904	2,748	303,067	243,639	258,987	562,054
1905	3,783	374,974	315,685	370,261	745,235
1906	3,741	297,430	283,400	314,048	611,478
1907	5,351	447,967	164,509	194,300	642,267
1908	3,367	308,997	115,890	140,137	449,134
1909	2,861	252,262	371,968	433,620	685,882
1910	5,232	459,814	372,504	460,220	920,034
1911	3,873	385,250	221,651	275,287	660,537
1912	2,278	296,519	311,254	380,699	677,218
1913	2,884	270,275	212,285	321,551	591,826
1914	4,178	409,060	254,167	350,794	759,854

Source: *JHA*, supplemented by information from the *Blue Books of Newfoundland*. The minor quantities of seal products from the Labrador coast were not regularly published prior to 1898 and are not included in this table prior to that year; beginning in 1898, the Labrador coast statistics were published regularly and are included in the above table. Up to and including 1864, the annual government accounts were recorded in sterling; beginning in 1865, accounts were recorded in dollars. Sterling was worth about $4.86 per £ and was frequently rounded off to $5.00 per £. Using the lower, more precise figure, one can calculate the dollar values for the export of all seal products for the brief period from 1861 to 1864 (inclusive) as follows: 1861 - $1,118,019; 1862 - $858,067; 1863 - $1,116,055; and 1864 - $462,380.

Table 2.8
Prices of Seal Fishery Exports: 1861-1914

Year	Oil per Tun (Stg.)	(Approx.$)	Skins (Stg.)	(Approx.$)
1861	£33	$160	40d	$0.80
1862	40	194	36	.72
1863	45	219	36	.72
1864	47 10s	231	36	.72

Year	Oil ($)	Skins	Year	Oil ($)	Skins
1865	230	.75	1890	90	1.00
1866	160	.75	1891	93	1.00
1867	160	.80	1892	75	1.20
1868	160	.80	1893	70	.66
1869	170	1.00	1894	68	.80
1870	170	1.00	1895	70	1.25
1871	140	1.00	1896	70	1.25
1872	140	1.00	1897	75	.60
1873	130	1.05	1898	?	.68
1874	140	1.30	1899	?	.49
1875	131	1.30	1900	?	.80
1876	136	1.30	1901	9	1.86
1877	128	.75	1902	96	.80
1878	120	.70	1903	104	.95
1879	96	.70	1904	110	1.06
1880	155	.80	1905	99	1.17
1881	120	.90	1906	80	1.11
1882	124	1.00	1907	84	1.18
1883	124	1.00	1908	92	1.21
1884	116	1.20	1909	88	1.17
1885	100	.90	1910	88	1.24
1886	72	1.00	1911	71	1.24
1887	68	1.00	1912	107	1.22
1888	80	1.00	1913	94	1.51
1889	84	.90	1914	98	1.38

Source: *JHA* and the *Blue Books of Newfoundland*. Where necessary, figures have been rounded off to the nearest whole number. This table, in conjunction with Table 2.6, can be compared with Table 1.8. Note the increase in seal oil prices during the period leading up to the mid 1850s and the decline after the mid 1860s. However, prices in Tables 2.6 and 2.8 are f.o.b., while prices in Table

1.8 are c.i.f. Unfortunately, for the purposes of this study, the decline in demand for seal oil (and other traditional oils) on the London Market by 1880 makes it impossible to extend Table 1.8 to 1914.

Table 2.9

Percentage of Newfoundland's Exports Consisting of Seal Products: 1850-1914 (Value)

Year	%	Year	%	Year	%	Year	%	Year	%
1850	27	1863	19	1876	16	1889	11	1902	8
1851	30	1864	9	1877	16	1890	9	1903	8
1852	?	1865	17	1878	18	1891	11	1904	5
1853	31	1866	16	1879	16	1892	15	1905	7
1854	?	1867	22	1880	15	1893	6	1906	5
1855	18	1868	24	1881	15	1894	9	1907	5
1856	21	1869	20	1882	8	1895	11	1908	4
1857	22	1870	18	1883	14	1896	9	1909	6
1858	22	1871	23	1884	12	1897	7	1910	8
1859	17	1872	14	1885	12	1898	7	1911	6
1860	16	1873	17	1886	11	1899	6	1912	5
1861	21	1874	15	1887	9	1900	7	1913	4
1862	15	1875	17	1888	9	1901	8	1914	5

Source: Calculated from statistics in the *JHA* and the *Blue Books of Newfoundland*.

Table 3.1

Newfoundland Seal Fishery: 1819-33

Year	St. John's		Conception Bay		Grand Total	
	Ships	Seals	Ships	Seals	Ships	Seals
1819	32	39,052	140	179,051	182	227,831
1820	40	24,132	170	152,502	227	187,214
1821	39	39,920	152	201,392	?	?
1822	?	105,504	164	209,158	?	?
1823	62	36,611	182	145,306	?	?
1824	54	53,038	168	70,931	?	?
1825	66	61,925	168	139,532	?	254,059
1826	70	75,024	?	180,759	?	270,543
1827	100	113,370	174	219,778	307	414,420
1828	95	58,141	177	117,409	295	256,392
1829	88	?	175	?	?	?
1830	95	?	179	?	?	?
1831	118	210,798	205	342,871	349	610,048
1832	154	178,341	221	233,845	420	469,649
1833	110	125,874	215	296,302	359	467,649

Source: CO 194/23-87. This table is drawn from the statistics in the annual governor's report each year. There are a number of slight discrepancies between the figures for Conception Bay vessels here and those in Table 3.3, some of which were taken from the local newspapers. Please note that the Grand Totals above also contain the number of vessels from outports north of Conception Bay and south of St. John's.

Table 3.2
St. John's Sealing Fleet: 1819-62

Year	Ships No.	Tonnage Total	Tonnage Average	Crew Members Total	Crew Members Average
1819	32	1,710	53	545	17
1820	40	1,951	48	666	17
1821	39	1,998	51	651	17
1822	?	?	?	?	?
1823	62	3,157	51	1,120	18
1824	54	2,874	53	996	18
1825	66	3,780	57	1,253	19
1826	70	?	?	1,322	19
1827	100	5,841	59	1,903	19
1828	95	6,002	63	1,950	21
1829	88	5,452	62	1,746	20
1830	95	6,659	70	2,578	22
1831	118	8,046	68	2,578	22
1832	153	11,462	75	3,294	22
1833	106	8,665	82	2,564	24
1834	125	11,029	89	2,910	23
1835	126	11,167	89	2,912	23
1836	126	11,425	91	2,965	24
1837	121	10,648	88	2,940	24
1838	110	9,500	86	2,826	26
1839	76	6,447	85	2,029	27
1840	75	6,190	83	2,058	27
1841	73	6,026	83	2,106	29
1842	74	5,142	69	2,018	27
1843	104	9,501	91	3,066	29
1844	121	11,088	92	3,775	31
1845	126	11,863	94	3,895	31
1846	141	13,165	93	4,470	32
1847	95	9,353	98	3,215	34
1848	103	10,046	98	3,541	34
1849	58	5,847	101	2,170	37
1850	72	6,842	95	2,616	36
1851	90	9,035	100	3,417	38
1852	97	10,236	106	3,861	40
1853	108	11,204	104	3,947	37
1854	89	9,144	103	3,396	38
1855	73	7,494	103	2,813	39
1856	72	7,988	111	2,855	40

1857	83	9,109	110	3,377	41
1858	80	9,958	124	3,831	48
1859	99	12,342	125	4,542	46
1860	80	10,418	130	3,989	50
1861	73	9,729	133	3,836	53
1862	48	6,173	129	2,513	52

Source: Most of the statistics for the period up to 1826 have been taken from CO 194/23-87. The statistics for the period beginning in 1827 have been taken from the local newspapers. The information on the fleet leaving port for the ice fields was usually reported in most newspapers in March as the ships cleared port. The *Public Ledger* and the *Newfoundlander* have been particularly useful for the purposes of this table. However, there are certain problems associated with drawing up this table. Not all St. John's ships cleared from this port for the ice fields; some cleared from more northerly ports. Therefore, it is possible that some ships have been missed. Similarly, not all ships clearing from more northerly ports belong to the port from which they cleared. Finally, it is apparent that some ships were counted twice; they were counted as coming from St. John's and from another port. (The same problems are associated with trying to construct a picture of the fleets from Harbour Grace, Carbonear, Bay Roberts and Brigus - plus the added problems in these Conception Bay ports, where there was more intermingling of the port of ownership, the port of supply and the port of departure. In addition, fewer statistics, especially during the earlier period, were published.) However, in the table above, in nearly all cases, it has been possible to identify St. John's vessels clearing from other ports because some newspapers identify them. Furthermore, particular care has been taken to ensure that the statistics are as accurate as possible, and while a small discrepancy will occasionally be found between the reports as they were published in the various newspapers, such discrepancies are insignificant. However, St. John's participation in the seal fishery in the 1850s and 1860s, was greater than this table indicates because whenever suppliers are listed for the various vessels, especially those in Trinity, Catalina and Greenspond, it is evident that St. John's firms were supplying many of these northern sealing vessels, and some were probably the actual property of St. John's owners. Furthermore, many northern vessels were bringing their seal catch to St. John's to dispose of it by the end of the period described in this table.

Table 3.3
Conception Bay Sealing Fleet: 1814-65

Year	Ships No.	Tonnage Total	Tonnage Average	Crew Members Total	Crew Members Average
1814	74	4,009	54	1,207	16
1815	68	4,091	60	1,127	17
1816	93	5,064	54	1,558	17
1817	?	?	?	?	?
1818	100	5,763	58	1,534	15
1819	140	7,528	54	2,203	16
1820	170	10,688	63	3,082	18
1821	152	8,778	58	2,382	16
1822	164	9,463	58	2,636	16
1823	182	10,485	58	2,917	16
1824	168	9,670	58	2,708	16
1825	168	9,788	58	2,720	16
1826	?	?	?	2,970	?
1827	174	?	?	3,226	19
1828	177	?	?	3,383	19
1829	175	11,204	64	3,342	19
1830	173	11,767	68	3,561	21
1831	205	?	?	4,350	21
1832	218	16,193	74	4,710	22
1833	212	16,436	78	4,680	22
1834	218	17,785	82	4,894	22
1835	204	17,349	85	4,558	22
1836	?	?	?	?	?
1837	206	17,790	86	4,937	24
1838	200	17,304	87	5,106	26
1839	?	?	?	?	?
1840	?	?	?	?	?
1841	?	?	?	?	?
1842	?	?	?	?	?
1843	?	?	?	?	?
1844	173	16,081	93	5,047	29
1845	?	?	?	?	?
1846	186	18,856	101	5,733	31
1847	156	15,846	102	5,042	32
1848	?	?	?	?	?
1849	158	?	?	?	?
1850	160	15,105	94	5,835	36
1851	161	14,974	93	5,794	36

1852	170	16,363	96	6,232	37
1853	184	19,175	104	6,744	37
1854	193	20,210	105	7,392	38
1855	193	20,627	107	7,744	40
1856	190	20,704	109	7,460	39
1857	185	20,794	112	7,743	42
1858	?	?	?	?	?
1859	?	?	?	?	?
1860	154	18,713	122	7,394	48
1861	152	18,644	123	7,416	49
1862	139	17,373	125	7,198	52
1863	113	14,073	125	5,836	52
1864	127	15,738	124	6,688	53
1865	117	14,907	127	6,015	51

Source: See note at end of Table 3.2. It is more difficult to gather statistics on the Conception Bay sealing fleet because fewer were published, and when they were, it was often in the form of scattered lists which one must combine to form an overall picture. In addition, some reports indicate the individual ports to which vessels belong, including such minor ones without customs houses as Harbour Main, Mosquito, Port de Grave, Spaniard's Bay and Cupids. Ships from ports without customs houses cleared from the nearby larger ports.

Table 3.4
St. John's and Conception Bay Sealing Fleets: 1851-1862

Port	1851	52	53	54	55	56	57	58	59	60	61	62
St. John's	92	97	101	88	73	72	83	80	99	80	73	48
Hr. Grace	63	62	73	71	62	57	48	?	?	50	57	60
Carbonear	33	32	32	39	37	41	41	27	31	35	32	29
Brigus	26	30	31	83	89	92	88	76	45	42	37	32
Bay Roberts	12	16	22	5	?	?	?	?	27	27	26	18
Port de Grave	11	14	12									
Cupids	7	7	11									
Hr. Main	1	1	3	?	?	?	7	8	7			
Spaniard's Bay	56											
Mosquito	2											
Total	252	265	285	286	261	262	260	*	*	234	225	187

Port	Total Tonnage (Selected Years and Major Ports)				
	1851	1852	1856	1861	1862
St. John's	9,200	10,236	7,998	9,729	6,173
Hr. Grace	5,834	5,918	6,388	7,273	7,633
Carbonear	3,293	3,341	4,724	4,013	3,728
Brigus	2,531	2,839	9,594	4,189	3,706

Port	Total Men Employed (Selected Years and Major Ports)				
	1851	1852	1856	1861	1862
St. John's	3,480	3,861	2,864	3,836	2,513
Hr. Grace	2,353	2,321	2,345	2,867	2,919
Carbonear	1,186	1,188	1,622	1,513	1,555
Brigus	953	1,107	3,493	1,772	1,704

Source: A variety of newspapers during the months of March and April of each year. See especially the *Newfoundlander, Royal Gazette, Patriot* and *Weekly Herald*. A report in the *Evening Telegram*, 30 March 1906, written by a local chronicler and poet, James Murphy, presents a complete picture of the seal fishery in 1853. In that year, according to his report, there were 392 vessels, employing 14,000 men, engaged in the Newfoundland seal fishery. *Because the figures for Harbour Grace are missing totals would be rather meaningless.

Table 3.5
St. John's and Conception Bay Sealing Fleets: 1853

Vessels	Captains	Tons	Men	Supplier
		St. John's		
Caledonia	Houlahan	114	45	W. & H. Thomas
Coquette	Houlahan	154	50	"
Margaret	Cummins	164	49	Job Brothers & Co.
Diana	White	130	54	"
Sarah	Weir	124	50	"
Elizabeth	Ashman	127	45	"
Arthur Leary	Halern	161	51	Stabb, Row & Co.
Scottish Chief	Delaney	96	40	"
Charles	Gerard	101	36	"
Mary	King	72	29	"
Clipper	Geran	132	40	"
Pursuit	Knight	135	43	"
Snipe	Murphy	95	33	Newman & Co.
Prosperity	Brien	84	32	Brocklebank & Anthony
Harmony	Brien	85	32	"
Active	Brien	75	31	"
Hecla	Bulger	121	45	"
Creole	Doyle	140	45	"
Empress	Kennedy	136	48	R. Alsop & Co.
Friends	Duff	90	34	"
Dolphin	Goss	79	33	"
Seamew	Carew	124	45	"
Margaret	Mealey	103	33	"
Brothers	Ryan	70	33	"
E.M. Dodd	Nurse	77	30	
Anna	Hally	163	48	McBride & Kerr
Scottish Lass	Fitzgerald	143	43	"
Avalon	Aspell	45	18	"
Elizabeth	Neill	81	30	"
Gannet	Cahill	52	19	"
Alice	Coady	65	27	"
John & Maria	Colbert	76	30	R. O'Dwyer
Glenara	Graham	126	48	P. Rogerson & Son
Gem	Mills	123	45	"
Eliza	Scott	97	37	Baine, Johnston & Co.
Hero	Pike	106	39	"
Trial	Kennedy	83	33	"
Triton	Keefe	94	37	"

Rover	Aylward	88	35	"
Native Lass	Silvey	116	43	"
Swift	Woodford	109	40	"
Speed	Pike	105	40	"
Emerald Isle	Reddy	149	45	"
Maggie	Gosse	93	38	"
Spartan	Cullen	57	26	"
Kingaloch	Burke	143	50	L. O'Brien & Co.
Normal	Lynch	114	40	"
Shamrock	French	143	40	"
Iris	Power	117	40	"
Nancy	Moore	74	30	"
Emily	Foran	102	40	P. Duggan
Skipwith	Walsh	129	35	Clift, Wood & Co.
Albatross	Hearn	108	35	"
Morning Star	Woods	40	20	"
Three Sisters	Rhodes	165	50	Ewen Stabb
Favourite Lass	Mountain	117	40	"
Georgina	Butler	107	35	"
Sisters	Chafe	79	35	"
Margaret	Young	60	25	"
Grace Darling	French	99	34	C. F. Bennett & Co.
Superior	Furneaux	122	38	"
Lena	Feehan	147	50	Bowring Brothers
Regina	Foster	138	45	"
Fanny Bloomer	Silvey	142	50	"
Roxana	Noseworthy	134	48	"
Swallow	Glinden	118	45	"
Defiance	Phean	74	30	"
Willaim	Withycomb	116	43	Mudge & Co.
Cora	Jackman	126	45	"
Mary	Staunton	150	50	J. H. Cozens
Escape	Cummins	116	40	Hunters & Co.
Alice Howard	Davidson	132	50	"
Primrose	Mullowney	90	32	"
Dartford	Jackman	70	30	D. Steele
Queen Victoria	Skinner	35	19	"
Alpha	Mullowney	131	45	"
Piscator	Lacey	129	47	W. Boden & Co.
Telegraph	Sheehan	135	48	"
Lioness	Jackman	93	40	Goodridge & Kelligrew
Ann	Hennessy	116	44	"

Kirtland	Lynch	160	45	J. & W. Stewart
Chedabucto	Cole	83	33	"
Resolution	Hally	122	40	"
Dash	Meaney	160	55	John Barron
Nisibis	Callahan	151	55	"
James	Barrington	40	20	"
Clara	McKay	174	50	McKay & McKenzie
Tweed	Callahan	94	35	"
Cadmus	Shea	101	43	P. & L. Tessier
Star	Ryan	118	42	J. Cusack & Sons
Iris	Knight	143	50	J. B. Barnes & Co.
Flirt	Davis	154	50	"
Mary Hounsell	Ryan	184	50	Hounsell & Co.
Orestes	French	154	45	"
Wyoming	McLaughlan	127	38	"
Juno	Priop	100	36	"
Ann	Vye	96	35	"
George	Stack	80	32	"
Billow	Comerford	64	26	"
Juno	Stafford	112	30	L. Macassey
Ann	Allen	106	40	Brooking & Son
Total 101 vessels		11,194	3967	

	Harbour Grace			
Vessels	Captains	Tons	Men	Suppliers
Providence	Taylor	161	48	Punton & Munn
John Munn	Munn	151	49	"
Thrasher	Hennebury	141	48	"
Maria	Keefe	182	45	"
Favourite	Butt	117	42	"
True Friend	Antle	112	42	"
Emily	Power	112	38	"
Rose	Avery	112	39	"
Superb	Taylor	111	38	"
James	Lynch	109	36	"
Elizabeth Margaret	Power	109	36	"
Jane & Mary	Parsons	101	34	"
Maria	Pike	100	30	"
Union	Parsons	100	34	"
Shannon	Avery	102	40	"
Rasselas	Keefe	99	34	"
Haidee	Kelly	93	36	"

J. Walmesley	Pike	93	30	"	
Elizabeth	Green	84	35	"	
Jane	Parsons	76	30	"	
William	Bransfield	91	36	"	
Hornet	Parsons	80	32	"	
Glide	Pike	75	36	"	
J. Nichole	Fitzgerald	62	28	"	
Nightingale	Callahan	61	28	"	
John Martin	Taylor	75	32	"	
William	Green	114	36	"	
Greyhound	Pike	152	50	Ridley & Sons	
Acastus	Smart	155	50	"	
Melrose	Pike	153	46	"	
Adamant	Murphy	129	45	"	
Linda	Baggs	121	48	"	
Brothers	Noel	116	42	"	
Cabot	Alcock	113	36	"	
Terra Nova	Pike	110	40	"	
Myrtle	Pike	109	42	"	
Alabama	Glavene	108	36	"	
Rosina	Brown	108	36	"	
Dart	Smart	100	34	"	
Suir	Ryan	100	40	"	
Princess	Heater	99	30	"	
Intrepid	Finn	95	32	"	
Wave	Johnson	94	38	"	
Imaun	Cleary	84	32	"	
William	Murphy	85	33	"	
Harp	Dean	59	28	"	
Isabella	Dean	74	30	"	
Margaret					
Trial	Noonan	72	30	"	
Osprey	Johnson	62	28	"	
Mary &	Hancock	59	30	"	
Elizabeth					
Meg Merrilies	Pike	53	25	"	
Malvina	Donnelly	112	45	William Donnelly	
Jemima	Stapleton	107	40	"	
Caroline	Corben	128	50	"	
Clarinda	Thompson	88	35	"	
Elizabeth	Phelan	82	32	"	
Scotia	Sheen	152	50	Patrick Devereaux	
Paragon	John Canty	124	45	"	
Gold	McCarthy	104	39	"	

Swordfish	Green	154	45	D. Green
Edward	Fitzgerald	110	37	"
Clio	Gordon	131	40	S. Gordon
Dove	Culliall	98	32	"
Funchall	Gordon	142	46	W. Gordon
Clipper	Gordon	108	42	J. Gordon
Orange	Thomey	118	40	A. Thomey
Scotch Lass	Thomey	123	45	"
Brothers	W. Davis	112	34	William Parsons
Eliza	Strapp	115	32	Patrick Strapp
Gazelle	Brien	75	33	Thomas Power
Romp	W. Gosse	97	30	Isreal Gosse
Leamon	Chandler	110	42	Robert Walsh
Elizabeth & William	Stevenson	134	48	John Stevenson
Total	73 vessels	7,757	2755	

		Carbonear		
Vessels	Captains	Tons	Men	Supplier
Princess Royal	W. Taylor	97	30	Pack, Gosse & Fryer
Trial	Humphry	135	48	"
Sir J. G. LeMarchant	T. Fitzgerald	131	48	"
Ann	T. Geary	94	27	"
Broadalbin	G. Joyce	113	38	"
Herald	E. Nichole	88	27	"
Morning Star	J. Vatcher	94	30	"
Orient	W. Bemister	133	37	W. Bemister & co.
Corfe Mullen	W. Ash	80	23	"
Victoria	W. Taylor	105	36	"
Briton	H. T. Forward	123	40	George Forward
Wm. IV	A. Forward	113	40	"
Samuel	S. Taylor	114	40	"
Thomas Ridley	Hanrahan	164	48	John Rorke
Echo	W. Pike	85	23	"
Caledonia	R. Bransfield	100	32	"
Adelaide	R. Davis	87	24	"
Bridgewater	C. Forward	106	38	"
Mary	T. Thistle	89	24	"
Margaret Jane	J. Forward	167	53	James Forward
Mary	J. Stapleton	122	36	Moses Wilshire
Sir John Harvey	Penny	125	42	"

The Ice Hunters

Walrus	M. Dwyer	131	44	Edward Dwyer
Sir H. Douglas	R. Stapleton	126	26	"
Lady of the Lake	Taylor	119	36	W. H. Taylor
Rosalie	T. C. Taylor	82	28	"
Belle	Thos. Oates Sr.	149	45	Thomas Oates Sr.
John Gibson	Thos. Oates Jr.	123	41	"
True Blue	J. Knox	100	30	Felix M'Carthy
John and Rachel	H. Taylor	92	36	John M'Niel
J.D.	W. Brown	126	45	W. Brown
Virgin Lass	J. Talbot	92	30	John Talbot
Total	32 vessels	3,605	1145	

Brigus

Vessels	Captains	Tons	Men	Supplier
John	Whelan	124	48	
Hound	Whelan	96	43	
Gleaner	Wilcox	104	42	Baine, Johnston & Co.
Lively Lass	Wilcox	129	45	
Rolla	Burk	83	37	
Sarah	Bartlett	91	39	W. & H. Thomas & Co.
Mary	Bartlett	109	40	
Theresa	Walker	60	26	
Alert	Munden	143	48	
Three Sisters	Pomery	122	43	
Atlas	Munden	115	43	
Stranger	Rabbits	77	32	
Jane Elizabeth	Norman	139	47	
Rose	Norman	103	40	
Mary	Wilcox	74	29	Job Brothers & Co.
Lady Norman	Norman	103	40	
Nymph	Norman	88	37	
Indian Chief	Norman	117	39	Job Brothers & Co.
Elizabeth	Rose	50	18	
Francis	Walsh	119	45	
Delmont	Clark	122	48	
Union	Clark	83	33	
Louisa Jane	Clark	70	32	Job Brothers & Co.
Terra Nova	Percey	106	40	
Jane	Curtis	93	37	
Hunter	Rabbits	118	43	Hounsell, Schenk&Hounsel

Victor	Spracklin	133	48	
Example	Wilcox	109	39	
Peal	Bartlett	127	38	
Witch	Bartlett	100	36	
Kingfisher	Spracklin	82	30	
Total	31 vessels	3,189	1205	

		Bay Roberts		
Vessels	Captains	Tons	Men	Supplier
Rosicus	Davis	120	45	James Cormack
Eclipse	Delaney	116	40	"
Harriet	Delaney	101	35	"
Billow	Delaney	114	42	"
Orator	Delaney	128	44	"
Jane Dalrymple	Mercer	106	40	"
Pactolus	Mercer	121	40	Pack, Gosse & Fryer
Haba	Rumson	114	40	"
Antigonish	Russell	95	32	"
Jane White	Russell	?	?	"
Symmetry	French	125	41	Sundry Suppliers
Falcon	Parsons	61	23	"
Eunice	Mercer	88	27	"
Brothers	Bartlett	107	40	"
Hunter	Bartlett	113	42	"
James Clift	Delaney	90	36	"
Hunter	Delaney	136	45	"
Rosebud	Daw	99	36	"
Glenfalloch	Daw	87	35	"
James White	Russell	118	40	"
Bloomer	Delaney	113	40	"
Charlotte	Russell	78	29	"
Total	22 vessels	2.230	792	(2 figures missing)

		Port de Grave		
Vessels	Captains	Tons	Men	Supplier
Rake	Batton	116	34	Baine, Johnston & Co.
Sea	Batton	104	34	
United Brothers	Halley	75	27	
Laurel	Daw	88	35	
Hound	Daw	92	36	McBride & Kerr
Pearl	Daw	85	32	

The Ice Hunters

Isabella	Andrews	106	34	
St. Patrick	Taylor	79	33	
Charles	Richards	84	27	
Volant	Richards	115	33	
Ann	Ledroe	71	29	
Sealer	Batten	107	32	Baine, Johnston & Co.
Total	22 vessels	1,122	386	

		Cupids		
Vessels	Captains	Tons	Men	Supplier
Sarah	W. Wells	110	40	
Isabella	Spracklin	81	32	
Sarah Grace	Spracklin	116	39	
Flora	Wills	172	46	
Joanna	Brine	45	18	
Billow	LeDrew	90	35	
Express	Whelan	103	35	
Isle of Skye	Hedderson	59	20	
Mary	Wills	76	31	
Ripple	Smith	108	38	
Racer	Smith	85	27	
Total	11 vessels	1,045	361	

		Harbour Main		
Vessels	Captains	Tons	Men	Supplier
Argo	Woodford	82	32	
New Packet	Woodford	73	25	Baine, Johnston & Co.
Nimrod	Fury	72	33	"
Total	3 vessels	227	90	

Source: The *Newfoundlander* and the *Public Ledger*, March 1853.

Table 3.6
Harbour Grace Sealing Fleet: 1867-1900

Year	Vessels No.	Tonnage Total	Tonnage Average	Men Total	Men Average
1867	50	5,923	119	2,504	50
1868	58	5,743	99	2,407	42
1869	52	5,743	110	2,574	50
1870	53	5,966	113	2,825	53
1871	52	6,292	121	2,930	56
1872	43	5,456	127	2,762	64
1873	35	4,385	125	2,401	69
1874	20	2,768	138	?	?
1875	21	2,869	137	?	?
1876	19	2,740	144	1,633	86
1877	18	2,640	147	1,418	79
1878	17	2,814	166	1,655	97
1879	12	2,085	174	1,374	115
1880	17	2,458	145	1,515	89
1881	11	1,884	171	1,185	108
1882	13	2,308	178	1,520	117
1883	10	1,884	188	1,305	131
1884	7	1,166	167	?	?
1885	5	945	189	585	117
1886	3	752	251	575	192
1887	3	826	275	600	200
1888	3	826	275	590	197
1889	3	826	275	630	210
1890	3	827	276	600	200
1891	3	826	275	620	207
1892	3	826	275	630	210
1893	2	581	291	516	258
1894	4	920	230	720	180
1895	4	1,113	278	885	221
1896	0	0	0	0	0
1897	0	0	0	0	0
1898	1	51	51	17	17
1899	0	0	0	0	0
1900	0	0	0	0	0

Source: "Book of Coasting and Fishing Ships Clearing from Harbour Grace: 1866-1918." Original in the Conception Bay Museum, Harbour Grace. This table includes steamers and sailers.

Table 3.7
Production of Seal Pelts: 1863-95

Year	St. John's	Hr. Grace	Steamers	Sail & Land	Total	%
1863	4,340	-	4,340	282,151	287,151	2
1864	1,059	-	1,059	124,891	125,950	1
1865	19,086	-	19,086	223,385	242,471	8
1866	28,058	23,400	51,458	217,571	269,029	19
1867	28,050	28,500	56,550	342,491	399,041	14
1868	57,800	31,200	89,000	246,858	335,858	26
1869	76,620	18,650	95,270	239,688	334,958	28
1870	102,310	42,700	145,010	120,179	265,189	55
1871	178,769	74,938	253,707	232,555	486,262	52
1872	76,261	42,435	118,696	112,548	231,244	51
1873	262,259	39,083	301,342	148,385	449,727	67
1874	175,792	24,908	200,700	191,528	392,228	51
1875	252,800	32,110	284,310	86,369	370,679	77
1876	193,990	30,531	224,521	116,771	341,292	66
1877	258,146	33,100	291,246	140,127	431,373	68
1878	249,716	24,136	273,852	145,368	419,220	65
1879	305,929	47,300	353,229	104,626	457,855	77
1880	124,968	8,712	133,680	127,828	261,508	51
1881	281,949	43,830	325,779	82,700	408,479	80
1882	137,864	10,650	148,514	30,298	178,812	83
1883	247,230	44,538	291,768	30,835	322,603	90
1884	178,198	20,886	199,084	67,206	266,290	75
1885	171,681	41,363	213,044	25,552	238,596	89
1886	169,890	30,375	200,265	72,391	272,656	73
1887	146,204	30,898	177,102	53,253	230,355	77
1888	195,191	18,078	213,269	73,195	286,464	74
1889	252,887	47,799	300,686	34,941	335,627	90
1890	165,052	42,034	207,086	13,777	220,863	94
1891	280,928	63,145	344,073	20,781	364,854	94
1892	303,196	45,628	348,824	41,350	390,174	89
1893	109,304	19,035	128,339	46,878	175,217	73
1894	132,792	20,029	152,821	131,239	284,060	54
1895	234,993	35,065	270,058	35,065	302,958	89

Source: Ryan, *Chafe*. Chafe recorded the annual catch of the steamer sealing fleets from St. John's and Harbour Grace. The 'Total' of 'Seal Pelts' above is actually the number of seal skins exported, as recorded in the Custom Reports. The numbers of seals killed by sailing ships and landsmen have been calculated from the other data. The 'percentage' in the last column refers to the percentage of the catch produced by the steamers. There was a large landsmen seal fishery in 1894, with 100,000 seals taken by the "People of Twillingate and Neighbourhood" alone; see *Evening Telegram*, 2 April 1894. This helps to explain the lopsided figures for that year.

Table 3.8

Newfoundland Sealing Steamers: 1863-1914

Wood and Steel — Class and Tonnage

Wood

Year	Steamer	Class	Tons
1863	Bloodhound I	3	153
1863	Wolf I	3	210
1864	Osprey	3	176
1866	Hawk	3	172
1866	Retriever	3	237
1867	Esquimaux	1	465
1867	Lion	2	293
1867	Mastiff	3	245
1867	Nimrod	3	226
1867	Panther	3	246
1869	Ariel	3	78
1869	Merlin	3	248
1870	Montecello	1	525
1870	Walrus	3	183
1871	Commodore	2	290
1871	Eagle I	2	343
1871	Hector	2	290
1872	Greenland	2	259
1872	Iceland	2	287
1872	Ranger	2	353
1872	Wolf II	2	353
1873	Micmac	1	463
1873	Neptune	1	465
1873	Bloodhound II	2	376
1873	Vanguard	2	322
1973	Tigress	3	217
1874	Bear	1	468
1874	Proteus	1	467
1874	Leopard	3	217
1877	Arctic	1	522
1877	Aurora	2	386
1877	Falcon	2	311
1877	Kite	3	112

Year	Steamer	Class	Tons
1878	Narwhal	2	362
1878	Tiger	3	147
1880	Resolute	1	424
1880	Xanthus	3	172
1881	Thetis	1	491
1884	Polynia	2	358
1885	Terra Nova	1	450
1885	Jan Mayen	2	283
1892	Eclipse	2	295
1892	Hope	2	307
1893	Newfoundland	1	568
1893	Algerine	3	223
1894	Harlaw	2	267
1894	Windsor Lake	2	293
1901	Virginia Lake	1	440
1901	Southern Cross	2	277
1902	Erik	1	412
1903	Grand Lake	1	463
1903	Windward	3	246
1904	Eagle II	1	394
1904	Bloodhound III	2	314
1904	Viking	2	256
1912	Lloydsen	3	247

Steel

Year	Steamer	Class	Tons
1906	Adventure	1	829
1906	Havana	3	190
1909	Florizel	1	1980
1909	Bellaventure	2	466
1909	Beothic	2	471
1909	Bonaventure	2	446
1910	Harlaw	3	267
1912	Nascopie	1	1004
1912	Stephano	1	2143
1912	Seal	3	277
1913	Sagona	3	420

Source: Michael Condon, *The Fisheries and Other Resources of Newfoundland* (St. John's, 1925), pp. 105-6. The *Fogota* is not included in this table.

Table 4.1
St. John's Sealing Fleet: 1834

Vessel	Captain	Tons	Men
Caledonia	H. J. Furneaux	120	25
Rasselas	Wm. Ready	106	18
Avalon	Jas. Maily	84	18
Metis Packet	Patrick Shea	76	18
Joseph	Dennis Murphy	60	17
Native	Thos. Phoran	130	29
Margaret Ellen	Stn. March	98	27
Piscator	Thos. Pinn	129	22
Sally Ann	Geo. Butt	77	14
Haberdine	Jas. Meagher	65	17
Oderin	Thos. Lea	82	23
Morning Star	John Knight	63	15
Juno	Edward Pyke	94	32
Cousins	Thos. Brien	80	19
Mary	Jeremiah Ryan	44	13
Notre Dame	Jas. Power	77	21
Actual	Garrett Dalton	68	15
Loyalty	Jas. Cummins	61	18
Rose & Thistle	F. Neagle	81	14
Nimrod	John Brennan	121	27
John & Horatio	Robt. Power	96	22
Trial	Thos. Power	76	18
Brazilian Patriot	T. Murphy	75	18
Brothers	Richard Cudihy	57	16
Malvina	Jeremiah Calahan	124	29
Water Lilly	William Gregory	172	29
Jane & Mary	Jas. Mulcahy	86	25
Eagle	Edward Purcell	99	26
Clondolin	John Roche	117	26
Clydesdale	George Corbin	117	29
Babe	John Glody	78	24
Elizabeth	Mat. Crawford	69	22
Shaver	Thomas Allen	132	29
Kate	James Canfield	60	20
Perseverance	Thos. Burke	72	19
Thistle	Thomas Hennesy	70	19
Auld	Nicholas Motley	84	26
Dido	Henry Charles	160	33
Royal Nigger	John Lamzed	75	21
Mary Jane	Nicholas Farrell	98	27

Aurora	Thomas Meally	157	30
St. Patrick	Garrett Dooley	96	24
Goose	Dennis Meally	106	27
Royal William	Thos. Casey	125	28
Belle Isle	George Pynn	95	19
Brittania	Maurice Cummings	145	28
Isabella	John Fitzgerald	118	29
Creole	Lawrence Geran	79	25
Dove	Joseph Houlahan	101	25
Mary	Patrick Murphy	74	25
Margaret Helen	B. Hagarthy	113	27
Ann	Thos. Barrington	75	22
Victory	Thomas Ashman	69	18
Belinda	James Pitts	110	26
Sir C. Hamilton	Peter Blake	118	28
Hope	John Bulger	76	18
Abeona	John Walsh	105	28
Scipio	Stephen Ryan	117	28
Despatch	Thomas McGrath	77	24
Duck	Timothy Shipton	108	26
Harriet Elizabeth	P. Brennon	114	24
Hero	William Dwyer	90	23
Perservance	Nicholas Power	70	18
Ocean	Garrett Hartery	121	24
Reindeer	William Shea	96	21
Edward	Richard Stephens	73	23
Gull	Patrick McKee	107	27
Sarah	Peter Kent	63	17
Trial	William Maxey	43	15
Charlotte	Pilley	48	17
Revenge	William King	55	16
Hope	John Parker	61	19
Industry	John Kenna	42	16
Speculation	William Burn	84	20
Daniel O'Connell	John Shea	75	18
Tweed	B. Maddocks	52	16
Shamrock	Cornelius Phoran	65	19
Eliza	James Hearne	97	27
Billow	Henry C. Hawson	103	26
Theresa	Neil McIsaac	57	19
James	John Breen	84	22
Norval	Stephen French	110	28
Nine Sons	Maurice Bolan	140	27
Only Son	Patrick Walsh	63	20

Ranger	William Walsh	128	30
Maria Ann	John Wood	53	17
Dolphin	John Cahill	79	23
Nightingale	Walter Walsh	91	23
Sophia	William French	67	19
James	Thomas Butler	144	30
Henry & Mary Ann	J. Chafe	99	27
Brothers	Edward Chafe	61	23
Annbella	Isaac Martin	66	19
Mary	James Bryan	93	23
Fury	Wm. Coady	72	17
Feronia	P. Kavanagh	83	20
Lady of the Lake	J. Casey	154	32
Sarah Mortimer	Thos. Butt	99	25
Amity	Edward Ryan	144	32
Hope	John Burke	126	26
Betsy	T. Knight	97	24
Mary Ann	Edward Riddy	74	17
Diana	George Pippy	72	17
Avon	James Silvey	93	23
Reliance	J. Ryan	76	22
Juliet	James Sullivan	48	14
Nancey	Thos. Ryan	56	14
George	Edward Pyer	73	18
Success	Jas. Axtell	123	24
Cambrian	Thos. MacKay	79	22
Catherine	Henry Tucker	104	24
Elizabeth	Charles Duttorel	61	18
Priviledge	Nicholas Power	66	18
Emulator	Jas. Carroll	96	24
Intrepid	Wm. Butt	111	28
Dingwell	George Carew	159	25
Phoebe & Jane	Wm. Kent	81	22
Active	Thomas Tracy	72	22
Orion	Edward Neal	63	15
Maria Louisa	James Penny	69	20
Huskisson	Isaac Warner	116	28
Peggy	David Barry	48	14
Total	122 vessels	10,992	2,847

Source: *Royal Gazette*, 18 March 1834. Average tons = 90; Average Crew = 23.

Table 4.2
Conception Bay Sealing Fleet: 1835

Vessel	Captain	Tons	Men
Eunice	Pike	83	22
Faith	Hopkins	106	22
Duncan & Margaret	Pike	122	28
Nancy	Kelly	75	20
Lady Ann	Pike	108	26
Friend	Simmons	73	19
Elizabeth	Johnston	92	25
Earl Grey	Donnelly	113	29
Dispatch	Phelan	100	26
Elizabeth	Delaney	71	20
St. Patrick	Hunt	131	29
William The 4th	Taylor	122	30
Joseph	Taylor	80	23
Fortitude	Pike	87	24
Curlew	Hanrahan	105	26
Trefoil	Bennett	78	18
Dewsbury	Nichol	107	26
Corfe Mullen	Finn	91	24
Benjamin	Howell	95	24
Fanny	Taylor	98	24
Rasselas	Keef	106	29
Lark	Pearce	98	27
Fox	Howell	74	21
Traveller	Brown	96	24
Britannia	Howell	93	26
Adelaide	Whelan	105	30
Sir Howard Douglas	Dwyre	134	22
Nancy	Pynn	94	25
Eagle	M'Carthy	67	20
Fanny	Glavine	88	21
Mary Frances	Dwyre	90	29
Emily	Coombs	98	30
Lavinia	Hudson	69	20
Philanthropy	Nicholl	92	26
Dart	Rearcy	109	28
Neptune	Hanrahan	60	17
13 Brothers & Sisters	Oats	96	25
Providence	Taylor	112	29
Reindeer	Guiney	96	24
Ferryland Packet	Blunden	54	17

Ann	Davis	94	30
Ann	Whelan	94	27
Waterloo	Ash	80	25
Ranger	Kennedy	65	20
Clinker	Nicholl	98	25
Elizabeth	Bemister	71	20
Sally	Forward	92	25
Hunter	Batt	68	20
Frederick	Kiely	62	21
Beginning	Taylor	51	15
Julia	Taylor	106	23
Hero	Barrett	83	27
Active	Scanlan	57	18
Greyhound	M'Carthy	104	24
Charlotte	Fillett	87	21
Alpha	Pearcey	105	27
Jane	Parsons	81	23
Orestes	Gosse	78	20
Alice	Bransfield	97	23
Mary	Bransfield	107	28
Codfish	Cole	63	19
Ann	Batt	122	31
Lavinia	Ridel	91	29
John	Crocker	64	16
William	Power	57	18
Sweet-Home	Moore	84	24
Sarah	Keef	93	28
Mary Ann & Martha	Taylor	94	28
Morning Star	Burdon	100	28
Minerva	Joyce	67	22
Tyro	Pike	63	22
Matilda	Parsons	54	14
Isabella & Margaret	Hearn	93	25
Cornelia	Parsons	90	27
Butler	Curtis	74	20
Willaim & Mary	Coony	71	20
Thomas & Hugh	Parsons	75	16
Edmund	Alcock	56	17
Rambler	Norman	63	16
William	Green	123	27
Shannon	Pike	124	28
Edward Piers	Batt	86	25
Jubilee	Simmons	86	20
Ambrose	Pelley	59	17

Eliza	Long	83	28
Trial	Pike	60	16
Susan	Moore	55	15
George	Oats	87	21
Maria	Howell	58	14
Caroline	Ash	86	21
Hope	Clark	54	14
Pandora	Horwood	75	21
Harriett	Pynn	46	16
Experiment	Sheppard	34	11
Wonderer	Davis	51	18
Phoenix	Barrett	63	11
Ianthe	Wills	126	30
Emily	Delaney	112	31
John & Maria	Burke	74	22
Jane Elizabeth	Munden	153	30
Mary	Wells	101	23
St. John's	Percy	149	24
Meg Merrilies	Newell	69	15
Nimrod	Cole	97	26
Nymph	Norman	88	24
Arabian	Percey	104	18
Abeona	Percy	94	24
5 Brothers	Antle	98	20
Jubilee	Percy	90	20
Terra Nova	Percy	119	31
Hebe	Rabbits	106	28
Dandy	Keating	70	19
Water Witch	Wilcocks	92	30
Dolphin	Whelan	56	13
True Blue	Whelan	109	24
Victory	Norman	49	15
John Alexander	Bartlett	96	23
John	Bartlett	85	20
Comet	Cole	126	30
Betsey	Le Drow	49	14
Alexander	Norman	90	24
Elizabeth & Maria	Le Drow	87	18
Margaret	Cahill	90	24
Isabella	Newell	57	17
Prosperity	Bryan	109	30

Surprise	Spraclin	61	15
Hunter	Edderson	38	12
Alligator	Wells	52	14
Sally	Keating	81	19
Oneas	Woodford	79	17
John	Saunders	53	17
Nancy	Cole	84	21
Tyro	Cole	73	18
Three Brothers	Burke	81	19
L'Avengeur	Gushue	82	20
Dove	Sheppard	70	18
Sarah	Sheppard	85	25
Isabella	Whelan	95	22
Venus	Le Drow	70	18
Nelson	Peyton	69	20
Indian Lass	Etchingham	44	12
Julia Ann	Kennedy	83	25
Indian Lass	Stabb	123	28
Agenoria	Hudson	91	26
Fair Cambrian	M'Carthy	98	26
Ethiopian	Parsons	87	26
Sylvanus	Davis	70	20
Herald	Gordon	104	25
Mary	Luther	85	25
Catherine & Margaret	Roach	74	22
Lord M'Donald	Webber	82	25
Wellington	Ryan	61	18
Jane & Mary	Parsons	88	22
Louisa & Frederick	Stevenson	132	30
John	Burt	70	17
Dolphin	Burke	79	18
Saint Anne	M'Carthy	93	21
Margaret	Lacy	105	29
Amelia	Pelley	64	17
Elizabeth	Ash	108	28
Relief	Davis	94	25
George Lewis	Snook	86	23
Maria	Heater	91	23
Ringwood	Goorney	111	32
Ranger	Mugford	88	23
Elizabeth Ann	Andrews	87	25
Henrietta	Kavanagh	75	15
Favourite	Richards	69	23
Swift	Batten	94	28

The Ice Hunters

Agenoria	Morgan	82	12
Lord Nelson	Spraclin	36	15
Sir T. Cochrane	Cowley	63	17
James	Hamilton	92	25
Revenge	Spraclin	71	14
Squirrel	Le Drow	63	14
Margaret & Ellen	Norman	98	22
Nonpareil	Williams	124	33
Active	Mercer	58	19
William	Snow	73	21
Samuel	Giles	110	32
Experiment	Davis	121	34
Montezuma	Russel	91	23
Nightingale	Russel	91	23
Margaret	Cave	104	24
Caroline	Mercer	68	22
Maria Louise	Penney	69	19
Joseph	Anthony	34	15
Hit or Miss	Roberts	93	20
Ann	Roberts	90	23
Naomi & Susannah	Munden	117	26
Highlander	Munden	125	25
4 Brothers	Munden	101	22
Bickley	Norman	94	24
Success	Shehan	56	16
Henry	Andrews	48	16
Calypso	Newell	54	16
Active	Kavanagh	62	20
Isaac & Elizabeth	Richards	105	30
Lady Ann	Richards	115	30
John & William	French	71	16
Success	Sheppard	55	14
Good Intent	Dawe	66	20
Rover	Herald	57	20
Total: 204 Vessels		17,349	4,558

Source: *Public Ledger*, 7 April 1835. This is reproduced as published; however, there are only 203 vessels in this list.
Total clearances for the seal fishery, from Conception Bay:
    1834   218 vessels   17,785 tons   4,894 men
    1835   204 vessels   17,349 tons   4,558 men

Table 4.3
Carbonear and Harbour Grace Sealing Fleets: 1836
Carbonear Sealing Fleet: 1836

Vessel	Captain	Tons	Men
Elizabeth	Wm. Roberts	108	29
Caledonia	Pat. Scanlan	113	27
Fanny	W. P. Taylor	98	25
Margaret	Daniel Lacey	105	29
Faith	Solomon Dean	106	27
Frederick	Stephen Blunden	92	26
Trial	Edward Pike	60	16
Sir Howard Douglas	Edward Dwyre	124	35
Curlew	E. Hanrahan	105	27
Alpha	Richard Parsons	105	29
Fortitude	George Pike	87	24
Dewsbury	Nicholas Nicholl	107	29
Julia	Richard Taylor	106	29
Dart	George Penny	109	28
St. Anne	Matthew George	93	26
Willaim IV.	Samuel Cleal	122	32
Joseph	William Clarke	80	21
Lavinia	Francis Taylor	91	26
Eliza & Ann	William Mahaney	67	19
Adelaide	William Udell	105	29
Lark	James Pearce	98	29
Benjamin	Francis Howell	95	27
Hero	Edward Barrett	83	20
Corfe Mullen	Thomas Finn	91	24
Traveller	Patrick Knox	96	24
Fox	James Howell	74	22
Neptune	J. Hanrahan	62	17
Minerva	George Joyce	67	19
James	C. Hamilton	92	25
Philanthropy	John Nicholl	92	27
Charlotte	James Jillett	87	23
13 Brothers & Sisters	Thomas Oats	96	25
Cod Fish	Henry Cole	63	18
Britannia	William Howell	93	27
Waterloo	Henry Ash	80	25
George Lewis	Nicholas Ash	86	24
George	James Keho	87	22
Ann	John Whelan	94	27
Ambrose	John Squires	66	18

Frederick	Patrick Meaney	62	21
Clinker	William Butt	98	28
Morning Star	William Burdon	100	28
Elizabeth	W. S. Bemister	71	18
Alice	R. Bransfield	97	27
Ann	Wm. Davis, Jun.	94	29
Mary	Rich Bransfield	107	28
Fair Cambrian	F. M'Carthy	90	26
Greyhound	C. M'Carthy	104	27
Reindeer	Edward Gulney	96	23
Agenoria	John Hudson	91	25
Mary Ann & Martha	Solomon Taylor	94	28
Julia Ann	John Kennedy	83	25
Sweet Home	John Moores	84	23
Experiment	Wm. Davis, Sen.	121	32
Dolphin	George Davis	120	30
Tyro	Thomas Pike	63	72
John	John Penny	70	17
Sally	James Forward	92	24
Ethiopian	John Parsons	87	25
Shannon	Francis Pike	124	29
Cornelia	Thomas Robbins	90	26
Wonderer	Clement Davis	51	16
Hunter	George Davis	68	21
Eliza	Clement Noel	91	29
Ranger	T. Kennedy	65	21
Amelia	John Pelly	64	17
Jubilee	Noah Perry	86	25
Eagle	C. M'Carthy	67	20
Pandora	Wm. Penny	75	21
Caroline	William Ash	86	22
Mary	Thomas Luther	85	24
Active	William Squires	57	18
Nancy	Robert George	50	19
Good Intent	Nicholas Howell	71	18
Catherine & Margaret	C. M'Carthy	74	24
Hope	David Clarke	54	14
Rambler	Richard Marshall	63	17
Venns	Henry Parsons	40	14
Maria	Wm. Beckett	58	17
Surprise	Richard Taylor	61	19
Total	80 vessels	6,889	1,918

Harbour Grace Sealing Fleet: 1836

Vessel	Captain	Tons	Men
Rasellas	Lawrence Keefe	106	29
Trefoil	Moses Pike	87	29
Intrepid	S. Johnson	111	32
Eunice	Edward Pike	83	23
Bustler	John Archer	74	21
William	David Power	57	20
Relief	John Murphy	106	29
Mary Francis	Thomas Dwyre	90	29
Sylvanus	Nathanael Davis	70	20
Lord M'Donald	William Ryan	82	26
Friends	Henry Davis	73	22
Nancy	Patrick Kelly	75	23
Lady Ann	Levi Pike	108	26
Wellington	James Kerby	61	20
Thomas & Hugh	R. Parsons	75	19
Jane & Mary	William Parsons	88	20
William	Daniel Green	123	30
Lavinia	Philip Adams	69	21
Nancy	Matthew Hudson	94	26
Brothers	George Parsons	51	17
Ann	Peter Hudson	122	31
Dart	John Parsons	77	21
Jane	William Parsons	81	23
Fanny	John Hogan	88	23
Isabella & Margaret	Patrick H. Hearne	93	30
William & Mary	James Cooney	71	21
Edward Piers	George Heater	86	25
Harriet	William Parsons	46	16
Edmund	William Ash	56	17
Louisa & Frederick	J. Stevenson	132	31
Experiment	William Sheppard	37	12
Defiance	Joseph Pynn	39	9
Total	32 vessels	2,611	741

Source: *Public Ledger*, 1 April 1836. Further information for both outports is given as follows: In 1835, 77 vessels with a total burthen of 6,554 tons and carrying 1,784 men sailed from Carbonear, while 36 vessels with a total burthen of 3,093 tons and carrying 822 men sailed from Harbour Grace. The only explanation for the descrepancy between the figures is the fact that ships from other neighbouring ports, including Mosquito, sailed from Carbonear.

Table 4.4
St. John's Sealing Fleet: 1838

Vessel	Captain	Tons	Men
Swan*	E. Chafe	95	31
Duck	T. Shipton	107	31
Drake	J. A. Francis	107	31
Goose	C. Grills	106	31
Prosperity	D. Mealy	109	29
Avalon	T. Mealey	84	22
Henry & Mary Ann	J. Chafe	99	30
Feronica	P. Kavanagh	84	21
Oderin	T. Lee	82	26
Metis Packet	J. Gushue	76	21
Margaret	J. Ryan	61	20
Brothers	H. Charles	61	22
Susanna Ford	T. Hughes	54	16
Hope	P. Walsh	51	15
Catharine Ann*	J. Warner	115	36
Christiana	T. Burke	110	31
Lady of the Lake	J. Power	89	30
Catharine Power	M. Power	105	27
Tryon	E. Ryan	85	27
Four Brothers	C. Colbert	79	18
Trial	M. Brien	76	24
John Stuart*	P. Feehan	95	31
Malvina*	L. Geron	86	31
St. Patrick	T. Casey	94	27
Juno	J. Pike	94	25
Nimrod*	J. Barron	93	31
Active	W. Shea	72	22
Loyalty	J. Lynch	60	20
Brothers	J. Glody	57	20
Ann	J. Geran	73	26
Jane	W. W. Boig	57	20
Waterlily*	J. Winser	100	34
Mary Jane	E. Prior	77	26
Rainbow	W. Walsh	77	26
Catherine*	R. French	65	30
Charlotte	F. Geary	67	19
Dan. O'Connell	M. Howlett	75	23
Joseph	P. James	60	20
Lady Young*	P. Houlihan	78	26
Eliza	S. French	105	30

Margaret Helen*	R. Rambury	92	29
Dove*	J. Roche	91	28
Antelope	T. Ebsary	93	25
Ann	T. Barrington	76	23
Privilege	N. Power	66	21
Sarah	G. Pippy	63	22
Active	S. Angel	58	17
Sarah	C. Harris	85	30
Hope	W. Maccassey	67	28
Charlotte*	H. J. Furneaux	99	37
Ranger*	J. Cahill	94	30
Speculation	W. Burn	84	24
Diana	R. Quidihy	72	20
Nancy	G. Hudson	56	16
Kitty	W. Pilly	53	18
Eliza Bunting	E. Purcell	117	35
Mary Ann*	G. Hartery	131	35
Margaret Ann	J. Hearn	139	36
Agenoria	H. Davis	128	32
Catherine	P. Brennan	75	21
Shaver	T. Allen	132	33
Eliza	W. Mullins	121	29
Harriet Elizabeth	T. Butler	114	30
Mary Jane	P. Mackay	108	32
John Fulton	J. O'Neil	94	23
Billow	P. Breenock	90	30
Argyle	M. Cosgrove	86	24
Revenge	C. Dutton	71	24
Preseverance	J. Pendergast	70	24
Trial	J. Holly	74	26
Jane & Mary	J. Coady	59	18
Lady Ann*	J. Crawley	40	29
Lord Nelson	J. Woods	37	15
Clondolin*	E. Pike	77	20
Scipio	J. Walsh	114	32
Annabella	R. Power	70	20
St. Patrick	J. Martin	94	24
Mary	J. Houlahan	91	23
Charles	W. Knight	79	22
Preseverance	J. Kenna	80	25
Revenge	J. Ennis	60	17
Dirk Hatterick*	J. Casey	103	34
Royal William*	W. Kent	80	32
Britannia*	M. Cummins	104	34

Sarah	J. Carty	80	24
United Brothers	D. Brien	130	28
Superb	S. Gordon	124	30
John & Horatio	D. Dwyer	96	30
Hero	J. Cooney	90	34
Daniel O'Connell	M. Burke	81	28
Sarah Isabel	W. Dwyer	63	17
Abeona	T. Williams	66	21
Theresa	J. Axtell	57	18
United Brothers	P. Lynch	113	29
Despatch	P. Manning	77	24
Victory	J. Fitzgerald	105	30
Kingaloch	W. Stanton	110	34
Isabella	R. Maher	94	33
Friends	T. Phoran	63	19
Orion	H. Ryan	63	18
Sir. C. Hamilton	P. Blake	78	29
Alpha	E. Morey	105	30
Nine Sons	J. Price	102	31
Mary	G. Carew	74	24
Alligator	G. White	52	15
Phoenix	B. Haggarty	89	26
Actual	J. Mace	63	20
Dart	D. M'Grath	90	23
Hunter	B. M'Grath	52	16
Hope	J. Walsh	76	22
Total 110 vessels		9,300	2,826

Source: *Newfoundlander*, 22 March 1838. * indicates new tonnage measurement.

Table 4.5
Sealing Vessels Clearing from Brigus: 1838

Sailing from **Bay Roberts**:			
Vessels	Captains	Tons	Men
William	Edward Snow	73	26
Newfoundlander	Isaac Mercer	92	27
Dolphin*	George Davis	86	35
Samuel	William Giles	110	33
Ann	William Davis	94	29
Nonpareil	Edward Russell	124	33
Montezuma	Stephen Russell	91	27
Margaret	Henry Cave	104	26
Nightingale	James Delany	91	28
Henrietta	Edward Williams	75	22
Despatch	James Goozenay	101	26
Caroline	Elijah Mercer	70	25
Total: 12 vessels		1,111	337

Sailing from **Port-De-Grave**:			
Vessels	Captains	Tons	Men
Active	Henry Andrews	62	19
Elizabeth Ann	Robert Andrews	87	26
Maria	Wm. H. Andrews	91	27
Mary	William Andrews	87	24
Good Intent	John Dawe	66	20
Glenfalloch	Isaac Dawe	101	25
Favorite	William Richards	69	23
Isaac & Elizabeth	John Richards	105	25
Lady Ann	William Richards	115	29
Swift	John Batten	94	24
Young Harp	Philip Corban	97	28
Ringwood*	William Taylor	103	34
John & William	Michael Keefe	71	25
Ranger*	Charles Mugford	77	85
Total: 14 vessels		1,225	414

Sailing from **Brigus:**

Vessels	Captains	Tons	Men
Saint John's*	Esau Percy	107	30
Ann	Stephen Roberts	90	26
Jane Elizabeth*	Nathl. Munden	115	34
Emily	Richard Walsh	112	33
Elizabeth & Maria	Edw. Kennedy	87	25
Water Witch	James Wilcocks	92	27
True Blue	William Whelan	85	33
Alexander	Caleb Whelan	90	23
Dolphin	Henry Whelan	56	15
John Alexander	Joseph Bartlett	96	24
Union*	Nathaniel Norman	77	30
Jane*	Johnathan Percy	120	36
Nymph	James Norman	88	28
Jubilee	John Wilcocks	90	22
Highlander*	Azariah Munden	102	34
Ianthe*	George Brown	117	32
Meg Merrilies	Daniel Bryant	69	23
Comet*	Moses Percy	104	35
Five Brothers	William Antle	98	27
George	William Walker	72	22
John	Abraham Bartlett	85	24
Agenoria	Thomas Delaney	82	26
Terra Nova*	Stephen Percy	101	32
Dandy	William Spracklin	70	18
William & Robert*	Wm. Munden	94	30
Jane	John Norman	100	33
Four Brothers	Reuben Munden	101	29
Hebe	William Rabbitts	85	30
Arabian	Nathan Percy	104	28
Hit or Miss	Thomas Roberts	93	32
John & Maria	William Burke	74	29
Bickley	William Norman	94	24
Margaret	Edmund Shehan	90	23
Rover	Thomas Spracklin	57	16
Total: 34 vessels		3,097	933

Sailing from **Cupids:**			
Vessels	Captains	Tons	Men
Dove	Thomas Snow	70	20
Victory	James King	69	18
Sir Thomas Cochrane	W. Spracklin	52	22
Venus	William Ledroe	71	20
Squirrel	William H. Ledroe	63	18
Nimrod	Abraham Ledroe	96	25
Margaret Ellen	John Norman	98	25
Success	James Ledroe	55	15
Nelson	William Smith	69	20
Liberty	Simon Spracklin	68	22
Isabella	John Whelan	76	30
Amphion	William Wells	132	33
Justin	Thomas Peyton	66	26
Orion	Henry Sheppard	68	24
Mary	George Wells	101	27
Elizabeth	John Noseworthy	74	20
Total: 16 vessels		1,228	365

Sailing from **Chapel Cove:**			
Sally	Michael Keating	81	21

Sailing from **Bacon Cove:**			
L'Avengeur	George Gushue	82	21

Sailing from **Colliers:**			
Three Brothers	Patrick Burke	81	21
Nimrod	William Cole	97	27
Tyro	Charles Saunders	73	20
Total: 5 vessels		414	110

Source: *Newfoundlander*, 29 March 1838. * indicates the new tonnage measurement.

Table 4.6
Newfoundland Sealing Fleet: 1869

St. John's:

Vessel	Captain	Tons	Men	Supplier
Hawk (SS)	Jackman	170	100	Bowring
Isabella	Ryan	140	60	"
Fanny Bloomer	Jackman	126	56	"
Sea Flower	Martin	117	55	"
Racer	Noseworthy	86	36	"
Bloodhound (SS)	Bartlett	153	90	Baine, Jn.
Primrose Bank	Joy	72	32	"
Mary Jane	Connors	67	30	"
Ellen	Gosse	49	15	"
Nimrod (SS)	White	226	140	Job Bros.
Otter	Joy	76	15	"
Mary Jane	Mansfield	51	19	"
Leader	Geran	125	51	Stabb, Row
Prima Donna	Knight	125	54	"
Sarah Grace	James	103	44	"
Gertrude	Mullowney	133	69	J.& W.Stewart
Convoy	Taylor	46	17	"
Lion (SS)	Graham	292	121	Walt. Grieve
Merlin (SS)	Ryan	248	128	W. H. Mare
Magic	Jackman	145	60	W. Kelligrew
Sterling Clipper	Jackman	98	55	A. Goodridge
Ariel (SS)	Hagan	78	37	P. Cleary
Mary Joyce	Blackler	60	15	E. Smith
Ebenezer	Taylor	59	39	P. Rogerson
Superior	Morris	68	33	E. Duder
Margaret Ann	Eclinton	44	20	L. O'Brien
May Flower	English	40	12	Ewan Stabb
Amerla	?	37	15	P.&L. Tessier
Total 28 vessels		3,034	1,418	

Vessel	Captain	Harbour Grace: Tons	Men	Supplier
Catherine	Morgan	44	19	Ridley & Sons
Flora	Davis	64	17	"
Kate	Henly	50	17	"
Robert Arthur	Sheppard	120	45	"
Retriever (SS)	Bartlett	237	115	"
Isabella Ridley	Thoomey	155	75	"
Medora	Pike	139	60	"
Lord Clyde	Pike	134	60	"
Greyhound	Alcock	153	60	"
Othello	Pike	53	20	"
Union	Pike	135	55	"
Trial	Cleary	67	40	"
Sophia	Heater	90	40	"
Elizabeth Jane	Ryan	131	60	"
Northern Light	Noel	49	17	"
Palestine	Noel	38	10	"
Mountaineer	Smart	177	100	Punton & Munn
William Whelan	Green	149	75	"
Vesta	Keefe	148	75	"
Curlew	O'Neille	168	80	"
Islay	Brine	134	70	"
Iona	Smalcombe	151	70	"
Emily	Geary	112	50	"
Penguin	Farrell	106	50	"
Glencoe	Delaney	133	65	"
Superb	Smart	109	55	"
Myrtle	Kennedy	76	35	"
Vulcan	Fitzgerald	59	25	"
John Nichole	McCarthy	57	20	"
Brilliant Star	Fleming	30	15	"
Elfrida	Jeffers	126	55	"
Friends	Avery	116	65	"
Jane & Mary	Parsons	101	50	"
Union	Parsons	105	55	"
Eclipse	Davis	146	60	"
Dolphin	Delaney	173	85	"
James Clift	Alcock	83	40	"
Saint Kilda	Parsons	70	25	"
Louisa	Morgan	49	25	"
Elizabeth & William	Stevenson	134	60	"

The Ice Hunters

Vessel	Captain	Tons	Men	Supplier
Rapid	Franey	85	30	"
Anastasia	Hennebury	177	80	"
Lizzie	Butt	81	25	"
George	McCarthy	137	60	WJS Donnelly
Hecla	Kielly	117	50	"
Topa	Stapleton	148	55	"
Emigrant	Corban	63	26	"
Rebecca	Moores	96	52	"
W. Donnelly	Harkness	125	50	"
Iona	Blundon	40	16	McBride & Co
Susan	Green	147	65	Daniel Green
Total 51 vessels		5,587	2,524	

Carbonear:

Vessel	Captain	Tons	Men	Supplier
Gleaner	Hopkins	127	55	Jos Hopkins
Bandit	Pearcy	71	34	Job Bros.
Cabot	Pike	126	54	Ridley & Sons
Gulnare	Vatcher	81	40	"
Sweet Home	Soper	67	30	"
True Blue	Joyce	152	56	"
Orient	Joyce	132	50	"
Emeline	Penny	135	58	Wm Penny
Thomas Ridley	Forward	164	53	John Rorke
Alarm	Mortimer	119	56	"
Hope	Thistle	79	30	"
Echo	Pike	115	43	"
Willaim	Fitzgerald	145	61	"
Adam Averel	Winsor	77	27	"
Dominion	Hopkins	80	40	"
Britannia	Hiscock	61	27	Aspey & Co.
Isobel	Howard	139	57	"
Walrus	Dwyer	131	53	Punton & Munn
Staff	Mahony	56	18	"
Experiment	Joyce	35	14	"
Ruby	Priddle	52	32	"
Cecilia	Pearce	60	19	"
Total 22 vessels		2,204	907	

Bay Roberts:

Vessel	Captain	Tons	Men	Supplier
Dundanna	Mercer	71	27	Bowring
Naomi	Parsons	125	64	McBride & Co
Ecliptic	Delany	149	65	J. Cormack
R.S.C.	Delany	131	54	"
Sneezer	Daw	128	65	W. S. Green
Jane White	Wilcox	118	48	"
Rolling Wave	Daw	152	68	R. Daw
Huntsman	Daw	120	57	"
Mary Ann	Batten	66	28	"
Louisa	Mercer	159	66	J&W. Bartlett
Brothers	Mosdell	136	62	"
Rescue	Daw	146	70	S. Daw
Total 12 vessels		1,501	674	

Brigus:

Vessel	Captain	Tons	Men	Supplier
Havelock	T. St. John	110	46	P&L. Tessier
Pearl	T. Wilcocks	132	51	Baine, Jn.
John Bull	M. Byrne	136	57	"
Susan	A. Smith	134	57	"
Bell Clutha	W. Wilcocks	117	46	"
Panther (SS)	A. Bartlett	238	105	"
Spy	J. Burke	111	50	Bowring
Abeona	J. Spracklin	135	55	"
Garland	G. Smith	130	60	Walt. Grieve
Deerhound	J. Bartlett	101	44	"
Breadalbane	R. Walsh	131	50	Punton & Munn
Atlanta	A. Munden	140	70	"
Eastern Packet	T. Kehoe	89	42	"
Gladiator	S. Wilcocks	116	47	Job Bros.
Topas	J. W. Norman	171	65	"
Sultana	I. Clarke	104	50	"
Mercury	S. Clarke	129	55	"
Maxim	M. Clarke	142	54	"
William	S. Whelan	133	68	McBride & Co.
Hunter	N. Rabbitts	114	50	Hounsell & Co
Herald	J. Bartlett	128	56	John Bond
Matilda	S. Dooling	110	44	John Munn
Total 22 vessels		2,851	1,222	

The Ice Hunters

Hants' Harbor:

Vessel	Captain	Tons	Men	Supplier
Ghoorka	Nicholas Short	80	38	Job Bros.
Lillian	John Harris	55	15	"
Lizzie	Timothy Pelly	45	17	"
Total 3 vessels		180	70	

Trinity:

Vessel	Captain	Tons	Men	Supplier
Wolf (SS)	Gent	210	100	Grieve & Brem
Gem	Facey	130	54	"
Young Prince	Doberty	53	11	"
Flash	Morris	121	55	Brooking & Co
Ariel	Ash	143	60	"
Hound	Rix	45	14	"
Othello	Coleman	47	14	Baine, Jn.
Abeona	Butler	57	12	Isaac Butler
Total 8 vessels		806	320	

Catalina:

Vessel	Captain	Tons	Men	Supplier
John & Elizabeth	Duffet	54	18	Ridley & Sons
Goldfinder	Martin	73	24	"
Rusina	Nowlan	125	60	"
Mastiff (SS)	Murphy	245	140	"
Margaret Ann	Murphy	140	66	"
Florence	Murphy	130	66	"
William	Stone	105	60	Job Bros.
Melena	Perry	116	60	J&W. Stewart
Young Prince	Snelgrove	70	28	Ben Snelgrove
Micmac	Perry	66	39	Bowring
Portia	Johnston	53	23	"
Ida	King	56	23	"
Soelia	Reed	45	23	"
Elizabeth	Joy	50	26	Baine, Jn.
Robert Grieve	Doody	47	25	"
Avalon	Fennell	46	26	A. Goodridge
Total 16 vessels		1,421	707	

Greenspond:

Vessel	Captain	Tons	Men	Supplier
Packet	Osborne	119	50	Brooking & Co
Selah Hutton	Batterton	89	39	"
Hebe	Blanford	121	65	J&W. Stewart
Renfrew	Blanford	124	65	"
Billow	Windsor	90	50	"
Brothers	Barber	134	65	"
Oban	Windsor	129	65	"
Argo	Easton	114	55	Ridley & Sons
Fanny Smallwoo Carter		45	20	"
Queen	Ham	48	25	"
Barbara	Kean	162	70	Baine, Jn.
Success	Kean	128	65	"
Clara Jane	Haines	68	29	"
Good Intent	Turner	60	23	"
Kingfisher	Cashin	51	20	"
Stella	Knee	126	57	Bowring
Glenara	Samsburg	126	57	P. Rogerson
Kitty Clyda	Davis	129	60	"
United Brothers	White	136	50	Punton & Munn
Ellen	Burton	57	21	W. Burton
Total 20 vessels		2,056	951	

La Poile:

Vessel	Captain	Tons	Men	Supplier
Golden Era	Hooper	44	12	D. Grouchy
Jessie	Poole	33	8	"
Vivid	Evans	29	8	"
Cornelian	Buffet	28	9	"
Necumtaw	Baker	38	8	"
Sisters	Garcin	39	8	"
Brisk	Cox	32	8	"
Romeo	Whittle	21	8	"
Leopard	Cains	33	8	"
Albert Mckay	Forward	19	6	"
Morning Light	Rose	43	10	Ben Rose
Volant	Bonnell	43	10	S. Bonnell
Ella	Bonnell	31	8	"
B. Wier	Hall	32	8	John Rose
Minnie	Hynes	14	6	Ridley & Sons
Harriet	Lea	14	6	"
Total 16 vessels		493	131	

The Ice Hunters

Burgeo:

Vessel	Captain	Tons	Men	Supplier
Albert Mckean	Forward	17	7	D. Grouchy
Leopard	Conez	31	10	"
Total 2 vessels		48	17	

Channel:

Vessel	Captain	Tons	Men	Supplier
President	Blanchard	47	10	J. Gillam
Isabel	Quiin	29	8	"
J. B. Fray	Bragg	48	10	J. Bragg
Susan	Mclean	44	8	J. Poole
Thetis	Poole	78	7	"
Express	Evans	68	10	J. Evans
Ripple	Evans	23	8	"
Hawk	Tobin	41	10	W. Pryor
John William	Carter	31	8	"
Petrel	Carter	15	6	"
E. Venno	Mclane	24	8	"
Lily Dale	Hall	24	8	"
British Lass	Forsay	29	8	A. Forsay
Total 13 vessels		501	109	

Fogo:

Vessel	Captain	Tons	Men	Supplier
Lily	Russell	43	21	Robt. Scott
Harp	Dwyer	36	13	"
Prince Alfred	Rolls	81	27	James Rolls
Edwin Duder	Keough	72	31	Edwin Duder
Gratitude	Downer	70	21	"
United Brothers	Downer	49	18	"
Total 6 vessels		351	131	

Grand Total 219 vessels;   21,033   9,181

		Abstract:	
Port or District	Vessels	Tons	Men
St. John's	28	3,034	1,418
Harbor Grace	51	5,587	2,524
Carbonear	22	2,204	907
Bay Roberts	12	1,501	674
Brigus	22	2,851	1,222
Hants' Harbor	3	180	70
Trinity	8	806	320
Catalina	16	1,421	707
Greenspond	20	2,056	951
Fogo	6	351	131
La Poile	16	493	131
Burgeo	2	48	17
Channel	13	501	109
Grand Total	219	21,033	9,181

Sources: 1869 Newfoundland Newspapers - *Courier*, 13, 17, 24 and 31 March; *Times*, 3, 7, 17 and 28 April; and *Morning Chronicle*, 16 April. Because of space restrictions, it was necessary to abbreviate slightly, or to alter to a small degree, the names of some suppliers. The following is a list of the names abbreviated or altered: Bowring Brothers; Baine, Johnston & Co.; Job Brothers & Co.; Stabb, Row & Co.; J. & W. Stewart; Walter Grieve & Co.; P. Rogerson & Son; W. J. S. Donnelly; Joseph Hopkins: J. & W. Bartlett; Grieve & Bremner; Benjamin Snelgrove; Allan Goodridge & Sons; W. Burton & Brothers; D. Grouchy & Co.; Benjamin Rose; and James Rolls & Sons.

Table 4.7
Newfoundland Sealing Steamer Captains: 1869/1909

St. John's: 1869

Bartlett, Abram: Born Brigus, Conception Bay, 1819; died 1889. Father of Capt. William; grandfather of Capt. Bob. *Panther* in 1869. First steamer, *Panther* in 1867; last, *Panther* in 1883. Part owner of this steamer.

Bartlett, John Jr: Born Brigus. *Bloodhound* in 1869 and 1870.,

Gent, George: Born Scotland. *Wolf* in 1868 and 1869.

Graham, Alexander: Born Scotland. *Lion* in 1869. First steamer, *Bloodhound* in 1863; last, *Bear* in 1874.

Hagen, John: Born Scotland. *Ariel* in 1869 and 1872.

Jackman, William: Born Renews 1837; died 1877. Brother of Capt. Arthur. *Hawk* in 1869. First steamer, *Hawk* in 1867; last, *Eagle* in 1876.

Ryan, William: Born St. John's. *Merlin* in 1869. First steamer, *Bloodhound* in 1866; last, *Eagle* in 1872.

White, Edward, Sr: Born Tickle Cove, Bonavista Bay, 1811; died 1886. Father of Capts. Edward Jr. and Richard. *Nimrod* in 1869. First steamer, *Hawk* in 1866; last, *Neptune* in 1882.

Harbour Grace: 1869

Bartlett, Isaac: Born Brigus; died 1906, aged 85 years. *Retriever* in 1869. First steamer, *Retriever* in 1867; last, *Tigress* in 1875.

Murphy, James: Born Catalina, Trinity Bay; died 1871. *Mastiff* in 1869. First steamer, *Retriever* in 1866; last, *Mastiff* in 1870.

St. John's: 1879

Adams, William, Sr: Born Scotland. *Arctic* in 1879. First steamer, *Arctic* in 1877; last, *Arctic* in 1883.

Ash, Francis: Born Trinity; died 1918, aged 84 years. *Lion* in 1879. First steamer, *Lion* in 1872; last, *Kite* in 1888.

Bannerman, James: Born Scotland. *Aurora* in 1879.

Barbour, Joseph: Born Cobbler's Island, near Newtown, Bonavista Bay. Son of Benjamin; eight brothers, including Capts. William, Thomas, George and James. *Walrus* in 1879. First steamer, *Walrus* in 1876; last, *Ranger* in 1890.

Bartlett, Abram: See 1869. *Panther* in 1879.

Blandford, Samuel: Born Greenspond, Bonavista Bay, 1840; died 1909. Brother of Capts. Darius, Jr., and James. *Eagle* in 1879. First steamer, *Osprey* in 1874; last, *Virginia Lake* in 1906.

Cummins, John: Born St. John's. *Nimrod* in 1879. First steamer, *Nimrod* in 1876; last, *Nimrod* in 1879.

Dawe, Charles: Born Port de Grave, Conception Bay 1845; d. 1908. Brother of Capt. Henry. *Iceland* in 1879. First steamer, *Greenland* in 1874; last, *Vanguard* in 1898.

Dawe, Henry: Born Port de Grave. Brother of Capt. Charles. *Leopard* in 1879. First steamer, *Leopard* in 1878; last, *Nimrod* in 1895.

Delaney, Mark: Born Bay Roberts, Conception Bay. *Ranger* in 1879. First steamer, *Walrus* in 1872; last, *Ranger* in 1881.

Delaney, Patrick: Born Bay Roberts, Conception Bay. *Kite* in 1879. First steamer, *Walrus* in 1875; last, *Eagle* in 1886.

Diamond, Levi: Born Catalina, Trinity Bay, 1833; died 1920. *Bear* in 1879. First steamer, *Wolf* in 1870; last, *Bear* in 1880.

Jackman, Arthur: Born Renews, 1843; d. 1907. Brother of Capt. William. *Falcon* in 1879. First steamer, *Hawk* in 1871; last, *Eagle* in 1906.

Joy, James: Born? *Tiger* in 1879. First steamer, *Tiger* in 1878; last, *Kite* in 1898.

Kean, Benjamin: Born Pool's Island, Bonavista Bay. *Greenland* in 1879. First steamer, *Greenland* in 1877; last, *Hector* in 1889.

Pike, Richard: Born Harbour Grace, Conception Bay; died 1893, aged 59 years. *Proteus* in 1879. First steamer, *Retriever*, in 1870; last, *Hope* in 1892.

Smith, Azariah: Born Scotland; with the Dundee Sealing and Whale Fishing Co. First Steamer, *Narwhal* in 1879; last, *Iceland* in 1884.

Smith, George: Born Conception Bay. *Wolf* in 1879. First steamer, *Wolf* in 1874; last, *Wolf* in 1888.

Walsh, Samuel: Born? *Merlin* in 1879. First steamer, *Merlin* in 1872; last, *Merlin* in 1882.

White, Edward, Sr: See 1869. *Neptune* in 1879.

White, Edward, Jr: Born St. John's 1840; died 1915. Son of Capt. Edward, Sr., and brother of Capt. Richard. *Hector* in 1879. First steamer, *Hector* in 1877; last, *Hector* in 1888.

Yule, Charles: Born Scotland; with the Dundee Sealing and Whale Fishing Co. *Esquimaux* in 1879. First steamer, *Esquimaux* in 1878; last, *Resolute* in 1880.

Harbour Grace: 1879

Dawe, Henry: Born Bay Roberts, Conception Bay; died 1921, aged 71 years. First steamer, *Mastiff* in 1879; last, *Adventure* in 1910.

Munden, Azariah: Born Brigus, Conception Bay 1813; died 1889. Son of Capt. William, who was the first to build a sealing vessel of over 100 tons. First steamer, *Commodore* in 1871; last, *Vanguard* in 1879.

Thomey, Henry: Born Mosquito, Conception Bay; died 1911, aged 91 years. Son of Capt. Arthur. *Commodore* in 1879. First steamer, *Commodore* in 1878; last, *Greenland* in 1889.

St. John's: 1889

Barbour, Joseph: See 1879. *Ranger* in 1889.

Bartlett, William, Sr: Born Brigus, Conception Bay. Son of Capt. Abram, who died in 1899; father of Capt. Bob. *Panther* in 1889. First steamer, *Panther* in 1884; last, *Thetis* in 1929.

Blandford, Samuel: See 1879. *Neptune* in 1889.

Bragg, Robert: Born Bonavista Bay. Brother of Capt. Daniel. *Walrus* in 1889. First steamer, *Walrus* in 1886; last, *Walrus* in 1903.

Dawe, Charles: See 1879. *Terra Nova* in 1889.

Dawe, Henry: See 1879. *Leopard* in 1889. (Resident of Port de Grave)

Dawe, Henry: See 1879. *Nimrod* in 1889. (Resident of Bay Roberts)

Guy, William: Born Scotland; with the Dundee Sealing and Whale Fishing Co. *Polynia* in 1889. First steamer, *Arctic* in 1885; last, *Eclipse* in 1892.

Jackman, Arthur: See 1879. *Eagle* in 1889.

Kean, Abram: Born Flowers Island, Bonavista Bay 1855; died 1945. Father of Capts. Joseph, Nathan and Westbury. First steamer, *Wolf* in 1889; last, *Beothic II* in 1936.

Kean, Benjamin: See 1879. *Hector* in 1889.

Knee, Job: Born Bonavista Bay. Brother of Capts. Kenneth and William. First steamer, *Falcon* in 1889; last, *Sagona* in 1923.

Knee, William: Born Bonavista Bay. Brother of Capts. Job and Kenneth. *Kite* in 1889. First steamer, *Kite* in 1877; last, *Kite* in 1893.

McKay, Henry: Born Scotland; with the William Stephens and Co. First steamer, *Aurora* in 1889; last, *Esquimaux* in 1900.

Milne, William: Born Scotland; with the Dundee Sealing and Whale Fishing Co. *Esquimaux* in 1889. First steamer, *Esquimaux* in 1885; last, *Esquimaux* in 1890.

Harbour Grace: 1889

Gosse, Robert: Born Spaniard's Bay, Conception Bay 1835; died 1899. *Vanguard* in 1889. First steamer, *Mastiff* in 1881; last, *Vanguard* in 1895.

Noel, J.F. Born Harbour Grace. Only steamer, *Mastiff* in 1889.

Thomey, Henry: See 1879. *Greenland* in 1889.

Winsor, William, Sr: Born Swain's Island, Bonavista Bay 1846; died 1907. Father of Capts. Jesse, Samuel and William, Jr. (Billy). *Iceland* in 1889. First steamer, *Vanguard* in 1881; last, *Vanguard* in 1906.

St. John's: 1899

Barbour, Alpheus: Born Bonavista Bay. *Diana* in 1899. First steamer, *Walrus* in 1896; last, *Bloodhound* in 1915.

Barbour, George: Born Cobbler's Island, near Newtown, Bonavista Bay, 1858; died 1928. Son of Benjamin; eight brothers including Capts. Joseph, William, Thomas and James. *Vanguard* in 1899. First steamer, *Walrus* in 1893; last, *Beothic II* in 1928.

Bartlett, William: See 1889. *Hope* in 1899.

Blandford, Darius: Born Greenspond, Bonavista Bay, 1843; died 1917. Brother of Capts. Samuel and James. *Iceland* in 1899. First steamer, *Iceland* in 1898; last, *Vanguard* in 1909, which was lost that year.

Blandford, Samuel: See 1879. *Neptune* in 1899.

Dawe, Henry: See 1889. *Ranger* in 1899. (Resident of Bay Roberts)

Farquhar, J. A. Only Nova Scotian to participate in the "Newfoundland" seal fishery. *Newfoundland* in 1899. First steamer, *Newfoundland* in 1893; last, *Sable Island* in 1919.

Hann, George: Born Swain's Island, Bonavista Bay 1850; died 1942. *Labrador* in 1899. First steamer, *Leopard* in 1890; last, *Labrador* in 1908.

Jackman, Arthur: See 1879. *Terra Nova* in 1899.

Kean, Abram: See 1889. *Aurora* in 1899.

Kean, Job: Born Brookfield, Bonavista Bay. Nephew of Capt. Abram. *Leopard* in 1899. First steamer, *Leopard* in 1896; last, *Diana* in 1917.

Knee, Job: See 1889. *Algerine* in 1899.

Mercer, Edward: Born Bay Roberts. First steamer, *Walrus* in 1899; last, *Walrus* in 1901.

Mercer, Isaac: Born Bay Roberts. *Greenland* in 1899. First steamer, *Mastiff* in 1894; last, *Greenland* in 1904.

Scott, D.A.: Born? *Harlaw* in 1899. First steamer, *Harlaw* in 1896; last, *Harlaw* in 1903.

Spracklin, Thomas: Born Brigus? *Nimrod* in 1899. First steamer, *Nimrod* in 1896; last, *Nimrod* in 1900.

Winsor, William, Sr: See 1889. *Panther* in 1899.

Young, James: Born? First steamer, *Kite* in 1899; last, *Kite* in 1900.

St. John's: 1909

Barbour, Alpheus: See 1899. *Neptune* in 1909.

Barbour, Baxter: Born Newtown, Bonavista Bay? *Labrador* in 1909. First steamer, *Nimrod* in 1905; last, *Neptune* in 1915.

Barbour, George: See 1899. *Beothic* in 1909.

Bartlett, Moses: Born Brigus? First steamer, *Nimrod* in 1893; last, *Southern Cross* in 1909.

Bartlett, William: See 1889. *Viking* in 1909.

Bishop, Noah: Born Swain's Island, Bonavista Bay. First steamer, *Algerine* in 1909; last, *Algerine* in 1912.

Bishop, Edward: Born Swain's Island, Bonavista Bay. *Terra Nova* in 1909. First steamer, *Algerine* in 1906; last, *Eagle II* in 1926.

Blandford, Joseph: Born? First steamer, *Diana* in 1909; last, *Diana* in 1912.

Carroll, William: Born King's Cove? First steamer, *Kite* in 1909; last, *Kite* in 1912.

Dawe, Henry: See 1889. *Adventure* in 1909. (Resident of Bay Roberts)

Green, Daniel: Born Newtown, Bonavista Bay. *Aurora* in 1909. First steamer, *Ranger* in 1891; last, *Aurora* in 1911.

Kean, Abram: See 1899. *Florizel* in 1909.

Kean, Edwin: Born Flowers Island, Bonavista Bay. Brother of Capt. Jacob. First steamer, *Iceland* in 1909; last, *Iceland* in 1910, which was lost at the ice.

Kean, Job: See 1889. *Erik* in 1909.

Kean, Joseph: Born in Brookfield, Bonavista Bay; died 1918. Son of Capt. Abram; brother of Capts. Nathan and Westbury. *Eagle* in 1909. First steamer, *Panther* in 1897; last, *Sable Island* in 1917. He was lost on the *Florizel*.

Knee, Job: See 1889. *Bellaventure* in 1909.

Parsons, John: Born Bay Roberts, Conception Bay. *Bonaventure* in 1909. First steamer, *Newfoundland* in 1906; last, *Terra Nova* in 1931.

Winsor, Jacob: Born? *Bloodhound II* in 1909; died 1912. First steamer, *Walrus* in 1907; last, *Bloodhound II* in 1911. He was lost on the *Erna*.

Winsor, Jesse: Born Swain's Island, Bonavista Bay. Son of Capt. William Sr; brother of Capts. Samuel and William C (Billy). *Newfoundland* in 1909. First steamer, *Panther* in 1906; last, *Bloodhound II* in 1914.

Winsor, Samuel: Born Swain's Island, Bonavista Bay. Son of Capt. William, Sr; brother of Capts. Jesse and William C (Billy). *Ranger* in 1909. First steamer, *Walrus* in 1904; last, *Ranger* in 1920.

Sources: *Chafe's Sealing Book*, four editions; and "CWA 36, II," which is a comprehensive alphabetical file on Newfoundland sealing captains and a separate smaller file on the Dundee sealing captains who operated from St. John's. Information is incomplete. Please note that the name 'Swain's Island' was changed to 'Wesleyville' in 1880 and that the community of Brookfield was named by Captain Abram Kean. Note also that there were two captains named Henry Dawe - one a resident of Bay Roberts and the other lived in Port de Grave.

Table 4.8
Harbour Grace Sealing Steamer Captains

Name	Number of Seasons	Owner and/or Supplier
Antle, William	2	Munns
Bartlett, Isaac	3	Ridleys
Curtis, H	1	Munns
Dawe, Charles	4	Munns
Dawe, Henry (BR)	10	Munns; and Paterson & Foster
Dawe, Henry (PG)	2	Munns
Dawe, Samuel	1	Paterson & Foster
Fitzgerald, T	1	Munns
Gosse, Robert	11	Munns; and Paterson & Foster

Name	Number	Firm
Green, Thomas	1	Munns
Hanrahan, N	1	Munns
Hicks, John	1	Ridleys
Jeffers, J	6	Munns
Kean, Benjamin	1	Munns
Keefe, J	2	Munns; and Paterson & Foster
Kennedy, J	2	Munns
Mercer, Isaac	2	Munns
Munden, Azariah	9	Munns
Murphy, Edward	1	Ridleys
Murphy, James	5	Ridleys
Noel, John F	1	Munns
Perry, J	1	Munns
Pike, Richard	4	Ridleys
Smith, Azariah	3	Munns
Thomey, Henry	7	Munns
Wilcox, William	2	Munns; and Paterson & Foster
Winsor, Jacob	1	Munns
Winsor, John	1	Paterson & Foster
Winsor, William	15	Munns

Source: Chafe (First edition), Mosdell, *Chafe* and Ryan, *Chafe*. Please note that in 1877, Captain Fitzgerald commanded the *Mastiff* on its first trip to the ice, while Captain Perry commanded it on its second trip.

Table 4.9
High Liner Captains and Cargoes
Sail in General

Year	Captain	Number
1822	Heighton Taylor	5,000
1826	Nathaniel Munden	6,666 second trip
1832	George Carew	silk flag for greatest number
1846	John Barron	9,646 first trip
1847	Burke	6,400 first trip
1848	John Barron	9,500 first trip
1852	White	6,000
1857	Andrews	9,000
1858	J. Houlahan	12,584

1859	Terry Halleran	9,500	first trip
1861	Terry Halleran	8,600	first trip
1866	John Bartlett	10,000	first trip, sail

Steam in St. John's

1863	Alexander Graham	3,000	one trip
1864	James Gulliford	800	"
1865	Patrick Skinner	7,286	two trips
1866	William Ryan	13,358	"
1867	William Ryan	9,600	"
1868	Alexander Graham	15,500	one trip
1869	Edward White, Sr.	24,000	three trips
1870	Edward White, Sr.	21,200	two trips
1871	Edward White, Sr.	31,996	three trips
1872	Pierce Mullowney	17,765	two trips
1873	Willaim Jackman	39,407	three trips
1874	William Jackman	26,491	two trips
1875	Richard Pike	44,377	"
1876	Edward White, Sr.	25,272	"
1877	Edward White, Sr.	40,018	"
1878	Willaim Adams, Sr.	33,678	"
1879	Benjamin Kean	27,361	"
1880	William Adams, Sr.	17,012	"
1881	Arthur Jackman	40,978	"
1882	William Adams, Sr.	24,662	"
1883	Charles Dawe	30,834	"
1884	Samuel Blandford	43,167	"
1885	Arthur Jackman	39,307	"
1886	Joseph Barbour	35,888	"
1887	Alex. Fairweather	26,134	"
1888	Samuel Blandford	42,242	one trip
1889	Joseph Barbour	34,373	"
1890	Samuel Blandford	21,949	"
1891	Charles Dawe	35,239	"
1892	William Barbour	41,101	two trips
1893	Arthur Jackman	12,770	one trip
1894	J. Brett	16,499	"
1895	Henry McKay	33,886	two trips
1896	Samuel Blandford	22,496	one trip
1897	Arthur Jackman	27,941	"
1898	Abram Kean	25,633	"
1899	Samuel Blandford	32,134	"

1900	Samuel Blandford	36,255	"
1901	Abram Kean	32,416	"
1902	George Barbour	25,707	"
1903	Abram Kean	26,069	"
1904	Abram Kean	34,849	"
1905	Arthur Jackman	32,064	"
1906	Henry Dawe	30,193	"
1907	George Barbour	30,985	"
1908	Henry Dawe	27,255	"
1909	George Barbour	34,837	"
1910	Abram Kean	49,069	"
1911	George Barbour	35,767	"
1912	William C. Winsor	34,561	"
1913	Abram Kean	37,882	"
1914	William C. Winsor	28,308	"

Steam in Harbour Grace

1866	James Murphy	23,400	two trips, only one vessel
1867	James Murphy	19,200	"
1868	James Murphy	18,700	"
1869	James Murphy	13,750	"
1870	Richard Pike	23,500	"
1871	Azariah Munden	26,658	"
1872	Azariah Munden	31,314	one trip
1873	N. Hanrahan	17,690	two trips
1874	J. Jeffers	16,500	one trip
1875	J. Jeffers	14,630	"
1876	Azariah Munden	15,200	two trips
1877	Azariah Munden	21,300	"
1878	Henry Thomey	14,386	"
1879	Henry Thomey	24,200	"
1880	Henry Thomey	5,606	"
1881	Benjamin Kean	23,332	"
1882	William Winsor, Sr.	8,000	"
1883	William Winsor, Sr.	18,888	"
1884	William Winsor, Sr.	16,015	one trip
1885	William Winsor, Sr.	25,256	two trips
1886	Henry Thomey	14,011	one trip
1887	Charles Dawe	13,453	two trips
1888	William Winsor, Sr.	16,389	one trip
1889	Robert Gosse	21,271	"

1890	Henry Dawe	14,236 " (Bay Roberts)	
1891	Henry Dawe	25,907 " "	
1892	William Winsor, Sr.	25,107 "	
1893	Robert Gosse	9,276 "	
1894	Robert Gosse	7,410 "	
1895	Henry Dawe	14,411 " (Bay Roberts)	

Source: Mosdell, *Chafe* and Ryan, *Chafe*. Information on the sailing vessels is very limited, as can be seen. The number of trips made is given when known because while one captain could be the high liner of the first trip, a different captain could claim that title of the overall spring voyage if his total catch after two or three voyages was the largest for the season. Furthermore, the above were the high liners in terms of the number of seals brought into port; they may not have been the high liners in terms of the weight of fat or in terms of the total value. In general, one can assume that they were, but there is at least one exception to this rule - and a famous exception at that, although it lies outside the period under study here. In 1933, while returning to St. John's with an adequate cargo of pelts, the *Ungava*, under Captain Peter Carter, chanced upon a herd of old seals (probably including bedlamers) on a patch of tight ice. The batsmen killed all of them because they could not escape, and the vessel brought in the heaviest load of seal pelts (gross tonnage) in the history of the industry - 1,254 tons, 19 cwts., 1 qr., 25 lbs. - and also broke the previous record set by Captain Abram Kean by bringing in a total of 49,285 pelts. However, the *Imogene*, under Captain Al Blackwood, set a new record that same year by bringing in 55,636 pelts, with an overall net tonnage slightly lower than that of the *Ungava's* cargo.

Table 5.1

Newfoundland Sealing Steamers Lost: 1863-1914

1871	Wolf I		
1872	Bloodhound I	Retriever	Montecello
1874	Osprey		
1875	Ariel	Tigress	
1876	Hawk		
1878	Micmac		
1880	Xanthus		
1882	Lion	Merlin	
1883	Commodore	Proteus	
1884	Narwhal	Tiger	
1886	Resolute	Jan Mayne	

1887	Arctic		
1891	Polynia		
1893	Eagle I		
1894	Falcon		
1896	Windsor Lake	Wolf II	
1898	Mastiff		
1901	Hope		
1907	Greenland	Leopard	
1908	Grand Lake	Panther	Walrus
1909	Vanguard	Virginia Lake	
1910	Iceland		
1912	Erna		
1913	Labrador		
1914	Kite	Southern Cross	

Source: Mosdell, *Chafe*, p. 47; and local newspapers. Note that the *Micmac* was lost in 1878, not in 1888, as reported by Chafe.

Table 6.1
Relief Subscriptions: 1830

"Account of the Subscriptions received by the Committee for the Relief of those who lost their Friends at the Seal Fishery last Spring [1830] - and its distribution."

Donated by:

	£	s	d
James Simms, Esq.	1	0	0
Mr. Robert Pearce	1	0	0
A Friend		9	0
Messrs. J.& W. Pitts	1	10	0
Rev. Charles Blackman		10	0
Joshua Green, Esq.	1	0	0
Miss H. Furneaux		5	0
Mr. T. Martin		5	0
Mr. Lind	1	0	0
Mr. E. Morgan		5	0
Mr. T. Parker and friends		15	0
Thomas Allen, per Baine, Johnston		10	0
Michael Dunn "		10	0
John Norman "		10	0
Dennis Nowlan "		10	0
Wm. Butt "		10	0
Wm. Mullins "		10	0
Thomas Butt "		10	0
Garrett Dalton "		10	0
Philip Kelly "		5	0
Wm. Reddy "		5	0

Collected by:

	£	s	d
Messrs. Gosse, Pack & Fryer, Carbonear	58	10	0
" T. Ridley and Co., H. Grace	4	10	0
" Newman and Co., St. John's	15	10	0
" Rennie, Stewart and Co.	10	2	7
" Robinson and Brooking	2	3	0
" Henderson, Bland and Co.	7	0	0
" W. and H. Thomas		3	0
" Bulley, Job and Co.		10	0
" Brine, March and Co.		5	0
" M'Bride and Kerr	3	7	0

The Ice Hunters 507

			£	s	d
"	John Howley			5	0
"	Robert Alsop and Co.			17	0
"	Benjamin J. Williams			3	0
"	James Stewart and Co.		6	0	10
"	Hunters and Co.		12	5	6
"	Wyse, Baker and Co.			8	10
"	John Ryan			5	0

Crew of Schooners (Baine, Johnston & Co.):

		£	s	d
	Renown	1	2	0
	Elizabeth		19	4
	Wellington	1	0	0
	Margaret	1	2	0
	Sally Ann		19	0
	Belinda	1	10	0
	Hannah		15	0
	Perseverance	1	5	0
	Fanny		18	0
Sundries		2	15	0

Amount of Subscriptions received £146 19 11

TO PAID (@ 47s 2d per person): (Persons in family)

		£	s	d
Mrs. Metcalfe, South Shore	3	7	1	6
" Metcalfe, Portugal cove	3	7	1	6
" Power, Belle Isle	3	7	1	6
" King, Quidi Vidi	8	18	17	4
" Tope, Belleisle	3	7	1	6
" Prendergast, Quidi Vidi	5	11	15	10
" M'Kay, St. John's	1	2	7	2
" Hurley, St. John's	2	4	14	4
" Codner, Torbay	6	14	3	0
" Foley, St. John's	3	7	1	6
" Williams, Bay Bulls	3	7	1	6
" Fry, Portugal Cove	2	4	14	4
" Ryan, St. John's	2	4	14	4
" Picco, Portugal Cove	2	4	14	4
" Angell, Petty Harbour	1	2	7	2
" Thompson, St. John's	1	2	7	2
" Copton, St. John's	1	2	7	2
" Walsh, St. John's	3	7	1	6
" Allen, Portugal Cove	3	7	1	6

"   Roach, St. John's   2   4   14   4
"   Doyle, St. John's   4   9   8   8

Number of persons relieved   61   £146   19   11
Signed by Henry Hawson and Wm. Johnston,
St. John's, Newfoundland, 10th Aug. 1830.

Source: *Newfoundlander*, 12 August 1830.
An additional £36 4 5 was collected later and distributed to the same families; see *Newfoundlander*, 21 April 1831.

BIBLIOGRAPHY

PRIMARY SOURCES

Manuscripts

1. The original correspondence from the convoy commanders and naval governors is the most important manuscript source in early Newfoundland. CO 194 series (Originals in PRO, Microfilm copy in CNS, MUN.)

    Great Britain, Colonial Office, Newfoundland. Original Correspondence, CO 194. The following volumes contain information on the origins of the Newfoundland seal fishery in the eighteenth century: 8 (1725), 9, 10, 18, 21 and 23.

    _____. CO 194. The following volumes are primarily useful in the study of the origins and early development of the Newfoundland *spring* seal fishery: 23, 39, 40-5, 47-9, 51, 53-5, 57, 59, 60-4, 72, 74, 78, 80-1, 83, 85, 87, 129, 131 and 139 (1853).

2. The original customs returns in the PRO provide the necessary information on imports to England and Great Britain. (These have been supplemented by printed records in the Board of Customs Library—see "Printed" below.)

    Great Britain, Customs Returns. CUST 4. British imports by place of origin. Vols. 5 (1809) to 54 (1859).

    _____. Customs Returns. CUST 5. British imports by item. Vols. 1A (1792) to 58 (1858), and 161 (1899).

    _____. Customs Returns. CUST 17. "State of Navigation, Commerce and Revenue" Imports into England (Great Britain beginning in 1798). Vols. 1 (1772) to 30 (1808).

3. Other British Manuscripts

    Great Britain. Board of Trade. Miscellanea, BT6/90, 91 and 92.

    _____. CO 195. Colonial Office. Newfoundland Entry Books. 1786-1867.

    Trinity House: Cash Book, 1778-1784; 1785-1792; 1820-1824. (Tower Hill, London.)

    _____: By Minutes, 1786-1791. (Tower Hill, London.)

    _____: Court Minutes, 1778-1789. (Tower Hill, London.)

    _____: Court Minutes, 1790-1797; 1816-1826; 1844-1848. (Goodge Street storage, London.)

4. Newfoundland

    "Annual Report of the Chamber of Commerce." 1834. PANL.

"Book of Coasting and Fishing Ships." Harbour Grace: Conception Bay Museum. 1866-1918.

John and William Boyd Letter Books, 1875-1878. PANL.

William Cox and Company Letter Book, 1858, 1865-1867. PANL.

Sir John T. Duckworth Papers. 1810/1812. PANL.

Government. Newfoundland. Colonial Secretary's Department. (GN2) 2, 1832. PANL.

Job Business Papers, 1810-1885. PANL.

Job Family Papers, 1864. PANL.

Mac Lee Collection. Munn, "Balance to December, 1873." Box 4B, No. 262. PANL.

Newman and Company Letter Books, #64, #67, #68, #69, #70; 1864-1900. PANL.

Ryan's "Sealers' Book: 1876-1877." Bonavista. MHA, MUN.

Printed

Great Britain. Customs and Excise Records. *Trade of the United Kingdom*. Annual Reports beginning in 1831. Title was changed to *Trade and Navigation of the United Kingdom* in 1853, and changed in 1912 to *Annual Statement of Trade*. London, 1831-1914. (Import statistics located in the Board of Customs Library, Mark Lane, London.)

_____. Parliament. *Parliamentary Papers*. 1817 (VI).

Chamber of Commerce Minute Book, 1866-1875, Vol. 5.

The Crown vs the Directors and Manager of the Commercial Bank of Newfoundland. St. John's, 1895.

The Crown vs the Directors of the union Bank of Newfoundland. St. John's, 1985.

Lambert, Sheila (ed.). *House of Commons Sessional Papers of the Eighteenth Century*. "Third Report from the Committee appointed to enquire into the State of the Trade to Newfoundland, 1793." Wilmington 1975, Vol 90.

Newfoundland. *Blue Books*. 1835-1915.

_____. *Census Returns*. 1836, 1845, 1857, 1869, 1874, 1884, 1891, 1901 and 1911.

_____. *Journals of the House of Assembly*. 1833-1915.

_____. "[Royal] Commission of Enquiry into the Sealing Disasters of 1914." Report. *Journal of the House of Assembly*. 1915.

_____. *Statutes of Newfoundland*. 1873-1916.

Canada. *Gazetteer of Canada: Newfoundland and Labrador.* 1968.

NEWSPAPERS

1. Newfoundland

    The local newspapers were extensively researched by Mr. Wayne Andrews and Mr. Michael O'Connell who were employed by Dr. Cater Andrews as research assistants and paid from a grant received from the Canada Council. A student, Mr. Jesse Fudge, also aided in this research. Research notes and photocopies of newspapers can be found in the Cater W. Andrews (CWA) Collection: 156 V, 158 V, 159 V and 975 VI (see below). The information referenced to the local newspapers was almost exclusively found in the above files. Readers are asked to note that there are some gaps in the CWA Collection because not every issue was researched and some photocopies are faded and useful only as guides—still an important use. (Tables of sailing vessels and captains were taken directly from the local newspapers.) The newspapers used are as follows:

    Courier, (St. John's) 1844-78. Began as the *Morning Courier and General Advertiser*, became the *Morning Courier* and, later the *Courier*.

    Chronicle (St. John's) 1865-81. Began as the *Morning Chronicle* in 1865 and later became the *Chronicle*.

    Daily News (St. John's) 1894-1914.

    Day Book (St. John's) 1862-65. Became the *Morning Chronicle*.

    Evening Herald (St. John's) 1890-1914. Began as the *Evening Mercury*.

    Evening Mercury (St. John's) 1882-1889. Became the *Evening Herald* 1890.

    Evening Telegram (St. John's) 1879-1914.

    Express (St. John's) 1851-76. Began as the *Newfoundland Express* and later became the *Express*.

    Mercantile Journal (St. John's) 1816-27. Full title: *Newfoundland Mercantile Journal*.

    Newfoundlander (St. John's) 1827-84.

    Patriot (St. John's) 1833-77. Began as the *Newfoundland Patriot* but changed its title several times, always retaining "*Patriot*".

    Pilot (St. John's) 1852-53.

    Public Ledger (St. John's) 1827-82. Began as the *Public Ledger* but underwent several name changes including a brief period as the *Daily Ledger* (1879-80).

    Royal Gazette (St. John's) 1807-1914. Full title: *Royal Gazette and Newfoundland Advertiser* although for a brief period it was called the *Royal Gazette*. (It became the *Newfoundland Gazette* in 1924.)

Sentinel (Carbonear) 1836-45. Began as the *Sentinel and Conception Bay Advertiser*, became the *Carbonear Sentinel and Conception Bay Advertiser*, and finally in 1845 the *Sentinel*.

Standard (Harbour Grace) 1859-1914. Began as the *Standard and Conception Bay Advertiser* and became the *Harbour Grace Standard and Conception Bay Advertiser* and later the *Harbour Grace Standard*.

Star (St. John's) 1840-47. Full title: *Star and Newfoundland Advocate*.

Times (St. John's) 1832-95. Full title: *Times and General Commercial Gazette*.

Weekly Herald (Harbour Grace) 1842-54. Full title: *Weekly Herald and Conception Bay General Advertiser*.

Weekly Journal (Harbour Grace) 1828-29. Full title: *Harbour Grace and Carbonear Weekly Journal and General Advertiser for Conception Bay*.

2. Great Britain

Mark Lane Express (London) 1832-1880. The most important source for prices of commodities on the London markets with some information on Bristol, Hull and Liverpool markets. After 1880 prices quoted for seal oil and other "common fish oils" were no longer published.

CATER W. ANDREWS COLLECTION

The CWA Collection was catalogued by Ms. J. M. Neeson. The following files, listed as they appear in her catalogue, were used in this study (except where noted, square brackets used as inserted by Ms. Neeson):

23 II *4 x 6 card file "MEN"*
Various sections:
i.Unfiled names of captains etc. with references from *ET* re orbits. [SR: obits.] or other biographical details.
ii.Names of captains etc. with note of their vessels (schooners, SS or m.v.) and occasionally other biographical information.
iii.Notes of ships arranged by name and date of building, including details of size, speed etc., various dates 1870-1968, mostly 20thC.

24 II *4 x 6 card file "The Schr. Men of B.N."* (The Schooner Men of Bonavista North
Various sections:
i.Men 1925-72, 23 names mostly *ET* source; also one from *Sydney Daily Post*, 1907. Photograph of a Labrador fishing schooner going north off Hopedale, 1953.

The Ice Hunters 513

iii. Towns. Details of the captains and vessels associated with these towns arranged by town, mostly from *ET*.

25 II *4 x 6 card file labelled "B.N."* [Bonavista North]
Various sections:
i. Unfiled notes re Nfld in general and St. John's especially (from McGregor, 1832 largely); also B.N. biographical notes.
ii. "B.N. HIST., Resources, & People" notes by subject: Churches, Educn, Health and Welfare, Industry, Communications etc. Various sources, often Prowse or similar, occasionally newspapers (*ET*, *NQ*). Several photographs of Wesleyville and CWA on fishing trip, 1940?.

33 II *4 x 6 card file "one"*
Various sections:
i. Unfiled cards for this index.
ii. "Ships of the Polar Seas"
Arranged by name, A—M. details of SF ships from various primary and secondary sources. Some photographs including primary and secondary sources. Some photographs including Bloodhound 2, Algerine 1, Arctic Prowler, Bellaventure, Blue Seal..., Eagle III, etc. Dates: 1870-1950s.

34 II *4 x 6 card file "two"*

i. Unfiled material
ii. Continuation of 33,II "Ships of the Polar Seas" arranged by ship N-Y. Photographs and details of ships as in 33,II.

35 II *4 x 6 card file "three"*
Various sections:
i. Unfiled material
ii. 3 x 5 cards indexed by subject: "Catches" [1892 only]; Communications; Conservation; Dictionary of Terms; Dundee [Chafe, *DN*, *Bowr.Mag.*]; Gulf SF etc. Sources: printed primary.
iii. 4 x 6 cards arranged by name of vessel, details of sailing ships in the SF. This is the major part of the file.
iv. 4 x 6 cards, arranged by name of sea captain or master or other, details of voyages and vessels from printed primary sources (*ET*, *Times*, *Roy.Gaz.*, *NQ*, Chafe, *DN*), also Prowse.
v. 4 x 6 cards, arranged by town, reference to "SB" or "SBk" [Chafe?] at various dates, 1832, 1834, 1864 etc; further notes from *ET* 1930's—'60's.

36 II	*4 x 6 card file "four"* Various sections: i. Unfiled material ii. "Captains". Notes from Condon, 1925; *Nfld Who's Who*; *DN*; Chafe; *ET*; Loomis C.C.—1971; *Polar Rec.*; *NQ*; Stefanson, 1921. Includes photograph of Ben Andrews; photograph of portrait of Capt. Wm Munden, 1776; photograph of Capt. Geo. Whiteley, 1897. Most notes are for the period 1860's—1960's with one or two dated earlier.
156 V	*Box file, not labelled* Notes taken from *ET* by Wayne Andrews, Michael O'Connell and CWA re SF for the period 1879-83; 1899; 1906-8; 1910-19; 1921-5. Some of this material is annotated by CWA "4x6" etc. suggesting that it was further filed in his card files of SF notes.
158 V	*Box file not labelled* Notes taken from newspapers: (as 156 and 157,V) *Morning Courier* 1857; 1864; 1845-56. *Daily News* 1920-1. *Evening Herald* 1892; 1900-5; 1909. *Evening Mercury* 1887-9 (superseded by *Evening Herald*). *Harbour Grace Standard* 1863-7; 1873; 1876-90. *Mercantile Journal* 1826-7. *Newfoundland Express* 1852-64. *Newfoundland Patriot* 1836-79. *Public Ledger* 1827-82. *Royal Gazette* 1828-45; 1846-82. *Sentinel and Conception Bay Adventurer* 1839-40; 1843-5. *Star* 1844-6. *Times* 1833-40; 1841-79. *Weekly Herald* (Harbour Grace) 1845-54; 1829. Notes taken by J. Fudge from documents from Trinity Historical Archives (loaned by K. Matthews).
159 V	*Box file, not labelled* Notes taken from the *Newfoundlander*, photocopies made from microfilm and placed in two ring binders, one for the period 1828-48 and 1849-79.
213 VI	*Ring binder labelled "4 Correspondence"* Correspondence re SF work: 1968-75. Again, much of the work of tracing individual careers, making contact with others researching the same or similar fields, surfaces here. One example: notes on the career of Capt. Joseph Jeffers, 1830-87, from Dr. George Jeffers. Also Capt. Nathan Norman, 1809-?. Also correspondence with the Trinity archivist Walter White. Two file folders at end with further correspondence as above.
361 III	*Envelope labelled "Dr. George Jeffers"* Biographical details of Captain Joseph Jeffers 1830-87 (commands and catches at seal hunts 1867-77). Also some IFAW literature for

	1974 and between Dr. George Jeffers and CWA re sealing (including note about the Munns of Harbour Grace).
375 III	*Envelope labelled "Lacey Alice"* Letter from Alice Lacey re photograph of the Erna
975 VI	*Large ring binder labelled "SF/DN 1920 Fudge..."* Notes taken from *ET* by M. O'Connell re SF for following years: 1880: 1883, 1885, 1886, 1890, 1891, 1893, 1894, 1896, 1899; and similarly from *DN* for 1895; and *Evening Herald* for 1897, 1900-5, 1909. All re SF.
980 Misc.	*Bound volume* Bound carbon copy of "Evidence taken before the Commission In the Matter of the Enquiry into [the] disaster at [the] Seal Fishery of 1917" [evidence given re Newfoundland and Southern Cross disasters].
1047 IV	*File folder, not labelled* i.Original ms: Sealers' Agreement, 1871, Dart. ii.Original ms: Sealers' Agreement, 1871, Gem. xi.Xerox copies: Sealing agreement, SS Bear, 1879.

SECONDARY

Books

Anspach, Reverend Lewis Amadeus. *A History of the Island of Newfoundland*. London: 1819.

Archibald, Samuel George. *Some Account of the Seal Fishery of Newfoundland*. St. John's: 1852.

Aspinall, J. *Liverpool: A Few Years Since*. Liverpool: 1885.

Bacot, H. Parrott. *Nineteenth Century Lighting: Candle-powered Devices, 1783-1883*. West Chester, Pennsylvania: 1987.

Baines, Thomas. *History of the Commerce and Town of Liverpool and of the Rise of Manufacturing Industry in the Adjoining Counties*. London: 1948.

Baker, Melvin, J. Miller Pitt and R. Pitt. *The Illustrated History of Newfoundland Light and Power*. St. John's: 1990.

Bartlett, Robert A. *The Log of Bob Bartlett*. New York: 1928.

Bergman, Gösta M. *Lighting in the Theatre*. Stockholm: 1977.

Bliss, Henry. *Statistics of the Trade, Industry and Resources of Canada and the other Plantations in British America*. London: 1833.

Bonnycastle, Richard Henry. *Newfoundland in 1842*. London: 1842.

Brooke, Richard. *Liverpool as it was During the Last Quarter of the Eighteenth Century: 1775-1800.* Liverpool: 1853.

Bullen, Frank T. *The Log of a Sea-Waif.* London: 1901.

Busch, Briton Cooper. *The War against the Seals: A History of the North American Seal Fishery.* McGill: 1985.

Brown, Cassie. *Death on the Ice.* Toronto: 1972.

Byrne, Cyril J. and Margaret Harry. eds. *Talamh an eisc: Canadian and Irish Essays.* Halifax: 1986.

Candow, James E. *Of Men and Seals: A History of the Newfoundland Seal Hunt.* Ottawa: 1989.

Carroll, Michael. *The Seal and Herring Fisheries of Newfoundland together with a Condensed History of the Island.* Montreal: 1873.

Cashin, Peter. *My Live and Times: 1890-1919.* ed. R. E. Buehler. St. John's: 1976.

Cell, Gillian. *English Enterprise in Newfoundland: 1577-1660.* Toronto: 1969.

_____, ed. *Newfoundland Discovered: English Attempts at Colonization, 1610-1630.* London: 1982.

Chafe, Levi George. *Report of the Newfoundland Seal-Fishery from 1863 ... to 1894.* St. John's: 1894.

_____. *Report of the Newfoundland Seal Fishery from 1863 ... to 1905.* St. John's: 1905.

Chappel, Edward. *Voyage of the HMS Rosamund to Newfoundland and the Southern Coast of Labrador.* London: 1818.

Coaker, W. F. compiler. *Twenty Years of the Fishermen's Protective Union of Newfoundland: from 1909-1929.* St. John's: 1930.

Connor, R. D. *The Weights and Measures of England.* London: 1987.

Cossons, Neil. ed. *Rees's Manufacturing Industry.* Vol. 5. London: 1819-20.

Cuff, Robert, Melvin Baker and Robert Pitt. eds. *Dictionary of Newfoundland and Labrador Biography.* St. John's: 1990.

Davenport, W. H. *Lighthouses and Lightships.* London: 1878.

Davis, Ralph. *The Rise of the English Shipping Industry in the Seventeenth and Eighteenth Centuries.* London: 1962.

_____. *A Commercial Revolution: English Overseas Trade in the Seventeenth and Eighteenth Centuries.* London: 1967.

Defoe, Daniel. *A Tour through the Whole Island.* London: 1788.

Devine, Patrick Kevin. *Ye Olde St. John's.* St. John's: 1936.

Ditchfield, H. H. *The City Companies of London and their Good Works*. London: 1904.

Doyle, Gerald S. *The Old Time Songs and Poetry of Newfoundland*. St. John's: 1927.

England, George Allan. *The Greatest Hunt in the World*. New York: 1924, reprinted Montreal: 1969.

Gillespie, Bill. *A Class Act: An Illustrated History of the Labour Movement in Newfoundland and Labrador*. St. John's: 1986.

Greene, William Howe. *The Wooden Walls Among the Ice Floes*. London: 1933.

Gunn, Gertrude. *The Political History of Newfoundland: 1832-64*. Toronto: 1966.

Handcock, W. Gordon. *Soe longe as there comes no women: Origins of English Settlement in Newfoundland*. St. John's: 1989.

Harrington, Michael. *Goin' To The Ice: Offbeat History of the Newfoundland Sealfishery*. St. John's: 1986.

Harvey, Moses and Joseph Hatton. *Newfoundland, The Oldest British Colony*. London: 1883.

Head, Grant. *Eighteenth Century Newfoundland*. Toronto: 1976.

Heath, George. *The New History, Survey and Description of the City and Suburbs of Bristol*. London: 1794.

Hudson, Derek and Kenneth W. Luckhurst. *The Royal Society of Arts*. London: 1954.

Hunt, Robert. *Ure's Dictionary of Arts, Manufacturers and Mines, containing a clear exposition of their principles and practices*. 6th ed. London: 1875.

Jackson, Gordon. *The British Whaling Trade*. London: 1978.

Jenkins, J. T. *A History of the Whale Fisheries*. London: 1921.

Jukes, Joseph Beete. *Excursions in and about Newfoundland During the Years 1839 and 1840*. London: 1842.

Kean, Abram. *Old and Young Ahead*. London: 1935.

Kemp, Peter. ed. *The Oxford Companion to Ships & the Sea*. Oxford: 1976.

Kent, Rockwell. *North by East*. New York: 1930.

Laing, Alastair. *Lighting: The Arts and Living*. London: 1982.

Lindsay, David Moore. *A Voyage to the Arctic in the Whaler 'Aurora'*. Boston: 1911.

Mannion, John. *Irish Settlements in Eastern Canada: A Study of Cultural Transfer and Adaptation*. Toronto: 1974.

_____. ed. *The Peopling of Newfoundland: Essays in Historical Geography.* St. John's: 1977.

Martin, Cabot. *No Fish and Our Lives: Some Survival Notes for Newfoundland.* St. John's, 1992.

Martin, Wendy. *Once Upon a Mine: Pre-Confederation Mines in Newfoundland.* Montreal, 1983.

Matthews, Keith. *Lectures on the History of Newfoundland: 1500-1830.* St. John's: 1988.

McDonald, Ian D. H. *To Each His Own: William Coaker and the Fishermen's Protective Union in Newfoundland Politics, 1908-1925.* ed. James K. Hiller. St. John's: 1987.

McGrath, Patrick Thomas. *Newfoundland in 1911.* London: 1911.

Mosdell, Harris Munden. *When was That: 5000 facts about Newfoundland.* St. John's: 1923.

_____. *Chafe's Sealing Book.* 3rd ed. With an Introduction by William Archibald Munn. St. John's: 1923.

Murphy, James. *Songs Sung by Old Time Sealers of Many Years Ago.* St. John's, 1925.

Murray, Peter. *The Vagabond Fleet: A Chronicle of the North Pacific Sealing Schooner Trade.* Victoria: 1988.

Neary, Peter and Patrick O'Flaherty. eds. *By Great Waters.* Toronto: 1974.

Noel, S. J. R. *Politics in Newfoundland.* Toronto: 1971.

O'Dea, W. T. *Darkness into Light: An Account of the Past, Present and Future of Man-Made Illumination.* London: 1948.

_____. *The Social History of Lighting.* London: 1958.

Peary, Robert E. *Nearest the Pole.* London: 1907.

_____. *The North Pole.* London: 1910.

Pedley, Reverend Charles. *The History of Newfoundland from Earliest Times to the Year 1860.* London: 1863.

Prowse, Daniel Woodley. *A History of Newfoundland from the English, Colonial, and Foreign Records.* London: 1895.

Quinn, David B. and Neil M. Cheshire. eds. *The New Found Land of Stephen Parmenius.* Toronto: 1972.

Robbins, F. W. *The Story of the Lamp and the Candle.* Oxford: 1939.

Rompkey, Ronald. *Grenfell of Labrador: A Biography.* Toronto: 1991.

Rowe, Frederick W. *A History of Newfoundland and Labrador.* Toronto: 1980.

Ryan, Shannon and Larry Small. *Haulin' Rope and Gaff: Songs and Poetry in the History of the Newfoundland Seal Fishery.* St. John's: 1978.

_____. *Fish out of Water: The Newfoundland Saltfish Trade, 1814-1914.* St. John's: 1986.

_____ assisted by Martha Drake. *Seals and Sealers: A Pictorial History of the Newfoundland Seal Fishery.* St. John's: 1987.

_____. ed. *Chafe's Sealing Book: A Statistical Record of the Newfoundland Steamer Seal Fishery, 1863-1941.* St. John's: 1989.

Sager, Eric. *Seafaring Labour: The Merchant Marine of Atlantic Canada, 1820-1914.* Kingston: 1989.

Scoresby, William Jr. *An Account of the Arctic Regions with a History and Descriptio of the Northern Whale Fishery.* Edinburgh: 1820.

Schumpeter, Elizabeth Boody. *English Overseas Trade Statistics.* Oxford: 1960.

Simmonds, P. L. *Science and Commerce: Their Influence on our Manufacturers.* London: 1872.

Smallwood, J. R. *Coaker of Newfoundland.* London: 1927.

Smith, Nicholas. *Fifty-two Years at the Labrador Fishery.* London: 1936.

Stevenson, D. Alan. *The World's Lighthouses Before 1820.* London: 1957.

Stevenson, Robert. *English Lighthouse Tours: 1801, 1813, 1818.* ed. D. Alan Stevenson. London: 1946.

Story, George M., William J. Kirwin and John D. A. Widdowson. eds. *Dictionary of Newfoundland English.* Toronto: 1982 and 1990.

Thompson, F. F. *The French Shore Problem in Newfoundland.* Toronto: 1961.

Thwing, Leroy. *Flickering Flames: A History of Domestic Lighting through the Ages.* London: 1959.

Tocque, Reverend Philip. *Wandering Thoughts or Solitary Hours.* London: 1846.

_____. *Newfoundland: As it was and as it is in 1877.* London: 1878.

Unwin, George. *The Gilds and Companies of London.* London: 1963.

Wallace, James. *A General and Descriptive History of the Ancient and Present State of the Town of Liverpool.* Liverpool: 1885.

Waterer, John W. *Leather in Life, Art and Industry.* London: 1946.

Whiteley, George. *Northern Seas, Hardy Sailors.* New York: 1982.

Winsor, Naboth. *Stalwart Men and Sturdy Ships: A History of the Prosecution of the Seal Fishery by the Sealers of Bonavista Bay North, Newfoundland.* Gander, Newfoundland: 1985.

Articles

Abbott, Edward. "Lighthouses." *Galaxy* VII. New York: 1869.

Alder, C. R. "The Manufacture of Toilet Soaps." *Journal of the Society of the Arts* XXXIII. London: 1884-85.

Blake, Edith. "On Seals and Savages." *Nineteenth Century* XXV. London: 1889.

Burnham, John C. "A Neglected Field: The History of Natural Disasters." *Perspectives: American Historical Association Newsletter.* April, 1988.

Busch, Briton Cooper. "The Newfoundland Sealers' Strike of 1902." *Labour/Le Travail* XIV, (Fall 1984).

Cadigan, Sean. "The Staple Model Reconsidered: The Case of Agricultural Policy in Northeast Newfoundland." *Acadiensis* XXI, 2 (Spring 1992).

Church, Professor. "The Manufacture of Soap." Edited by G. Phillips Bevan. IV. *British Manufacturing Industries.* London: 1876.

de Beer, E. S. "Early History of London Street-Lighting." *History* New Series, XXV, London: 1941, 316-8.

Dunfield, Brian. ed. "The *Dash,*" *Newfoundland Law Reports, 1846-1853: Decisions of the Supreme Court of Newfoundland.* St. John's: 1915.

Fogarty, James J. "The Seal-Skinners' Union." *The Book of Newfoundland*, II, 1938.

Fyfe, Andrew. "On the Comparative Expense of Light derived from different sources...." *Journal of the Franklin Institute* V. Philadelphia: 1863.

Grenfell, Dr. Wilfred. "The Seal Hunters of Newfoundland." *Leisure Hours* (1897-98). (Photocopy in CNS).

Hawes, William. "On the Manufacture of Soap." *Journal of the Society of the Arts* IV. London: 1856.

Hiller, James K. "The Newfoundland Seal Fishery: An Historical Introduction." *Bulletin of Canadian Studies* VII, 2 (Winter 1983-84).

Howley, R. "The Fisheries and Fishermen of Newfoundland." *The Month* LXI. London: 1887.

Imellos, Stephanos D. "Hard Tack as Popular Food." In *Food in Change: Eating Habits from the Middle Ages to the Present Day,* ed. Alexander Fenton and Eszter Kisbán. Edinburgh: 1986.

"Impressions From Seals." *Chambers' Journal.* Edinburgh: 1854.

Kisbán, Eszter. "Food Habits in Change: The Example of Europe." In *Food in Change: Eating Habits from the Middle Ages to the Present Day*, ed. Alexander Fenton and Eszter Kisbán. Edinburgh: 1986.

"Light-House Construction and Illumination." *Putnam's Monthly Magazine of American Literature, Science and Art* VIII. New York: (1857).

"Lighthouse Illuminants." *Van Nostrand's Eclectic Engineering Magazine* XXXI. New York: 1884.

"Lighthouses." *Harper's New Monthly Magazine* XXXVIII. New York: (1868-69).

"Liverpool Fifty Years Ago." *Liverpool Daily (Evening) Albion*. Reprinted from *Albion* 1825-33.

"Lubricants for Machinery," *The Practical Magazine* VI, no. 19. (Translated from a report by M. D. Grothe in "Bulletin of the Netherlands Industrial Society".) London: 1874.

Lux, Herr. "Lubricants: Proceedings of the Society of German Engineers." *Van Nostrand's Eclectic Engineering Magazine* XXIX. New York: 1883.

Munn, William Archibald. "History of Harbour Grace." *Newfoundland Quarterly*. St. John's: 1934-39.

"Newfoundland." *Blackwood's Magazine* CXIV. Edinburgh: 1873.

Olds, Dr. John M. "Seal Finger or Speck Finger: A Clinical Condition Observed in Personnel Handling Hair Seals." *Canadian Medical Association Journal* LXXVI. Toronto: 1957.

Parsons, Alexander A. "Newfoundland Tragedy and the loss of the *Southern Cross*." *Newfoundland Quarterly* XIV, no. 1. St. John's: 1914.

_____. "Our Great Sealing Industry." *Newfoundland Quarterly* XIV. St. John's: 1915.

Paul, B. H. "Artificial Light and Lighting Materials." *Journal of the Franklin Institute* XLVIII (July-December). Philadelphia: 1864.

"Report from the Exhibition at the Crystal Palace, New York." *New York Freeman's Journal*. (Quoted in the *Newfoundlander* XX December 1853.)

Rodahl, Kaare. "Speck-Finger or Sealer's Finger." *Arctic* V, no. 4, 1952.

Sanger, Chesley W. "The Evolution of Sealing and the Spread of Permanent Settlement in North-eastern Newfoundland." Edited by John J. Mannion, *The Peopling of Newfoundland: Essays in Historical Geography*. St. John's: 1977.

_____. "The 19th Century Newfoundland Seal Fishery and the Influence of Scottish Whalemen." *Polar Record* XX, no. 126. Cambridge: 1980.

_____. "The Dundee-St. John's Connection: Nineteenth Century Interlinkages Between Scottish Arctic Whaling and the Newfoundland Seal Fishery." *Newfoundland Studies* IV, no. 1. St. John's: 1988.

Scoffern, J. "Artificial Illumination." *St. James Magazine* XV. London: 1866.

"The Seal Fishery of Labrador." *Hunts' Merchants Magazine* XLV, no. 5. New York: 1861.

"Seal Hunting." *The Penny Magazine* IV. London: 1835.

Shortis, H. F. "Sealing in the Old Days." *Newfoundland Quarterly* I, no. 4. St. John's: 1902.

Thornton, Patricia A. "The Evolution of Sealing." Edited by John J. Mannion, *The Peopling of Newfoundland: Essays in Historical Geography*. St. John's: 1977.

Tocque, Reverend Philip. "The Seal Fishery of Newfoundland." *Littell's Living Age* XXVII. Boston: 1850.

Tyndall, John. "A Story of our Lighthouses." *Nineteenth Century*, XXIV. London: 1888.

Wardle, Arthur. "Liverpool and the Newfoundland Trade." A paper submitted to the Liverpool Society for Nautical Research, 11 January 1939. *Liverpool Nautical Research Publications* I. 1933-44.

Williams, Mattieu. "Oils and Candles." Edited by G. Phillips Bevan. IV. *British Manufacturing Industries*. London: 1876.

MISCELLANEOUS

Cadigan, Sean. "Economic and Social Relations of Production on the Northeast Coast of Newfoundland with Special Reference to Conception Bay, 1785-1855." Ph.D. thesis. MUN, 1991.

Christmas Review, (St. John's) 1901.

Coates, Kenneth S. and W. R. Morrison. "Towards a Methodology of Disasters: The Case of the *Princess Sophia*." Presented to the Canadian Historical Association Meeting, Victoria, BC: 1990.

Co-Partners' Magazine (The Liverpool Gas Co.). IX, no. 2. April 1948.

Crewe, Nimshi. "A Descriptive Monograph on the Slades." William White Collection. PANL.

Davis, David. "The Bond-Blaine Negotiations: 1890-1891." MA thesis. MUN, 1970.

Dictionary of Canadian Biography.

Disasters. "In the Matter of Enquiry into Disasters at Seal Fishery of 1914: Evidence taken before the Commission," 1915. (Unpublished mss.) CWA Collection, CNS. Copies in CNS library and PANL.

Encyclopedia Americana. 1993. "Leather" and "Oils."

Encyclopedia Britannica. 11th ed. "Leather," "Oils" and "Soap."

Encyclopedia of Newfoundland and Labrador. Newfoundland Book Publishers and Harry Cuff Publications, St. John's: 1967. "Agriculture" by Catherine F. Horan; and "Lighthouses" by Malcolm MacLeod.

Feltham, John. "The Development of the F. P. U. in Newfoundland." MA thesis. MUN, 1959.

Greene, John P. "The Influence of Religion in the Politics of Newfoundland: 1850-1861." MA thesis. MUN, 1970.

Harris, Leslie. "The First Nine Years of Representative Government." MA thesis. MUN. 1958.

_____. "Independent Review of the State of the Northern Cod Stock: Final Report." Ottawa, 1990.

Hattenhauer, Ralph. "A Brief Labour History of Newfoundland." (CNS), Typescript, 1970.

Hickey, Patrick Joseph. "The Immediate Impact of the 1894 Bank Crash." BA Honours Dissertation. MUN, 1980.

Hiller, James K. *The Newfoundland Railway: 1881-1949.* Newfoundland Historical Society Pamphlet. St. John's: 1981.

Joy, John Lawrence. "The Growth and Development of Trades and Manufacturing in St. John's, 1870-1914." MA thesis. MUN, 1977.

LeMessurier, H. W. *Old Time Newfoundland.* Edited by C. W. Fay. Typescript. CNS, 1955.

Little, Linda. "Plebian Collective Action in Harbour Grace and Carbonear, Newfoundland." MA thesis. MUN, 1984.

MacKinnon, Robert. "The Growth of Commercial Agriculture around St. John's, 1800-1935: A Study in Local Trade in Response to Urban Demand." MA thesis. MUN, 1981.

MacWhirter, W. D. "A Political History of Newfoundland: 1865-1874." MA thesis. MUN, 1963.

Matthews, Keith. "History of the West of England-Newfoundland Fishery." D.Phil. thesis, Oxford, 1968.

_____. "The Class of '32: The Newfoundland Reformers on the Eve of Representative Government." Typescript, 1974. CNS.

_____. "Profiles of Water Street Merchants." Typescript, 1980. CNS.

Moulton, E. C. "The Political History of Newfoundland: 1861-1869." MA thesis. MUN, 1960.

Murphy, James. *The Old Sealing Days*. St. John's: *The Evening Herald*, 1916. Reprinted as, "Project of the Newfoundland Archives," 1971.

Neary, Peter. "The French Shore Question: 1865-1878." MA thesis. MUN, 1961.

Ralph, Elizabeth. *The Streets of Bristol*. Bristol Branch of the Historical Association, Pamphlet #49, 1981.

Rees, Abraham. "The Cyclopedia; or Universal Dictionary of Arts, Science and Literature." London: David and Charles Reprints, 1972.

Reeves, William. "The Fortune Bay Dispute; Newfoundland's Place in Imperial Treaty Relations under the Washington Treaty, 1871-1885." MA thesis. MUN, 1971.

_____. "Our Yankee Cousins: Modernization and the Newfoundland-American Relationship, 1898-1910." Ph.D. thesis. Maine, 1987.

Reid, John. "Warrior Aristocrats in Crisis: The Political Effects of the Transition from the Slave Trade to Palm Oil Commerce in the Nineteenth Century Kingdom of Dahomey." Ph.D. thesis. University of Stirling, Scotland, 1986.

Regular, Donald K. "The Commercial History of Munn and Company, Harbour Grace." Typescript, 1973[?]. CNS.

Ryan, Shannon. "Abstract of C.O.194 Statistics." Unpublished manuscript, MUN, 1969. CNS.

_____. "The Newfoundland Cod Fishery in the Nineteenth Century." MA thesis. MUN, 1972.

_____. "The Seal in Newfoundland Culture." Presented to the American Folklore Society, Philadelphia: 1976.

_____. "The Origin and Early Growth of Newfoundland's Seal Fishery." Presented to the annual meeting of the Canadian Historical Association, Guelph: 1984.

Sanger, Chesley W. "Technological and Spatial Adaptation in the Newfoundland Seal Fishery in the Nineteenth Century." MA thesis. MUN, 1973.

_____. "The Newfoundland Seal Fishery and Scottish Whalemen in the Nineteenth Century." Presented to the 15th General Assembly of the International Geographical Union, Tokyo: 1980.

_____. "Dundee Steam-Powered Whalers and the Newfoundland Harp Seal Fishery." Presented to the Annual Conference, British Association for Canadian Studies, Southampton: 1988.

Scott, John Roper. "The Function of Folklore in the Inter-relationship of the Newfoundland Seal Fishery and the Home Communities of the Sealers." MA thesis. MUN, 1975.

Seals and Sealing in Canada: Report of the Royal Commission on Seals and the Sealing Industry in Canada. Albert H. Malouf, Chairman. Ottawa: 1986.

Shortis, Henry Francis. "From the well-stored Mines of the Tradition of Newfoundland." In *Fugitive History of Newfoundland.* PANL.

_____. "The Brigantine *Fanny Bloomer*." (Typescript. CNS) Vol. IV, 243, PANL, n.d.

_____. "Hard Winters of the Past when the Mercury almost Froze." In *Fugitive History of Newfoundland.* (Typescript. CNS) Vol. V, 380, PANL.

_____. "Loss of the Brigantine *Eric*: 1878." In *Fugitive History of Newfoundland.* (Typescript. CNS) Vol. V, 71, PANL.

Wells, E. A. "The Struggle for Responsible Government in Newfoundland: 1846-1855." MA thesis. MUN, 1967.

White, Edward. 1866. Letter to Jobs. Private collection of Dr. George Story, MUN, St. John's.

Films

Frissell, Varick. *The Great Arctic Seal Hunt.* 1927.

_____. *The Viking.* 1930.

Index

A

Aberdeen, 150, 383
"Accounts of the Spring Seal Fishery", 127, 128
Act to Regulate the Prosecution of the Seal Fishery, 113, 354
Adams, William Sr. (Capt.), 261, 280n, 375, 402n
Agriculture, 43, 92, 94-7, 108, 117, 118n, 119n, 135
American Embargo Act, 1807, 41
American Revolutionary War, 27, 27, 32, 37, 60n, 70, 78
Anglo-American War, 1812-14, 36, 42-3, 122, 137, 284
Anglo-French agreement, 1904, 165
Anspach, Lewis Amadeus, 122-5, 129, 132, 202, 203n, 204n, 230, 234-5, 243-5, 249-53, 274n, 275n, 276n, 277n, 283, 291, 318, 321n
Antarctic, 149, 195, 254, 279n, 403n
Archibald, Samuel George, 148, 204n
Arctic, 147-8, 150-1, 153, 165, 175, 186, 188, 195, 197, 206n, 375, 393-6, 402n
Argand lamp, 73-6, 78, 80, 82
see also Lighting; Oil lamps
Association of Newfoundland Fishermen and Sharemen, 397

B

Baccalieu Island, 55, 180, 213, 215, 284
Baine, Johnston and Company, 135, 136, 138, 145-7, 150-9, 175, 177, 179, 183, 184, 189, 191, 193, 195, 197, 206n, 207n, 211n, 220, 258, 310, 325, 348, 361, 414, 458, 463-5, 487, 490-2, 494, 506-7
Baird, James, 164, 165, 183-5, 193, 348, 363
Bank crash, 108, 170, 179, 210n, 399n, 408
Bank fishery, 27-9, 34-6, 59n, 108, 166, 318
Barbour, George (Capt.), 191-2, 195, 196, 207, 229, 241, 498, 499, 503
'Barrel man', 150, 206n, 257
Barron, John (Capt.), 139, 140, 370, 481, 501
Bartlett, Abram (Capt.), 55, 142, 152, 174, 207n, 220, 221, 241, 485, 490, 495-6
Bartlett, Isaac (Capt.), 155, 221, 258, 273n, 382, 393, 402n, 495, 500
Bartlett, Moses (Capt.), 226, 242, 394, 499
Bartlett, Robert A. (Capt.), 55, 196, 198, 210n, 384, 394, 402n, 403n, 447, 465, 475, 487, 488
Bartlett, William J. (Capt.), 189, 191, 196, 223, 242, 447, 497-9
'Batsman', 197, 230-1, 247-8, 312, 315, 318, 330, 332-4, 340, 360, 395, 397n, 504
see also Ice hunters

Bay Bulls, 128, 371, 418, 425-6, 507
Bay de Verde, 52, 144, 360, 418
Bay Roberts, 56, 124, 131-2, 134, 138, 14, 154-5, 159-60, 178, 180, 214, 218, 220, 222, 225-6, 232, 241, 255, 258, 272n, 342, 346, 381, 419, 454, 457, 464, 484, 490, 494, 496-500, 504
'Beaver Hat Man', 125
'Bedlamers', 48, 53, 148, 197, 504
see also Harp seal; Hood seal
Bennett, C.F., and Company, 155, 145, 414, 459,
Berth, 161, 163, 166, 173-5, 177-9, 181, 186, 188, 210, 217, 218, 224-7, 230-3, 238, 239, 243, 248, 251, 262, 272n
'Betty' lamp, 66
see also Lighting; Oil lamps
Bird, Richard E. (Admiral), 403n
Blake, Lady Edith, 390-2, 402n
Blake, Sir Henry, 392, 394, 402n
Blanc Sablon, 208n, 241, 382
Bland, John, 52, 54
Blandford, Darius (Capt.), 189, 268, 496, 496
Blandford, Samuel (Capt.), 115, 171-2, 179, 181-2, 191, 222, 227, 240-41, 299, 344, 364, 382, 392, 496, 497, 502, 503
'Blue back' (hood seal), 49, 192
Bonaventure, 52, 426
Bonavista, 29, 30, 49-53, 55-7, 59, 60n, 121, 123-4, 128, 134, 136-7, 143-4, 173-4, 179-80, 202n, 204n, 205n, 213, 219, 225, 229, 234, 237-8, 242, 265-8, 271-2, 284, 290-4, 296, 302,
306, 308, 327, 382, 418, 425-8, 495, 497-9, 500
Bonavista Bay, 53, 56, 57, 124, 134, 136, 144, 179, 180, 202n, 219, 234, 265, 268, 271n, 284, 290, 293, 296, 306, 308, 342, 345, 368, 377, 392, 425, 495, 496-500
Bond, Sir Robert, 110, 115, 166, 364
Bonnycastle, Sir Richard, xvii, xix
Bowring Brothers, 135, 145, 150-9, 170-1, 179, 184, 185, 189, 190-3, 195-7, 206n, 211n, 212n, 220, 241, 257, 278n, 345, 347, 349, 358, 363, 371, 459, 487, 490, 491, 492, 494
Boyle, Sir Cavendish, 342, 363, 365
Brigus, 44, 55-6, 102-3, 124, 125, 130-4, 138-9, 142, 144, 155, 158, 159, 160, 1613, 174, 201, 202, 213-5, 217-8, 220-2, 224, 258-9, 272, 335-6, 338-40, 342, 346, 353, 378, 380, 293-4, 397n, 418, 454, 463, 484, 485, 490, 494, 495, 497, 499
Brookfield, 499, 500
Brooking, Son and Company, 135-6, 155-6, 159, 460

Brown, Cassie, 351
Bye boat fishery, 27, 34-6

C

Calloway, Simeon, 342
Canada, 50, 107, 109, 115, 119, 199, 354, 408, 414
Candles, 65, 67, 68, 70, 71, 74, 76, 78, 82, 84, 86, 88, 89n
  see also Lighting
Cape Freels, 179, 295, 306, 324n
Cape Race, 200, 310, 311, 413
Cape Ray, 59n, 116, 413
Cape St. Francis, 224, 300, 302
Cape St. John, 45, 58, 60n, 266, 377
Cape Spear286, 320n, 371
Carbonear, 30, 44, 52, 56, 95-7, 124-5, 130-4, 136, 138, 141, 142, 146, 156-8, 160, 162-3, 179, 180, 201-2, 204, 214, 216-7, 222, 224, 237, 255, 258, 259, 269, 273n, 280n, 284, 288-9, 291, 293-4, 306, 315, 317, 324n, 325n, 327n, 388, 396n, 404, 409
Carcel lamp, 74
  see also Lighting; Oil lamps
Carson, William, 95, 406
Casey, Tom (Capt.), 215, 269, 272, 385
'Cat' (harp seal), 48, 114, 354, 401
  see also 'Whitecoats' (harp seal)
Catalina, 136-7, 154, 156, 159, 160, 162-3, 174, 196, 202, 204n, 205n, 218, 220-1, 225-6, 229, 255, 265-66, 296, 297, 301, 326n, 327n, 368, 371, 374, 394
Chafe, Levi George, xvii, xix, 130, 139, 152, 154, 163, 174, 187, 211, 214, 237-8, 240, 266, 272n, 273, 275n, 282, 291, 292-3, 299, 318-9, 320n, 321n, 384, 396, 402n
Change Islands, 266, 301
Channel-Port aux Basques, 16-8, 174, 180, 186, 193, 196, 226, 299
Clarke, George (Capt.), 310, 325n
Clarke, John (Capt.), 223
Cleary, Philip (Capt.), 145, 153, 154, 206
Coaker, Sir William, 197, 212, 230, 346, 353, 355, 366, 400n
'Coaling charge', 231, 239, 345, 348
  see also Ice hunters
Coconut oil, 83-5
  see also Oil
Cod fishery, xiii, xiv, xvii, xxiv, 65, 70, 71, 76, 81, 84, 92-4, 97, 98, 100, 104-6, 108, 114, 117-8, 123, 125, 130, 135, 137-8, 142-3, 146, 156-7, 164-5, 178, 183, 201, 208n, 216-7, 219, 230, 232, 235-7, 239, 244, 254, 270-1, 275n, 331, 346, 350, 352-3, 368, 370, 382, 385, 396, 404, 405, 407, 408, 410, 411
  see also Migratory fishery; Saltfish industry
Cod liver oil, 84
Cod oil, 49, 54, 67, 70, 71, 76-7, 79, 80-1, 84, 87, 88n, 89n, 99, 104, 183, 185, 201, 202n, 278n
  see also Fish oil; Oil; Train oil
Collins, Jesse, 315, 367
Colza oil, 82-3
  see also Oil; Rape seed oil
Commercial Society (St. John's), 100
Committee for the Relief of Those Who Lost Their Friends at the Seal Fishery Last Spring (1830), 317
  see also Sealing disasters
Conception Bay, 95-6, 110, 119, 121-35, 137, 139, 141-4, 158, 161, 173-4, 179, 196, 201, 203n, 204n, 213-9, 222, 224-5, 231-6, 242, 245-6, 271, 274n 283, 284, 285, 292, 293, 296, 298-9, 321n, 336, 337, 340-2, 346, 353, 356, 381, 397n, 399n, 404-9, 411
Coopers' Union, 398n
Cordwainers' guild, 69, 87n
Cox, William, and Company, 136-7, 156-7, 207
'Crop', 173, 177, 185, 191-2, 195, 196, 199, 209n, 210n, 212n, 239, 343, 345, 347
  see also Ice hunters
Crosbie and Company, 195, 197-8
'Cruises', 68
  see also Oil lamps
Cupids, 124, 131-2, 159, 214, 287-8, 293, 325n, 339
Curriers' guild, 69
Curtis, Tim, 375-6

D

Dawe, Charles (Capt.), 178, 241, 381, 401n
Dawe, Henry (Capt.) (Bay Roberts), 189, 192-2, 241, 263
Dawe, Robert (Capt.), 152-4, 220, 300
Dawson, Thomas, 312-5
Devine, Maurice A., xvii, xix, 384, 401n
Devine, Patrick Kevin, xvii, xix
'Dog' (gunner's assistant), 197, 248
'Dog' (harp seal), 48
'Dog' (hood seal), 48, 248, 390
Donnelly, W. J. S., 156, 158-9
Doyle, Gerald S., 401n
Duckworth, Admiral John Thomas, 42
Dundee Seal and Whale Fishing Company 147, 152, 175
Dundee Shipbuilders Company, Ltd., 186

E

Electricity, 82-84, 409, 413
  see also Lighting
England, 25-8, 30, 33-4, 39, 43, 49, 57, 59n, 60n, 62n, 65, 68-71, 76, 77, 85, 88n, 142, 145, 155-6, 183-4, 195, 199, 207n, 214, 237, 243-4, 246, 273n, 352, 360
England, George Allan, xvii, xx, 253, 257, 259, 264, 273n, 279n, 282, 318, 386
Evening Herald, 272n, 361, 369, 397n
Evening Mercury, 172

Evening Telegram, 115, 120, 163, 172-4, 176-8, 206n, 208n, 209n, 210n, 211n, 212n, 228, 233, 261, 264, 272n, 308, 322n, 345, 356-62, 365, 368, 370, 398

F

Fairweather, James (Capt.), 263-4
Farquhar, J. A. (Capt.), 150, 176-8, 180-2, 190, 195, 197-8, 210n, 226, 232
Feehan, Peter (Capt.), 214, 219, 372
Ferryland, 29, 52, 57, 123, 127-8, 137, 214, 218, 232, 327n
Fish oil, xvii, 69-71, 76-7, 82, 84, 105, 115, 404
  see also Cod oil; Oil; Seal oil; Train oil; Whale oil
Fish Out of Water, xvii, xix, 62n, 413
Fishermen of Carbonear and Harbour Grace, 396n
Fishermen's Mutual Protective Society of Newfoundland, 335, 397n
Fishermen's Protective Union, 197, 230, 346, 347, 353, 363, 365, 397n, 398n, 400n, 411
Fishing room, 25, 27, 30, 32, 34, 36, 42, 45, 56, 144, 216, 222, 405
Fishing ships, 25-8, 30, 34-6, 43, 50, 59, 71, 95, 104, 328, 404
  see also Migratory fishery
Fishing stages, 25, 29, 40, 62, 96, 122
Fishing station, xvii, xix, 35, 37, 40, 104, 122, 130, 328, 352, 395, 409
Fogo, 49, 52, 54, 55, 57, 124, 127-8, 134, 143, 156-7, 180, 195, 198-9, 200-1, 204n, 213, 265-8, 327n
Forest resources, 29, 97, 110
Fort Amherst, 320n
France 28, 31, 33, 35, 37, 41, 49, 58, 65, 67, 73, 78, 82, 88n, 93, 108, 110, 133, 408
French fishery, 43, 58, 110
French Revolutionary War, 34, 328, 409
French Shore, 35, 43, 45, 56-8, 60, 94, 96, 110, 165, 328, 404-5, 409
Frissell, Varick, 278n 320n
'The Front', 165, 178, 185
'Frosty Spring', 321n
Funk Island, 55, 213, 295-6, 306

G

Gaff, 247
Gambo, 225, 227, 228, 374
Gas lighting, 78-9, 81-4, 409
  see also Lighting
Gent, George (Capt.), 220
Gillam, John (Capt.), 165, 167-8, 342
Gosse, Pack and Fryer, 125, 317
  see also Pack, Gosse and Fryer
Gower, Sir Erasmus, 40, 41, 44-5, 58, 61n-64n
Graham, Alexander (Capt.), 151, 153, 154, 219-21
Great Britain, 41, 54, 67, 101, 115, 145, 199, 283, 405-6, 408-9

Anglo-French agreement, 110
Colonial Office, 39, 58, 130, 363
commercial revolution, 65-81
Commission of Government (Nfld.), 408
  imports, 429, 431, 432-5, 439
  oil Markets, xviii, 76-81, 83, 84, 143-4, 182-3, 297, 404, 406, 409, 410, 413
  oil Prices, 439-41
Passenger Act, 1803, 44
Peninsula War, 41-2
Great Northern Peninsula, 45, 59n, 60n, 268, 404-5, 412
Greely, A. W., 394, 403n
'Green Bay Spring', 265-6, 295-301
Greene, William Howe, xviii, xxn
Greenland, 76-7, 147-8, 175, 201, 394
Greenock, Scotland, 310
Greenspond, 52, 57, 124, 128, 136-7, 143, 156-7, 159, 160, 162-3, 167, 174, 179, 180, 213, 226-8, 288, 292-3, 295, 299, 305, 324n, 365, 382
Grenfell, Sir Wilfred, 269-70, 393, 412
Grey Islands, 196, 229, 293, 377
Grieve and Bremner, 145, 156-7
Grieve, Walter, and Company, 145, 147, 150-3, 155-6, 159, 175, 178, 345, 363
Guilds, 68-9, 85
'Gunner', 197, 216, 230-1, 247-50, 330-4, 340, 392, 395
  see also Ice hunters
Gulf of St. Lawrence, 41, 45, 47, 51, 116-7, 160, 164, 165, 167, 168, 173-7, 182-3, 189, 196, 198, 199, 223, 226, 305, 308-11, 357-8
Gulliford, James (Capt.), 219-20
Gushue, Moses, 226-7
Guy, William (Capt.), 263-4

H

Hagen, John (Capt.), 220-1
Halifax, 46, 176-8, 190, 195, 323, 382
Halifax Herald, 176
Hall Arctic expedition, 273n, 393
Hall, Charles, 393
Halleran, Terry (Capt.), 152-3, 188, 220-1, 371, 385
Hamilton, Sir Charles, 58, 64n
Hanrahan, Nicholas (Capt.), 155, 222
Hant's Harbour, 136, 143, 156, 160, 165, 293
Harbour Grace, 29-30, 44, 52, 55-6, 122, 124, 130-48, 154-63, 169-74, 177, 186, 189, 201, 202, 204n, 206n, 207n, 208n
Harbour Grace Standard, 173
Harbour Main, 124, 132-3, 144, 336, 338-40, 346
Hard bread, 243-7, 252-4, 259
Harp Seal (Phoca groenlandica), 47-9, 53, 70, 148, 163-4, 171, 184, 196, 247-8, 254, 266, 269-70, 375-6, 404, 410-1
  see also 'Whitecoats'
Harrington, Michael, xviii, xxn

Harris, Leslie, 407
Harvey and Company, 146, 153, 155-6, 187, 192-3, 196, 198, 311, 347-8
Harvey, Moses, xvii, xixn, 124
Hattenhauer, R., 333, 397
Hatton, Joseph, 124
Head, C. Grant, 95-6
Herring Neck, 303, 365
'High liner' sealing captains, 153, 190-2, 197-9, 241-2
see also Ice hunters; Sealing captains
Hiller, James K., xviii, xixn
Hiscock, John E., 315
Holloway, Admiral John, 41-2
Holyrood, 124, 338
Hood seal (Cystophora cristata), 48-9, 53, 148, 163-4, 172, 192-3, 248, 358, 375-6, 390
Horners, 67
Horwood, Warrick, 279n
Hudson Bay, 199, 394
Hudson's Bay Company, 199-200, 393
Hull, England, 111
'Hummocky' ice, 254
Hussey, Joseph, 361

I

Ice
    'hummocky' ice, 254
    'running' ice, 140, 251, 291, 296-7, 304
    'seal meadows', 101, 250
    'sheet' ice, 51, 306
    'slob' ice, 188
    'whelping' ice, 180, 199, 268
Ice blindness, 254
Ice breaker, 187-9, 200
'Ice captain', 177-8
Ice Cutting Act, 126
Ice hunters
    Arctic exploration, 393-5
    income, 234-43, 353, 368, 411
    labour movement and strikes, 328-53
    landsmen, 264-71
    living and working conditions, 243-64
    steamer captains, 219-223
    stowaways, 227-9
    see also Berth; Landsmen seal fishery; Seal fishery; Sealers' Agreements; Sealers' shares; Sealers' strikes
'Ice master', 256-7
Industrial revolution, 74
Inshore fishery, 28-9, 36, 45-6
    see also Cod fishery
Ireland,
    emigration to Newfoundland, 33, 43-4
    famine, 102-3
Iron clad steamers,
    see Steel vessels

J

'Jack Ass Brig', 125

Jackman, Arthur (Capt.), 214, 221, 227, 228, 232, 241, 257, 272n, 279n
Jackman, Peter, 310, 325n
Jackman, Thomas (Capt.), 221, 273n, 371
Jackman, William (Capt.), 153-5, 207n, 221, 240-1, 371-2
Jackson, Gordon, 76-7, 84
Jeffers, Joseph (Capt.), 222, 273n, 294, 322
Job Brothers and Company, 130, 135, 138, 145, 150, 152-7, 159, 167, 172, 179, 181, 184, 185, 189, 191-8, 206n, 208n, 211n, 212n
Jukes, Joseph Beete, xvii, xix, 131, 231, 244 5, 250-3, 256, 260, 261, 274n, 276n, 386, 389-92, 402n

K

Kean, Abram (Capt.), 221-2, 241, 253, 265, 277n, 306, 215, 326n
Kean, Jacob (Capt.), 191, 195-6, 364
Kean, Job (Capt.), 222
Kean, Joseph (Capt.), 192-3, 222, 311-2
Kean, Nathan (Capt.), 222
Kean, Westbury (Capt.), 198, 199, 222, 242, 311-315
Kean, William (Capt.), 151, 219, 221, 253
Kearney, Michael, 141
Keats, Richard Goodwin, 42-3, 62n
Kent, Rockwell, 316
Kerosene lamp, 82, 84
Kerosene oil, 109
    see also Oil
Kielty's Long Room, 335, 340
King's Cove, 124, 128, 136, 143, 146
Knee, Job (Capt.), 192, 196, 198, 222-3, 227
Knee, William (Capt.), 170-2, 223, 305
Kyle (SS), 311

L

Labrador, 37, 41', 47, 50-1, 52-4, 55, 58, 63n, 110, 207n, 208n, 270, 272n, 328, 352, 381, 383,
393, 405, 407-9, 412-3, 415n
Labrador fishery, 57-9, 98-9, 107-8, 117, 123-6, 135, 137-9, 142, 146-7, 157, 158, 205n, 209n, 240, 342, 346, 352-3, 382, 385, 396
    see also Cod fishery; 'North Shore' fishery
Lady Franklin Bay, 394
'Landsmen' seal fishery, 223, 264-71, 301-5
    see also Seal fishery
Leather, 65, 68-9, 70, 77, 84, 86, 87n, 126, 201, 244, 335, 410
LeMarchant, John Gaspard, 54
LeMessurier, Henry William, 369
Lighthouses, 74-6, 78-9, 82-3, 86, 204n
Lighting, 65, 67-8, 70-9, 81-9
    see also Argand lamp; Gas lighting; Oil lamps
Lindsay, David Moore, 259-63

Linklater, L. M. (Capt.), 258-9, 325n
Linseed oil, 70-2, 80-1
   see also Oil
Little, Linda, 204n, 215-7, 330, 396n
Liverpool, 67, 70-1, 81-2
London, 40, 43, 65-9, 81-3
Longshoremen's Protective Union, 352

M

MacGregor, William, 363-5, 368
MacKay, Alexander M., 145, 153, 182
Mackey, Pat, 286, 320n
MacKinnon, Robert, 96, 119n
Maddigan, Mick, 257
'Manus'
   264, 342, 347, 350-2, 398n, 399n
   see also Ice hunters;
   Sealers' strikes
March, Stephen, 146, 158, 295
'Marconimen', 259
   see also Seal fishery;
   Wireless communication
Martin, Emma Jane, 399n
Matthews, Keith, 26-28, 34, 36, 59n, 60n, 62n
McBride and Company, 155, 156, 159
McBride and Kerr135, 158, 159, 350, 372, 414
McCallum, Henry E., 363, 364, 365, 368
McCarthy, John, 261, 279n, 280n, 375-6
McGrath, Patrick Thomas, xvii, xix
McLea, Kenneth, and Sons, 135, 155, 158, 159
Mechanics Society, 352
Mercantile Committee, 340
Mercer, Albert, 342, 346
Mercury
   see *Evening Mercury*
Migratory fishery xvii, 25-37
   see also Cod fishery; West of England fishery
Mineral oil, 83-4, 90n, 91n, 413
   see also Oil
Morine, Alfred Bishop, 115, 343, 366, 385
Morris, Patrick, 95-96, 118n, 212n 233, 406, 414n
Mosdell, Harris Munden, xvii, xixn, 63n, 333, 397n
Mouland, Arthur, 312, 314-5
Muir and Duder, 155-6
Mullowney, Pierce (Capt.), 154, 221
Munden, Azariah (Capt.), 155, 221-2, 238, 240-1, 278n, 340, 375-8, 379, 383-4, 401n
Munden, William (Capt.), 125, 142, 155, 214
Munn, John, and Company, 130-1, 136, 138, 142, 146, 17, 151, 154-61, 170, 173, 179, 189, 206n, 207n
Murphy, Edward (Capt.), 222
Murphy, James, xvii
Murphy, James (Capt.), 218, 221, 238, 273
Murray and Crawford, 211n, 310

Mutual Insurance Association of St. John's, 139, 285
Mutual Insurance Society of Conception Bay, 139, 285

N

Napoleonic War, 32, 39-41, 43, 44, 55, 48, 59
New Perlican, 124, 135, 326n, 327n
New World Island, 124, 128
Newfoundland
   exports, 419, 446, 448, 450-1
   Select Committee on Agriculture, 108
   imports, 429-33, 436, 438, 440, 441
   industrialization, 110
   Irish emigration to Newfoundland, 43-6
   labour movement, 353
   living conditions, 243-7
   mineral resources, 110
   Newfoundlanders in the United States, 93
   populations and occupations, 425, 441-2
   prices, 420-3
   reciprocity with the United States, 110, 115
   settlement, 25, 29-31, 33, 35, 43, 46, 49, 51, 59
   wages, 421, 422
Newfoundland Blue Books, 98
Newfoundland S.S. Sealing Company Limited, 185, 193, 211n
Newfoundland Steam Sealing and Whale Fishing Company, 178-9
Newfoundlander, 113-4, 168-9, 317, 320
Newman and Company, 81, 157, 208n, 414n
Newtown, 179, 225, 324n, 326n, 327n
'Noggins', 173, 204n
Norman, John (Capt.), 215, 340
Norman, Nathaniel (Capt.), 142
Norman, William (Capt.), 394
North Pacific fur seal industry, 318, 393, 402n, 410
'North Shore' fishery, 45-6, 56-9, 63n
   see also Cod fishery; Labrador fishery
Notre Dame Bay, 174, 266, 365, 368
Nova Scotia, 64
Nowlan, Jeremiah, 237, 289-90, 332-3

O

O'Brien, Lawrence, 108, 296, 331-2, 347n
O'Brien, Lawrence, and Company, 135-6, 139-40, 155-6, 159, 240, 272n
Oil, 26, 29, 38, 40, 43, 45, 47, 49-58, 63
   prices, 184
   production, 71, 72
   trade, 76-86
   see also Coconut oil; Cod liver oil; Cod oil; Colza oil; Fish oil; Kerosene oil; Linseed oil; Mineral oil; Olive oil; Palm oil; Rape seed oil; Seal oil; Sperm oil; Train oil; Vegetable oil; Whale oil

Oil lamps, 65-8, 71, 73-4, 81-2, 86
    see also Argand lamp; Carcel lamp; Lighting
Old Perlican, 52, 124, 146, 299
Olive oil, 67-70, 80-3
see also Oil

P

Pack, Gosse, and Fryer, 125, 317
    see also Gosse, Pack and Fryer
Palm oil, 80-1, 83-5
    see also Oil
'Pan', 170-2, 182, 247, 250, 265, 289, 297, 303-5, 313
'Pan flag', 171-2, 181-2
'Panning' of seals, 171-2, 181-2, 259, 261, 268-9, 356, 368, 376-7
Parsons, Alexander A., xviii, xxn
Parsons, John (Capt.), 192
Patriot, 317, 330-5
Peary, Robert E., 394-5
Petroleum, 83-4, 86, 90n, 91n
Piccot, John (Capt.), 284-6
Pike, Richard (Capt.), 221-2, 393-4, 403n
Plantation, 25, 30-2
Planter, 27-8, 30, 33-5, 38-40, 44-5, 47, 51, 54-6, 59-60, 328-9, 331, 352, 384-5, 401n
    see also Resident fishery
'Pokers', 250
Polaris (SS), 393-4
Port de Grave, 121, 127, 131-2, 144, 160
Potato crop, 31-2, 38, 92, 95-7, 102-3, 109-10
    see also Agriculture
Prowse and Sons, 151-2, 184, 188
Prowse, Daniel Woodley, 46-7, 95, 103, 143-4, 146, 151, 186, 221, 236, 245-6, 248-9, 303
Public Ledger, 159, 163, 330-1
Punton and Munn, 142, 146-7, 383

Q

Quebec, 50-1
Quidi Vidi, 230, 259
Quinquet lamp, 88n
    see also Argand lamp

R

Rabbits, William (Capt.)[or Robbitts], 378
Ragged Harbour, 265
'Raggedy jackets' (harp seal), 48, 247
Railway, 93, 110, 117, 174, 179, 225, 354, 357, 408, 409
Randell, Isaac R. (Capt.), 198, 315
Rape seed oil, 65, 70-1, 74, 76, 78, 79, 81, 83, 85, 90
    see also Colza oil; Oil
Reardon, Stephen, 280n
Red Island, 327
Reeves, John, 36-7
Reid, H. D., 348-9
Reid, R. G. , 364
Reid Newfoundland Company, 382
Rendell, Stephen, 130
Renews, 29, 245
Rennie and Stewart, 146
Resident fishery, 24, 30-5, 47, 49, 51, 56-7, 59
    see also Migratory fishery; Planter
Ridley and Sons, 133, 136, 138, 142, 146, 148, 154-6, 158-63, 202, 204n
Rogerson, P., and Son, 136, 158-9
Roosevelt (SS), 394-5
Rorke, John, 136, 138, 141, 156-8
Royal Gazette, 159, 160
'Running' ice, 140, 251, 291, 304
Ryan, William (Capt.), 152-3, 219-21

S

Saddlers' guild, 69
Sailing vessels
    see Sealing vessels, sail
St. John, Oliver, 130
St. John's, 38-47, 49-59, 95-103, 121-9, 134-51, 154-202, 213-238, 255-62, 283-311, 330-5, 340-54, 404, 414
St. John's Chamber of Commerce, 111, 113-4, 160, 162, 169, 175, 285, 288-9, 299, 407-8
St. John's Fire, 1817, 46-7
St. John's Fire, 1846, 102, 139, 289
St. John's Fire, 1892, 119n
St. Pierre and Miquelon, 35, 43
Saltfish industry, 33, 34, 37, 40-3, 47, 49, 54, 56-8, 62n, 92-3, 119, 137, 147, 328, 369
Sanger, Chesley W., 259, 261
Scotland, 145, 147, 150, 175, 176, 219, 221, 230, 231, 258, 259, 310
'Sculp', 48, 248, 259
'Scunner', 150, 253, 257
'Seal finger', 254
Seal fishery,
    competition, 178-83
    conservation, 111-7
    cruelty to seals, 388-93
    decline, 105-6, 111-8, 168-73, 218, 269-71, 353-6, 368, 395
    exhibitions, 111-2
    impact on Newfoundland society, and culture ,328, 368-96
    legislation, 111-7, 165-7, 178-80, 182, 186, 190, 194, 259, 351, 35405, 357-8, 362, 366-8, 412
    songs, 215, 233-4, 263, 308, 309, 384-86
    wireless communication, 191-4, 196, 198, 259, 311, 314, 316
    women, 226-7, 365-7
    see also Ice hunters; Sealers' strikes; Sealing captains; Sealing disasters; Sealing firms; Sealing fleets; Sealing ports; Sealing vessels

Seal flippers, 239-4, 247, 252, 254, 260, 387-8
Seal hunters
   see Ice hunters
'Seal meadows', 101, 250
Seal oil, 49-52, 54, 56-8, 63n, 64n, 99, 104-5, 108, 115, 119, 215, 236-8, 240-3, 262, 283, 297, 329, 353-4, 372, 397n
   processing plants, 134-5, 180-1, 184-5, 186-7, 230, 368
   see also Fish oil; Oil; Train oil
Seal skinners, 58, 178, 216, 349, 352
Seal Skinners' Union, 349
Sealers
   see Ice hunters
Sealers' Agreements, 231, 347-9
Sealers' shares, 138-9, 164, 183, 185-6, 189, 191-6, 199, 215, 236-8, 240-3, 263, 329, 345
Sealers' strikes
   1832, 329-30
   1838, 397n
   1842, 330-5
   1843, 397n
   1845, 336-41
   1853, 341-2
   1902, 342-7
Sealing captains, 213-5, 219-23, 240-42, 284-300, 458-65, 470-93, 495-504
Sealing disasters, 282-319
   victims, 323n-327n
Sealing firms, 136-8, 142, 145-7, 151-61, 168-70, 175, 176, 178, 183-7, 191, 293, 406-7, 414n, see also names of firms
Sealing fleets, 443-44, 453-66, 470-94
Sealing ports, 121-202
Sealing vessels
   sail
     Active (Brigus), 284
     Active (Carbonear), 289
     Adelaide, 224
     Afton, 219
     Ajax, 295
     Albert F., 164
     Alder Davis, 237
     Alert, 383
     Alice May, 164
     Alma, 297
     Amazon, 164
     Ambrose, 291
     Amy Ann, 292
     Anastasia, 255
     Ann, 288, 294
     Anna, 288
     Annie, 165
     Annie Laurie, 370
     Argyle, 295
     Arthur O'Leary, 370
     Atlanta, 376
     Atlas, 383
     Azariah, 287
     Balaclava, 297
     Beatrice, 165
     Belisarius, 284
     Billow, 295
     Brighton, 300
     Britannia, 296
     Brothers (Hr. Grace), 142
     Brothers (St. John's), 285
     Cabot, 222
     Caledonia, 139, 295
     Candid, 164
     Carolina, 284
     Caroline, 140, 297
     Cecilia, 300
     Cedella, 238
     Charles, 293
     Charlotte, 291
     Charming Lass, 164
     Christianna, 295
     Christina, 297
     Comet, 165
     Confidence, 285
     Coquette, 295, 299
     Corfe Mullen, 293, 296
     Cornelia, 296
     Curlew, 255
     Cyrus, 300
     Dart, 291
     Dash, 139, 295, 296, 370
     Despatch, 291
     Devonport, 286
     Dingwell, 289
     Dolphin, 255, 300
     Dorothy, 164
     Drake, 219
     Duck, 219
     Dundanah, 300
     Echo, 294
     Eclipse, 299
     Ecliptic, 300
     Eldora, 164
     Elfrida, 299
     Eliza, 133, 297
     Elizabeth, 297, 370
     Elizabeth (brig.), 295
     Elizabeth and William, 136
     Elizabeth Jane, 297
     Elizabeth Margaret, 297
     Elizabeth Margaret (brig.), 292
     Emily Tobin, 297
     Endeavour, 136
     Eneas MacIntyre, 300
     Enthusia, 167
     Envy, 285
     Eric, 373-5
     Escape, 296
     Escort, 167
     Evanthes, 370
     Experiment, 284
     Fanny, 284
     Fanny Bloomer, 299, 369-72

Favourite, 284
Florence B., 165
Fortitude, 296
Four Brothers, 125
Frances, 255
Friends, 293
Funchel, 296
G.M. Johnson,
Gannet, 295
Gardiner, 167
Gem, 295
Georgina, 300
Gertrude, 255, 300
Gipsy, 383
Glance, 370
Gleener, 295
Glencoe, 300
Glengarry, 300
Glide, 237
Goodship Jubilee, 167
Goorkha, 227
Goose, 219
Greyhound, 142, 300, 378
Guitar, 370
Gull, 219
Haidee, 294
Hammer,
Hare, 294
Harvest Home, 165
Hebe, 382
Helen, 295
Henrietta, 295
Herald, 294
Hibernia, 293
Highland Laddie, 285
Highlander, 240, 383
Hope (Carbonear), 288
Hope (Conception Bay), 296
Hope (St. John's), 297
Hornet, 297
Hound, 139
Hunter, 297
Huntress, 136
Huntsman, 220, 300, 381
Imauna, 295
Industry, 285
Iona, 255
Isabella, 220
Isabella Ridley, 133, 255, 382
Island Gem, 164
Jane (St. John's), 299
Jane (Trinity), 136
Jasper, 136
Jessie Brown, 297
Jessie Louis, 296
John, 291
John and Maria, 297
John and William, 292
John Martin, 263, 294
Joseph, 287

Jubilee, 167
Jura, 297
Kate, 152
Kate Cummins, 255
Kingaloch, 139, 295, 296
Kitty Clyde, 220
Lady Margaret, 284
Lark (Greenspond), 136
Lark (St. John's), 288
Liberator, 294
Livingstone, 152
Louisa, 255
Louise Stuart, 292
Maggie, 383
Maggie A, 167
Margaret (Harbour Grace), 293
Margaret (St. John's), 297, 372
Margaret Ellen, 293
Mary (Harbour Grace), 291
Mary (St. John's), 291, 295, 297
Mary Anne Rossiter, 297
Mary Belle, 371
Mary Francis, 293
Mary Joyce, 152, 300
May Queen, 165, 167
Mayflower, 285, 293
Melrose, 297
Mildred, 238
Moonlight, 299
Mountaineer, 255
Nancy, 285
Nancy Lee, 370
Nautilus, 299
Nine Suns, 219
Nora Creina, 370
North Star, 136
Ocean, 391
Olive Branch, 288
Orlando, 167
Oresta, 296
Packet, 299
Peerless, 292
Perseverance, 329
Placid, 295
Portland, 164
Prince Edward, 297
Princess, 291
Princess Royal, 294
Pursuit, 295
Rachel and Ellen, 237
Rake, 139, 296
Rebecca (Harbour Grace), 291
Rebecca (St. John's), 292
Relief, 291
Renfrew, 299, 382
Rescue, 300
Ringwood, 292
Rise Over, 164
Robert Brine, 288
Rolling Wave, 255, 381

Ronana, 297
Rosebud, 297
Sally (Conception Bay), 296
Sally (St. John's), 284
Sally (Trinity), 136
Sarah Ann, 371
Sarah Jane, 294, 371
Seaflower, 398, 300
Selah Hutton, 299
Selina, 288
Silver Stream, 300
Sir Howard Douglas, 294
Sir John Harvey, 293
Sisters, 165
Sonora, 370
Speedwell, 284
St. Margaret, 136
St. Patrick, 215
Stella Jessie, 300
Success, 293
Swan (Channel), 167
Swan (St. John's), 292
Terra Nova, 289
Thomas Ridley, 139, 255
Three Sisters, 155, 383
Topa, 255
Topaz, 250
Trial (Bay Bulls), 289
Trial (Harbour Grace), 291
Triumph, 219
True Blue, 286
Tyro, 292
Union (Harbour Grace), 399n
Union (Trinity), 288
Velocipede, 300
Vesta, 295
Vesta (Harbour Grace), 148
Victoria, 297
Village Belle, 300
Visiter, 284
Waterlily, 293
Western Trader, 295
William, 136, 296
William (Carbonear), 163
William L. Black, 292
William Stairs, 297
William Whelan, 255
Winged Arrow, 167
Winnie L., 165
Young Prince, 220
Zambesi, 152, 370
steam (first reference)
  Admiral
    see Vanguard
  Adventure, 187
  Algerine, 179
  Arctic, 151
  Ariel, 145
  Aurora, 175
  Bear, 150

  Bellaventure, 191
  Beothic I, 191
  Bloodhound I, 151
  Bloodhound II, 150
  Bonaventure, 191
  Camperdown, 147
  Commodore, 155
  Diana, 153
  Eagle I, 150
  Eagle II, 190
  Eclipse, 469
  Erik, 190
  Erna, 195
  Esquimaux, 150
  Falcon, 170
  Florizel, 191
  Fogota, 195
  Grand Lake, 190
  Greenland, 154
  Harlaw, 180
  Havana, 190
  Hawk, 151
  Hector, 153
  Hope, 179
  Iceland, 154
  Kite, 179
  Labrador, 179
  Leopard, 179
  Lion, 150
  Lloydsen, 195
  Mastiff, 155
  Merlin, 153
  Micmac, 150
  Montecello, 150
  Narwhal, 175
  Nascopie, 195
  Neptune, 150
  Newfoundland, 150
  Nimrod, 150
  Osprey (or Ospray), 151
  Panther, 150
  Polynia, 147
  Proteus, 150
  Ranger, 150
  Resolute, 151
  Retriever, 149
  Sagona, 195
  Seal, 183
  Southern Cross, 150
  Stephano, 195
  Terra Nova, 149
  Thetis, 151
  Tiger, 305
  Tigress, 258
  Vanguard, 155
  Viking, 189
  Virginia Lake, 190
  Walrus, 150
  Windsor Lake,
  Windward, 185

Wolf I, 150
Wolf II, 150
Xanthus, 469
see also Sailing vessels;
Sealing disasters;
Steam vessels; Steel vessels;
Wooden walls.
Seals, 47-9
Shallop, 51, 55
Shipbuilding, 51, 142-3, 186
Ship's biscuit
see Hard bread
Shipwrights' Union, 352-3
Shortis, Henry Francis, 261, 350, 369
Skinners' guild, 68
Slade, J., and Company, 156
Slade, R., and Company, 136
Smallpox, 234
Smith, Nicholas, 142, 174
Soap making, 69
Society for the Prevention of
Cruelty to Animals, 390-91
'Speck finger'
see 'Seal finger'
Sperm oil, 71, 77, 78, 80, 82, 440-41
see also Oil; Train oil; Whale oil
'Spring of the Growlers', 291
'Spring of the Wadhams', 295
'Spring of White Bay', 321n
'Spy master'
see 'Barrel man'
Stabb, Nicholas and Sons, 146
Stabb, Row and Holmwood, 135
Steam vessels
see Sealing vessels, steam
Steamboat Labourers' Union
see Longshoremen's Protective Union
Steel vessels, 187-200, 469
see also Sealing vessels, steam;
Steam vessels
Stephen, Alexander, and Sons, 147-50
Stewart, J. and W., and Company, 145-6
Supple, Henry (Capt.), 333-40, 344, 346, 397
Swain's Island
see Wesleyville
'Swile finger'
see 'Seal finger'

T

Tallow Chandlers, 67
Temperance Band (Harbour Main), 336, 339
Terra Nova Advocate, 369
Tessier, P. and L., 135, 155, 156, 371
Thomas, W. and H., and Company, 135, 158, 414
Thomey, Arthur (Capt.), 133-34, 138, 162, 214-15, 383
Thomey, Henry (Capt.), 133, 163, 176-7, 204n, 214-5, 221-2, 273n, 383

'Ticket'
see Berth
Times (London), 112
Times (St. John's), 133, 351
Tin Plate Workers, 67
Tizzard, John, 313
Tocque, Rev. Philip, 111, 332-3, 388-9, 390-1
Torbay, 227, 230, 324, 326, 361, 383
Train oil, 76, 79-81, 83-4, 86
see also Cod oil; Fish oil; Oil; Seal oil;
Whale oil
Trinity, 45, 50, 52, 55-6, 121, 123, 124, 128, 129, 136, 137, 143, 145, 156, 160, 164, 172-4, 196, 202n, 213, 225-6, 229n 235, 266, 271n, 284, 288, 292, 303, 305, 323-4
Trinity Bay 45, 49, 50-1, 55-7, 123-4, 127-8, 134, 136, 137, 144, 157, 202, 205n, 213, 223, 232, 234-5, 291, 298, 303-4, 368, 385
'Trinity Bay Disaster', 304
Trinity House Corporation, 75-6, 78, 83
Tuff, George, 307, 312-5
Twillingate, 49, 52, 56-7, 124, 127-8, 134, 136, 143, 156-7, 170, 172, 207n, 289, 299, 301, 302, 356, 358, 365

U

Union Bank, 179
Unionist Party, 355, 368
United States of America, 33, 40-44, 77, 79, 82, 86, 93, 110, 115, 117, 153, 183-4, 199, 219, 317, 329, 393-4, 402n, 408, 409, 410, 412, 414-5

V

Vegetable oil, 77, 79, 82-3, 85, 409
see also Oil
The Viking (Movie), 320n

W

Waldegrave, William, 37-39, 60n, 61n
Wallace, Sir James, 37, 60n, 61n
Walrus, 85, 318, 402n, 410, 413
Walsh, William, 286, 331-2
Warren, M. H., and Company, 136, 155, 156
Waterford, 43-4, 59
Weekly Herald, 132, 319n
Wesleyville, 179, 198, 225-6, 259, 324n, 325n
West Country, 26, 30, 31, 35, 36-7, 40, 43-4
West Country merchants, 26, 30, 36-7, 43, 50-1, 104
West Indies, 33, 39, 41
West of England fishery, 25-6, 34, 57, 65, 70
'Western Adventurers'
see West Country merchants
Western Charter, 26
Whale fishery, 50-1, 70-1, 76-7, 79, 84, 103, 108, 147-8, 175, 201, 283, 410
Whale oil, 70-2, 74, 76-80, 82, 88n, 155, 410
see also Fish oil; Oil; Sperm oil; Train oil

Whelan, William (Capt.), 215, 378, 380
White Bay, 51, 188, 215, 267-9, 299, 300, 350, 376, 384, 385
White, Edward Sr. (Capt.), 151, 153-4, 220-1, 238, 240-1, 276, 299, 379-80, 384, 401
'Whitecoats' (harp seal), 48-9, 51, 113-6, 126, 177, 189, 191, 193-4, 197, 199, 215, 247-8, 252, 260, 265-6, 311, 348, 371, 393
  see also Harp seal
Whiteway, William, 110, 176, 354, 257, 359
Williams, Sir Ralph, 364, 368
Winsor, Jacob (Capt.), 222, 259, 310
Winsor, James (Capt.), 220-1
Winsor, William Sr. (Capt.), 116, 228, 392
Winsor, William C. (Capt.), 195, 198-9, 241
Wireless communication, 191-4, 196, 198, 259, 311, 314, 316
Wooden walls, 116-7, 121, 144-87, 187-200, 202, 226, 242, 306-7, 312
  see also Sealing vessels, steam
World War I, 200, 270, 349, 412, 413
Wyndus, Edward, 65

www.ingramcontent.com/pod-product-compliance
Lightning Source LLC
Chambersburg PA
CBHW022054150426
43195CB00008B/137